Xunren Yang

Atmospheric Acoustics

Xunren Yang

Atmospheric Acoustics

DE GRUYTER

Science Press
Beijing

Physics and Astronomy Classification Scheme 2010
43.28.+h, 42.25.Dd, 92.60.Cc, 92.60.H-

Author
Xunren Yang
Chinese Academy of Sciences
Institute of Acoustics
yangxunren@tsinghua.org.cn

ISBN 978-3-11-031152-5
e-ISBN (PDF) 978-3-11-031153-2
e-ISBN (EPUB) 978-3-11-038302-7
Set-ISBN 978-3-11-031154-9

Library of Congress Cataloging-in-Publication Data
A CIP catalog record for this book has been applied for at the Library of Congress.

Bibliographic information published by the Deutsche Nationalbibliothek
The Deutsche Nationalbibliothek lists this publication in the Deutsche Nationalbibliografie; detailed bibliographic data are available on the Internet at http://dnb.dnb.de.

Cover image: HT Pix/iStock/Thinkstock
Printing and binding: CPI books GmbH, Leck
♾ Printed on acid-free paper
Printed in Germany

www.degruyter.com

In memory of my mother. Via this book I seemed to hear her voice from heaven.

Preface

It is inconceivable that for a long time, no monograph had been dedicated to atmospheric acoustics, such an important discipline with a long history. This situation did not change until 1997, when the first Chinese edition of this book was published by Science Press. Hereafter, quite a few publishers expressed their interest in publishing its English version. Some "sample chapters" were reviewed and approved by expert groups and earned favourable comments. However, all projects miscarried purely due to the dereliction of duty of some staff members there.

The second Chinese edition was published in 2007, with much new material added. Some supplements, mainly in Chapter 10, were completed by Prof. Chen Yu at Tsinghua University. A couple of years later, Dr. Bao Ming and my other young colleagues at The Institute of Acoustics, Chinese Academy of Sciences, suggested restarting the plan to publish an English version. Subsequently a publishing agreement was reached between Science Press and the German publishing house De Gruyter.

This English version is based on the translation from the 2nd Chinese edition. Thanks to Zhao Nan, Ao Lin and Lin Jianheng for their translating drafts of the corresponding parts of this book.

Though there is no other book dedicated to atmospheric acoustics till now, several similar publications exist indeed. In the process of writing, we took relevant publications into consideration, and those materials are reflected in the corresponding chapters in the book.

Based on over 30 years of experience in atmospheric acoustics, the book provides readers with a fundamental understanding of this field. Attention is paid to both theoretical and experimental aspects, with emphasis on the former.

The foundation for writing this book is the author's thirty-year experience of researching atmospheric acoustics. The main aim of this book is to give the readers a fundamental understanding concerning this very field. Though both theoretical and experimental materials have been presented, the former is dominant.

Xunren Yang
June 1st, 2015, Beijing

Foreword

Acoustics is a branch of physics with distinctive features. Firstly, it is one of the earliest developed branches in physical sciences. Ancient Chinese and Greeks understood the basic concepts of "sound" without diversions – in this regard, acoustics is much more fortunate as compared to its companion discipline, optics. As a systematic sub-branch, the fundamental theory was perfectly developed by the middle of the 19th century, the golden age of classical physics. Secondly, acoustics is also the most active branch in physics today. Due to the interlocking connections with so many fields in natural sciences, and even the humanities, many research frontiers have emerged. The number of sub-branches in acoustics has exceeded twenty so far, with new fields forming every year. This is rare in physical sciences, and even in natural sciences.

Among the sub-branches of acoustics, atmospheric acoustics turns out to be the most ancient and most important one. In the beginning of the 18th century, acoustics was preliminarily formed as a discipline, and in the 19th century a lot of important fundamental phenomena had been studied, with theories constructed. Thereafter, research in this field proceeded uninterrupted, and great developments have been achieved in recent years. As the material on which humanity relies for its existence, and as one of the three vast media (along with the oceans and the earth's crust) through which sound waves propagate, the atmosphere embodies numerous and varied wave phenomena. Humanity would do well to develop a thorough understanding of the basic nature of wave phenomena, which are closely related to many other natural phenomena and human activities. Thus it is argued that atmospheric acoustics is not only an important theoretical discipline, but also an important applied field.

However, what puzzles us is the fact that there was not a monograph devoted to such an important and peculiar discipline in the worldwide literature until now. There are two books of a similar nature, i.e., T. Beer's *Atmospheric Waves* and E. E. Gossard & W. H. Hooke's *Waves in the Atmosphere* (references [30] and [202] in this book). As shown by their titles, however, they do not belong solely to the category of atmospheric acoustics. Needless to say any monographs, or even general texts regarding acoustics with contents concerning atmospheric acoustics, are scarcely found. A. D. Pierce's *Acoustics: An Introduction to Its Physical Principles and Applications* (reference [56]) is perhaps the only exception. Besides, there are merely a couple of reviews, such as the paper *Advances in Atmospheric Acoustics* written by E. H. Brown & F. F. Hall, Jr. (reference [1]) plus items in some encyclopedias of science and technology, such as those written by H. E. Bass (reference [174]) and by E. H. Brown, respectively. All the above-mentioned material has been of great help in writing the book, and the content in several sections and/or chapters is quoted directly from them.

In view of the above-mentioned situation, to publish a high-level monograph of atmospheric acoustics thus becomes a task of top priority to filling in the gaps in this very field. This book is nothing but an attempt to "cast a brick to attract jade".[1] The aim is to provide readers with a systematic and rigorous general picture that reflects the highest academic level in this field, which is as up-to-date and as comprehensive as possible. Consideration has been given to both theory and practice, though more emphasis has been placed on the former. Special attention was paid to the discussion of fundamental concepts and the treatment of physical processes. We did not go too far in mathematical derivations, as the key of the matter is pointing out the origin and development rather than rigidly adhering to rigorousness.

In this book, essential descriptions of structural properties and acoustical behaviors of the atmosphere as a propagation medium of sound waves will be given first. Based on fundamental equations of fluid dynamics, together with the practice of the medium atmosphere, a series of acoustic wave processes such as reflection, refraction, scattering, diffraction and absorption are discussed. Therein, alternative methods of wave acoustics and geometric acoustics are used in accordance with different situations. Regarding the discussion of influences from the earth's gravity field and rotation, we transit from classical atmospheric acoustics to modern atmospheric acoustics. In order to make the discussion more coherent and more comprehensive, fundamental acoustic relations were derived in detail. Just through the description of these processes, an understanding of atmospheric acoustics is established. In the chapter on computational atmospheric acoustics, further treatments of basic problems in atmospheric acoustics are supplemented by extending traditional computational methods to this field. Finally, in the chapter on remote sensing for the atmosphere, the application aspects of atmospheric acoustics are discussed from the angle of inverse problem.

The references directly related to the text are arranged in serial numbers listed at the end of the book, while those involving secondary problems (e.g., the derivation of a less important equation) are indicated as footnotes.

Non-linear phenomena are inevitable in atmospheric acoustics; the fundamental problems with this aspect are discussed in detail in Chapter 9. The problems concerning sound sources in atmospheric acoustics are also arousing general interest and recent advances are described in the last chapter.

Xunren Yang
Jan, 2007

1 This is a "literal" translation of a famous Chinese idiom (usually a self-deprecating remark), whose actual implication is "to issue some opinion, suggestion or writings of lower level in order to inspire the corresponding ones of higher level". Thus, the author expects that a book of much higher level will be contributed to this field in the near future.

Contents

Chapter 1
Introduction

1.1 Scope of the discipline and historical review

The earth's atmosphere, on which humanity relies for its existence, is ubiquitously flooded with acoustic waves of various hues. These waves include both audible sound and infrasound, where the latter embodies a series stretching from "true" infrasonic waves to "acoustic gravity waves", "internal gravity waves" down to "planetary waves" and "atmospheric tides", whose behaviors becoming increasingly different to acoustic waves under ordinary concepts. As for ultrasonic waves, they occupy very less important position in atmospheric acoustics due to their rapid attenuation in the air. Atmospheric acoustics is a discipline that studies the generation, propagation (including the processes of reflection, refraction, diffraction, scattering and attenuation), reception and the various effects and applications of these waves. It is one of the most ancient and most important sub-branches in acoustics.

Obviously, the behavior of acoustic waves in the atmosphere depend strongly on the properties of the atmosphere itself. The variations of atmospheric properties in both space and time present very complicated relationships. The interaction of acoustic waves with this complex atmosphere, as well as with the variety of ground surfaces, requires a wide spectrum of physical phenomena be understood in order to completely describe a specific sound field.

Although the atmosphere consists of air, there exists qualitative differences between acoustic phenomena in the atmosphere and the "airborne-acoustic" phenomena at small scales, especially in closed spaces. This difference represents two distinct features of atmospheric acoustics. At first, it is a fundamental discipline involving the influences on acoustic waves from the variations and non-homogeneity of the earth's atmosphere, which are difficult to forecast; secondly, it is an application of acoustic waves to probe the atmosphere. The second feature can be considered as the "inverse problem" of the first one, and in a certain sense this also expresses the notion that atmospheric acoustics can also be regarded as a sub-branch of atmospheric physics. Of course, due to the difference between the main disciplines subordinated to, the focus of attention and the set-up system for these two sub-branches are quite different.

Indeed, the traditional motive for studying atmospheric acoustics turns out to be the need to communicate audibly in the atmosphere – and therefore the problem of acoustical probing of the atmosphere and then acquiring information of the atmosphere itself (i.e., the inverse problem) emerges. In recent years, studies of atmospheric acoustics have also been motivated by concerns of environmental noise. Consequently, it was suggested that "aero-acoustics" should be included in atmospheric acoustics as a sub-branch, whose main object of study is the strong atmospheric flows acting as sources of sound and noise.

According to the view of, atmospheric acoustics can be divided into two categories, i.e. as "classical" and "modern"; where the former takes the audible sound in the lower atmosphere as the main object of study; while the latter takes the infrasound located in the upper atmosphere as the main object.

The main contributions to acoustics made by ancient Chinese and ancient Greece are concentrated in phonologic and musical aspects, and this emphasis continued during the "rebirth" of science in the occident in the 17th century. The measurements of the speed of sound in the free atmosphere made in 1635 may be counted as the first atmospheric acoustic experiment, but to move the experiment outdoors at that time was only to obtain sufficiently long propagation path lengths.

It is generally acknowledged today that the initiative work in atmospheric acoustics was performed in a famous experiment carried out by English clergyman Derham and his Italian acquaintance Averrani in 1704. In order to clarify a circulated ambiguous statement that "the outdoor audibility in England is better than that in Italy", they conducted outdoor experiments in their respective countries which showed that after properly accounting for the effects of wind, sound propagation did not differ in these two countries. The physical phenomena that affect the acoustic effects of the atmosphere were generally identified at the beginning of 19th century, although many of the details and mathematical techniques for dealing with these phenomena only evolved more recently, and the sustained development of atmospheric acoustics did not begin until the last half of the 19th century. During this period, Reynolds [2] studied the variation of atmospheric temperature with altitude (referred to the modern term as "profile") and their effects on the refraction of sound. Stokes [3] briefly considered the refractive effects of winds and wind gradients. Tyndall [4] investigated the scattering of sound by atmospheric fluctuations. This problem subsequently evoked a contention, which saw it shelved for nearly an entire century, before it was eventually solved in the 1970s (see 5.1 in Chap V). The huge horn used by Tyndall at that time (Fig. 1) is an excellent example of early experimental work in atmospheric acoustics.

Most of the turbulence-dependent processes in the atmosphere take place in the troposphere. Via balloon measurements, Teisserenc de Bort [5] discovered the stratosphere, an isothermal layer lying just above the troposphere, in 1902. At about the same time, the so-called "anomalous" sound propagation phenomenon was observed [6]. Emden [7] explained this phenomenon theoretically using the concept of "temperature inversion", i.e., increasing temperature with height, that existed at a certain altitude. Later on, observations of meteor trails proved that at the "stratopause" (with an altitude of about 50 km) did indeed have an increasing temperature, caused by the absorption of ultraviolet light in solar radiation by atmospheric ozone. Basing on this evidence, Whipple [8] described the anomalous propagation of sound further by refraction caused by this temperature inversion. In the following years,(1920s–1930s), numerous tests regarding the anomalous propagation of sound from explosions were

Fig. 1: The apparatus for studying sound propagation in the atmosphere used by Tyndall in 1874.

carried out. During this process, in 1929 Gowan [9] noted hitherto unrecognized very low-frequency sound waves, now called "infrasound". Subsequently, starting from 1938, Gutenberg [10] made extensive investigations into the sources and propagation of such infrasonic waves, and thereupon the field of "modern atmospheric acoustics" was created.

Research balloons appear to be powerless at high altitudes owing to the thin air there, thus the technique of using rocket grenades was developed in 1950s, which provided measurements of both temperature and wind in the upper atmosphere, up to heights of 80 km [11]. The richening of information concerning the upper atmosphere considerably promoted studies of the "extra-long-distance" propagation of infrasound.

Regarding the turbulent atmosphere, the gradual introduction of turbulence theory to studies of the small-scale dynamic structure of the atmosphere beginning with Taylor [12], who revived interest in the nature of sound scattering. In 1941, Obukhov [13] related the acoustic scattering cross section to a general expression for the spectral density of turbulent kinetic energy. His derivation, however, neglected the fact that acoustic refractivity and its gradient appear in the "source term" of the inhomogeneous wave equation. Nevertheless, his analysis represents the first attempt at a quantitative explanation of sound scattering by the atmosphere. In the case of non turbulent "smooth" atmospheres, Bergmann [14] showed that the presence of the static density gradient produces dispersions in the propagation of sound, i.e., a dependence of phase velocity on frequency. At the same time, he also studied scattering from the point of view of correlation functions [15].

For the turbulent atmosphere, Blokhintzev [16] introduced the assumption that only wave numbers in the inertial range of turbulence affect the propagation of audible and high-frequency sound, thus permitting the use of the Kolmogorov spectrum in the scattering cross section. In 1953, Kraichnan [17] and Lighthill [18] simultaneously found the correct form of the velocity-dependent part of the scattering cross section. Four years later, Batchelor [19] obtained the scattering cross sections used today, finding that in addition to the above-mentioned velocity-dependence, there is an additional dependence on the spectral density of the fluctuations of a passive additive, such as acoustic velocity or temperature. Although he did not express the total cross section as the sum of these two parts, such a summation is implicit since the cross section represents a normalized energy flow and thus "partial cross sections" are additive.

Evidently unaware of the earlier work, in 1961 Tatarskii [20] obtained the total scattering cross section expressed as the sum of a velocity-dependent part and a temperature-dependent part. Unfortunately, he made the same mistake as Obukhov by neglecting gradients in his derivation, and thus omitted a factor of $\cos^2\theta$, which is dependent on the scattering angle θ. The striking divergence in scattering-angle dependence near 90° between his theoretical expression and the extensive experimental results prompted him and Monin [21] to re-examine their derivations.

Along with infrasounds coming from various natural sources in the atmosphere being detected successively, and nuclear testing frequently carried out in the atmosphere, the technique of passive remote sensing developed rapidly since the 1960s. Whereas the active remote sensing is developed under the enlightenments from several companion disciplines. For example, the precursor of echosonde, the so-called "SODAR" (for "SOnic Detecting and Ranging") was invented by Gilman et al. [22] in 1946. McAllister introduced three major improvements in 1968 [23]: a vastly improved transmitting antenna permitting operation just below 1000 Hz and at ranges up to 1.5 km; use of the transmitting antenna itself as the receiver; and use of the facousticsimile recorder adopted from sonar. The invention of the echosonde to such a great extent reawakened interest in acoustics of atmospheric physical researchers, and brought about great advances in developing modern atmospheric acoustics.

The developments in recent years include: new ways to use refraction, studies of phase and amplitude fluctuations during the propagation of sound along a path, insights into large-scale atmospheric processes gained from infrasound, non-linear effects near high-powered acoustic antennas, problems related to noise, applications dependent on the Doppler effect, hybrid devices using both acoustic and electromagnetic waves, and so forth. What should be pointed out emphatically is that, along with the rapid development of computational methods and technology, as a hallmark of the publishing of a monograph [171] in the early 21st century – a brand new "sub-branch" — computational atmospheric acoustics was formed and it is an important promotion for the development of atmospheric acoustics. In the hopes of inspiring even further developments, the concrete contents will be discussed minutely in respective chapters, especially in the last four chapters.

1.2 Structure and acoustic properties of the atmosphere

1.2.1 Stratification structure of the atmosphere

The main physical parameters describing the state of atmosphere are its density, pressure and temperature. The atmospheric density, and thus the pressure, decreases with increasing altitude due to the effect of gravity. The variation of temperature with altitude is much more complicated, which results from different responses of the atmosphere to solar radiation at different altitudes. The atmospheric absorption of radiation in well-defined and quite distinct regions is approximately like this: at heights above about 100 km, solar radiation of very short wavelengths is absorbed mainly by molecular oxygen (O_2); at heights between 35 and 70 km, the absorption of radiation in the near ultra-violet by ozone (O_3) is dominant; in the lower 15 km of the atmosphere, the re-radiated infra-red part of the spectrum is absorbed by water vapor (H_2O) and carbon dioxide (CO_2). At ground level, radiation in the visible part of the spectrum is absorbed by land and sea surfaces. The differences between these absorption mechanisms forms the altitude-dependence of atmospheric temperature, as shown in Fig. 2.

To simplify the research, meteorologists divide the structure of the atmosphere into a succession of layers according to their characteristics. In the lowest layer, the troposphere, the temperature decreases with increasing altitude up to the tropopause, where a local minimum is reached. The tropopause varies in height from about 10 km near the poles to about 20 km over the equator. This change with latitude, however, is not smooth but is jumping: the height jumps from that of the polar tropopause to the subtropical tropopause, and again from that to the tropical tropopause. The regions just below these two gaps contain (at least in winter; in summer the strong westerlies at certain heights change to weak easterlies) the high-velocity flows called jet streams.

The structure of the troposphere depends mainly on the presence of, and phase changes in, atmospheric water content as well as the vertical convection of water vapor. The lowest part of the troposphere, the atmospheric boundary layer, which varies in thickness from 1 or 2 km during sunny daytime to a few hundred meters at night, departs from the general temperature decrease at night and during smog conditions. The boundary layer represents the furthest extent of the influence of the ground on atmospheric fluxes of heat and momentum. The boundary layer can have either an unstable, "open" structure, acting as a pathway for fluxes to affect higher levels (as temperature decreases with altitude), or a stable, "closed" structure, acting as an impenetrable container disconnected to large-scale events, in the opposite case. An increasing temperature in the boundary layer (called an inversion) acts as a "lid" that prevents the dispersion of atmospheric pollutants.

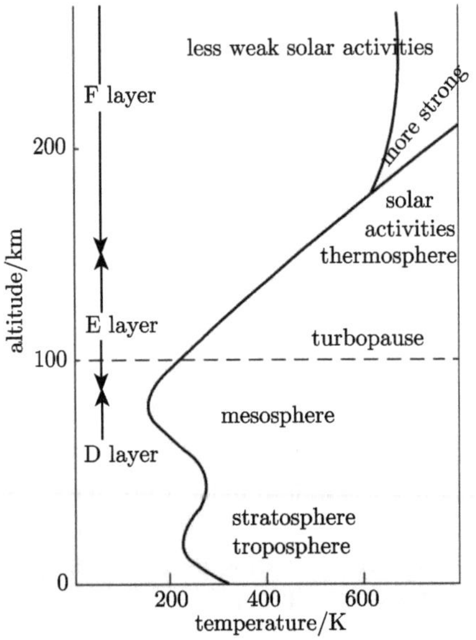

Fig. 2: Temperature profile of the atmosphere.

The part of the atmosphere above the tropopause is called the "upper atmosphere", the lowest part of which is the stratosphere, which contains very little water vapor. As for its temperature, its discoverer Teisserenc de Bort (1902) originally considered it as a constant, thus giving it another name: the "isothermosphere". However, later research showed that the upper part of the stratosphere contains an increasing amount of ozone that changes with altitude. Since ozone strongly absorbs ultraviolet light, the temperature in the upper part of the stratosphere increases with altitude, reaching a maximum at the stratopause (altitude about 45∼55 km). In the upper layer, the mesosphere, temperature again decreases with altitude, and the molecular composition of the atmosphere begins to change. Above the second minimum at the mesopause (altitude about 80∼85 km), atmospheric temperature begins a second (and the final) "climb", and thus forms the thermosphere up to altitude about 200 km. The neutral gas temperature remains constant up to heights exceeding 1000 km. The isothermal behavior of the upper thermosphere arises because its thermal conductivity is so high that most of the energy absorbed by the gas is removed downwards. The limiting thermospheric temperature is determined by the incoming ultraviolet radiation, so that it varies with season, solar cycle, as well as the diurnal and latitude differences in temperature.

Above 600 km, the gas density has diminished to such low values that the air no longer acts as a fluid, therefore fluid dynamics are inapplicable and this height will be considered here to mark the upper limit of the earth's atmosphere. There are various names given to the region above 600 km, such as metasphere (atom hydrogen dominating sphere), exosphere, protonosphere, geocorona, etc.

Another important effect of solar radiation is its ability to ionize neutral particles, so that a region of free electrons, ions and neutral particles exists in the upper atmosphere, which is known as the ionosphere. It consists of three major regions: D region (altitudes from 60 to 90 km), E region (from 90 to 150 km) and F region (above 150 km, and up to the top of the atmosphere at higher altitudes). In the D and E regions, the temperatures of all the atmospheric constituents are equal, whereas in the F region, the electron temperature exceeds the ion temperature, which in turn exceeds the natural gas temperature.

The dash line in Fig. 2 represents the height (about 105 km) of the turbopause, which is the boundary between the turbulent regions of the atmosphere. The transition region between 85 and 105 km may also contain laminar flows. Below 85 km, laminar flows are always present, which causes convection and mixing of atmospheric constituents and keeps the mean molecular weight constant. Above this height, the separation of oxygen and nitrogen produces a steady decline in the value of the mean molecular weight.

1.2.2 Turbulence structure of the atmosphere

Besides the "macro" stratification structure, the atmosphere also has a "micro" turbulence structure. This means that the motion of atmospheric particles is random and obeys certain statistical rules. The analysis of turbulence is based on the postulate for the existence of vortices over a wide range of scales. These vortices superpose and interact with each other, and energy continuously flows from vortices with larger scales to those with smaller scales. The lower limit reached eventually depends on the minimum fluctuation from which vortices can no longer be created. The atmospheric vortices with time scales of tens of minutes are referred to as atmospheric turbulences.

In realistic viscous fluids, the generation of turbulence is closely related to the ratio of the inertial force and viscous force acting on the fluid (i.e., the Reynolds number[1] $Re = UL/\nu$, where U is the flow velocity, $\nu = \mu/\rho$ is the kinematic viscosity of the fluid; and L is a characteristic length). When the fluid is perturbed, the inertial force causes the perturbation to absorb energy from the main flow, while the viscous force causes the perturbation to be damped. Therefore $Re > 1$ forms the necessary condition for the creation of turbulence. In addition to this, the corresponding thermodynamic and dynamic conditions are also necessary, and therefore two origins of

1 This definition, which originates from pipe-flow in the laboratory, might not be available to a vast medium like the atmosphere – at least the viscous stresses introduced by the solid walls of the pipes will be lacking. Therefore the "geophysical" Reynolds number $Re_g = \omega_B L^2/\nu$ is sometimes introduced, where ω_B is the V-B frequency defined in Eq. (1.28). For the large-scale turbulence phenomena, $Re_g \gg Re$, at least, while Re_g is pertinent for explaining the behavior of small-scale turbulences.

atmospheric turbulences are respectively formed: the inhomogeneous temperature distribution, especially thermal plumes produced by the heating and cooling of the ground surface, which leads to the formation of "thermal turbulence"; and the wind velocity gradient caused by friction with the ground surface that leads to wind shear, which develops eventually into turbulent vortices and thus leads to the formation of "wind turbulence". The latter is dominant in most cases.

A quantity that indicates whether turbulent motions will persist or decay in a stratified fluid is the Richardson number $Ri = \omega_B^2/(\partial_z U)^2$, according to Eq. (1.28″) shown below, which can be rewritten as (for the definition of potential temperature θ see section 3.3 in this chapter):

$$Ri = (g/\theta)\left[\partial_z\theta/(\partial_z U)^2\right]$$

This number provides a measure of the stabilizing influence of gravity, modified by temperature gradients, in comparison with the destabilizing effects of wind shear. In fact, it is the ratio between the two parts in the energy of turbulent motion: the work done by buoyancy and by Reynolds stress. $Ri < 0$ represents the instable condition, in this case the buoyancy works to strengthen the turbulence; $Ri > 0$ represents the stable condition, in this case the turbulence will expend a part of its energy in doing work against the buoyancy, and thus its strength will weaken. Ri takes its critical value (roughly between 0.25 and 1), which is a necessary condition for instability in an incompressible and/or compressible fluid (Miles-Howard criterion). At this time the turbulence will be completely controlled and will transfer to the laminar flow.

In general, stronger atmospheric turbulences appear in the following three regions: the atmospheric boundary layer, jet flows at the upper part of troposphere, and the body of convective clouds. The corresponding turbulence is referred to as boundary-layer turbulence, clear-air turbulence (CAT) and convective-cloud turbulence, respectively. Atmospheric turbulence has different scales and strengths in the three different directions (along the wind, across the wind, and vertical), and thus it is anisotropic. In general cases, the vertical component is smaller than the other two horizontal components. Boundary-layer turbulence is mainly controlled by ground–surface conditions; for CAT, the vertical dimension (from tens to hundreds meters) is always smaller than the horizontal dimension (from tens of hundreds to hundreds of hundreds km); as for convective cloud turbulence, there are different situations.

The vortex energy spectrum of the atmospheric turbulence can be divided into two frequency bands: the large-scale energy-containing range and the medium- and small-scale equilibrium ranges. In the equilibrium range, a part of the turbulence energy gained from a preceding large-scale vortex is transferred to the next vertex with a smaller scale, while another part is dissipated by the molecular viscosity. The equilibrium range can be further divided into two sub-ranges: the inertial sub-range in which the molecular viscous dissipation is disregarded and the molecular viscous dissipation is sub-range (Fig. 3). In the energy transmission process influences

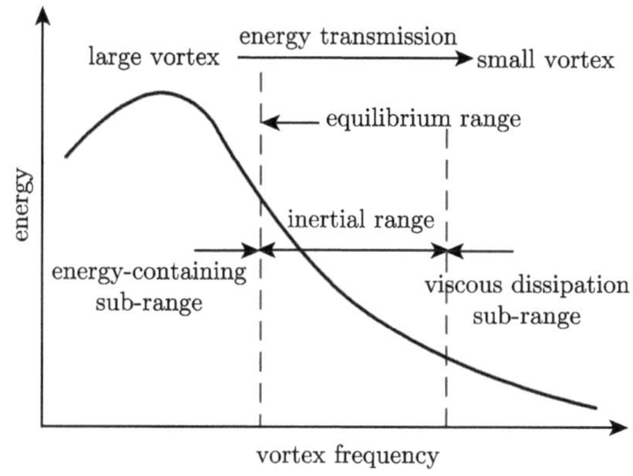

Fig. 3: Energy distribution of atmospheric turbulence.

from external condition diminish gradually. Therefore the anisotropy of the large-scale vortices vanishes gradually, and the isotropy of the small-scale vortices appears gradually. As a result, in a realistic atmosphere, the turbulences are locally isotropic, on the whole. Such local isotropy can be expressed in terms of statistical functions basing on dimensional analysis. For example, the following structure function of the turbulent velocity fluctuations

$$D(r_1, r) = \langle |v(r_1 + r) - v(r_1)|^2 \rangle \tag{1.1}$$

(where brackets $\langle\ \rangle$ denote ensemble averaging) satisfies both conditions of isotropy and local homogeneity $D(r_1, r) = D(r)$, and the Kolmogorov-Obukhov inertial law (i.e., the famous "law of two-thirds power") [13, 24]

$$D(r) = C_V^2 r^{2/3} \tag{1.2}$$

Eq. (1.2) also defines the turbulent velocity fluctuation "structure constant" C_V^2, which varies with altitude in unstable cases.

1.2.3 The acoustic properties of the atmosphere

From the point of view of acoustics, the atmosphere is a type of moving inhomogeneous medium, whose non-homogeneity is expressed, as stated above, firstly in its stratification, i.e., its main parameters vary only with altitude (this functional relationship is called a "profile"). This stratified non-homogeneity leads to continuous refractions of acoustic waves (in this case, reflection can often be ignored), and a definite profile corresponds to a definite distribution of the acoustic field.

The atmosphere is constantly in uninterrupted motion, thus making the acoustic phenomena in it more complicated. What is fortunate, however, is the fact that its motion (the wind) depends also mainly on vertical coordinates only. In other words, the atmosphere is basically a kind of moving stratified medium. This situation makes the problem relatively easy to handle.

The acoustic properties of the atmosphere are characterized by the sound velocity in it, which varies monotonically with the temperature.[2] In most cases, wind effects can be accounted for by introducing the concept of "effective-sound velocity", i.e., to consider the atmosphere to be "quiet", but the sound velocity in it would be replaced by the vector sum of the sound and wind velocities. Thereby the temperature and wind profiles are combined. A typical situation for such a profile is shown in Fig. 4. In general, this profile varies with latitude and season, and even with daily time. The variations are more violent in the near-ground surface region, but relatively stable at upper altitudes. It can be noticed that there are two minimal effective sound velocities that arise at altitudes of about 15 and 80 km, respectively. No matter how the profile varies with altitude, season, etc., these two minima always exist. Sometimes there may be a third minimum near the ground surface. Each of these minima can form "wave guides" (or "sound channels"), which are of benefit to the long-range propagation of sound waves in the atmosphere.

Secondary random non-homogeneities, however, include either the non-homogeneity in atmospheric ingredients, i.e., water droplets, fog-drops and other small particles existing in the atmosphere, or the non-homogeneity in the atmospheric motion itself, i.e., the atmospheric turbulences mentioned in the last sub-section. There are intense interactions between acoustic waves and atmospheric turbulences, and the theoretical analysis of such interactions requires the corresponding one- and three-dimensional spatial spectra of the velocity fluctuations, rather than the structure function $D(r)$ defined by Eq. (1.2). These spectra result from a Fourier transformation of the coordinate space x_i to the associated Fourier space of the variables κ_i. In terms of its magnitude, $\kappa = (\kappa_i\kappa_i)^{1/2}$ (repeated indices imply summation over the squares of components), the kinetic energy spectrum associated with $D(r)$ becomes [25]

$$E(\kappa) = 0.76C_V^2\kappa^{-5/3} \tag{1.3}$$

And the corresponding three-dimensional velocity spectrum becomes

$$\Phi_{ij}(\kappa) = \left[\delta_{ij} - \kappa_i\kappa_j/\kappa^2\right] E(\kappa)/(4\pi\kappa^2) \tag{1.4}$$

where δ_{ij} is the Kronecker delta, equal to 1 for $i = j$, and 0 otherwise.

Of course, the stratification structure that leads to acoustic refraction is not irrelevant to the turbulence structure which leads to acoustic scattering; interactions

2 In general, the atmosphere can be considered as a perfect gas and thus the sound speed in it is proportional to the square root of the absolute temperature. If a higher precision is required, an empirical formula expressing the relationship between sound speed and temperature can be adopted, and this is just a monotonic relation. More details can be seen in the following text.

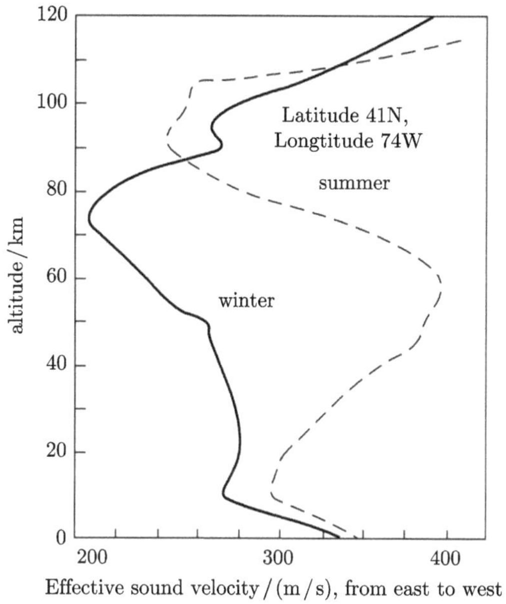

Fig. 4: Typical profile of the effective-sound velocity for the atmosphere at middle latitudes.

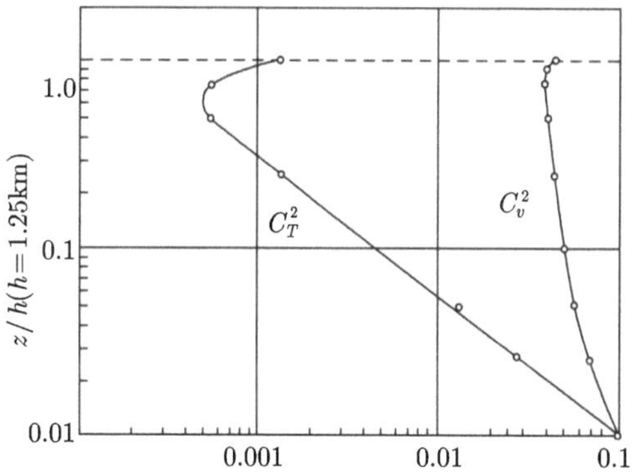

Fig. 5: Typical daytime profiles of C_T^2 and C_v^2.

also exist between them. For instance, the turbulent velocity fluctuation field acting on the mean vertical gradient of the acoustic refractivity n in the atmosphere can derive the fluctuations in n. Similarly, the structure function of refractivity fluctuations $D_n(r) = \langle | n(r_1 - r) - n(r_1) |^2 \rangle$ also satisfies the Kolmogorov-Obukhov law

$$D_n(r) = C_n^2 r^{2/3} \tag{1.5}$$

where

$$C_n^2 = C_T^2/4T_0^2 + 2(0.307)C_{eT}/4p_0T_0 + (0.307)^2 C_e^2/4p_0^2 \tag{1.6}$$

is the structure constant for refractivity fluctuations [26], where T_0, p_0 are the equilibrium temperature and pressure respectively. C_T^2 is a structure parameter for temperature fluctuations and C_{eT} is a parameter related to the correlation of humidity and temperature fluctuations.

In Fig. 5, the typical variations of the parameters C_T^2 and C_v^2 (both describing the turbulence strength) with altitude (the profiles) under daytime unstable conditions are shown [27].

1.3 Thermodynamic relationships in the atmosphere [30]

1.3.1 Equation of state and adiabatic equation

The properties of the atmosphere, just as any fluid, can be described also by any two from the density ρ, pressure p, temperature T and entropy s. Here we choose the first two variables, since they give the most intuitive concepts of the physical processes in the atmosphere.

To a high-degree approximation, the atmosphere can be considered as a perfect gas. Thus all of its molecules are identical and only interact with each other via collisions, and move along straight lines between two adjacent collisions. The specific heats at constant pressure and at constant volume of a perfect gas are kept constant. Therefore the density and the pressure are related with the temperature via the *equation of state*

$$p = (\rho/M)RT = (\rho/m)kT \tag{1.7}$$

where M is the average molecular weight (kg/kmol), m is the average mass of the molecule (kg), thus $M = N_A m$, where N_A is the Avogadro number = 6.02×10^{26} molecules/kmol; R is the universal gas constant = 8.31×10^3 J/kmol·K, and k is the Boltzmann constant = $R/N_A = 1.38 \times 10^{-23}$J/K.

Throughout the lowest 80 km of the atmosphere, the mean molecular weight remains at a constant value if 28.966 kg/kmol, whilst above 80 km it starts to decrease with height. This is due to the dissociation of O_2 and the diffusive separation of O_2 and N_2 from atomic oxygen up to 400 km. Above 400 km, atomic oxygen is predominant. This variation of M with altitude z can be expressed by an analytic expression as follows (z in km) [28]

$$M = 28.9 - 6.45 \left\{1 + \tanh\left[(z - 300)/100\right]\right\} \tag{1.8}$$

As any motion process of a perfect gas is adiabatic, from the first law of thermodynamics, the internal energy increment of a parcel of atmospheric gas is equal to the work done by the gas by changing its volume: d ε–dW, i.e.,

$$C_V \mathrm{d}T = (p/\rho^2)\mathrm{d}\rho \tag{1.9}$$

where C_V is the specific heat per unit mass at constant volume. By using the equation of state (1.7), we obtain the adiabatic equation after integration

$$p\rho^{-\gamma} = \text{const} \tag{1.10}$$

where $\gamma \equiv C_p/C_V$, and the Mayer equation $R = M(C_p - C_V)$ was applied, while C_p is the specific heat per unit mass at constant pressure.

In the region of the upper atmosphere where molecular oxygen and nitrogen gives way to atomic oxygen, γ will increase from its value of 1.4 for a diatomic gas to a value of 1.67 for a monatomic gas. In a similar manner to Eq. (1.8), an analytical form of γ may be used (z also in km)

$$\gamma = 1.4 + 0.135\left\{1 + \tanh\left[(z - 300)/100\right]\right\} \tag{1.11}$$

As the same as any form of motion in a perfect gas, the acoustic process is also adiabatic, and thus infinitesimal changes in pressure p_1 and in density ρ_1 are connected by the following relation:

$$p_1 = \left(\partial p/\partial \rho\right)_s \rho_1 = c^2 \rho_1 \tag{1.12}$$

where $c = (\partial p/\partial \rho)_s^{1/2}$ is the sound speed.

Except for processes in which heat exchange exists (e.g., when taking heat conduction into account such that acoustic attenuation is caused), acoustic waves with extra-low frequencies also cannot be regarded as adiabatic. For these cases the above discussion must be modified.

1.3.2 Barometric equation and scale height, isothermal atmosphere and atmosphere with constant temperature gradient

The atmosphere up to considerable altitudes is governed by a hydrostatic balance between the vertical pressure-gradient force and the gravitational force, which can be expressed as the hydrostatic equation

$$\partial p/\partial z = -\rho g \tag{1.13}$$

where g is the gravitational acceleration, the minus sign arises because g is directed downwards, whilst altitude z is measured upwards. Using the equation of state (1.7) to eliminate ρ, the barometric equation can be obtained after integration

$$p = p_0 \exp\left(\int_0^z \left[\frac{Mg}{RT}\right] \mathrm{d}z\right) \tag{1.14}$$

where p_0 is the pressure at $z = 0$.

One of the simplest atmospheric models is the isothermal atmosphere, in which T is independent of z, and thus the above equation can be simplified. A similar relation concerning the density can be obtained in the same way, and both are referred to as Halley's law

$$p/p_0 = \rho/\rho_0 = \exp(-z/H) \tag{1.15}$$

where an important parameter, the so-called "*scalar height*" is introduced

$$H = \frac{Mg}{RT} \tag{1.16}$$

which, in general, is a function of z. The physical meaning of H lies in that the scale height at any particular altitude z represents the height that a column of air with uniform density ρ_0 would have if it produced the uniform pressure p_0 at $z = 0$, provided that the column had the same temperature as at altitude z, and the variation in g can be ignored.

H may be re-written in terms of the ratio of specific heats γ and the sound speed c. Because we can get

$$c^2 = \gamma p/\rho = \gamma RT/M \tag{1.17}$$

by the perfect gas law (1.7), so that

$$H = c^2/\gamma g \tag{1.17a}$$

A model atmosphere with a constant temperature gradient is somewhat more complex than the isothermal atmosphere, but it does possess basic theoretical significance, since the combination of both models create a more realistic atmosphere.

In this case we can assume

$$T = T_0 + \alpha z \tag{1.18}$$

where α is a constant. In this case, the solution of Eq. (1.14) is, instead of Halley's law in case of isothermal atmosphere,

$$p = p_0(T/T_0)^{-Mg/R\alpha} \tag{1.19}$$

At the same time, in this case ρ/ρ_0 is no longer equal to p/p_0, and in fact

$$\rho = \rho_0(T/T_0)^{-1-(Mg/R\alpha)} \tag{1.20}$$

In the case of an adiabatic or isoentropic atmosphere, by using (1.4) we have

$$T\rho^{1-\gamma} = \text{const} \tag{1.21}$$

$$Tp^{(1-\gamma)/\gamma} = \text{const} \tag{1.22}$$

Differentiating with respect to z and using the hydrostatic equation (1.13) and the equation of state (1.7), the adiabatic temperature gradient can be obtained

$$\alpha_S = \partial_z T|_S = \left[(\gamma-1)/\gamma\right](T/p)\partial_z p = -\left[(\gamma-1)Mg\right]/\gamma R \cong -9.8\ \text{K/km} \tag{1.23}$$

which is an important criterion for judging the stability of the atmosphere: when the negative temperature gradient exceeds this value, the atmosphere would be unstable. The quantity $-\alpha$ is called the lapse rate in meteorology, and correspondingly, $-\alpha_S$ is called the adiabatic lapse rate.

After substituting Eq. (1.18) and (1.23) into (1.19) and (1.20) respectively, we get the profiles of both pressure and density in adiabatic atmosphere

$$p = p_0 \left\{1 - \left[(\gamma - 1)Mg/\gamma RT_0\right] z\right\}^{\gamma/(\gamma-1)} \tag{1.24}$$

$$\rho = \rho_0 \left\{1 - \left[(\gamma - 1)Mg/\gamma RT_0\right] z\right\}^{1/(\gamma-1)} \tag{1.25}$$

Substituting the related numerical values into (1.25), we can see $\rho = 0$ when $z = 27.5$ km, which can be regarded as the upper limit of the adiabatic atmosphere.

Adiabatic equilibrium tends to be set up in an atmosphere possessing convective motion (e.g. turbulence) because convection in gases proceeds more quickly than the condition necessary to establish isothermal equilibrium. So the atmosphere below the tropopause at an altitude of 100 km should exhibit adiabatic equilibrium. To a certain extent, this is true of the whole troposphere, though in fact it is not in perfect adiabatic equilibrium: it has a lapse rate of 5 K/km but not 9.8 K/km as given by (1.23); while the tropopause lies between 8 and 20 km but not at 27.5 km as given by (1.19). However, only in the troposphere is the temperature distribution controlled by convection, while in the stratosphere it is controlled by radiation. This is because the atmosphere allows visible and infra-red radiation emitted by the sun to penetrate with very little absorption. Therefore the incoming radiation does not actually heat the atmosphere, except at very high altitude, but instead it heats the ground, which then re-radiates the energy in the far infra-red part of the spectrum. The minor constituents such as water vapor, carbon dioxide, ozone, etc. absorb infrared strongly. Therefore, if there were no convective motions, the temperature distribution would fall very rapidly near the ground where the water vapor and carbon predominate. This rapid fall would make the atmosphere unstable, and this instability will be balanced eventually by convective motions.

1.3.3 Potential temperature and Vaisala-Brunt frequency

Static equilibrium can occur only when forces are properly balanced, but the converse does not hold. A balance of forces in no way guarantees that the equilibrium configuration does in fact occur. In the atmosphere, there will always be small extraneous forces to disturb a strict equilibrium condition. The manner in which a system responds to small disturbing influences is characterized by its stability, and this stability in the atmosphere is controlled by the temperature distribution.

Besides the lapse rate, another meteorological parameter is introduced, i.e. the *potential temperature* θ. This is the temperature that a parcel of air would attain if it was brought adiabatically to a reference pressure (in general it is chosen as the pressure over the sea level p_0, in practice being 1000 mbar). By this definition we obtain Poisson's equation

$$\theta = T(p/p_0)^{(1-\gamma)/\gamma} \cong T(1000/p)^{0.286} \quad (1.26)$$

It can then readily be seen that an atmosphere in adiabatic equilibrium has a constant potential temperature throughout. The practicality of the potential temperature lies in that its gradient implies that heat is transfered upwards or downwards, and thus represents whether an air parcel is "stable" or "unstable", i.e., an accident displacement of the parcel results in either a restored motion which diminishes the displacement or inversely increases the displacement. Therefore the potential temperature 0 is a very proper quantity in discussing atmospheric states.

When a parcel is displaced vertically in a stable atmosphere, then a restoring force will act. If the frequency of the resultant oscillation is high enough, then the effects due to gravitation are negligible and the wave propagates as an ordinary sound wave with the compressibility of the medium providing the restoring force. At lower frequencies, however, the gravitational restoring forces are comparable to those due to compressibility and thus can no longer be ignored.

Assume that an air parcel at altitude z has experienced an infinitesimal displacement Δz, its density has thus changed from initial value ρ to $\rho + \Delta\rho$, where $\Delta\rho = (\Delta\rho/\Delta p)\Delta p$. If the displacement takes place adiabatically, then $\Delta p/\Delta\rho = c^2$. In its initial state the density of the parcel ρ equals the density ρ_0 of the external atmosphere. After being displaced, in general these two densities are no longer equal. As the density decreases with increasing altitude, after displacing the environmental pressure the parcel is also incrementally displaced: $\Delta\rho_0 = \mathrm{d}_z\rho_0\Delta z$. Thus the parcel is acted upon by a buoyancy force that must balance the inertial force

$$\rho_0 \mathrm{d}t^2 \Delta z = g(\Delta\rho_0 - \Delta\rho) = g(\mathrm{d}_z\rho + \rho_0 g/c^2)\Delta z$$

Providing that the hydrostatic equilibrium is not disturbed by the displacement. Provided

$$(g/\rho_0)\left(\mathrm{d}_z\rho_0 + \rho_0 g/c^2\right) < 0 \quad (1.27)$$

the solution of the above equation is a free oscillation with the frequency[3]

$$\omega_\mathrm{B} = \left[-(g/\rho)\mathrm{d}_z\rho + (g/c)^2\right]^{1/2} \quad (1.28)$$

This intrinsic frequency of an atmospheric oscillation is called *Vaisala-Brunt frequency* (abbreviated as V-B frequency), which is one of the most important parameters in atmospheric dynamics. For a perfect gas, obviously

$$\omega_\mathrm{B}^2 = g(\alpha - \alpha_S)/T \quad (1.28a)$$

3 In general this quantity is denoted by the letter v in literatures, but we adopt symbols ω_B and ω_g to reflect the systematic nature.

For an isothermal atmosphere ($\alpha = 0$) further,

$$\omega_g^2 = (\gamma - 1)(g/c)^2 \tag{1.28b}$$

So we can obtain

$$\omega_B^2 = \omega_g^2 + (g/c^2)\partial_z c^2 = \omega_g^2 + (g/T)\partial_z T \tag{1.28a'}$$

By using some of the aforementioned relations and definitions it is easy to see that the V-B frequency can be expressed in terms of the sound speed entirely

$$\omega_B^2 = -g\left[(g/c^2)(1-\gamma) - (2/c)\partial_z c\right] \tag{1.28'}$$

As well as in a more concise form, as expressed in terms of its potential temperature:

$$\omega_B^2 = (g/\theta)\partial_z\theta \tag{1.28''}$$

1.4 Fundamental relations of atmospheric dynamics

As is well-known, the motion of a fluid can be described by two systems: the Lagrange description, in which the coordinate system is attached to the particles of the moving fluid; and the Euler description, in which the coordinate system is fixed in space. These two systems are connected with each other by the *Stokes' operator*[4]

$$D_t = \partial_t + \boldsymbol{v} \cdot \nabla \equiv \partial_t + v_i \partial_i \tag{1.29}$$

Where $\boldsymbol{v}$ is the moving velocity of the fluid, in our case it is the wind velocity in the atmosphere.

By using this operator "related to both space and time", the fluid-dynamic fundamental equations describing the motion of the atmosphere can be written in the most succinct form.

1.4.1 Equation of motion

The implication of Newton's second law of motion "the time- and space-rate of change in momentum equal to force acted on unit volume" stated by the equation of motion

4 The notations for derivatives adopted in this book are: ∂_t represents the time derivatives $\partial/\partial t$; ∂_x, ∂_y, ∂_z and ∂_i represent the space derivatives $\partial/\partial x$, $\partial/\partial y$, $\partial/\partial z$ and $\partial/\partial x_i . \equiv \nabla$ respectively. Besides the obvious advantage of succinctness, this representation is superior also in the unification with tensor notations. We specify that, any written-out subscript implies that its three values will be taken all over and these values correspond to the three components in rectangular coordinates; thus vector $\boldsymbol{v}$ will be written as v_i. The subscript appeared twice implies that after the three values are taken they will be added together, e.g., $\boldsymbol{A} \cdot \boldsymbol{B} = A_1B_1 + A_2B_2 + A_3B_3 = \sum A_iB_i \equiv A_iB_i$. On the other hand, the unit tensor (or the Kronnecker's notation, as introduced in Eq. (1.4)) will be introduced as $\delta_{ik}A_k = A_i$. A second-order tensor, A_{kl}, will have $\delta_{ik}A_{kl} = A_{il}$, $\delta_{ik}A_{ik} = A_{il}A_{ii}$, etc.

(the *Euler equation*) can be expressed in terms of the above notations as

$$\rho D_t \boldsymbol{v} = \boldsymbol{f} \tag{1.30}$$

Where $\boldsymbol{f}$ has different meanings in different cases, it is just $-\nabla p$, if:
If what is under consideration are the ordinary acoustic waves with additional forces and dissipative forces being neglected; at this time Eq. (1.30) becomes

$$D_t \boldsymbol{v} = \partial_t \boldsymbol{v} + (\boldsymbol{v} \cdot \nabla)\boldsymbol{v} = -(1/\rho)\nabla p \tag{1.30a}$$

An additional term $\boldsymbol{g}$ must be added if the gravitational force is taken into account; an extra term, the Coriolis force, should further be added for the planetary waves and the atmospheric tides with even more longer wavelengths. See the discussions in chapter VI for details.

As can be seen from Eq. (1.30a), the acceleration of a given fluid particle is composed of two parts: the first term in the left-hand side represents the local acceleration of a specific point in the fluid, namely, the time-rate of change for the velocity at a point fixed in space (non-stationary flow); the second term, referred to as "convective" acceleration, is due to the difference between the velocities (at the same instant) at two points $\mathrm{d}\boldsymbol{r}$ apart, where $\mathrm{d}\boldsymbol{r}$ is the distance moved by the given fluid particle during the time $\mathrm{d}t$.

For the isoentropic motion such as the acoustic wave motion, Eq. (1.30a) can be written in another form. From the second law of thermodynamics

$$\mathrm{d}w = T\mathrm{d}s + (1/\rho)\mathrm{d}p$$

where $w = \varepsilon + p/\rho$ is the enthalpy per unit mass (ε is the internal energy per unit mass), $\mathrm{d}w = (1/2)\,\mathrm{d}p$ when s = const., thus (1.30a) becomes

$$\partial_t \boldsymbol{v} + (\boldsymbol{v} \cdot \nabla)\boldsymbol{v} = -\nabla w \tag{1.30a$'$}$$

Using the well-known formula in vector analysis

$$(1/2)\nabla v^2 = \boldsymbol{v} \times \nabla \times \boldsymbol{v} + (\boldsymbol{v} \cdot \nabla)\boldsymbol{v}$$

the above equation can be rewritten as

$$\partial_t \boldsymbol{v} + (1/2)\nabla v^2 - \boldsymbol{v} \times \nabla \times \boldsymbol{v} = -\nabla w \tag{1.30a$''$}$$

If taking the operation $\nabla\times$ (curl) in both sides of the above equation, because $\nabla \times \nabla = 0$, we obtain the equation involving the velocity $\boldsymbol{v}$ only

$$\partial_t(\nabla \times \boldsymbol{v}) = \nabla \times (\boldsymbol{v} \times \nabla \times \boldsymbol{v})$$

1.4.2 Equation of continuity, equation of state, tensor presentation

The *equation of continuity* representing the conservation of mass can be written as

$$D_t\rho + \rho\nabla \cdot \boldsymbol{v} = 0 \tag{1.31}$$

Or as

$$\partial_t\rho + \nabla\cdot(\rho\boldsymbol{v}) = 0 \tag{1.31'}$$

While the equation of state (1.12) representing the conservation of energy can be written as

$$D_t p = c^2 D_t\rho = -c^2\rho\nabla\cdot\boldsymbol{v} \tag{1.32}$$

In some situations, expressing the continuity equation (1.31′) and Euler's equation (1.30a) in tensor presentations is more convenient

$$\partial_t\rho = -\rho v_k, \quad \partial_t v_i = -v_k\partial_k v_i - (1/\rho)\partial_i p$$

From the last two equations we have

$$\begin{aligned}\partial_t\rho v_i &= -\rho v_k\partial_k v_i - \partial_i p - v_i\partial_k\rho v_k = -\partial_i p - \partial_k\rho v_i v_k\\ &= \delta_{ik}\partial_k p - \partial_k\rho v_i v_k \equiv -\partial_k\Pi_{ik}\end{aligned} \tag{1.30b}$$

where

$$\Pi_{ik} \equiv p\delta_{ik} + \rho v_i v_k \tag{1.33}$$

is called the *momentum flux-density-tensor* (sometimes, this name is referred to as the second term only, while the first one is called the stress tensor).

The set of equations (1.30) to (1.32) represent the fundamental equations which describe all kinds of fluid motions including acoustic waves and other atmospheric waves. Among them only the continuity equation remains unchanged, whereas the other two will be modified at varying degrees corresponding to different conditions.

1.4.3 Conservation laws

The above three fundamental equations may be reduced to three conservation laws. For doing so, we define the amount a certain physical quantity φ passing through a unit area perpendicular to the fluid velocity $\boldsymbol{v}$ in unit time as the *flux density* of φ

$$\boldsymbol{j} = \varphi\boldsymbol{v} \tag{1.34}$$

If φ represents the fluid density ρ, then the above equation represents the conservation of mass

$$\boldsymbol{j}_{\mathrm{m}} = \rho\boldsymbol{v} \tag{1.34a}$$

Which is equivalent to the continuity equation; at the end of the last subsection we have already seen that Euler's equation is equivalent to the conservation of momentum, and a similar relation holds for energy flux.

The energy per unit volume of the fluid is the sum of the kinetic energy $\rho v^2/2$, and the internal energy $\rho\varepsilon$ (the absence of potential energy is due to us ignoring the gravitational force here). Thus the variation of the kinetic energy with time can be written as

$$\begin{aligned}\partial_t(\rho v^2/2) &= (v^2/2)\partial_t\rho + \rho\boldsymbol{v}\cdot\partial_t\boldsymbol{v}\\ &= -(v^2/2)\nabla\cdot(\rho\boldsymbol{v}) - \boldsymbol{v}\cdot\nabla p - \rho\boldsymbol{v}\cdot(\boldsymbol{v}\cdot\nabla)\boldsymbol{v}\end{aligned}$$

In which the continuity equation (1.31′) and Euler's equation (1.30a) are applied. The last term can be transformed into $-\frac{1}{2}\boldsymbol{j}_{\mathrm{m}}\cdot\nabla v^2$ by using the formulae in vector analysis; and ∇p can be transformed according to the second law of thermodynamics (see the discussion preceding Eq. (1.30a′)), then we have

$$\partial_t(\rho v^2/2) = -(v^2/2)\nabla\cdot\boldsymbol{j}_{\mathrm{m}} - \boldsymbol{j}_{\mathrm{m}}\cdot\nabla\left[w + (v^2/2)\right] + \boldsymbol{j}_{\mathrm{m}}\cdot T\nabla s$$

We now calculate the variation of internal energy with time. Also from the second law of thermodynamics, we obtain

$$\mathrm{d}(\rho\varepsilon) = \varepsilon\mathrm{d}\rho + \rho\mathrm{d}\varepsilon = \varepsilon\mathrm{d}\rho + \rho T\mathrm{d}s + (p/\rho)\mathrm{d}\rho = w\mathrm{d}\rho + \rho T\mathrm{d}s$$

thus

$$\partial_t(\rho\varepsilon) = w\partial_t\rho + \rho T\partial_t s = -w\nabla\cdot\boldsymbol{j}_{\mathrm{m}} - \boldsymbol{j}_{\mathrm{m}}\cdot T\nabla s$$

here in the second step we have used Eq. (1.31′) and the continuity equation for entropy

$$\partial_t(\rho s) + \nabla\cdot(\rho s\boldsymbol{v}) = 0 \tag{1.35}$$

which can be seen directly from Eq. (1.34) with φ representing the fluid's entropy density ρs. Subtracting Eq. (1.31′) from Eq. (1.35), we obtain the adiabatic equation for the motion of the ideal fluid

$$\partial_t s + \boldsymbol{v}\cdot\nabla s = 0 \tag{1.35a}$$

Therefore the rate of change in time for the total energy is

$$\begin{aligned}\partial_t\left[(\rho v^2/2) + \rho\varepsilon\right] &= -[w + (v^2/2)]\nabla\cdot\boldsymbol{j}_{\mathrm{m}} - (\boldsymbol{j}_{\mathrm{m}}\cdot\nabla)[w + (v^2/2)]\\ &= -\nabla\cdot\left\{\boldsymbol{j}_{\mathrm{m}}[(v^2/2) + w]\right\}\end{aligned} \tag{1.36}$$

This is the *continuity equation for energy flux density*, which is very similar in form to the continuity equation for mass (1.31′), only that the ε in the left-hand side changes to w in the right-hand side. The physical meaning of this can be interpreted as follows.

Integrating Eq. (1.35) over some volume, and using Gauss' theorem to transform the volume integral into the surface integral, we obtain

$$\partial_t\int[(\rho v^2/2) + \rho\varepsilon]\mathrm{d}V = -\oint[(v^2/2) + w]\boldsymbol{j}_{\mathrm{m}}\cdot\mathrm{d}S \equiv -\oint\boldsymbol{j}_{\mathrm{E}}\cdot\mathrm{d}\boldsymbol{S} \tag{1.37}$$

where

$$\boldsymbol{j}_\mathrm{E} \equiv \boldsymbol{j}_\mathrm{m}[(v^2/2) + w] \tag{1.38}$$

is called the energy flux density vector (Poynting-Umov vector). After substituting the definition of w, the right-hand side of Eq. (1.36) reduces to

$$-\oint \left[(v^2/2) + \varepsilon\right] \boldsymbol{j}_\mathrm{m} \cdot \mathrm{d}\boldsymbol{S} - \oint p\boldsymbol{v} \cdot \mathrm{d}\boldsymbol{S}$$

The first term represents the kinetic energy and internal energy transported directly through the surface in unit time by the fluid mass, while the second term represents the work done by the pressure force on the fluid within the surface. The physical meaning of (1.36) is thus obvious: the decrease of fluid energy within a certain volume in unit time is equal to the sum of the energy flowing out of this volume in unit time and the work done by the pressure force – i.e., the law of conservation of energy.

To summarize, the flux densities and continuity equations are universal concepts in fluid dynamics. Mass, entropy and energy themselves are all scalars, while the flux density of mass (in another sense, it is just the momentum per unit volume) $\boldsymbol{j}_\mathrm{m} = \rho\boldsymbol{v}$, of entropy $\boldsymbol{j}_S = s\boldsymbol{j}_\mathrm{m}$ and of energy $\boldsymbol{j}_\mathrm{E} = [(v^2/2) + w]\boldsymbol{j}_\mathrm{m}$ are all vectors; momentum itself is a vector, while its flux density is determined by a second-order tensor[(1.33)]. These quantities satisfy "continuity equations" very similar in form, respectively (compare Eqs. (1.31′), to (1.35), (1.36) and (1.30b)).

1.4.4 Geopotential altitude and Coriolis force

The gravitational acceleration $\boldsymbol{g}$ possesses an importance which cannot be ignored when describing the behavior of the atmosphere. Although it can be regarded as a constant at small scales, in fact it is a function of both latitude and altitude: Taking the sea level value as the criterion, it is about 0.5 percent higher at the poles than at the equator; its dependence on altitude can be written as

$$g(z) = \frac{g(0)}{[1 + (z/R_\mathrm{E})]^2} \tag{1.39}$$

here we assume that the effects of earth's rotation may be neglected, where $g(0)$ is the value at sea level and R_E is the radius of the earth.

The above dependence is often merged into a quantity called "geopotential altitude", which is defined as

$$\chi = \frac{1}{g_0} \int_0^z g(z)\mathrm{d}z \tag{1.40}$$

Where g_0 is a dimensionless quantity, and is chosen to be 9.80 $\mathrm{m/s^2}$ in order to make the numerical value of χ similar to that of the corresponding geometric altitude. As g_0 is taken to be dimensionless, the dimension of χ is energy per unit mass, thus its physical meaning is: the potential energy that would be gained by a unit mass lifted from the earth's surface to the altitude z.

The earth's surface is not fixed in space but rotates about an axis with an angular velocity $\boldsymbol{\Omega}_E$. For wave motions with very long wavelengths, the "apparent force" or inertial force (Coriolis force) acting on a body (e.g., an air parcel) in motion that arises due to the rotation of the observer on the earth's surface must be included. Such a force acts everywhere at right angles to the motion of the object and is proportional to the speed, and thus it can change the direction of motion only but not the speed.

Between a vector measured in an inertial coordinate system fixed in space (represented by subscript s) and the vector measured in a rotating coordinate system (represented by subscript r), there exists the relationship

$$(\mathrm{d}_t)_s = (\mathrm{d}_t)_r + \boldsymbol{\Omega}_E \times \boldsymbol{r} \tag{1.41}$$

From which the relationship between velocities in these two systems can be easily obtained

$$\boldsymbol{v}_s = \boldsymbol{v}_r + \boldsymbol{\Omega}_E \times \boldsymbol{r} \tag{1.42}$$

where $\boldsymbol{r}$ is the radius vector from the earth's center to the body in motion, $\boldsymbol{v}_r = (\mathrm{d}_t\, \boldsymbol{r})_r$; using (1.41) once again, the relationship between accelerations is obtained

$$\boldsymbol{a}_s = (\mathrm{d}_t \boldsymbol{v}_s)_r + \boldsymbol{\Omega}_E \times \boldsymbol{v}_s = \boldsymbol{a}_r + 2(\boldsymbol{\Omega}_E \times \boldsymbol{v}_r) + \boldsymbol{\Omega}_E \times (\boldsymbol{\Omega}_E \times \boldsymbol{r}) \tag{1.43}$$

where the last term is the centrifugal acceleration, while the second term is the *Coriolis acceleration* of interest. Since $\Omega \approx 7.29 \times 10^{-5}$ rad/s, $R_E \approx 6370$ km, so that the maximum value of the centrifugal acceleration (roughly in opposite direction to $\boldsymbol{g}$) $\Omega_E^2 R_E \approx 3.38 \times 10^{-5}$ km/s$^2 \ll g$, and thus will be always neglected in the discussions later.

What will be neglected also is the horizontal component of the Coriolis force (this neglect is often referred to as the "*traditional approximation*"); since if this is not done, then the equations of motion when solved would indicate that an airflow that starts out by being horizontal would be purely vertical six hours later. This is not what actually happens because the airflow follows the spherical motion of the earth's surface. On the other hand, quite obviously, the vertical component of the Coriolis force will be a function of the latitude. Just this dependence forms the restoring force for the planetary waves (see also chapter 6 section 4).

1.5 Types of atmospheric waves [30]

The characteristic of wave motion is merely energy transferring from one point to another without producing any permanent displacement of the medium as a whole, therefore the wave may be considered as a perturbation on the steady slowly changing background. The mathematical analyses of wave motion deserve to be one of the greatest triumphs of human powers in synthesis. Various types of wave motion can be reduced to the solutions of corresponding differential equations.

The specialties of the earth's atmosphere are represented as a compressible, rotating, spherical fluid that is permeated by density- and temperature-gradients, which is capable of sustaining a large number of wave phenomena. All of these waves (electromagnetic waves are excluded of course) are objects of study in "generalized" atmospheric acoustics.

All of these atmospheric waves (total number exceeds twenty, as shown in Table I [5] can be resolved into three fundamental types: vertical transverse waves which propagate horizontally and are composed of vertical displacements; horizontal transverse waves which propagate horizontally with horizontal displacements perpendicular to the propagation direction; and longitudinal waves whose displacements are in the same direction as the propagation. These three types are depicted in Fig. 6 [29]. All of these atmospheric waves can either be referred directly to these types, or be regarded as combinations of them.

In most cases, these waves exist in the form of infinitesimal perturbations superposed on a steady state of the atmosphere, and thus satisfy linear equations, namely small disturbances of different amplitude, wavelength or frequency can then be superimposed without interactions. Inversely, a wave, however complex its shape is, can also be analyzed into regular sinusoidal components (Fourier components). The amplitude of the wave will attenuate during its progressing, in some cases, however, it may actually grow up also. It may then eventually grow to such an amplitude for which nonlinear effects become important; at this time the interactions of one wave with another and with itself can no longer be ignored.

The best known, and most important, atmospheric waves is the principal part of the objects studied in this book – the sound wave. Sound waves, or acoustic waves, are longitudinal waves formed by a balance between the compressibility (resistance to changes in its volume) and inertia (resistance to a change of velocity) of a fluid. In a homogeneous stationary fluid in the absence of an external force, acoustic waves are the only type of waves that can exist. Besides an ordinary audible sound whose frequency spectrum covers the response range of human hearing, the atmosphere also contains an abundance of infrasound whose frequencies lie below the lower limit of hearing. Ultrasound, whose frequencies lie above the threshold of human hearing, can only exist at very small scales due to their rapid damping in air.

The earth's atmosphere is a fluid continually being acted upon by gravity, therefore its density will inevitably decrease with an increase in altitude, and thus a gradient is formed. This density gradient endows the atmosphere with a stability that is completely lacking in a homogeneous fluid. When the magnitude of the stabilizing

5 The basic idea of this table is cited from [30], some of the main items in which will be mentioned more or less in successive chapters.

Tab. I: Atmospheric wave system

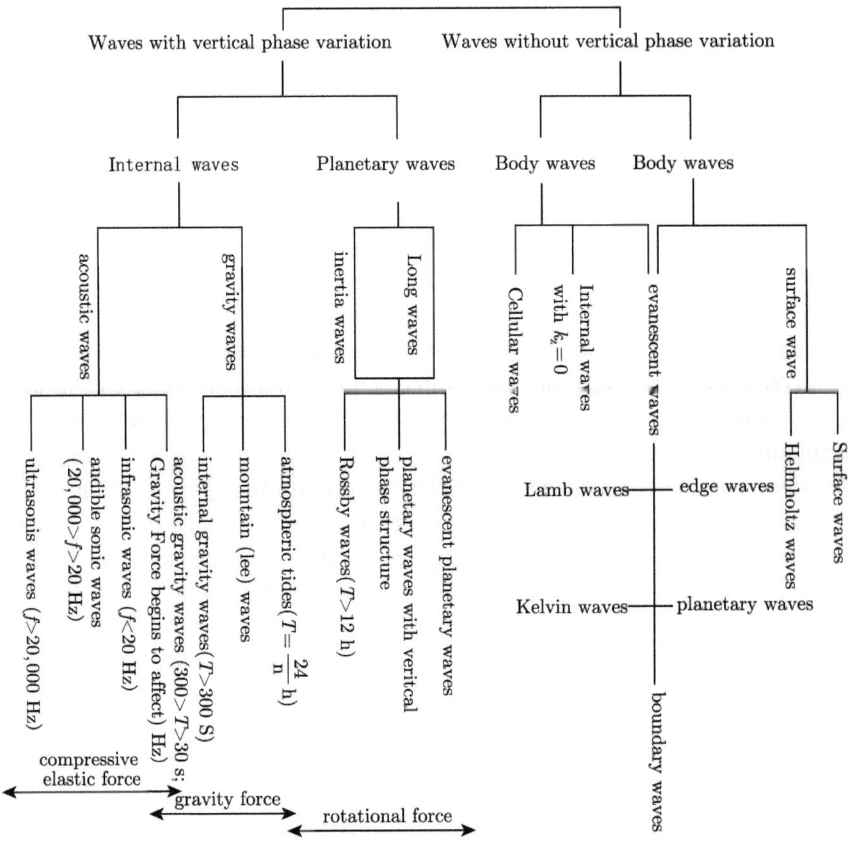

restoring force caused by this density gradient becomes comparable with the compressive force on which the ordinary acoustic waves rely for propagating, the resulting waves are referred to as acoustic gravity waves (AGW). Such waves are no longer purely longitudinal (except when they propagate vertically) because gravity has produced a component of air particle motion that is transverse to the propagation direction.

When the above stability-restoring force has been strengthened further such that it causes the dominant motion in the atmosphere, internal gravity waves (IGW) are created. The mechanism of such waves is the same as those of ordinary (external) gravity waves existing at the interface between two different media, merely the former is existing at the "stratum of density stratification" inside the same medium.

In general, the Froude number Fr is introduced to measure the significance of the gravitational force. The definition is

$$Fr = |\boldsymbol{U}|/(gL)^{1/2}$$

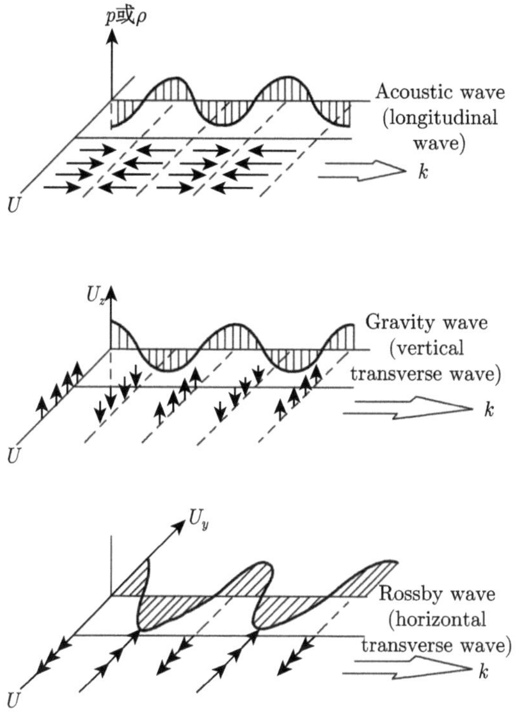

Fig. 6: Three principal types of atmospheric waves.

Where $\boldsymbol{U}$ is a characteristic velocity and L is a characteristic length. In the case of atmospheric gravity waves, the wind speed is chosen as $\boldsymbol{U}$, while L is taken as the height required for the density to drop to one half of its original value. Hydrostatic forces operate in the range of $F \ll 1$ when gravity is the force of dominant importance compared with the compressibility. Conversely, $F \gg 1$ represents the case when the gravitational force is dwarfed by pressure and inertia – most aerodynamic and hydrodynamic flows are in this category. In our case: acoustic waves have $F > 1$, AGWs have $F \sim 1$, and IGWs have $F < 1$.

In addition to the gravitational force, the curvature and the rotation of the earth also influence wave motions. As pointed out in the above section, the strength of the Coriolis effect varies with altitude, and this variation acts as an external force field which results in horizontally transverse waves with extra-long wavelengths (up to thousands of kilometers). These waves are referred to as planetary waves, or as Rossby waves after its discoverer. The phase velocity of such waves is always directed towards the west and is often directed opposite to the background wind. Thus the concepts of these kinds of waves was primarily of meteorological interest in regions of prevailing westerlies, and provided a theory for describing the pressure distribution associated with moving wave-like high- and low-pressure systems.

Shearing waves can occur also in the atmosphere, and it is of special importance at the interface (called the "front") between two adjacent air parcels with different physical properties.

Atmospheric shear flows are always unstable, and any regular wave is rolled up by nonlinear effects into discrete vortices. IGWs and surface gravity waves, whose wavelengths are much smaller than Rossby waves, can develop unstable vortices as well. An important typical case are gravity waves produced by shear flows at the lee side of a mountain, i.e., the so-called mountain lee waves, which will be studied in some detail in chapter 8.

Atmospheric motions whose periods are sub-multiples of the solar or lunar day are called atmospheric tidal oscillations. Differing from sea tides, which are produced mainly by gravitational effects from the moon, atmospheric tides are produced mainly by heating effects from the sun, but the gravitational effects from both the sun and the moon are of secondary importance. Another difference lies in that sea tides can be gauged accurately by measuring the changing height of the water surface, while air tides cannot be measured in this way since the atmosphere does not have such clearly defined boundaries. The alternative is to use a barometer at the bottom of the aerial "ocean". The vertical accelerations of the air are so small that the barometer can effectively measure the weight of the overlying air. Therefore an above-average barometer reading implies a high tide of air, whereas an under-average reading corresponds to a low tide.

There is common ground between atmospheric tides and atmospheric gravity waves: the amplitudes of both increase as $\rho^{-1/2}$. This increase in amplitude is a direct consequence of the conservation of energy. If the wave amplitude is represented by A, then the kinetic energy $\rho A^2/2$ remains constant. In a real atmosphere, the density ρ decreases almost exponentially with altitude so that the amplitude A increases correspondingly. Such an "upward amplification" phenomenon of the tides and gravity waves indicates that there are very considerable components of the wind velocity in the upper atmosphere, provided that these waves can actually propagate to these heights. In fact, both of these types of waves actually do exist at ionospheric heights (60$\sim$500 km altitudes). Atmospheric tidal winds can also produce small-scale variations in the geo-magnetic field. This due to the fact that tidal winds set charged particles in the ionosphere into motion, thus producing an electric field plus a current (dynamo effect), and eventually variations in the magnetic field. These variations are around 4×10^{-4} of the background field.

Since many of the important geophysical forces, such as the gravitational force and the Coriolis force, have specific directions, waves propagating in the earth's atmosphere are anisotropic, i.e. the wave properties are not the same in all directions. On the other hand, these waves are also dispersive, namely that wave frequency depends on wavelength (or wave number). In general, dispersion can arise from two distinct effects: one is structural in origin and depends only on the properties of the medium; the other one is geometric in origin, and arises from interference effects due to reflections at the boundaries of the medium. The dispersion of atmospheric waves is referred to the former, and thus is connected with internal resonant frequencies possessed by the atmosphere itself: V-B resonant frequency for gravity waves; acoustic resonance for acoustic waves; and a gyroscopic resonance due to precessional motions in a rotating fluid for planetary waves. One consequence of dispersion is that the direction for energy flux (group velocity) is different from that for phase propagation (phase velocity).

Chapter 2
Basic Concepts and Processing Methods

The generation and propagation of sound waves are always associated with a medium. The medium of interest in this book is of course the atmosphere, which is a very complex medium as pointed out in Chapter I. All of the complex factors cannot be considered simultaneously in our analysis. Hence, as a basic model, the atmosphere is treated as an ideal medium. Firstly, the curvature and rotation of the earth are ignored, meaning the atmosphere is flat and non-rotational; secondly, the atmosphere is assumed to be static (windless), lossless, and homogeneous, except the motion caused by sound waves; finally, only sound waves with high enough frequency are considered so that the effects of gravity can be ignored. With those assumptions, the atmosphere acts as an ideal medium and the ordinary wave equation is applicable.

The derivation of the wave equation is based on the three basic principles in Fluid Mechanics discussed in Chapter I. Pressure p, density ρ, and particle velocity v are chosen to describe the state of the atmosphere.

When there is a fluctuation in the atmosphere, the three aforementioned quantities (as well as other related quantities) will change to

$$p \to p_0 + p_1, \quad \rho \to \rho_0 + \rho_1, \quad v \to v_0 + v_1 \tag{2.1}$$

Where the subscript 0 denotes static quantities, and subscript 1 denotes the perturbation quantities. In this Chapter, $v_0 = 0$ is assumed to be as discussed previously. All of the quadratic terms of the perturbation quantities are disregarded for small amplitude acoustic waves, which results in the linearization of the related equations.

2.1 Wave equation in homogeneous atmosphere

2.1.1 Derivation of the wave equation

In principle the basic coupled equations, Eq. (1.30) to (1.32), from hydrodynamics should completely describe fluid movement, including sound waves. However, these equations are generally nonlinear and difficult for analysis. For small amplitude acoustic waves, all of the equations can be linearized, and elementary elimination process can be applied to unify them into a single equation.

Substituting the sound wave perturbation expression given by Eq. (2.1) into Eqs. (1.30a), (1.31′), and (1.32), and considering now $v_0 = 0$, and neglect the quadratic terms of small quantities, the three equations become

$$\partial_t v_1 = -\frac{1}{\rho_0} \nabla p_1 \tag{2.2}$$

$$\partial_t \rho_1 + \rho_0 \nabla \cdot v_1 = 0 \tag{2.3}$$

$$\partial_t p_1 = c^2 \partial_t \rho_1 \tag{2.4}$$

Applying an elementary elimination process to those equations, we obtain the wave equations for sound pressure p_1, density perturbation ρ_1, and particle velocity v_1, respectively. For example, substituting Eq. (2.4) into Eq. (2.3) yields

$$\frac{1}{c^2}\partial_t p_1 + \rho_0 \nabla \cdot v_1 = 0 \tag{2.5}$$

Eliminating v_1 by taking the derivative of Eq. (2.5) with respect to time, and taking the divergence of Eq. (2.2), the wave equation for p_1 is obtained

$$\nabla^2 p_1 = \frac{1}{c^2}\partial_t^2 p_1 \tag{2.6}$$

Substituting Eq. (2.4) into (2.6), the wave equation for ρ_1 is obtained immediately

$$\nabla^2 \rho_1 = \frac{1}{c^2}\partial_t^2 \rho_1 \tag{2.7}$$

As for the particle velocity caused by sound waves, the circumstance is somewhat different. In the one-dimension case, it is easy to see that the particle velocity also obeys the wave equation. However, in the three-dimension case, the wave equation can only be derived with the condition that sound wave motion is irrotational. Taking the derivative of Eq. (2.2) with respect to time, together with using Eq. (2.5) to eliminate p_1, the following equation is acquired

$$\partial_t^2 \mathbf{v}_1 = c^2 \nabla(\nabla \cdot v_1)$$

Applying the familiar formula in vector analysis $\nabla(\nabla \cdot \mathbf{v}_1) = \nabla^2 \mathbf{v}_1 + \nabla \times \nabla \times \mathbf{v}_1$, when $\nabla \times \mathbf{v}_1 = 0$ (irrotational motion), the wave equation for $\mathbf{v}_1$ is obtained

$$\nabla^2 \mathbf{v}_1 = \frac{1}{c^2}\partial_t^2 \mathbf{v}_1 \tag{2.8}$$

From now on, for simplicity, subscript 1 will be omitted for all quantities relating to sound waves to avoid any confusion.

2.1.2 Velocity potential (acoustic potential) and wave equation including quantities of second order

According to the principle that a potential must exist when an acoustic wave is irrotational, equation $\nabla \times \mathbf{v} = 0$ implies the existence of a velocity (acoustic) potential (because the rotation of any scalar gradient is always zero)

$$\mathbf{v} = -\nabla\phi \tag{2.9}$$

With the introduction of an acoustic potential, the three components of the Euler equation, Eq. (2.2), can be expressed in one equation as

$$\nabla p = -\rho_0 \partial_t v = \rho_0 \nabla(\partial_t \phi)$$

This avoids the difficulties of solving vector differential equations. By obtaining the integral of the above formula, the following important relation is obtained

$$p = \rho_0 \partial_t \phi + p_0 \tag{2.10}$$

Thus, not only the three components of particle velocity, but also the sound pressure, can be calculated from the acoustic potential via a simple differentiation. By substituting Eq. (2.10) into (2.6), and integrating Eq. (2.6) over time, after ignoring the insignificant constant of integration, the wave equation for an acoustic potential is given as

$$\nabla^2 \phi = \frac{1}{c^2} \partial_t^2 \phi \tag{2.11}$$

Therefore, the acoustic field can be expressed by a scalar function with spatial coordinates.

Wave equation (2.11) is a linear equation accurate only to the first order. If we substitute Eq. (2.9) and (2.1) into the non-linearized Euler equation, and disregard the small third-order quantity, $\rho \mathbf{v} \cdot \nabla v$, and perform the transforms and integrals several times, the acoustic pressure expression, which includes second-order terms, is obtained:

$$p = \rho_0 \partial_t \phi - \frac{1}{2} \rho_0 \left[|\nabla \phi|^2 - \frac{1}{c^2} (\partial_t \phi)^2 \right] \tag{2.10a}$$

where the constant of integration is omitted.

Applying the state equation (2.4) and the continuity equation (2.3) again, and repeating the elimination process used in the first paragraph of this section, the wave equation with second-order quantities of p and v is obtained and given below.

$$\partial_t^2 \phi - \frac{1}{2} \partial_t \left[|\nabla \phi|^2 - \frac{1}{c^2} (\partial_t \phi)^2 \right] = c^2 \nabla^2 \phi \tag{2.11a}$$

From this, it can be seen that even some second-order effects such as viscosity and heat conductivity are disregarded in the simple wave motion equation (2.11), but can be rectified by including second-order quantities.

2.1.3 Helmholtz equation

The description of many acoustic phenomena can be significantly simplified by applying a Fourier integral transform. Fourier temporal and spatial transforms, as well as their inverse transforms, are defined by the following integrals

$$F(\mathbf{x}, k) = F_t f(\mathbf{x}, t) = (2\pi)^{-1} c_0 \int dt \exp(ikc_0 t) f(\mathbf{x}, t) \tag{2.12a}$$

$$f(\boldsymbol{x}, t) = F_t^{\dagger} F(\boldsymbol{x}, k) = \int \mathrm{d}k \exp(-\mathrm{i}kc_0 t) F(\boldsymbol{x}, k) \tag{2.12b}$$

$$F(\boldsymbol{\kappa}, t) = F_{\kappa} f(\boldsymbol{x}, t) = (2\pi)^{-3} \int \mathrm{d}^3 x \exp(-\mathrm{i}\boldsymbol{\kappa} \cdot \boldsymbol{x}) f(\boldsymbol{x}, t) \tag{2.13a}$$

$$f(\boldsymbol{x}, t) = F_{\kappa}^{\dagger} F(\boldsymbol{\kappa}, t) = \int \mathrm{d}^3 \boldsymbol{\kappa} \exp(\mathrm{i}\boldsymbol{\kappa} \cdot \boldsymbol{x}) F(\boldsymbol{\kappa}, t) \tag{2.13b}$$

Where $\boldsymbol{x} = (x, y, z)$ are the spatial coordinates, and $\boldsymbol{\kappa}$ are the Fourier spatial coordinates (spatial wave-vector). In the Cartesian coordinate system, the intervals of all integrals' are $(-\infty, +\infty)$. The superscript "+" denotes the complex conjugate of the functions and operators. When c is a constant, or varies slowly in time (relative to the acoustic wave period), when we apply the Fourier operator shown in Eq. (2.12a) to wave equation (2.6), the Helmholtz equation is obtained.

$$\nabla^2 P + k^2 P = 0 \tag{2.14}$$

Where P represents the Fourier transform of the acoustic pressure (or other corresponding physical quantities related to an acoustic wave), and $k(= \omega/c)$ is the wavenumber. The Helmholtz equation can be derived from another point of view. If we assume that the time relation of the wave motion is sinusoidal, (i.e. a pure periodical process), the time factor $\mathrm{e}^{-\mathrm{i}\omega t}(\partial_t = -\mathrm{i}\omega)$ can be detached. For example, by substituting $p = P(x)\mathrm{e}^{-\mathrm{i}\omega t}$ into Eq. (2.6), we obtain the pressure equation that only relates to the spatial coordinates, i.e., Eq. (2.14). Since any process can be decomposed into a series of sinusoidal processes through a Fourier analysis, the above two derivation methods are consistent. Because the number of independent variables is reduced, the advantage of using the Helmholtz equation becomes obvious. Therefore, it is often used as the starting point of analyses in linear acoustics.

In addition, the acoustic potential can be replaced by the acoustic pressure via the multiplication of an imaginary constant of purely periodical processes. Hence, the particle velocity can be directly expressed from the acoustic pressure, which gets rid of the need of introducing an uninteresting physical quantity ϕ. The following equations are obtained from Eq. (2.10) and (2.9) respectively ("$\sim$" denotes periodical quantity)

$$\tilde{p} = -\mathrm{i}\omega\rho_0\tilde{\phi} \tag{2.15}$$

$$\tilde{v} = -(\mathrm{i}/\omega\rho_0)\nabla\tilde{p} \tag{2.16}$$

2.2 Energy relations in acoustic waves

2.2.1 Energy and energy flow density in acoustic waves

It is pointed out in Section 4.3 of Chapter I that the total energy of a fluid per unit volume is $\rho\varepsilon + \frac{1}{2}\rho v^2$. When acoustic waves enter static fluids, perturbations are introduced to the density $\rho \to \rho_0 + \rho_1$ and the internal energy $\varepsilon \to \varepsilon_0 + \varepsilon_1$. Substituting these perturbations into the expression of the total energy per unit volume, and expanding it into a second-order Taylor series, yields

$$\rho_0\varepsilon_0 + \rho_1 \partial_\rho(\rho\varepsilon) + \frac{\rho_1^2}{2}\partial_\rho^2(\rho\varepsilon) + \frac{\rho_0}{2}v^2$$

Because the acoustic process is adiabatic, we take the differential quotient with entropy as constant. According to the second law of thermodynamics, $d\varepsilon = Tds + \frac{p}{\rho^2}d\rho$, we obtain $(\partial_\rho(\rho\varepsilon))_s = \varepsilon + \frac{p}{\rho} = \omega$, $(\partial_\rho^2(\rho\varepsilon))_s = (\partial_\rho\omega)_s = (\partial_p\omega)_s(\partial_\rho p)_s = \frac{c^2}{\rho}$. Then the fluid energy per unit volume is

$$\rho_0\varepsilon_0 + \omega_0\rho_1 + \frac{c^2}{2\rho_0}\rho_1^2 + \frac{\rho_0}{2}v^2$$

Where the first term represents the internal energy of a static fluid, which has nothing to do with acoustic waves. The second term denotes the energy variation introduced by the change of fluid mass per unit volume, which will not affect the total energy when the integral is over all the volume of fluid, because the total quantity of fluid does not change, i.e., $\int \rho_1 dV = 0$, whose physical meaning is that the denseness and sparseness of the fluid introduced by acoustic waves cancel each other out in the whole fluid. Therefore, the variation of the total energy resulting from the existence of an acoustic wave is

$$\int \left(\frac{\rho_0 v^2}{2} + \frac{c^2\rho_1^2}{2\rho_0}\right) dV$$

The integrand can be regarded as the acoustic energy density. Using Eq. (2.4) and the relation between the sound speed and the fluid compression coefficient $\kappa \left(= \frac{1}{\rho}\partial_p\rho \right)$, i.e., $c = \left(\frac{1}{\kappa\rho}\right)^{1/2}$, and replacing ρ_1 by p_1, we obtain a more symmetrical expression

$$E = \frac{1}{2}\rho_0 v^2 + \frac{c^2}{2\rho_0}\rho_1^2 = \frac{1}{2}\rho_0 v^2 + \frac{1}{2}\kappa p_1^2 \tag{2.17}$$

Where the first term represents the kinetic energy density, and the second term represents the potential energy density resulting from the compressibility (elasticity) of the fluid.

Applying Eq. (1.38) to an acoustic wave, and ignoring small, third-order quantities, we obtain the average energy flow density

$$\overline{\rho_0 \boldsymbol{v}\omega} = \omega_0\overline{\rho_0\boldsymbol{v}} + \overline{\rho_0\omega_1\boldsymbol{v}}$$

Where ω_0 is the enthalpy per unit mass of fluid before being perturbed, while ω_1 is the perturbation value of the enthalpy caused by the acoustic waves. For the latter, we have

$$w_1 = \partial_p(w)_s p_1 = \frac{p_1}{\rho_0}$$

Substituting this result into the above expression, and integrating it, we obtain the total energy flow passing through a given surface

$$\oint (w_0\overline{\rho_0 \boldsymbol{v}} + \overline{p_1 \boldsymbol{v}}) \cdot \mathrm{d}\boldsymbol{S}$$

Because the total quantity of fluid in given volume is unchangeable, the average value of the mass flow passing through a close surface over time is 0. The acoustic energy flow becomes

$$\oint \overline{p_1 \boldsymbol{v}} \cdot \mathrm{d}\boldsymbol{S}$$

Then we introduce the acoustic energy flow vector

$$\boldsymbol{I} = p_1 \boldsymbol{v} \tag{2.18}$$

Evaluating the differential quotient of Eq. (2.17) over time, and applying the relations shown in Eq. (2.2)–(2.4), we easily prove that

$$\partial_t E + \nabla \cdot \boldsymbol{I} = 0 \tag{2.19}$$

The above equation actually describes the law of acoustic energy conservation.

2.2.2 Momentum in acoustic waves and time-averaged values of acoustic pressure

Substituting Eq. (2.1) into Eq. (1.34a), and applying Eq. (2.4) and Eq. (2.18), the momentum per unit volume of acoustic waves is given as

$$\boldsymbol{j}_{\mathrm{m}} = \rho_0 \boldsymbol{v} + \frac{1}{c^2}\boldsymbol{I} = \frac{1}{c^2}\boldsymbol{I} - \rho_0 \nabla \phi \tag{2.20}$$

The total momentum is obtained via the integral of the above equation over the volume. However, because $\phi = 0$ outside of the region occupied by the acoustic wave (wave packet)

$$\int \nabla \phi \mathrm{d}V = \oint \phi \mathrm{d}\boldsymbol{S} = 0$$

Thus, the total momentum of an acoustic wave becomes

$$\int \boldsymbol{j}_{\mathrm{m}} \mathrm{d}V = \frac{1}{c^2} \int \boldsymbol{I} \mathrm{d}V \tag{2.21}$$

A conclusion can then be made that the propagation of an acoustic wave packet is accompanied by the transfer of fluid substance. Because $\boldsymbol{I}$ is a second-order quantity, this is a second-order effect.

In the first-order approximation, which corresponds to the motion equation being linearized, the acoustic pressure p_1 is a quantity that changes its sign periodically. Hence, its average is 0. However, in a high-order approximation, this conclusion is not valid. If the equation is accurate only to the second-order approximation, p_1 can be expressed by the quantities obtained from the linear equations, and there is no need to solve the nonlinear equation that involves high-order quantities directly.

Substitute Eq. (2.9) into Euler's equation (1.30a″), the following equation is obtained

$$\nabla\left(-\partial_t\phi+\frac{v^2}{2}+\omega\right)=0$$

We can get the single integral of Euler's equation (Lagrange integral) for a motion with potential and under the adiabatic condition (as acoustic wave is, in such a situation)

$$-\partial_t\phi+\frac{v^2}{2}+\omega=C(t) \tag{2.22}$$

The above equation is the generalized Bernoulli equation.[1]

When taking the average of Eq. (2.22) over time, we notice that the average of $\partial_t\phi$ over time is 0 ($\phi(t)$ should be bounded for all t). Instead, when we replace ω with $\omega_0+\omega_1$, and combine ω_0 into the constant term on the right-hand side of equation, we obtain

$$\overline{\omega_1}+\frac{\overline{v^2}}{2}=\text{const}$$

As long as there is some absorption in the fluid, all of the quantities related to acoustic waves, such as v and ω_1, will be 0 at infinity. Therefore, the constant should be zero in the above equation

$$\overline{\omega_1}+\frac{\overline{v^2}}{2}=0 \tag{2.23}$$

Next, by expanding ω_1 into the series of the power of p_1, which is accurate to the second-order term, we get

$$\omega_1=(\partial_p\omega)_s p_1+\frac{1}{2}(\partial_p^2\omega)_s p_1^2=\frac{p_1}{\rho_0}-\frac{p_1^2}{2\rho_0^2c^2}(\partial_p\rho)_s=\frac{p_1}{\rho_0}-\frac{p_1^2}{2\rho_0^2c^2}$$

And when we substitute Eq. (2.23) into it, we obtain the average of the acoustic pressure

$$\overline{p_1}=-\frac{\rho_0\overline{v^2}}{2}+\frac{\overline{p_1^2}}{2\rho_0c^2}=-\frac{\rho_0}{2}\overline{v^2}+\frac{c^2}{2\rho_0}\overline{\rho_1^2} \tag{2.24}$$

Which is evidently a small quantity of the first or second order.

Correspondingly, we can also express the average value of the density fluctuation caused by acoustic waves as

$$\overline{\rho_1}=(\partial_{p_0}\rho)_s\overline{p_1}+\frac{1}{2}(\partial_{p_0}^2\rho)_s\overline{p_1^2} \tag{2.25}$$

1 It is different to the Bernoulli equation by general implication: the latter is tenable under the condition of a finite constant flow, so that $v^2/2+w=C$ can be obtained by singly integrating Eq. (1.30a″) when $\partial_t\mathbf{v}=0$, where C is a constant taken variously along various flow lines. But the $C(t)$ term in Eq. (2.22) is a function of time, and as such is constant in the whole fluid range.

2.2.3 Lagrange density in acoustic waves[2]

All of the above derivations are based on Newton's equations of motion. If the concepts and methods of these analytical mechanics are applied, many results, especially those regarding the density of energy flow and momentum flow, can be obtained more simply.

The Lagrange density of a system, i.e., the Lagrange function, L, in unit volume (motion potential), is defined as the difference between the densities of kinetic energy and potential energy. According to Hamilton's principle, the actual motion of the system should make the integral of L over time (Hamilton function quantity $\int_{t_0}^{t_1} L dt$) be at its extremes (in most cases the minimum), meaning $\delta \int L dt = 0$. Thus, the motion equation in Lagrange form is obtained[3]

$$\frac{\partial}{\partial t}\left(\frac{\partial L}{\partial \phi_t}\right)+\frac{\partial}{\partial x}\left(\frac{\partial L}{\partial \phi_x}\right)+\frac{\partial}{\partial y}\left(\frac{\partial L}{\partial \phi_y}\right)+\frac{\partial}{\partial z}\left(\frac{\partial L}{\partial \phi_z}\right)-\frac{\partial L}{\partial \phi}=0 \tag{2.26}$$

For acoustic waves, and referring to Eq. (2.17), we have

$$L=\frac{1}{2}\rho_0 v^2-\frac{1}{2}\kappa p_1^2 \tag{2.27}$$

When applying Hamilton's principle, v and p_1 must be expressed as a common function with coordinate x and time t. The acoustic potential ϕ is the more precisely suitable function. Substituting Eq. (2.9) and (2.10) into Eq. (2.27) yields

$$L=-\frac{1}{2}\rho_0\left[|\nabla\phi|^2-\frac{1}{c^2}\left(\frac{\partial\phi}{\partial t}\right)^2\right]=-\frac{1}{2}\rho_0\left[\phi_x^2+\phi_y^2+\phi_z^2-\frac{1}{c^2}\phi_t^2\right] \tag{2.28}$$

In the above equation, we have to change the sign of L because such expression is somewhat "abnormal", where the spatial differentiation of ϕ corresponds to kinetic energy, while the time differentiation of which corresponds to potential energy.

Putting Eq. (2.28) into (2.26) results in the wave equation (2.11). From this we can see that wave equation satisfies the following requirement: on average the difference of the total kinetic and total potential energy of a system under certain initial and boundary conditions should be as small as possible. This result is very enlightening and useful, based on which many other relations can be derived.

2 Please see reference [31] for more details.

3 Here (only in this section) we introduce another sign system, in one respect, to restore traditional expression for partial differential quotient, for example, $\partial/\partial t$, $\partial/\partial x$ (∂_t, ∂_x were used previously); in another respect, we still retain the subscripts to denote the partial differential quotient over the quantity, for example, $\Phi_t \equiv \Phi f/\partial t$, $\Phi_x \equiv \partial\Phi/\partial x$, and so forth.

Because the potential energy density in Eq. (2.27) is only accurate to the 2nd order (p_1 used is only accurate to the 1st order in Eq. (2.10)), so L in Eq. (2.28) can only be accurate to the 2nd order. If the expression of p_1 (2.10a), which includes the second-order quantity (being able to denoted as $p_1 = \rho_0\phi_t + L$ with currently used symbols), is used, and we also consider the third-order quantity, Eq. (2.28) becomes

$$L^{(1)} = \left[1 + \frac{1}{c^2}\phi_t\right] L \tag{2.29}$$

By substituting the above equation into Eq. (2.26), we obtain the wave equation that includes second-order effects, Eq. (2.11a). What is interesting is that the modified term is proportional to the differentiation of the unmodified L over time: L_t/ρ_0.

It is well-known that the total energy (the sum of kinetic and potential energy) of a system being expressed by canonical variables is the Hamilton function of the system. In acoustics, the corresponding canonical coordinate is ϕ, and hence the canonical momentum is $\partial L/\partial\phi_t$, (equal to $\frac{1}{c^2}p_1$ observed from Eq. (2.28)). Because ϕ depends not only on time t, but also on coordinate x, the scenario is more complex than that in general analytic mechanics. That is to say, the relationship between the momentum, field, and the gradient of the field is more complex than that given by the ordinary Hamilton canonical equation. In analytic mechanics, it is proven that the energy of a particle is the time (temporal) component of a four-element vector, and its spatial components are proportional to the momentum of the particle. Applying this principle to acoustics, a four-element parallel vector Ξ can be introduced, whose components are

$$W_{ij} = \phi_{x_i}\left(\frac{\partial L}{\partial\phi_{x_j}}\right) - L\delta_{ij} \tag{2.30}$$

Where, δ_{ij} is unit tensor, $x_1 = x$, $x_2 = y$, $x_3 = z$, $x_4 = ict$.

Apparently, the time-time component is the Hamilton density H

$$W_{44} \equiv W_{tt} = \phi_t\left(\frac{\partial L}{\partial\phi_t}\right) - L = H \tag{2.31}$$

Substituting Eq. (2.28) into (2.31), we find that the above expression is just the acoustic energy density E defined in Eq. (2.17).

The three time-space components form a three-element vector. Its x component is

$$W_{tt} = \phi_t\left(\frac{\partial L}{\partial\phi_x}\right) = -\rho_0\phi_t\phi_x = p_1v_x$$

Which is just the acoustic energy flow density $\boldsymbol{I}$ defined in Eq. (2.18). Because the product of the acoustic pressure and the particle velocity represents the power flux in unit area caused by the acoustic waves, $\boldsymbol{I}$ is also referred to as the acoustic intensity.

Next, let us check the space-time components of the four-element parallel vector Ξ. If its x component is given as

$$W_{xt} = \phi_x\left(\frac{\partial L}{\partial\phi_t}\right) = \frac{\rho_0}{c^2}\phi_x\phi_t = -\frac{1}{c^2}p_1v_x = -\frac{1}{c^2}W_{tx}$$

So it forms the momentum density vector of the acoustic wave [corresponding to the vector of fluid momentum density $\boldsymbol{j}_{\rm m} = \rho\boldsymbol{v}$ defined in (Eq. (1.34a)]

$$\boldsymbol{M} = -\frac{1}{c^2}p_1\boldsymbol{v} = -\frac{1}{c^2}\boldsymbol{I} \tag{2.32}$$

Since $\frac{1}{c^2}$ is usually very small, M is a first- or second-order small quantity which satisfies the continuity equation for the vectors given in Eq. (1.30b)

$$\frac{\partial}{\partial t}M_i = -\frac{\partial}{\partial x_k}\Xi_{ik} \tag{2.33}$$

Where Ξ_{ik} is a tensor consisting of nine space-space components (the acoustic wave stress tensor), which is a measure of the momentum flux of the acoustic waves

$$\Xi_{ik} = \begin{bmatrix} W_{xx} & W_{xy} & W_{xz} \\ W_{yx} & W_{yy} & W_{yz} \\ W_{zx} & W_{zy} & W_{zz} \end{bmatrix} \tag{2.34}$$

Where

$$W_{xx} = \phi_x\left(\frac{\partial L}{\partial \phi_x}\right) - L = \frac{\rho_0}{2}(-v_x^2 + v_y^2 + v_z^2) - \frac{\kappa}{2}p_1^2$$

$$W_{xy} = \phi_x\left(\frac{\partial L}{\partial \phi_y}\right) = -\rho_0 v_x v_y$$

And so on.

By the way,

$$c\boldsymbol{M} = -\frac{1}{c}\boldsymbol{I} \tag{2.35}$$

It is the radiation pressure of acoustic waves.

2.3 Wave equation in inhomogeneous atmosphere

2.3.1 Wave equation and solution-defining conditions

The actual atmosphere is inhomogeneous. It means both c in the wave equation (2.6) and k in the Helmholtz equation (2.14) are functions of spatial coordinates. For harmonic waves, these equations can be derived from the coupled equations of fluid mechanics directly. Starting from Eq. (1.30a′) and (2.5), and by noticing that both ρ_0 and c are functions of coordinates $\boldsymbol{x} = (x, y, z)$, and by eliminating v in the two equations by applying $\partial_t = -i\omega$, the following equation is yielded

$$\rho_0\nabla\cdot\left[(1/\rho_0)\,\nabla p\right] + k^2 p = 0$$

Expanding the operator, the equation becomes

$$\nabla^2 p + k^2 p - (1/\rho_0)\,\nabla\rho_0\cdot\nabla p = 0 \tag{2.36}$$

Using ψ defined in the following to replace p [32]

$$\psi = \rho_0^{-1/2} p \tag{2.37}$$

After the simple transformation, the Helmholtz equation about ψ is obtained

$$\nabla^2 \psi + K^2 (\mathbf{x}) \psi = 0 \tag{2.38}$$

Where

$$K^2 = k^2 + (1/2\rho_0) \nabla^2 \rho_0 - (3/4) \left[(1/\rho_0) \nabla \rho_0\right]^2 \tag{2.39}$$

When density ρ_0 varies slightly within one wavelength, the second and third terms in the above equation can be neglected.

It is well-known that in order to determine the solution of a differential equation describing a physical process, the corresponding solution-defining conditions, including the initial condition the and boundary condition, must be given.

The initial condition describes the state of the system at the initial time $t = 0$, which is rather simple, and is not necessary at all for harmonic steady-state processes. The boundary condition describes the effects of boundaries on the wave field, which specifies the relation the wave function ψ should satisfy at the boundary S due to the confinement of the physical condition at the boundary. Then the properties of the total wave field can be described (not only at the boundary). The commonly used boundary conditions can be divided into several categories, and are shown in the following:

(1) An ideal absolute soft boundary – unable to bear any pressure. That is, the following relation is satisfied at the boundary

$$p|_S = 0 \tag{2.40}$$

This is called the boundary condition of the first kind, which actually can be used at the boundary between an atmosphere and water.

(2) An ideal absolute hard boundary, which allows no normal particle velocity, i.e.

$$v_z|_S = 0, \text{ or } \partial_z p|_S = 0 \tag{2.41}$$

This is called the boundary condition of the second kind, which can be used for very rigid ground surfaces.

(3) An impedance boundary condition, which means certain impedance relations must be satisfied at the surface S[4]

4 Of course, Eq. (2.42) can also be written as $p|_S - Z_{Sv_x}|_S = 0$, and with Eq. (2.40) and (2.41) they are equally named inhomogeneous boundary conditions. If the right-hand sides of these three equations are not 0, but have certain distributions of pressure, particle velocity (or combination of both), then the inhomogeneous boundary can be obtained.

$$\left.\frac{p}{v_z}\right|_S = Z_S \tag{2.42}$$

This is called the boundary condition of the third kind. It can be regarded as the intermediate situation of the first (corresponding to $Z_S = 0$) and the second (corresponding to $Z_S \to \infty$) kind of boundary conditions. Therefore, it is the most useful in practice.

In ordinary cases, Z_S is not only related to the lower medium parameter (such as the density ρ and the sound speed c), but also to the characteristics of the acoustic field itself (such as the incident angle θ) at the boundary.

Therefore Z_S cannot be specified as a known quantity in advance, so it is not able to be regarded as a solution-defining condition. However, when Z_S is independent of the angle θ, the normal velocity of each particle at the surface is determined by the acoustic pressure at the point completely, and has nothing to do with the pressure at other positions. This kind of surface is called a local response surface,[5] which will be met frequently later on.

Besides the boundary conditions discussed, the solution-defining conditions should also include the so-called radiation condition, which requires the field to be 0 (vanished) at infinity if the acoustic field involves the field at infinity,

2.3.2 Review of the existing solutions

As demonstrated in the above discussions, many fundamental problems in atmospheric acoustics come down to the solving of the Helmholtz equation (2.38) under certain solution-defining conditions. If the incident plane of the wave coincides with the $x-z$ plane, ψ is the function of only two variables, x and z. Applying the method of the separation of variables, the following solution form is obtained

$$\psi(x,z) = \Psi(z)\exp(\mathrm{i}\xi x), \quad \xi = k_0 \sin\theta_0 \tag{2.43}$$

For the new function $\Psi(z)$, the differential equation it satisfies can be obtained by substituting Eq. (2.43) into (2.38),

$$\Psi'' + \left[K^2(z) - \xi^2\right]\Psi = 0 \tag{2.38a}$$

where the primes denote the differentiation over variable z. Notice that the atmosphere was treated as being stratified. Therefore, $K^2(\boldsymbol{x})$ is simplified as $K^2(z)$.

5 The mechanism for forming this kind of surface can be due to the fact that the sound velocity in the lower medium is much smeller than that in the upper medium, or due to the case when there is strong absorption or some anisotropism (e.g. “comb surface”) in the lower medium. For further discussions see section 7.3 in Chapter 3.

Mathematically, the solutions of the second-order ordinary differential equation in the form of Eq. (2.38) correspond to some special functions, and only when $K^2(z)$ takes some particular form. The close-formed solutions of the known functions can be obtained, or already-made function tables can be checked. Up to now, there are less than ten such strictly solvable situations. We are to review them briefly in the following

(1) $K^2(z) = k_0^2(1 + az)$

This is the simplest, but very important, linear case. The function can be solved by applying Airy functions (whose properties will be discussed in some detail in the following sections).

(2) $K^2(z) = k_0^2(1 + az)^m$

The normal incidence case ($\xi = 0$), with arbitrary a and m, was studied [33]. The plane wave case with m being an integer and at arbitrary incident angles ($\xi = 0 \to k_0$) has also been studied [34, 35].

(3) $K(z) = \dfrac{k_0 a}{z + d}$

The normal incidence case was studied by Rayleigh [36]. After making the replacement of $z + d = \xi$, the solution to Eq. (2.38a) (where $\xi = 0$) takes the form of

$$\Psi = A\xi^{\frac{1}{2}\pm im}, \quad m^2 = (k_0 a)^2 - \frac{1}{4}$$

(4) $K^2(z) = k_0^2\left(a^2 - \dfrac{b^2}{z^2}\right)$

In this case, Eq. (2.38a) has two linearly independent solutions [37]

$$z^{1/2}H_P^{(1)}(\beta z) \text{ and } z^{1/2}H_P^{(2)}(\beta z)$$

where $p = \left(k_0^2 b^2 + \dfrac{1}{4}\right)^{1/2}$, and $\beta^2 = k_0^2 a^2 - \xi^2$.

The above third case discussed by Rayleigh can be treated as a special scenario of this case, where $a = 0$.

(5) $K^2(z)/k_0^2 = 1 - P(z)$

$$P(z) = c_2 z^2 + c_3 z^3 + c_4 z^4 + \cdots$$

After a series of transformations, Eq. (2.38a) can be solved [39] using parabolic cylinder functions [38].

It is worth to point out that if only the first term c_2z^2 of polynomial $P(z)$ is counted, and when $c_2 > 0$, the Schrödinger Equation for a harmonic oscillator is obtained. When $c_2 < 0$, study [40] related radio ground wave theory in an atmosphere, which is helpful in the study of acoustic wave propagation in a waveguide.

(6) $K^2(z)/k_0^2 = 1 - N\dfrac{e^{mz}}{1 + e^{mz}} - 4M\dfrac{e^{mz}}{(1 + e^{mz})^2} = n^2(z)$

This case is also known as the famous Epstein profile, which obtained its name from the scientist who first studied the reflection of electromagnetic waves [41]. This case can be divided into a symmetrical layer ($N = 0, M \neq 0$) and a transition layer ($N = 0, M \neq 0$) according to the distribution of the square of the refraction index $n^2(z)$. The solution can be expressed in terms of hyper-geometric functions. The generalized Epstein profiles are discussed in ref. [42], [43].

(7) $K(z) = k_0 e^{az}$

After substituting $v = \dfrac{k_0}{a}e^{az}$, Eq. (2.38a) can be solved by using cylindrical functions $H_q^{(1)}(v)$ and $H_q^{(2)}(v)$, where $q = (k_0 \sin\theta)/a$.

(8) $K(z) = \dfrac{\left(z^2 - a^2\right)^{1/2}}{c_0 z}$

In this case, the solution to Eq. (2.38a) is expressed in terms of the Bessel function.

(9) $K^2(z)/k_0^2 = p^2 + \left(1 - p^2 + q\right)e^{-az} - qe^{-2az}$

In this case, the solution to Eq. (2.38a) can be expressed in terms of the Whittaker function [46] $M_{k,\mu}$ and $W_{k,\mu}$ [38].

Finally, it is necessary to point out that, in some cases, although we cannot directly obtain the strict closed-form solutions for Eq. (2.38a) as the above situations, we can make some proper transformations to get the solutions related to other solvable situations.

In summary, it is very limiting to find strict solutions for the wave equation (Helmholtz Equation) in an inhomogeneous atmosphere (medium). Firstly, strict solutions can only be obtained for very specific profiles – profiles that cannot be met in the real world. Secondly, even if the strict solutions are obtained, they are usually in the form of advanced transcendental functions, whose physical images are not clear. Therefore, in many actual cases, we have to depend on approximate and/or numerical solutions. The later will be discussed in a special Chapter.

2.4 WKB approximation[6]

The most important approximate method is the WKB (Wentzel-Kramers-Brillouin) approximation, or geometric optical approximation.

2.4.1 General remarks

First, we will discuss the simplest case where an acoustic wave propagates along the z axis. At this time, $\zeta = 0$ in Eq. (2.38a). Let $K(z) = k_0 n(z)$, where $k_0 = \omega / c_0$, and c_0 being the sound speed at an arbitrary point where $z = z_0$. When $n(z_0) = 1$, we have

$$\psi''(z) + k_0^2 n^2(z) \psi(z) = 0 \tag{2.38b}$$

In a homogeneous atmosphere, $n(z) = 1$, the solution to Eq. (2.38b) is the exponential expression $e^{\pm ik_0}$; while in the inhomogeneous case, the solution can be regarded as having the form

$$\psi = \exp[ik_0 M(z)] \tag{2.44}$$

Where $M(z)$ must be a function in the form of a power series of $1/k_0$ (assuming ω and the corresponding k_0 are large enough; the criterion will be shown in the following paragraph). Let

$$M(z) = \int_{z_0}^{z} dz \sum_{\nu=0}^{\infty} y_\nu(z) / k_0^\nu \tag{2.45}$$

Substituting it into (2.38b), and letting the coefficients of k_0^2, k_0 and k_0^0 be equal each other, the following can be obtained in succession

$$y_0 = \pm n, \quad y_1 = i\left(\ln n^{1/2}\right)', \quad y_2 = \pm\frac{1}{2} n^{-1/2}\left(n - 1/2\right)''$$

Retaining only the first three terms and noticing that $n(z_0) = 1$, the following formula can be obtained based on Eq. (2.45) and (2.44)

$$\psi = n^{-1/2} \exp\left[\pm ik_0 \int_{z_0}^{z} (1 + \varepsilon) n dz\right] \tag{2.46}$$

where

$$\varepsilon \equiv n^{-3/2}\left(n^{-1/2}\right)'' \Big/ 2k_0^2 \tag{2.47}$$

when

$$y_z / k_0^2 \ll y_1 / k_0 \ll y_0$$

6 See the intensive discussion in related monographs, such as reference [49].

or

$$\varepsilon \ll 1, \quad \left(\lambda_0^2/2\pi n^2\right) n' \ll 1 \tag{2.48}$$

When it is tenable, the result has little difference on the precise result. If $\varepsilon = 0$, expression (2.46) is regarded as the WKB approximation

$$\psi = n^{-1/2} \exp\left[\pm ik_0 \int n(z)\,\mathrm{d}z\right] \tag{2.46'}$$

Apparently, here the following condition must be satisfied

$$k_0 \int_{z_0}^{z} n\varepsilon \mathrm{d}z \ll 1 \tag{2.49}$$

This condition (restricting the second-order differentiation over n and the difference of $z - z_0$), together with the two conditions shown in Eq. (2.48) (the second condition restricts n') constitutes the necessary conditions for the existence of Eq. (2.46′). This expression represents the combination of two waves propagating in opposite directions without any interaction between them. Therefore, in the geometric optical approximation, there are no wave reflections. The exponential expression denotes the phase shift from z_0 to z, and the factor in front of the expression ensures that the energy is conserved.

In the case of oblique incidence (with incident angle being θ_0), $k_0 n$ in Eq. (2.46′) should be replaced by $k_0\left(n^2 - \sin^2\theta_2\right)^{1/2}$. However, for arbitrary z, we can always have $\left(n^2 - \sin^2\theta_0\right)^{1/2} = n(z)\cos\theta(z)$. Therefore, the general solution of the geometric optical approximation can be demonstrated to be

$$\psi(x,z) = (n\cos\theta)^{-1/2}\left[C_1 \exp\left(ik_0 \int_{z_0}^{z} n\cos\theta \mathrm{d}z\right) + C_2 \exp\left(-ik_0 \int_{z_0}^{z} n\cos\theta \mathrm{d}z\right)\right] \exp(ik_0 \sin\theta_0 x) \tag{2.50}$$

Where C_1 and C_2 are arbitrary constants. The selection of z_0 will only affect the values of the two constants, and not the solution itself. The criterion expressions, specified by Eq. (2.48) and (2.49), to guarantee the validity of the solution, will also use $\left(n^2 - \sin^2\theta_0\right)^{1/2} = n\cos\theta$ instead of n, and it is not difficult to see that the two criteria are not satisfied in the vicinity of the “turning point” which is determined by $n(z) = \sin\theta_0$, $\theta = 0$.

2.4.2 Airy functions

Before studying the field in the vicinity of the turning point, first we will briefly discuss the Airy function due to its particular importance not only to the problem at this moment, but also to other propagation problems (for example, the field near caustic surfaces, which is especially the diffraction problem to be discussed in Chapter IV).

Consider the typical equation in the following form

$$F'' - \eta F = 0 \tag{2.51}$$

Without considering the constant factor, the unique solution of Eq. (2.51) is the Airy function which possesses the characteristics of approaching zero when $\eta \to \infty$. The Airy function is defined as follows:
For η being a real number.[7]

$$Ai(\eta) = \frac{1}{\pi}\int_0^{\infty} \cos\left(\frac{s^{\varepsilon}}{3} + \eta s\right) \mathrm{d}s \tag{2.52}$$

For η being a complex number,

$$Ai(\eta) = \frac{1}{2\pi}\int_{C_{Ai}} \mathrm{e}^{\mathrm{i}(s^3/3+\eta s)} \mathrm{d}s \tag{2.52'}$$

The integration contour C_{Ai} starts from the point $|s| = \infty$ on the curve where the phase of s is $5\pi/6$ and ends at the point $|s| = \infty$ on the curve where the phase of s is $\pi/6$. It can be proved immediately by transforming the shape of C_{Ai} to the real axis that Eq. (2.52′) is equivalent to Eq. (2.52). For any real number η, A_i can be expressed by 1/3-order cylindrical functions whose value can be found in special function tables [51].

When $|\eta|$ is very large, the asymptotic expression of $A_i(\eta)$ can be derived from Eq. (2.52′). When the phase of η, ϕ is between $-\frac{2}{3}\pi$ and $\frac{2}{3}\pi$, transforming the shape of C_{Ai} to the steepest descent path $s = s(l)$, which passes through the saddle point at $s = \mathrm{e}^{\mathrm{i}\pi/2}\eta^{1/2}$(at this point, $\mathrm{d}s/\mathrm{d}l = \mathrm{e}^{-\mathrm{i}\phi/4}$). Because the integrand has a very sharp peak at the saddle point, $s^3/3 + \eta s$ can be approximated with $\mathrm{i}\left(\frac{2}{3}\right)\eta^{3/2} + \mathrm{i}|\eta|^{1/2} l^2$, where l is the distance from the saddle point to the interesting point along the path. Then we obtain

$$Ai(\eta) \to \frac{\mathrm{e}^{-(2/3)\eta^{3/2}}}{2\pi^{1/2}\eta^{1/4}}, \quad -2\pi/3 < \phi < 2\pi/3 \tag{2.53a}$$

When $2\pi/3 < \phi < 4\pi/3$, C_{Ai} is extended so that its middle point arrives at the point $-\mathrm{i}\infty$ on the negative virtual axis. Its left part is deformed to the steepest descent path passing through the saddle point at $s = \mathrm{e}^{\mathrm{i}\pi/2}\eta^{1/2}$ (where $\mathrm{d}s/\mathrm{d}l = \mathrm{e}^{-\mathrm{i}\phi/4}$), while its right part is deformed to the steepest descent path passing through the saddle point at

7 There will be many different definitions in different references. The definition here is adopted from reference [50].

$s = -e^{i\pi/2}\eta^{1/2}$ (where $ds/dl = e^{i\pi/2}e^{i\phi/4}$). Therefore, after making an approximation similar to that mentioned above, we have

$$Ai(\eta) \to \frac{1}{2\pi^{1/2}\eta^{1/4}}\left(e^{-(2/3)}\eta^{3/2} + ie^{(2/3)}\eta^{3/2}\right), \quad 2\pi/3 < \phi < 4\pi/3 \tag{2.53b}$$

Comparing Eq. (2.53b) with (2.52b), there seems to be existing discontinuities along the line $\phi = 2\pi/3$ and the line $\phi = 4\pi/3$. In reality, the discontinuity does not exist because when $|\eta|$ is very large, the value of $e^{(2/3)}\eta^{3/2}$ along the lines is so small that it can be neglected; and secondly, because of the two terms in Eq. (2.53b), the value of the second term at $\phi = 4\pi/3$ is exactly the same as that of the first term at $\phi = 2\pi/3$.

When η is a negative number, let $\eta = |\eta|\, e^{i\pi}$ in Eq. (2.53b), we obtain $\eta < 0$

$$\begin{aligned} Ai(\eta) &\to \frac{e^{i\pi/4}}{2\pi^{1/2}\eta^{1/4}}\left(e^{-i(2/3)|\eta|^{3/2}} - ie^{(2/3)|\eta|^{3/2}}\right) \\ &= \frac{1}{\pi^{1/2}|\eta|^{1/4}}\cos\left(\frac{2}{3}|\eta|^{3/2} - \frac{\pi}{4}\right) \end{aligned} \tag{2.54}$$

Therefore, when η is real number, $Ai(\eta)$ oscillates while $\eta < 0$ (as shown in Fig. 1). $Ai(0) = 0.355, \cdots$; For a series of negative values of η, at first $Ai(\eta)$ reaches the highest peak 0.536 at $\eta = -1.019$, then the first zero at $\eta = -2.338$, the minimum -0.419 at $\eta = -3.324$, the second maximum 0.308 at $\eta = -4.088$, the second zero point at $\eta = -4.82$. The nth zero occurs at $\eta = -\left(\frac{3}{2}\right)^{2/3}\left[\left(n - \frac{1}{4}\right)\pi\right]^{2/3}$.

2.4.3 The wave field in the presence of a turning point

In order to discuss the acoustic field in the vicinity of a turning point (at this point, $\cos\theta = 0$, the solution of Eq. (2.5) is meaningless) we will make a transformation on Eq. (2.38b) first. A new variable t is introduced to replace z, i.e., $z = z(t)$. Thus, Eq. (2.38b) can be expressed by (for the oblique incidence case, just replace n^2 with $n^2 - \sin\theta_0$)

$$\psi''(z) - \left(z''/z'\right)\psi' + k_0^2 n^2 z'^2 \psi = 0$$

where the prime denotes the differentiation with respect to t. In order to get rid of the terms containing ψ', we replace ψ with by Ψ,

$$\psi = \left(z'\right)^{1/2}\Psi \tag{2.55}$$

Substituting Eq. (2.55) into the above equation, we have

$$\Psi'' + k_0^2(z)\, n^2 \Psi = \left[3\left(z''\right)^2/4\left(z'\right)^2 - z'''/2z'\right]\Psi \tag{2.56}$$

Next, we specify the detailed form of function $z(t)$ by assuming

$$t = ik_0 \int_{z_0}^{z} n\mathrm{d}z \tag{2.57}$$

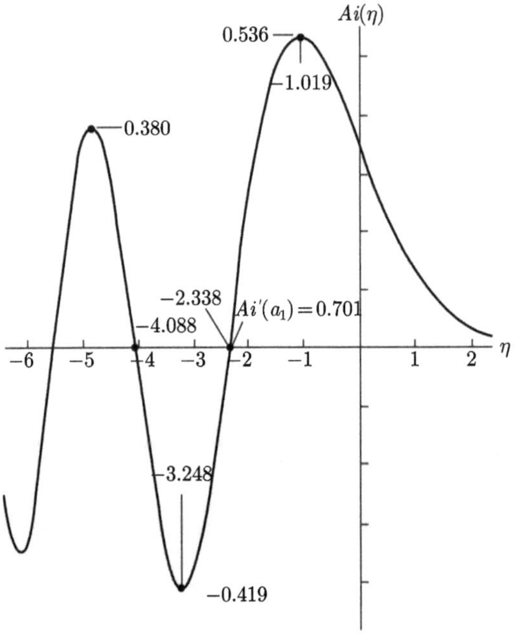

Fig. 7: Schematic diagram of the Airy function for the real argument *Ai*(*t*).

Then Eq. (2.56) is transformed into

$$\Psi'' - \Psi = \left[3(\dot{n})^2/2n - \ddot{n}\right] \Big/ 2k_0^2 n^3 \tag{2.56'}$$

where the dots on the top of n denote the differentiation with respect to argument z. When k_0 is very large, the right-hand side of the equation can be neglected, and the equation becomes

$$\Psi'' - \Psi = 0 \tag{2.58}$$

That is called the standard equation of Eq. (2.56′), whose solution is the uniform asymptotic solution of Eq. (2.54) at $k_0 \to \infty$. The two linearly independent solutions of Eq. (2.58) are $\Psi = e^{\pm t}$. If transforming Ψ and t back into ψ and z, we can find immediately that they are just the approximate expression of the geometrical optics in Eq. (2.46′).

Next we assume that the turning point corresponds to $z = 0$. At this time Eq. (2.58) is not the standard equation of Eq. (2.56′), because when $z \to 0$, the right-hand side of Eq. (2.56′) is not very small, but increases without limit. Assuming $n^2(z)$ in (2.38b) can be expressed by the following series in the vicinity of $z = 0$

$$n^2 = -az + bz^2 + \cdots, \quad a > 0 \tag{2.59}$$

Then, $z = 0$ is called the first-order turning point; $z = 0$ while $b \neq 0$ is called the second-order turning point, and so on.

For example, only retaining the first term in Eq. (2.59) (corresponding to the case that z is very small, i.e., in the vicinity of the turning point at $z = 0$), the solution can be expressed by the Airy function [52]. Hence, the equation satisfied by the Airy function is the standard equation for this case. In order to transform Eq. (2.56) into the standard equation, the following variable is introduced

$$t = \left(\frac{3ik_0}{2} \int_{z_0}^{z} n\mathrm{d}z \right)^{2/3} = (\pm 1)^{2/3} z/H, \quad H = \left(ak_0^2 \right)^{-1/2} \tag{2.60}$$

Replacing the right-hand side of Eq. (2.56) with 0, we obtain the Airy equation, which has the same form as in Eq. (2.51)

$$\Psi'' - t\Psi = 0 \tag{2.61}$$

Thus, we can constitute the solution of Eq. (2.38b) for any z. Therefore, we define a region in the vicinity of $t = 0$ (corresponding to $z = 0$) with $|t| \leqslant t_1, t_1 \gg 1$, so that only the first term can be retained in the expansion of Eq. (2.59). Using the equivalent steps described previously, we obtain the acoustic field expression for different ranges

$$t < -t_1 : \psi = C_1 n^{-1/2} \exp\left(ik_0 \int_0^z n\mathrm{d}z \right) + C_2 n^{-1/2} \exp\left(-ik_0 \int_0^z n\mathrm{d}z \right)$$

$$|t| \leqslant t_1 : \psi = 2C_1 \mathrm{e}^{-i\pi/4} \left(-t/n^2 \right)^{1/4} Ai(t)$$

$$t > t_1 : \psi = C_1 \mathrm{e}^{-i\pi/4} \left(-n^2 \right)^{-1/4} \exp\left(-k_0 \int_0^z \left(-n^2 \right)^{-1/2} \mathrm{d}z \right) \tag{2.62}$$

It can be seen from the above equation that at positions far from the turning point, the acoustic field in the approximation of geometrical optics is a direct wave (incident wave) and a return wave (reflecting wave), respectively. In summary, near the vicinity of the turning point, the field can be described by the Airy function; and far away from the turning point, the field decays exponentially with increasing z.

For the situation of two turning points, Ref. [54] can be consulted.

2.5 Normal mode solutions

Another basic approach to Eq. (2.38a) is to express its solution in the form of the summation of all of the normal modes; each of which satisfies the wave equation and boundary conditions independently but propagates at different speeds along the boundary.

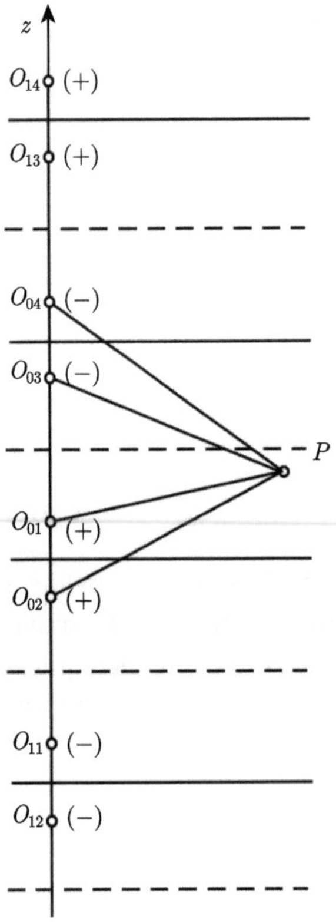

Fig. 8: Chain of virtual sources resulting from atmospheric layers when the top boundary surface is a free boundary face ($R = -1$), but the bottom boundary surface is a rigid boundary surface ($R = 1$).

2.5.1 Image of virtual sources

To express the solution in terms of normal modes, we take an ideal simplified model into account where the atmosphere layer is homogeneous and located between the $z = 0$ and $z = h$ planes. A point source is placed at $z = z_0$, and both the top and the bottom boundary surfaces reflect perfectly, i.e., being a free and a rigid boundary surface, respectively. The acoustic potential ψ satisfies the wave equation (2.38) inside the layer with $K \to k$ being a constant. At the top and bottom boundaries, ψ satisfies the following boundary conditions, respectively: the normal velocity is restricted and the acoustic pressure is 0:

$$\begin{aligned} \partial_z \psi &= 0, \quad z = 0 \\ \psi &= 0, \quad z = h \end{aligned} \tag{2.62a}$$

The multi-reflection of the acoustic source by the boundary surfaces forms an infinite chain of virtual sources (Fig. 8). The acoustic pressure at any point $P(r, z)$ can be treated as the sum of the wave from the source directly, and those from the virtual sources.

For example, using the field of a virtual source O_{02}, the virtual source created by the bottom boundary reflection complements the field of the original sound source O_{01}, and the resultant field becomes

$$\psi = \frac{e^{ikR_{01}}}{R_{01}} + \frac{e^{ikR_{02}}}{R_{02}} \tag{2.63}$$

Where $R_{01} = \left[r^2 + (z - z_0)^2\right]^{1/2}$ and $R_{02} = \left[r^2 + (z - z_0)^2\right]^{1/2}$ are the distances of O_{01} and O_{02} to the receiving point, respectively. Eq. (2.63) satisfies both the wave equation (2.38) and the boundary conditions on the bottom boundary. Because O_{01} and O_{02} are symmetric, and refer to the plane $z = 0$, $\partial_z\psi = 0$ that is satisfied at this boundary can be deduced directly from the symmetrical condition. Nevertheless, it does not satisfy the boundary condition at $z = h$.

By complementing another pair of virtual sources, O_{03} and O_{04}, which are created due to the reflection of O_{01} and O_{02} by the top boundary surface, then the solution satisfies both the wave equation and the boundary condition of the top boundary, but not that of the bottom boundary. If the virtual source pair O_{11} and O_{12}, created by the reflection of O_{03} and O_{04} due to the bottom boundary, are complemented, the resultant field satisfies the bottom boundary condition, but not the top boundary condition, and so on. Extending the length of the chain by including more virtual sources, the two scenarios, one boundary condition being satisfied while the other being not, appear alternately. Nevertheless, the newly included virtual source pair is increasingly farther away from the receiving point, and its contribution to the field becomes increasingly smaller. In the case of an infinite long virtual source chain, both the top and bottom boundary conditions can be satisfied simultaneously.

Notice that because the two boundary conditions are different, $\partial_z\psi = 0$ corresponds to the reflection coefficient $\Re = 1$, while $\psi = 0$ corresponds to reflection coefficient $\Re = -1$. Thus the virtual sources created by the top boundary have an opposite phase to the original source (as well as the virtual sources produced by the reflection of the lower boundary). The virtual source chain that corresponds to the sources being reflected an odd number of times by the top boundary is shown in Fig. 8, where the actual source and the virtual sources in phase with the actual source are denoted by "+" sign, while the sources with opposite phase denoted by "–" sign. It is not hard to observe that the straight line connecting the latter sources and the receiving point P intersects the top boundary and all its mirror images (denoted by a group of horizontal dotted lines) odd number of times. Finally, the total acoustic field at the receiving point, P, can be expressed as

$$\psi = \sum_{l=0}^{\infty} (-1)^l \left[\frac{e^{ikR_{l1}}}{R_{l1}} + \frac{e^{ikR2}}{R_{l2}} + \frac{e^{ikR_{l3}}}{R_{l3}} - \frac{e^{ikR_{l4}}}{R_{l4}}\right] \tag{2.64}$$

2.5.2 Integral representation of the field

It is well-known that when dealing with the reflection by a plane of the spherical wave produced by a point source, the spherical wave should be first decomposed into plane waves, which gives

$$\frac{e^{ikR}}{R} = \frac{ik}{2\pi} \int_0^{\frac{\pi}{2}-i\infty} \int_0^{2\pi} e^{i(k_x x + k_y y \pm k_z z)} \sin\theta d\theta d\varphi \tag{2.65}$$

Where R is the distance between the sound source and the receiving point; θ and φ are the angles between the wavenumber vector $\boldsymbol{k}$ and the z axis, and the angle of the projection of $\boldsymbol{k}$ on the (x, y) plane and x axis, respectively. The sign "±" corresponds to the $z > 0$ and $z < 0$ cases.

Considering the integral with respect to φ, it becomes the 0th order Bessel function J_0 (which can also be expressed as the sum of the 1st kind Hankel function $H_0^{(1)}$ and the 2nd kind Hankel functions $H_0^{(2)}$). The integral with respect to θ turns into a sum of two integrals having the same integrand (notice $H_0^{(2)}(e^{-\pi i}u) = -H_0^{(1)}(u)$) with different integral limits, one from 0 to $\frac{\pi}{2} - i\infty$, and the other from $-\frac{\pi}{2} + i\infty$ to 0. The two integrals can be combined into one along the path Γ from $-\frac{\pi}{2} + i\infty$ to $\frac{\pi}{2} - i\infty$. Thus, Eq. (2.65) will be transformed into

$$\frac{e^{ikR}}{R} = \frac{ik}{2} \int_\Gamma e^{\pm ikz\cos\theta} H_0^{(1)}(kr\sin\theta)\sin\theta d\theta \tag{2.66}$$

Apply the above equation to each term of Eq. (2.64), and set the quantity z cover all of the values of $z = z_{lj}$, $l = 0, 1, 2, \cdots, j = 1, 2, 3, 4$; the following relations can be obtained based on the similar relation between R_{01} and R_{02} given in Eq. (2.63)

$$\begin{aligned} z_{l1} &= 2lh + z - z_0 \\ z_{l2} &= 2lh + z + z_0 \\ z_{l3} &= 2(l+1)h - z - z_0 \\ z_{l4} &= 2(l+1)h - z + z_0 \\ R_{lj} &= \sqrt{r^2 + z_{lj}^2} \end{aligned} \tag{2.67}$$

$z_{lj} > 0$ is suitable to all l and j except for $z_{01} = z - z_0$.

By inserting the expression of $\frac{e^{ikR_{lj}}}{R_{lj}}$ specified by Eq. (2.66) into Eq. (2.64), and exchanging the order of the summation and integral, and introducing the notation $b \equiv ik\cos\theta$, and by making a simple transformation, such as applying the identity of

$$e^{b(z-z_0)} + e^{b(z+z_0)} - e^{b(2h-z-z_0)} - e^{b(2h-z+z_0)} \equiv (e^{-bz_0} + e^{bz_0})(e^{bz} - e^{b(2h-z)})$$

While the infinite geometric series satisfies

$$\sum_{l=0}^{\infty} (-1)^l e^{2lbh} = \frac{1}{1 + e^{2bh}}$$

The following expressions for $z > z_0$ and $z < z_0$ are obtained

$$\begin{aligned} z > z_0, \quad \psi &= 2ik \int_{\Gamma} \frac{\cosh bz_0 \sinh b(h-z)}{\cosh bh} H_0^{(1)}(kr \sin\theta) \sin\theta d\theta \\ z < z_0, \quad \psi &= 2ik \int_{\Gamma} \frac{\cosh bz \sinh b(h-z_0)}{\cosh bh} H_0^{(1)}(kr \sin\theta) \sin\theta d\theta \end{aligned} \tag{2.68}$$

It is shown that the difference between the above two expressions is the replacement of z with z_0, which conforms with the Principle of Reciprocity. It also demonstrates that the integral in Eq. (2.68) is convergent everywhere, except at the source point $(0, z_0)$, where the required singularity in the form of $\frac{1}{R}$ is given.

2.5.3 Normal modes

The integral expression in Eq. (2.68) can be transformed into the sum of normal modes. For this purpose, a new integral variable $\xi \equiv k \sin\theta$ is introduced to replace θ. The corresponding integral limit for the new variable is from $-\infty$ to $+\infty$. Therefore, Eq. (2.68) becomes

$$\begin{aligned} z > z_0, \quad \psi &= 2ik \int_{-\infty}^{+\infty} \frac{\cosh bz_0 \sinh b(h-z)}{b \cosh bh} H_0^{(1)}(\xi r) \xi d\xi \\ z < z_0, \quad \psi &= 2ik \int_{-\infty}^{+\infty} \frac{\cosh bz \sinh b(h-z_0)}{b \cosh bh} H_0^{(1)}(\xi r) \xi d\xi \end{aligned} \tag{2.69}$$

The integral obtained via this method can be transformed into the sum of the residues of the poles of the integrand. The positions of poles are determined by the roots of

$$\cosh bh = 0 \tag{2.70}$$

They are

$$b_l h = i\pi \left(l + \frac{1}{2}\right), \qquad l = 0, \pm 1, \pm 2, \cdots \tag{2.71}$$

Or, since $b = \sqrt{\xi^2 - k^2}$, the locations of the poles are at

$$\xi_l = \pm \sqrt{k^2 - \left(\frac{\pi}{h}\right)^2 \left(l + \frac{1}{2}\right)^2} \tag{2.71a}$$

When $\left(l + \frac{1}{2}\right)\pi < kh$, these poles are positioned along the real axis of ξ plane; when l is very large, the poles are at the image axis (the sign of l has no effect here and only positive values of l are considered). If the absorption of the atmosphere is counted, k

can be regarded as having a small positive imaginary part (this imaginary part could be treated as zero in final results). In this case, the poles are not located exactly on the coordinate axes, but at points departing small distances from the axes in the 1st and 3rd quadrants.

Next, we extend the integral path in the upper half plane from the real axis to infinity. The Hankel function $H_0^{(1)}(\xi r)$ approaches 0 at infinity only if $r \neq 0$. It can be proved that the integral along the infinite part of integral contour will vanish.[8]

Thus, the whole expression of ψ can be simplified to the sum of all the residues (expressed by sign Res) taken at the poles in the first quadrant

$$\mathrm{Res}\left[\frac{\xi \cosh bz_0 \sinh b(h-z)}{b\cosh bh}\right]_{\xi=\xi_l} = \left[\frac{\xi \cosh bz_0 \sinh b(h-z)}{b\dfrac{\mathrm{d}}{\mathrm{d}\xi}(\cosh bh)}\right]_{\xi=\xi_l}$$

Because $[\cosh bh]_{\xi=\xi_l} = 0$, $[\sinh b(h-z)]_{\xi=\xi_l} = \sin bh \cosh bz$, and also because $\frac{\mathrm{d}}{\mathrm{d}\xi}(\cosh bh) = \frac{\xi h}{b}\sinh bh$, the residues become

$$\mathrm{Res}[\ ] = \frac{1}{h}\cosh b_l z \cosh b_l z_0$$

Hence, the expression of acoustic field given by Eq. (2.68) is transformed to

$$\psi = \frac{2\pi \mathrm{i}}{h}\sum_{l=0}^{\infty}\cosh b_l z_0 \cosh b_l z H_0^{(1)}(\xi_l r) \tag{2.72}$$

Where b_l and ξ_l are given by Eq. (2.71) and (2.71a), respectively.

Eq. (2.72) satisfies all of the conditions of the problem, which can be validated using the substitution method – it satisfies the wave equation (2.38), as well as the boundary condition (2.62) term-by-term. Each term in Eq. (2.72) is in the so-called normal mode.

When the distance of the field point to the sound source is longer than the wavelength, the Hankel function can be replaced with its asymptotic form

$$r \to \infty: \quad \psi = \frac{2\mathrm{e}^{\mathrm{i}\pi/4}}{h}\sqrt{\frac{2\pi}{r}}\left\{\sum_{l=0}^{\infty}\cosh b_l z_0 \cosh b_l z \frac{1}{\sqrt{\xi_l}}\mathrm{e}^{\mathrm{i}\xi_l r}\right\} \tag{2.73}$$

It is not difficult to see that the velocity of each normal mode is

$$V_l = \frac{\omega}{\xi_l} = \frac{c}{\sqrt{1-\left[\left(l+\frac{1}{2}\right)\frac{\lambda}{2h}\right]^2}} \tag{2.74}$$

8 Generally speaking, at this time the total sum is not the correct solution of the question, in a certain condition, but is only an approximation to the real circumstance.

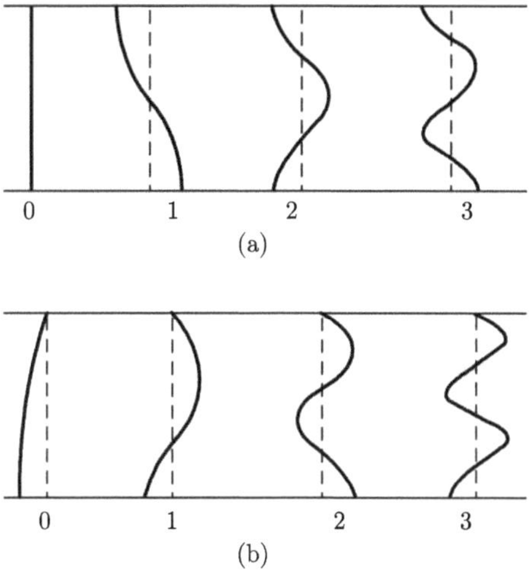

Fig. 9: In the atmosphere when the top boundary is free or when the bottom boundary is rigid, respectively, the changes of the amplitude of the first four order normal modes along the thickness of the layer, (a) reflection coefficient of top boundary = 1, (b) reflection coefficient of bottom boundary = −1.

Where c is the sound speed, and λ is the wavelength. The above equation specifies that when $\left(l+\frac{1}{2}\right)\lambda \leqslant 2h$, $V_l \geqslant c$; if $\left(l+\frac{1}{2}\right)\lambda > 2h$, V_l turns into an imaginary number. The physical meaning is that the corresponding normal modes are non-uniform waves whose amplitudes decay exponentially along the range r. The existence of these waves is necessary for obtaining the required singularity at the sound source.

When $r \geqslant h$, only a few terms, that satisfy $\left(l+\frac{1}{2}\right)\lambda < 2h$, of the summation in Eq. (2.73) need to be considered. The number of these terms equals the number of half-wavelengths contained in thickness h.

Each term in Eq. (2.73) denotes a standing wave in the z direction, which still travels in the r direction. The dependency of the amplitude of each standing wave on z is given by factor $\cosh b_l z$. Fig. 9 demonstrates the dependent relationships for the first four order normal modes ($l = 0, 1, 2, 3$).

2.5.4 Cases of arbitrary boundaries

Finally, we should point out that the total field equation (2.64) for perfectly reflecting interfaces is also valid for any arbitrary boundaries. In those cases the reflection coefficients of the top and bottom boundaries, $\Re_1$ and $\Re_2$, are functions of the incidence angle. Although the virtual source picture described in Section 2.5.1 is not correct any more, a similar equation corresponding to Eq. (2.64) can still be derived [55].

$$\psi = \sum_{l=0}^{\infty} (\Re_1\Re_2)^l \left[\frac{e^{ikR_{l1}}}{R_{l1}} + \Re_1\frac{e^{ikR_{l2}}}{R_{l2}} + \Re_2\frac{e^{ikR_{l3}}}{R_{l3}} + \Re_1\Re_2\frac{e^{ikR_{l4}}}{R_{l4}}\right] \tag{2.75}$$

Accordingly, we can replace Eq. (2.68) with (only the expression for $z > z_0$ is given because the expression for $z < z_0$ can be obtained by exchanging the positions of z and z_0)

$$\psi = \frac{\mathrm{i}k}{2} \int\limits_{\Gamma} \frac{(\mathrm{e}^{-bz_0} + \Re_1 \mathrm{e}^{bz_0})(\mathrm{e}^{-b(h-z)} + \Re_2 \mathrm{e}^{b(h-z)})}{\mathrm{e}^{-bh}(1 - \Re_1 \Re_2 \mathrm{e}^{2bh})} H_0^{(1)}(kr \sin\theta) \sin\theta \mathrm{d}\theta \tag{2.76}$$

B transforming the integration path Γ, and calculating the sum of residues at the poles (let it is θ_l) of the integrand, we can obtain the normal mode equation for the general situation

$$\psi = \pi k \sum_l \left[\frac{(\mathrm{e}^{-bz} + \Re_1 \mathrm{e}^{bz})(\mathrm{e}^{-bz} + \Re_2 \mathrm{e}^{bz})}{\Re_l (\partial/\partial\theta)(\Re_1 \Re_2 \mathrm{e}^{2bh})} \right]_{\theta_l} H_0^{(1)}(kr \sin\theta_l) \sin\theta_l \tag{2.77}$$

The above equation is valid for $0 \leqslant z \leqslant h$, because it does not change when the positions of z and z_0 are exchanged.

2.6 Basic concepts of geometrical (ray) acoustics [56]

The particularity of an atmosphere as a medium makes geometrical acoustics of great importance in the study of atmospheric acoustics. Therefore, it is necessary to discuss it in detail.

2.6.1 Wave fronts, rays and eikonal

The concept of wave fronts is the backbone of ray acoustics. Wave fronts are the curved surface formed by the wave at a given moment. They progress forward as time goes on. For example, if the acoustic pressure has an obvious peak arriving at point x at time $\tau(x)$, the ensemble of all the points satisfying the condition $t = \tau(x)$ describe the corresponding wave front at time t. For a perturbation with a constant frequency, the wave fronts are a series of curved surfaces where the phase of the oscillating acoustic pressure is same everywhere. However, theories of geometrical acoustics all assume that the amplitude of vibration is small compared with the wavelength, and that the radii of curvatures of the wave-fronts are much larger compared with their wavelengths.

A wavefront moves at speed c relative to a fixed coordinate system. If the ambient medium moves at a speed $\boldsymbol{v}$, the wave speed $c\boldsymbol{n}$ (where $\boldsymbol{n}$ is the normal unit vector of the wavefront) observed from the moving coordinate system will be transformed into $\boldsymbol{v} + c\boldsymbol{n}$ observed from the static coordinate system. In moving coordinates, and traveling at the same speed as the ambient medium, $\boldsymbol{n}$, the propagation direction are the same. Nevertheless, it is not always this case for static observers, because $\boldsymbol{n}$ does not have anything to do with the velocity of the reference system, while the propagation direction does.

Let $x_p(t)$ be a moving point on the wave front of $t = \tau(x)$ at the initial moment (as shown in Fig. 10). According to the above discussion, $x_p(t)$ is always on the moving wave front if its speed satisfies

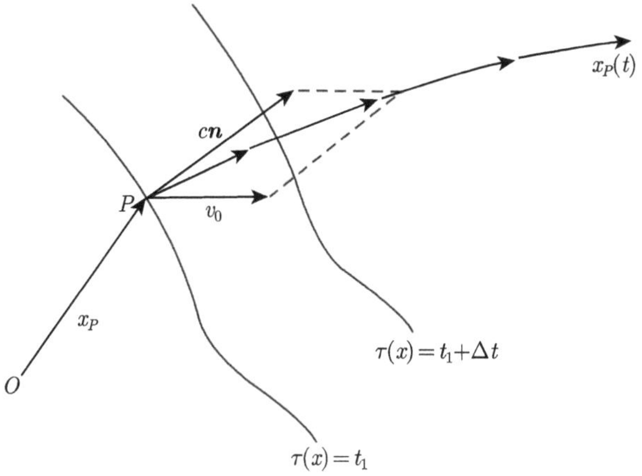

Fig. 10: The concept of wave fronts and rays: point $x_p(t)$ moves with velocity $c\boldsymbol{n} + \boldsymbol{v}$, so it is always in front of the wave front $c(x)$, and continuously repeating the course will "trace" a ray.

$$d_t\boldsymbol{x}_p = \boldsymbol{v}(\boldsymbol{x}_p, t) + \boldsymbol{n}(\boldsymbol{x}_p, t)\, c(\boldsymbol{x}_p, t) \equiv \boldsymbol{v}_{\text{ray}} \tag{2.78}$$

Here we allow the possibility that $\boldsymbol{v}$ and c change in space and time. The curves owing to the variation of $\boldsymbol{x}_p(t)$ over time t in the space are called rays. The function $\boldsymbol{x}_p(t)$ describes the ray's trajectory. The speed of the wave front in the normal direction is the dot product of the right-hand side of Eq. (2.78) and $\boldsymbol{n}$. The product is equal to $c + \boldsymbol{v}\cdot\boldsymbol{n}$, which is less than the value $|c\boldsymbol{n} + \boldsymbol{v}|$ of the ray's speed $\boldsymbol{v}_{\text{ray}}$.

Eq. (2.78) is a generalized expression of the Huygens principle, which forms the base for deciding the wave front's properties at a subsequent time. However, for inhomogeneous media, Eq. (2.78) is difficult to be applied directly, because it requires $\boldsymbol{n}$ along the ray path to be known at every moment (which requires knowing the wave front structure in the vicinity of rays during a very short time). To avoid this, we introduce an additional differential equation that can predict the rate of change in time of $\boldsymbol{n}$, rather than dealing with $\boldsymbol{n}$ directly. The wave delay vector[9] $\boldsymbol{s}(\boldsymbol{x}) = \nabla\tau(\boldsymbol{x})$ is adopted. Since $\nabla\tau$ is perpendicular to the curved surface $t = \tau(\boldsymbol{x})$, it is parallel to $\boldsymbol{n}$.

The reason why the nomenclature of "wave delay" is used is because the reciprocal of $|\boldsymbol{s}|$ is the wave rate $c + \boldsymbol{n}\cdot\boldsymbol{v}$, and the wave front moves forward at this velocity along its own normal direction. This can be proved by considering the wave fronts at the closely spaced time interval t and $t + \Delta t$. For a given ray trajectory $\boldsymbol{x}_p(t)$, its position at time $t + \Delta t$ can be approximated by $\boldsymbol{x}_p(t) + \dot{\boldsymbol{x}}_p(t)\Delta t$ (the dot on the top of $\boldsymbol{x}_p$ denotes taking the total differential quotient). Therefore, $t + \Delta t \approx \tau(\boldsymbol{x}_p(t) + \dot{\boldsymbol{x}}_p(t)\Delta t) \approx$

9 This term originates in reference [57]. For a plane wave with constant frequency, $\boldsymbol{s}$ is $\boldsymbol{k}/w$, so its direction is parallel to the phase velocity and its value is the reciprocal of the phase velocity.

$\tau(\boldsymbol{x}_p)+\Delta t\dot{\boldsymbol{x}}_p\cdot\nabla\tau$. Because $t=\tau(\boldsymbol{x}_p)$, and $\nabla\tau=\boldsymbol{s}$, hence $\nabla\tau\cdot\dot{\boldsymbol{x}}_p=1$ should be satisfied; or according to Eq. (2.78), for any given point on the wave front at any given moment, we have

$$\boldsymbol{s}\cdot(c\boldsymbol{n}+\boldsymbol{v})=1,\quad c\boldsymbol{s}\cdot\boldsymbol{n}=1-\boldsymbol{v}\cdot\boldsymbol{s} \tag{2.79}$$

Because $\boldsymbol{s}$ is parallel to $\boldsymbol{n}$, so $\boldsymbol{s}=(\boldsymbol{s}\cdot\boldsymbol{n})\,\boldsymbol{n}$. Considering $\boldsymbol{n}=\boldsymbol{s}/(\boldsymbol{s}\cdot\boldsymbol{n})$, the above equation gives

$$\boldsymbol{s}=\frac{\boldsymbol{n}}{c+\boldsymbol{v}\cdot\boldsymbol{n}},\quad \boldsymbol{n}=\frac{c\boldsymbol{s}}{\Omega} \tag{2.80}$$

Where

$$\Omega=1-\boldsymbol{v}\cdot\boldsymbol{s}=1-\boldsymbol{v}\cdot\nabla\tau=\frac{c}{c+\boldsymbol{v}\cdot\boldsymbol{n}} \tag{2.81}$$

Eq. (2.80) confirms the assertion that $|\boldsymbol{s}|^{-1}=c+\boldsymbol{n}\cdot\boldsymbol{v}$. In the same way, because $\boldsymbol{n}\cdot\boldsymbol{n}=1$ and $\boldsymbol{s}=\nabla\tau$, the previous relation gives

$$s^2=\frac{\Omega^2}{c^2},\quad (\nabla\tau)^2=\frac{\Omega^2}{c^2} \tag{2.82}$$

The above equation is called the Eikonal equation. $\tau(\boldsymbol{x})$ is called an eikonal.

2.6.2 Ray-tracing equations

The differential equation regarding the rate of change in time of $\boldsymbol{s}$ along the ray trajectory can be derived from the following expression[10]

$$\mathrm{d}_t\boldsymbol{s}(\boldsymbol{x}_p)=(\dot{\boldsymbol{x}}_p\cdot\nabla)\,\boldsymbol{s}=c\,(\boldsymbol{n}\cdot\nabla)\,\boldsymbol{s}+(\boldsymbol{v}\cdot\nabla)\,\boldsymbol{s} \tag{2.83}$$

Where all of the quantities should be regarded as taking their values at $\boldsymbol{x}_p(t)$. Because $\boldsymbol{n}$ has the same direction as $\boldsymbol{s}$, the first term contains the factor $(\boldsymbol{s}\cdot\nabla)\,\boldsymbol{s}$, which can be expressed as

$$(\boldsymbol{s}\cdot\nabla)\,\boldsymbol{s}=-\boldsymbol{s}\times(\nabla\times\boldsymbol{s})+\frac{1}{2}\nabla s^2=0+\frac{1}{2}\nabla\frac{\Omega^2}{c^2}=-\frac{\Omega}{c^2}\nabla(\boldsymbol{v}\cdot\boldsymbol{s})-\frac{\Omega^2}{c^2}\nabla c \tag{2.84}$$

Where we apply $\nabla\times(\nabla\tau)=0$ and the expression of s^2 is given in Eq. (2.82). Substituting Eq. (2.84) and the second equation in Eq. (2.80) into Eq. (2.83), sequentially, yields

$$\mathrm{d}_t s=-\frac{\Omega}{c}\nabla c-\nabla(\boldsymbol{v}\cdot\boldsymbol{s})+(\boldsymbol{v}\cdot\nabla)\,\boldsymbol{s} \tag{2.85}$$

Taking use of the following vector relation equation (Eq. (2.84) is a particular case of this relation), the above equation can be further reduced to

$$\nabla(\boldsymbol{v}\cdot\boldsymbol{s})=\boldsymbol{v}\times(\nabla\times\boldsymbol{s})+\boldsymbol{s}\times(\nabla\times\boldsymbol{v})+(\boldsymbol{v}\cdot\nabla)\,\boldsymbol{s}+(\boldsymbol{s}\cdot\nabla)\,\boldsymbol{v} \tag{2.86}$$

10 Although there are many different derivations, among these, the earliest must be in reference [58]. With regards to the ray trajectory analysis of motion stratified fluid, some early work can be consulted in reference [59]~[61].

Here the first term is zero because $\boldsymbol{s}$ (= $\nabla\tau$) is a gradient.

Eq. (2.78) and (2.85) are the so-called ray-tracing equations. Substituting the second expression in Eq. (2.80), as well as Eq. (2.86) into them, they can be expressed as (henceforward the suffix P is omitted)

$$\mathrm{d}_l \boldsymbol{x} = \frac{c^2}{\Omega} + \boldsymbol{v} \tag{2.87}$$

$$\mathrm{d}_l \boldsymbol{s} = -\frac{\Omega}{c}\nabla c - \boldsymbol{s} \times (\nabla \times \boldsymbol{v}) - (\boldsymbol{s} \cdot \nabla)\,\boldsymbol{v} \tag{2.88}$$

Or, expressed using the symbols of Cartesian coordinates

$$\mathrm{d}_l \boldsymbol{s} = -\frac{\Omega}{c}\partial_i c - \sum_{j=1}^{3} s_j \partial_i v_j \tag{2.88$'$}$$

These equations do not depend on the spatial differential quotient of $\boldsymbol{s}$. Therefore, if $c\,(\boldsymbol{x}, t)$ and $\boldsymbol{v}\,(\boldsymbol{x}, t)$ are specified, and so too are the ray's position $\boldsymbol{x}$ and the wave delay vector $\boldsymbol{s}$ at moment t_0, we can integrate the equation group (2.87)~(2.88) over time to determine $\boldsymbol{x}$ and $\boldsymbol{s}$ at any subsequent moment without needing any information about the adjacent rays. Although these equations are non-linear, they are all first-order ordinary differential equations, which can be solved using standard numerical integral technology, where references [62]~[63] can be consulted. Further details will be discussed in Chapter 7.

For more general cases, we can consider the propagation of wave packets with a slowly changing frequency $\omega\,(\boldsymbol{x}, t)$ and wavenumber $\boldsymbol{k}\,(\boldsymbol{x}, t)$ in an anisotropic media that depends on time. If $F\,(\omega, \boldsymbol{k}, \boldsymbol{x}, t) = 0$ demonstrates the dispersion relation in the vicinity of point $\boldsymbol{x}$ at time t, the ray equation is given by the following equation group [64, 65]

$$\mathrm{d}_t\omega = -\frac{\partial_t F}{\partial_\omega F}, \quad \mathrm{d}_t x_i = -\frac{\partial_{k_i} F}{\partial_\omega F}, \quad \mathrm{d}_t k_i = -\frac{\partial_{x_i} F}{\partial_\omega F} \tag{2.89}$$

In this particular case, the dispersion relation can be derived using the Eikonal equation (2.82): $F = (\omega - \boldsymbol{v} \cdot \boldsymbol{k})^2 - c^2 k^2 = 0$.

2.6.3 Fermat's principle

If l denotes the distance along the ray path, $\mathrm{d}_l x$ (simply expressed by $\boldsymbol{x}'$) denotes the direction of the ray. The speed of the ray, v_{ray}, as defined by Eq. (2.78), satisfies

$$c\boldsymbol{n} = v_{\mathrm{ray}}\boldsymbol{x}' - \boldsymbol{v} \tag{2.90}$$

As $\boldsymbol{n} \cdot \boldsymbol{n} = \boldsymbol{x}' \cdot \boldsymbol{x}' = 1$, v_{ray} satisfies the following quadratic equation

$$v_{\mathrm{ray}}^2 - 2v_{\mathrm{ray}}\boldsymbol{v} \cdot \boldsymbol{x}' - \left(c^2 - v^2\right) = 0$$

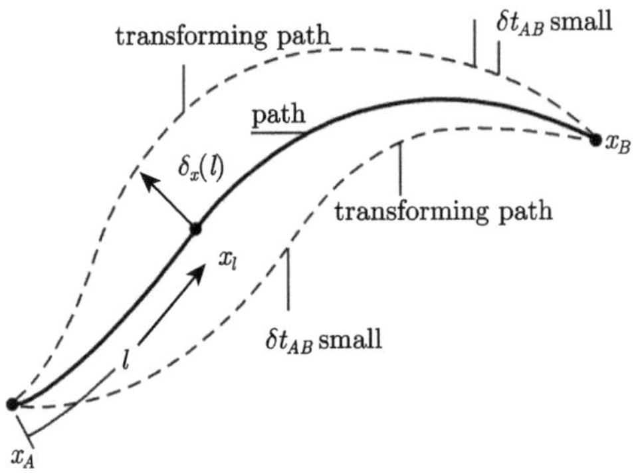

Fig. 11: Fermat's principle.

Its positive solution (for $c^2 > v^2$) is

$$v_{\text{ray}} = \boldsymbol{v} \cdot \boldsymbol{x}' + \left[c^2 - v^2 + \left(\boldsymbol{v} \cdot \boldsymbol{x}'\right)^2\right]^{1/2} \tag{2.91}$$

Therefore, the time required for the ray propagating from $\boldsymbol{x}_A$ to $\boldsymbol{x}_B$ is

$$t_{AB} = \int_{t_A}^{t_B} \frac{\mathrm{d}l}{\boldsymbol{v} \cdot \boldsymbol{x}' + \left[c^2 - v^2 + (\boldsymbol{v} \cdot \boldsymbol{x}')^2\right]^{1/2}} \tag{2.92}$$

Here c and $\boldsymbol{v}$ are assumed to only be functions of the coordinates. Therefore, for a given ray path, the time can be treated as a function of the path distance l.

Fermat's principle[11] specifies that the actual path connecting $\boldsymbol{x}_A$ and $\boldsymbol{x}_B$ makes the travel time integral t_{AB} steady for a small virtual variation of the path. By adding a small change to the actual path, $\boldsymbol{x}(l) \to \boldsymbol{x}(l) + \delta\boldsymbol{x}(l)$ (as shown in Fig. 11), and being accurate to the first-order quantity of $\delta\boldsymbol{x}$ the alteration of the travel time, δt_{AB}, should be zero.

In order to prove that when the path is being travelled there is no reflection, we can: transform the integral variable into the projection of the ray path on the straight line connecting $\boldsymbol{x}_A$ with $\boldsymbol{x}_B$, q. Hence $\mathrm{d}l$ is replaced with $(\boldsymbol{x}_q \cdot \boldsymbol{x}_q)^{1/2}\mathrm{d}q$, and $\boldsymbol{x}'$ with $\boldsymbol{x}_q / (\boldsymbol{x}_q \cdot \boldsymbol{x}_q)^{1/2}$, where $\boldsymbol{x}_q$ is the differential quotient of $\boldsymbol{x}$ over q. Therefore, the travel time t_{AB} becomes the integral of $L(\boldsymbol{x}_q, \boldsymbol{x})$ over q from 0 to $|\boldsymbol{x}_B - \boldsymbol{x}_A|$ with

$$L(\boldsymbol{x}_q, \boldsymbol{x}) = \frac{x_q^2}{\boldsymbol{v} \cdot \boldsymbol{x}_q + \left[(c^2 - v^2)\, x_q^2 + (\boldsymbol{v} \cdot \boldsymbol{x}_q)^2\right]^{1/2}} \tag{2.93}$$

11 Fermat originally "guessed" that the travel time of light is a minimum (1657, "the shortest time principle"). But later Hamilton recognized that this formulation including some exceptions, and its correct presentation should be: the actual path being referred to other adjacent paths is steady. The contribution that proved that this principle is also applicable to sound waves in a moving medium belongs to Ugincius (1972) [66].

The requirement for the travel time being stable results in the Euler-Lagrange equation[12]

$$\mathrm{d}_q \partial_{x_q} L - \partial_x L = 0 \tag{2.94}$$

Applying the relations and definitions derived in the first two paragraphs of this section, and after many algebraic operations, the partial differential quotient of function $L(\boldsymbol{x}_q, \boldsymbol{x})$ is transformed into

$$\partial_{x_q} L = \frac{\boldsymbol{n}}{\boldsymbol{n} \cdot \boldsymbol{v}_{\mathrm{ray}}} = \boldsymbol{s}$$

$$\partial_x L = -\frac{\mathrm{d}_q l}{v_{\mathrm{ray}}} \left[\frac{\Omega}{c} \nabla c + \boldsymbol{s} \times (\nabla \times \boldsymbol{v}) + (\boldsymbol{s} \cdot \nabla) \boldsymbol{v} \right] \tag{2.95}$$

Consequently Eq. (2.94) corresponds to the ray tracing equation (2.88), and Fermat's Principle can be derived from the ray equation.

For generalization, Fermat's Principle is valid not only for a smooth successive ray path, but also for a ray path with sudden changes of directions. As a result, all of the ray paths, regardless of being caused by reflection, refraction, or diffraction, can be predicted using Fermat's Principle.

12 About the concerned concept of variational method, please consult J. Mathews and R. L. Walker, *Mathematical Methods of Physics*, New York: Benjamin. 1965; 304~326; S. H. Grandall et al. *Dynamics of Mechanical and Electromechanical Systems*, New York: McGraw-Hill, 1968;1~35; 417~424. We may notice the similarity between Eq. (2.94) and the Lagrange Equation in analytical mechanics, the similarity between $L(\boldsymbol{x}_q, \boldsymbol{x})$ and Lagrange density, and the similarity between Fermat's principle and that of Hamilton.

Chapter 3
Sound Propagation in Atmosphere – Refraction and Reflection

Sound propagation is the most fundamental task in atmospheric acoustics. In general, it includes almost all of the physical processes experienced by acoustic waves, such as reflection from boundaries, refraction caused by temperature- and wind-gradients, diffraction resulting from object edges and boundary impedance changes, scattering caused by temperature- and turbulence-fluctuations as well as by particles of rain, snow and fog, absorption caused by all kinds of loss processes, and so on. Certainly, all of the above processes may actually happen simultaneously, provided that corresponding conditions are satisfied. However, in order to facilitate our study, these processes have to be dealt with separately. The meaning of "propagation" discussed in this Chapter is now limited to a narrower sense, which now mainly involves problems of reflection and refraction under the premise that the atmosphere possesses only "macroscopic" layered structures while the "microscopic" turbulent structures can be ignored for the time being. At the same time, all of the loss processes have been omitted.

One of the main characteristics of the atmosphere lies in the fact that it is always in ceaseless motion. For the winds that do not vary in height (impossible in meteorology), the solution can be obtained immediately by transforming the problem into a coordinate system moving with the wind (Galileo transformation). However, wind profiles (varying with height) are actually very complicated, and are especially more changeable in boundary layers. Roughly speaking, wind speed increases logarithmically with the height from ground to the top of the boundary layer due to friction, while the wind direction changes according to the Ekman spiral rule. At the top of this boundary layer, the geostrophic wind determined by the horizontal pressure gradient and the Coriolis force is dominant. The methods for solving refraction problems with complicated profiles can be reduced to two categories: (1) to obtain the whole solution of generalized wave equations with simple boundary conditions, where the method of normal mode expansion is usually available; (2) to apply the ray method, which is restricted by certain criterion (roughly speaking, some scales, such as the diameter of ray beams and the reciprocal of the gradient of the refractive index and the like should be much greater than wavelength; the strict criteria will be discussed in some details in Chapter 7). For this category, one can consider the path of energy flow and energy density flow only, and leave the wave properties aside. In this Chapter, we mainly discuss the reflection and refraction problems of acoustic waves in the atmosphere via the viewpoint of the ray method.

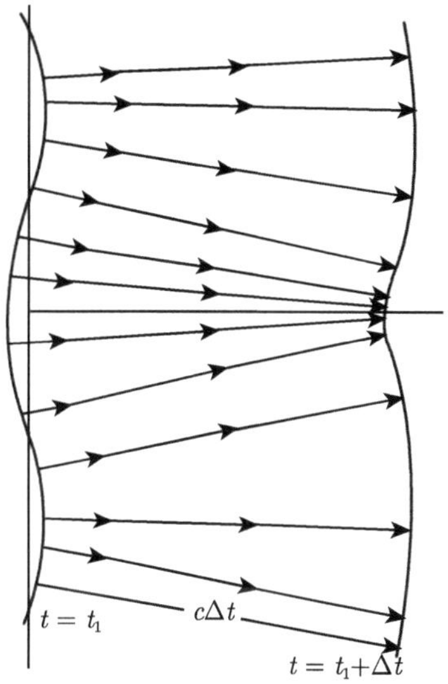

Fig. 12: Wave front at moment t deduced from the given wave front at $t = 0$.

3.1 Sound propagation in quiescent homogeneous media [56]

Although, in any conditions, the atmosphere cannot be treated as a quiescent homogeneous medium, but the analysis of this simplest case is helpful for understanding the "large scale behavior" of acoustic waves, and lays a foundation in conception for studying more complicated cases.

Because $v = 0$ and $c = \text{const}$, the ray speed $d_t x$, $\boldsymbol{s}$, and refraction index n are all constants, as can be seen from ray-tracing equations (2.87) and (2.88). Therefore all of the ray paths are straight lines, and the law of "straight line sound propagation" will be adhered in this case, even if $v = \text{const} \neq 0$.

3.1.1 Parametric description of wave fronts

Assume that a wave front moving along the positive direction of z is given by $z = f(x, y)$ at $t = 0$. Now we need to describe the wave front at a certain moment t afterwards (as shown in Fig. 12).

A ray passing through point $\boldsymbol{x}_P$ on the initial wave front moves along the normal direction $\boldsymbol{n}$, where $\boldsymbol{n}$ satisfies

$$\boldsymbol{n} = \left\{ \frac{\nabla[z - f(x, y)]}{|\nabla[z - f(x, y)]|} \right\}_{x = x_P} = \frac{\boldsymbol{e}_z - f_x \boldsymbol{e}_x - f_y \boldsymbol{e}_y}{(1 + f_x^2 + f_y^2)^{1/2}} \tag{3.1}$$

where $f_x = \partial f / \partial x$, $f_y = \partial f / \partial y$; each $\boldsymbol{e}$ represents the unit vector along the increasing direction of the axis specified by its subscript. The ray is located at $\boldsymbol{x} = \boldsymbol{x}_p + ct\boldsymbol{n}$ at moment t. If we let $x_P = \alpha$, $y_P = \beta$, the position can be written as

$$\boldsymbol{x}(\alpha, \beta, t) = \alpha \boldsymbol{e}_x + \beta \boldsymbol{e}_y + f(\alpha, \beta)\boldsymbol{e}_z + \frac{ct(\boldsymbol{e}_z - f_\alpha \boldsymbol{e}_x) - f_\beta \boldsymbol{e}_z}{(1 + f_\alpha^2 + f_\beta^2)^{1/2}} \tag{3.2}$$

Hence, we can give the parametric description of the wave front at the moment t in terms of parameters α and β. Any selection of α and β decides a point on the wave front. Consequently the method of picture composition of the Huygens principle can be replaced with an analytic expression.

3.1.2 Variation of principal radii of curvature along a ray

Any curved surface, whether it seems to be an ellipsoid bowl (concave and convex) or a saddle locally has two principal radii of curvature. If we take any point on the curved surface as the origin, and let the z-direction be normal to the surface at this point, the nearby curved surface can always be described by the combination of the square of x and y with the proper selection of the x and y axes[1]

$$z = \frac{x^2}{2r_1} + \frac{y^2}{2r_2} \tag{3.3}$$

where r_1 and r_2 are the two principal radii of curvature, which can be negative.

The variation of r_1 and r_2 along a ray traveling in a quiescent and homogeneous medium can be derived using Eq. (3.2). The coordinate system is chosen to make the ray pass through the origin at $t = 0$ along the $+z$-direction, where $f(\alpha, \beta)$ is equal to $\alpha^2/2r_1^0 + \beta^2/2r_2^0$ (accurate up to the quadratic terms of α and β; where the superscript 0 denotes initial values). Therefore, the z-component that is accurate up to the quadratic terms of α and β of Eq. (3.2) is given as

$$z = ct + \frac{\alpha^2}{2r_1^0}\left(1 - \frac{ct}{r_1^0}\right) + \left(\frac{\beta^2}{2r_2^0}\right)\left(1 - \frac{ct}{r_2^0}\right) \tag{3.4}$$

However, the x- and y- components of Eq. (3.2) that are accurate to the first order of α and β, are $\alpha(1 - ct/r_1^0)$ and $\beta(1 - ct/r_2^0)$, respectively. To obtain an accuracy up to the quadratic terms of x and y, we have

$$z = ct + \frac{\frac{1}{2}x^2}{r_1^0 - ct} + \frac{\frac{1}{2}y^2}{r_2^0 - ct} \tag{3.5}$$

Because Eq. (3.5) has the same form as Eq. (3.3), the direction related with the principal radius of curvature is kept constant along any given ray, and the radius itself reduces

1 This can be verified simply as follows: The equation of a circle with a radius of r_1 in the xy plane, which is tangent to the plane $z = 0$ is $(z - r_1)^2 + x^2 = r_1^2$. It can be reduced to $z = x^2/2r_1$ when $z \ll r_1$ and $|x| \ll r_1$. A similar relation is also applicable for a circle with a radius of r_2 in yz plane.

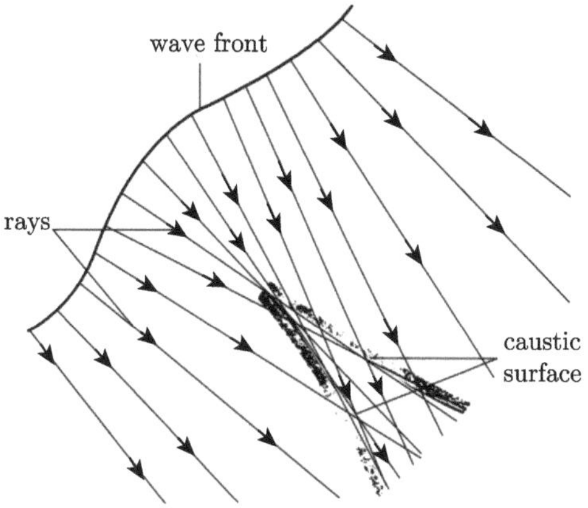

Fig. 13: Formation of caustic surface rays, wave front, and a caustic surface.

a quantity of ct in the duration of t, which, in other words, the ray reduces to Δz after proceeding for a distance of Δz. The above discussion is the result of when the wave front is concave along the ray direction. If it is convex or like a saddle, consequently (for example) $r_1^0 < 0$, and with an increase in the propagation distance, the increasing quantity of $|r_1|$ is equal to the increase of the distance along the ray. The decrease of the curvature radius of the wave front is related to focusing, whereas an increase of the radius is related with defocusing.

3.1.3 Caustic surface

Eq. (3.5) shows that, if (for example) $r_1^0 < 0$ and $r_2^0 < r_1^0$, the wave front at $t = r_1^0/c$ will develop into a sharp spire (i.e. a point of tangency). The points where this happens are also the points where neighboring rays intersect. Each point corresponds to a given ray starting from the initial wave front, and the trajectories of all the points consist of a caustic surface, as shown in Fig. 13. Because the wave front has a spire at the point contacting the caustic surface, the assumption that the wave front is acting locally like a proceeding plane wave everywhere is not valid, and the basic rule of geometric acoustics is not applicable any more.

However, geometric acoustics is very important in essence for predicting the position of caustic surfaces, because it shows where abnormally high intensities (large amplitudes) appear. Moreover, the concept of the caustic surface is also applicable to the rays in inhomogeneous media. Geometric acoustics has a unique meaning in atmospheric acoustics. For example, for some noisy activities (such as the static test of big rocket engines, etc.), a proper location and meteorological environment should be chosen well so that distant residential areas can avoid the formation of possible caustic surfaces [67].

Sometimes, the characteristic spire formed by the intersection of two caustic surfaces is called an arête [68]. Outside the arête, there are three rays (rather than one ray) passing through each point between the two caustic surfaces, while the wave front folds [69].

3.2 Sound refraction in stratified in homogeneous media

As previously pointed out a few times, the atmosphere is a type of stratified medium. That is, not only the parameters that represent its properties, such as density, temperature, and consequently sound speed, but also the parameters that represent its movement such as wind speed, depends on the vertical coordinate z. Therefore, acoustic waves are reflected and refracted at the interfaces between layers, just as they are at interfaces between different media. The difference is that the reflection is so small that it can be neglected, whereas the refraction happens gradually and continuously, and the corresponding refraction line is a continuous and gradually bending line, rather than a sharp "broken straight line" that occurs at a definite interface.

3.2.1 Refraction caused by sound-speed gradients

Assume a wind speed of $v = 0$, and a sound speed c that is independent of time. According to the first expression of Eq. (2.80), the wave-slowness vector can be obtained as $\boldsymbol{s} = \boldsymbol{n}/c$, and the tracing equations given by Eq. (2.87)~(2.88) are reduced to

$$\mathrm{d}_t x = c^2 \boldsymbol{s}, \quad \mathrm{d}_t \boldsymbol{s} = -\frac{\nabla c}{c} \tag{3.6}$$

To decide the effect of the sound speed gradient on the bending of rays, let us consider a ray that passes through the origin along the positive x-direction initially. Therefore, $\boldsymbol{s} = \boldsymbol{e}_x / c(0)$, when $t = 0$. Hence, accurate to the first order of t, the second expression in Eq. (3.6) gives

$$\boldsymbol{s} = \frac{1}{c}(\boldsymbol{e}_x - \nabla c t) \tag{3.7}$$

where c and its three partial derivatives (c_x, c_y, c_z) take the values at the origin of $(0, 0, 0)$. From the equations about $\mathrm{d}_t y$ and $\mathrm{d}_t z$, we can make the corresponding conclusion that both y and z are directly proportional to t^2 for very small t. Therefore, because $x = ct$ (accurate to the lowest order), the first expression of Eq. (3.6), which is accurate to the lowest power ($x \neq 0$), gives

$$y = -\frac{1}{2}\frac{c_y}{c}x^2, \quad z = -\frac{1}{2}\frac{c_z}{c}x^2 \tag{3.8}$$

Both expressions are parabolic equations.

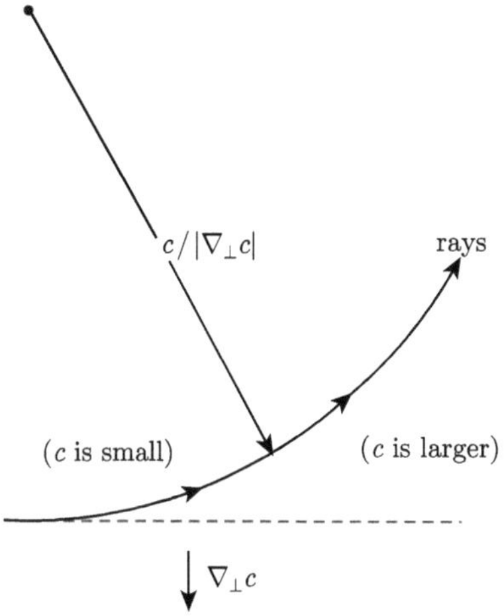

Fig. 14: Bending of ray paths in a medium where the sound speed varies in space.

In addition, assuming that a new coordinate system is chosen that makes $c_z = 0$ at $x = 0$ and ∇c be parallel to $\boldsymbol{e}_y$, the ray path will bend to the negative y-direction when $c_y > 0$, and to the positive y-direction when $c_y < 0$. The curvature radius of the ray path is $c/|c_y|$ in both cases, as shown in Fig. 14.

The following conclusions can be drawn from the above discussions: if acoustic rays move in a medium with a varying sound speed, and when the component of ∇c is perpendicular to the propagating direction, $\nabla_\perp c$, is not 0, the rays will bend and deviate away from the propagation direction. The bending of the rays occurs on the plane determined by both $\nabla_\perp c$ and local ray path. However, it deviates from the direction of $\nabla_\perp c$, and bends towards the direction where the sound speed is smaller. The curvature radius of a ray's path is $c/|\nabla_\perp c|$ or $c/(|\nabla c| \sin\theta_0)$, with θ_0 being the inclination angle between the ray's direction and the direction of ∇c. Based on this, a common-sense explanation can be given to the phenomenon: (take an extreme case, for instance) the effect of outdoor hearing is obviously better during a winter night than at summer noon. This is because the specific heat capacity at ground level is less than that in the atmosphere, and the former has a positive temperature (or sound speed) gradient, i.e., the temperature increases with height, resulting in the bending of the acoustic ray downward to the ground. The situation is opposite for the latter case.

The phenomenon that rays always bend to the area where the sound speed is smaller can also be illustrated by the concept of the wave front. Because wave fronts move slowly in the area where the sound speed is smaller, the entire wave front consequentially inclines to this side. Therefore, rays must also bend to this side in order to keep themselves perpendicular to the wave fronts ($v = 0$ is already given).

In the special case everywhere ∇c is zero, the corresponding rays are all circular arcs. To prove this, let us assume $c = c_0 - az$, thus $\nabla c = -a\boldsymbol{e}_x = \text{const}$. The rays only move in the xz plane at this time, thus $s_y = 0$. Taking into account that fact that $v = 0$ in Eq. (2.82), we then obtain $s_z^2 = c^{-2} - s_x^2$. Therefore, the relation $s_z/s_x = \mathrm{d}_x z$ given in Eq. (3.6) yields

$$(\mathrm{d}_x z)^2 - \frac{1}{c^2 s_x^2} = -1 \tag{3.9}$$

Furthermore, from the second expression in Eq. (3.6) we know that s_x is a constant. Then based on the algebraic properties of the circle equation it is not difficult to verify that the integral of Eq. (3.9) is a circle with a radius of $1/as_x$, where the center is on the height line $z = c_0/a$, and the sound speed is extrapolated to zero. Of all the potential rays passing through this point, those rays that bend the most are moving in the direction vertical to the sound speed gradient.

3.2.2 Refraction caused by windspeed gradients

Even in the same medium, reflection and refraction can also take place at the interface between two sections having relative movements. In this section, the refraction caused by wind speed gradients is investigated.

Let us consider rays passing through the origin at $t = 0$, where the normal direction of the wave front is n_0. The corresponding starting value of the wave-slowness vector is determined by Eq. (2.80). Therefore, accurate to the first-order of t, the integral of Eq. (2.88′) gives

$$\boldsymbol{s} \approx (c + \boldsymbol{v} \cdot \boldsymbol{n}_0)^{-1} \left[\boldsymbol{n}_0 - t\nabla_\perp (c + \boldsymbol{v} \cdot \boldsymbol{n}_0)\right] \tag{3.10}$$

Correspondingly, Eq. (2.87) gives the following power-series expansion

$$\boldsymbol{x} \approx (c\boldsymbol{n}_0 + \boldsymbol{v})t + \frac{1}{2}t^2 \left[(\boldsymbol{v}_{\mathrm{ray}} \cdot \nabla)(c\boldsymbol{n}_0 + \boldsymbol{v}) - c\nabla_\perp (c + \boldsymbol{v} \cdot \boldsymbol{n}_0)\right] \tag{3.11}$$

Where $\nabla_\perp = \nabla - \boldsymbol{n}_0(\boldsymbol{n}_0 \cdot \nabla)$ denotes the (transverse) gradient perpendicular with $\boldsymbol{n}_0$, and $\boldsymbol{v}_{\mathrm{ray}}$ is $c\boldsymbol{n}_0 + \boldsymbol{v}$ (shown in Eq. (2.78)). All coefficients and derivatives take their values at the origin.

The plane where a ray's bend contains two vectors, $\dot{\boldsymbol{x}}$ and $\ddot{\boldsymbol{x}}$, which is shown in Eq. (3.11) as the coefficients of terms t and $\frac{1}{2}t^2$. The ray bends to the direction of the component of $\ddot{\boldsymbol{x}}$ and $\ddot{\boldsymbol{x}}_\perp$, the latter is vertical to $\dot{\boldsymbol{x}}$. The curvature radius is $\dot{\boldsymbol{x}} \cdot \frac{\dot{\boldsymbol{x}}}{|\ddot{\boldsymbol{x}}_\perp|}$.

In many actual situations, the variation of wind speed along its blowing direction can be neglected, i.e., $(\boldsymbol{v} \cdot \nabla)\boldsymbol{v} = 0$, which is approximately valid. Further omitting the small difference between the directions of $\boldsymbol{n}_0$ and $\dot{\boldsymbol{x}}$, Eq. (3.11) gives

$$\ddot{\boldsymbol{x}}_\perp \approx c\left[(\boldsymbol{n}_0 \cdot \nabla)\boldsymbol{v} - \nabla(c + \boldsymbol{v} \cdot \boldsymbol{n}_0)_\perp\right] \approx -c\nabla_\perp c - c\boldsymbol{n}_0 \times (\nabla \times \boldsymbol{v}) \tag{3.12}$$

The above expression is especially suitable to the scenario of $|\boldsymbol{v}| \ll c$ or the scenario when $\boldsymbol{n}_0$ is parallel to $\boldsymbol{v}$. Based on those relations, we can arrive at the conclusion that the bending direction of a ray is opposite to the direction of $\nabla_\perp c + \boldsymbol{n}_0 \times (\nabla \times \boldsymbol{v})$, and the curvature radius is approximately equal to c divided by the value of this vector.

As an example, assume $\boldsymbol{n}_0 = \boldsymbol{e}_x \cos\theta + \boldsymbol{e}_x \sin\theta$, where c, v_x and v_y all depend on z; and $v_z = 0$. Thus, Eq. (3.12) is simplified to

$$\boldsymbol{x}_\perp = -c(\mathrm{d}_z c \sin\theta + \mathrm{d}_z v_x)\boldsymbol{e}_2 + c(\mathrm{d}_z v_y \cos\theta)\boldsymbol{e}_y \tag{3.12a}$$

Where $\boldsymbol{e}_2 = \boldsymbol{e}_z \sin\theta - \boldsymbol{e}_x \cos\theta$ is the unit vector vertical to $\boldsymbol{n}_0$ in the xz -plane; the y-component of $\ddot{\boldsymbol{x}}_\perp$ connects with the transverse drift of the rays, which is caused by the transverse wind. Usually this effect is secondary, because maybe the rays we are interested in are close to the horizontal plane ($\cos\theta$ is very small), or maybe the net offset of a ray's direction caused by this component is, on average, almost 0. Omitting this component, the curvature radius of a ray is given as [70]

$$r_c = \frac{c}{\mathrm{d}_z c \sin\theta + \mathrm{d}_z v_x} \tag{3.13}$$

The positive value of Eq. (3.13) means bending down, while the negative value means bending up.

In the case when a ray's travel in a direction is nearly horizontal, a further approximation can be made. In this case, using 1 instead of $\sin\theta$, $c\sin\theta + v_x$ can be replaced by $c + v_x$. This yields the following simple law: the refracted rays appear to move in a windless medium that has an effective sound speed[2] of $c_{\text{eff}} = c + v_x (v_x$ is the x-component of the wind speed in the vertical plane where the rays exist). From this viewpoint, both the wind speed and the sound speed gradients have the same effect. However, if θ is smaller than (for example) 30°, the effect of the wind speed gradient is much greater than that of the sound speed gradient of the same magnitude.

After the concept of the effective sound speed is introduced, the concept of the ordinary Snell law of refraction can be presented: the ratio of the sound speed over the sine of the angle between a ray and the normal of the interface is an invariant,

$$c/\sin\varphi = \text{const} \tag{3.14}$$

This can be naturally extended to the case of a moving medium by replacing v in Eq. (3.14) with $c_{\text{eff}} = c + v\sin\varphi$

$$(c/\sin\varphi) + v = \text{const} \tag{3.14a}$$

2 See Introduction Section 2.3.

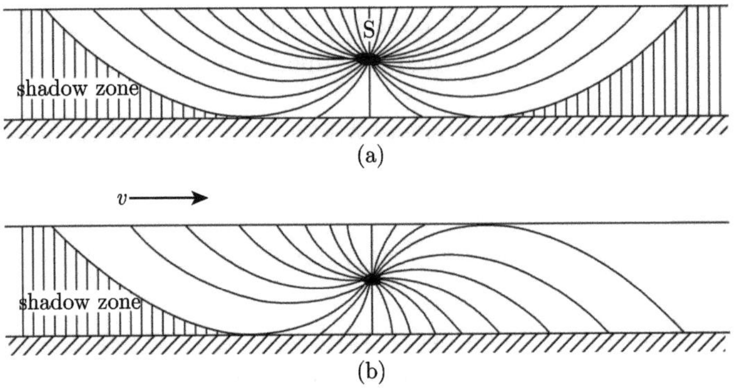

Fig. 15: Ray bending and shadow zone caused by a temperature-gradient and a wind speed-gradient. (a) Temperature progressively decreases with height; (b) Horizontal wind speed progressively increases with height.

As a brief summary, the typical behaviors of ray bending caused by a temperature-gradient and a wind speed-gradient are shown in graph (a) and graph (b) in Fig. 15, respectively.

3.3 Acoustic rays in the atmosphere

Under certain atmospheric profiles (as shown in Fig. 4), the integral of the ray-tracing equation is a fundamental subject in geometric atmospheric acoustics. Nevertheless, closed-form solutions can be obtained only for a few types of "ideal" profiles. For general cases, we can only use numerical integration methods. Some basic phenomena and concepts of stratified atmospheric models will be discussed in this section.

3.3.1 Ray integrals

For stratified media, the ray-tracing equation (2.88′) requires that s_x and s_y are constant along any given ray, which actually conforms to the generalized law of refraction (3.14a). Furthermore, once s_x and s_y are specified, s_z can be determined using the Eikonal equation as a function of height z, i.e.,

$$s_z = \pm\left[\left(\frac{\Omega}{c}\right)^2 - s_x^2 - s_y^2\right]^{1/2} \tag{3.15}$$

(Notice: Because $\boldsymbol{v}$ does not contain a z-component, $1 - \boldsymbol{v} \cdot \boldsymbol{s}$ has nothing to do with s_z). Therefore, Eq. (2.88) can be regarded as being solved. Thus, from Eq. (2.87) we have $\left(\mathrm{d}_z x = \frac{\mathrm{d}_t x}{\mathrm{d}_t z}\mathrm{d}_z t = \frac{1}{\mathrm{d}_t z}\right)$ [61]

$$\mathrm{d}_z x = \frac{c^2 s_x + \Omega v_x}{c^2 s_z}, \quad \mathrm{d}_z t = \frac{\Omega}{c^2 s_z} \tag{3.16}$$

And similar equations regarding $\mathrm{d}_z y$ can also be obtained.

Because the right-hand side of Eq. (3.16) is only a function of z, we can take the integral directly to determine x, y and t as functions of z (also s_x and s_y). Such as,

$$x = x_0 + \int_{z_0}^{z} \frac{c^2 s_x + (1 - \boldsymbol{v} \cdot \boldsymbol{s}) v_x}{c^2 s_z} \mathrm{d}z \tag{3.17}$$

where $x_0 \equiv x(z_0)$.

3.3.2 Rays in waveguides

In Sec. 2.3 of the introduction, we mentioned that there exists a minimum in an atmospheric sound speed profile, which corresponds to an “atmosphere waveguide (or sound channel)”. Now we can deal with the sound propagation in a waveguide by applying ray concepts.

An adjacent area to the minimum of the sound speed in an atmospheric profile is shown in the left graph of Fig. 16, while the right graph gives the ray’s path that corresponds to the profile. It is shown that the rays are restricted between two heights, which are named “turning points”. At turning points, $s_z^2 \to 0$, and in the area between the turning points, $s_z^0 \geqslant 0$. In this case x, y and t of an actual ray are all not single-valued functions of z, because every time the ray reaches a turning point, s_z will change its sign. A ray’s initial position and direction determines the initial signs of s_x, s_y and s_z. Assuming $1 - \boldsymbol{v} \cdot \boldsymbol{s} > 0$, s_z has the same sign as $\mathrm{d}_z t$ (see Eq. (3.16)). Therefore, for the rays traveling in a slanting upward direction, the sign of s_z is positive, and for the rays traveling in a slanting downward direction, the sign is negative.

Assuming the initial sign is positive, Eq. (3.17) and the similar relations for y and t describes the ray’s trajectory as traveling upward before it reaches “the height of the upper turning point” z_U. At $z = z_U$, s_z^2 becomes 0 for the first time. The ray becomes flat and starts to bend downward, corresponding to $s_z < 0$ (totally reflective), until the ray reaches “the height of lower turning point” z_L. Let x_{U1}, y_{U1}, t_{U1} be the corresponding values of x, y, and t when the ray arrives at z_U for the first time, respectively. Then, the subsequent value of x along the next ray trajectory is given by

$$x = x_{U1} + \int_{z}^{z_U} \frac{c^2 s_x + \Omega v_x}{c^2 |s_z|} \mathrm{d}z \tag{3.18}$$

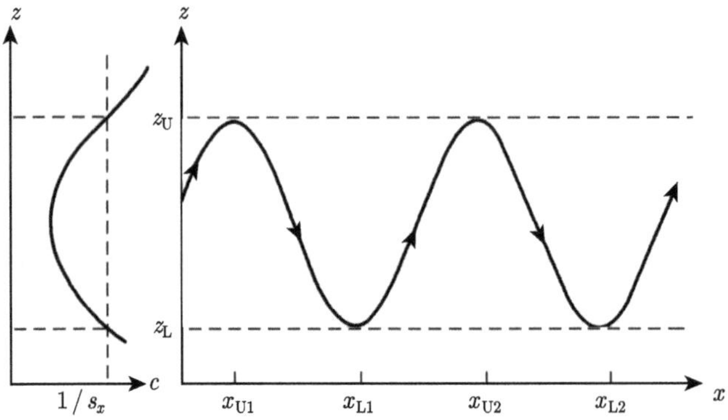

Fig. 16: A ray near the minimum of the sound speed in a sound channel.

And the corresponding y and t also follow a similar relation. When the ray arrives at $z = z_L$, s_z^2 becomes 0 again, and thus changes its sign.

It should be pointed out that, although $s_z(z_U) = 0$, an integral as that in Eq. (3.18) is still bounded. This is because when $z \to z_U$ the factor, $|s_z|$, in the denominator tends to approach to 0 according to $(z_U - z)^{1/2}$. Therefore, the integral is still integratable.

Therefore, the rays (thus acoustic energy) are, certainly only from the view point of ray acoustics, totally restricted to a region between two 'perfect reflective' planes, analogous to rays propagating in a pipeline. The propagation trajectories of these rays are periodic in both time and horizontal displacement (x_U and x_L occur alternately with a constant interval, as shown in Fig. 16). As for the displacement along the y-direction and the traveling time, they are in a similar situation. The average horizontal speed of a ray is

$$\boldsymbol{v}_H = \frac{(\Delta x)_{L\to U}\boldsymbol{e}_x + (\Delta y)_{L\to U}\boldsymbol{e}_y}{(\Delta t)_{L\to U}} \tag{3.19}$$

Where $(\Delta t)_{L\to U}$ denotes the net time needed for the ray to propagate from the upper turning point to the lower turning point.

3.3.3 "Abnormal" propagation

Actually, complicated and ever-changing atmospheric profiles can cause correspondingly complicated acoustic ray distributions. Sometimes the so-called "abnormal propagation" phenomenon will occur. The "abnormality" refers to a phenomenon that goes against the regular rule that the sound can be heard in many sites 300 km away, but could not be heard within 100 km away from the source – a "silent zone" was formed with a width of more than 100 km. In the subsequent seeking for theoretical explanation [72], an assumption that there should be a temperature inversion layer

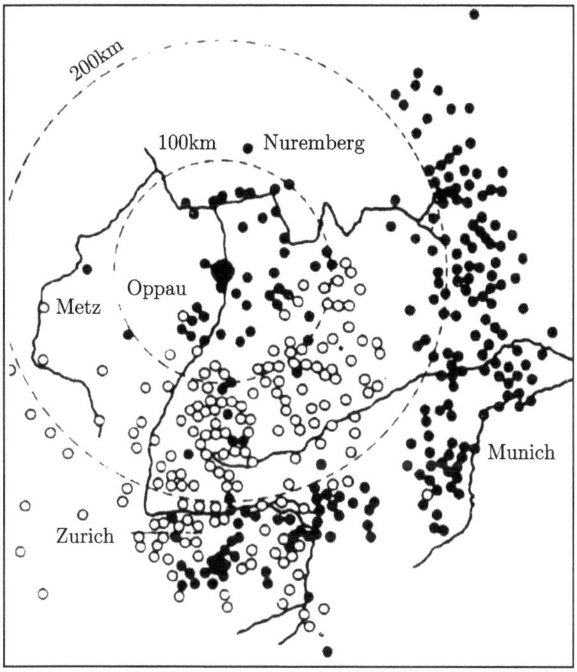

Fig. 17: Map of hearable and non-hearable zones during the Oppau arsenal explosion in Germany on Sept. 21, 1921. Black dots (·) represent sites where explosion can be heard; Circle dots (○) represent sites where explosion cannot be heard.

above the convection layer was firstly made. And soon, the advection layer (isothermal layer) was found in actual detection.

To explain this phenomenon, lateral winds were neglected for the sake of simplicity. So that, the rays starting from the source will all stay in the vertical plane, and the included angle θ between the normal unit vector of wave front n and plumb line satisfies

$$s_x = \frac{\sin\theta}{c + v_x \sin\theta} = \text{const} \tag{3.20}$$

Nonetheless in general, the direction of the rays differs slightly from the direction of $\boldsymbol{n}$ (see Fig. 10).[3] However, it is horizontal when $\boldsymbol{n}$ is horizontal. Hence, rays with initial

---

3 Although this difference is tiny, it is an important sign for the different cases between with wind and without wind. Historically, it is because of ignoring this difference that made Rayleigh obtain an incorrect ray equation by regarding that the path is catenary rather than circular arc in his study of sound propagation when wind speed increases linearly with the height and sound speed keeps constant everywhere. Although this mistake had been pointed out by many scholars in the beginning of twenty century, it still causes chaos in a relative long period because of the prestige of Rayleigh.

angle θ_0 will bend to the ground when they reach the height of turning point, z_{tp}, and z_{tp} is decided by

$$c(z_{tp}) + v_x(z_{tp}) = \frac{c_g}{\sin\theta_0} \tag{3.21}$$

where c_g is the ground sound speed (the ground wind speed is so small that it can be ignored).

According to Eq. (3.21), if there exist such rays that start from the ground, arrive at the advection layer, and then return to the ground, it must be satisfied that $c + v_x$ is larger than the ground sound speed c_g at certain height. In fact, this condition can be met from the profile shown in Fig. 4. In addition, it also can be observed that in convection layer $c + v_x$ decreases progressively with the increase of height typically. Hence, the zone within a medium distance from the sound source can form a silent zone (shadow zone). And the corresponding rays are drawn in Fig. 18 [73]. Because of the reflection of sound caused by grounds, sound paths can be circular, and form the second and third shadow zones. However, the paths intersect more and more frequently in the later zones, making those shadow zones be less and less obvious. The shadow effect sometimes can be canceled out by local meteorological conditions near the ground (the profile has less regular patterns in the height less than 3 km from the ground). But, in general the rays leave the sound source with an angle of elevation $\geqslant 10°$ mostly reach a height of 3 km or higher before they start to return the ground.

From the measurement of the horizontal propagation velocity of wave front $1/s_x$, sweeping across a microphone array, we can find the incident angle of arrival acousticalray θ_0, thus obtain $c(z_{tp}) + v_x(z_{tp})$ by applying Eq. (3.21). The arrival time is always much later (usually about 1 minute) than that of creeping waves propagating directly along the ground with sound speed (see Chap. IV). These creeping waves which are often neglected in geometrical acoustics, have very weak amplitudes, but they are often detected by sensible instruments.

The notable characteristic of "hearing zone" (where abnormally propagated sound can be received in this zone) is that it begins suddenly at the range with the order of 200 km (see Fig. 18). The existence of such a critical range is the result of the "jumping distance" $R(\theta_0)$ calculated by ray theory, which denotes the horizontal range traveled by a ray before it returns to the ground. When the profile of $c + v_x$ decreases monotonously to a minimum, and then increases with the height to a maximum ($> c_g$), at a height of the range corresponding to the grazing angle of $\theta = \pi/2$, $R(\pi/2)$ can be in the order of 200 km or more. With the decrease of θ_0, R also decreases until it approaches a minimum R_{min}. Then, it increases to $R(\theta_{0,m})$. Here $\theta_{0,m}$ is the value of θ_0 making $z_{tp} = z_m$ from Eq. (3.21). When θ_0 reduces to a value less than $\theta_{0,m}$, the range suddenly appears a big jump [the turning point locates at a much higher height, when $c + v_x$ becomes $c(z_m) + v_x(z_m)$ again]. Thus abnormal hearing zones are located between R_{min} and $R(\theta_{0,m})$. Because $R(\theta_0)$ has a minimum, there

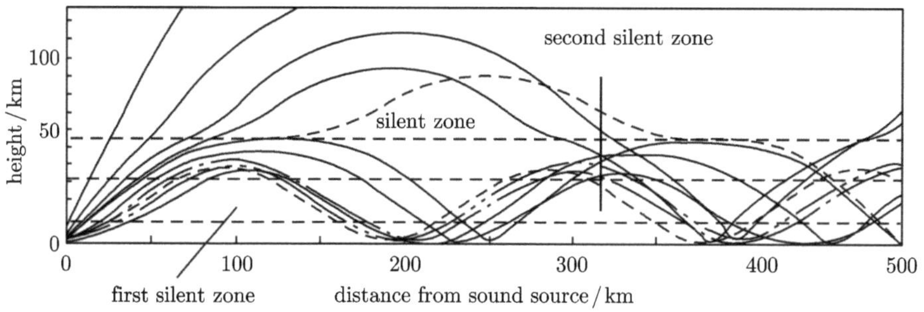

Fig. 18: Typical acoustical rays from east to west in the Northern Hemisphere in summer.[4]

should be a caustic surface that contacts the ground at the reach with $R_{\min}$. Therefore, the abnormal sound is most strong at the place just outside the inner border of the abnormal hearing zone.

Actually, Derham noticed the abnormal propagation phenomenon early in 1708: when French and Netherlandish navies encountered and engaged in a fight off the coast of Holland, the cannon sound could be heard in Welsh, several hundred kilometers away. Nevertheless, he was not aware of the silent zone in much nearer ranges.

3.4 Amplitude variations in quiescent media

3.4.1 Wave amplitude in quiescent and homogeneous media

Firstly, let's consider the simplest case where sound waves with a constant frequency propagate in a quiescent medium with constant density and sound velocity. In this case, acoustic pressure satisfies wave equation (2.6). Its complex amplitude $\hat{p}(x)$ satisfies Helmholtz Equation (2.14). Introducing the Fourier transformation[5] of p (i.e., $\hat{p}(\boldsymbol{x}) = P(\boldsymbol{x}, \omega)e^{i\omega\tau(x)}$), From Eq. (2.14), we obtain

$$\nabla^2 P + i\omega(2\nabla P \cdot \nabla\tau + P\nabla^2\tau) - \omega^2 P\left[(\nabla\tau)^2 - \frac{1}{c^2}\right] = 0 \tag{3.22}$$

4 See Malone, T. F. (ed.), *Compendium of Meteorology*, American Meteorological Society, Boston, 1951:374.

5 This treatment originates from optical principles. For details, refer to the following paper: A. Sommerfeld and J. Runge, *Application of vector calculus to the fundamentals of geometrical optics*, Ann. Phys., 1911, (4)35: 277–298.

To solve the above equations in high frequency limit, an asymptotic expansion of P is assumed

$$P(\boldsymbol{x}, \omega) = P_0(\boldsymbol{x}) + \frac{1}{\omega}P_1(\boldsymbol{x}) + \frac{1}{\omega^2}P_2(\boldsymbol{x}) + \cdots \tag{3.23}$$

Substitute Eq. (3.23) into (3.22), and let the coefficients of every power of ω be identical to 0, an infinite series of equations can be derived, the first two of which contain only τ and P_0. Assume that P_0 is a proper approximation of P, therefore we can just keep the first two equations, and replace P with P_0. Hence, the following equations will be obtained

$$(\nabla\tau)^2 = \frac{1}{c^2}$$

$$2\nabla P \cdot \nabla\tau + P\nabla^2\tau = 0 \text{ or } \nabla \cdot (P^2\nabla\tau) = 0 \tag{3.24}$$

We may notice that the above two equations can also be obtained by letting the coefficients of ω^2 and ω in Eq. (3.22) be 0. The latter form in the second expression of Eq. (3.24) comes from the first form being multiplied by P.

The first equation of (3.24) is just the Eikonal equation (2.28) when the medium motion speed is 0. Thus its solution can be given in the form of rays. Once any wavefront curved surface and its associated τ value are specified, $\tau(\boldsymbol{x})$ at any point $\boldsymbol{x}$ can be determined by the ray connecting the original designated wavefront with point $\boldsymbol{x}$. If the ray passes through the point $\boldsymbol{x}_0$ on the original designated wavefront, and satisfies $\tau(\boldsymbol{x}_0) = \tau_0$, $\tau(\boldsymbol{x})$ will be the sum of τ_0 and the travel time along the ray from $\boldsymbol{x}_0$ to $\boldsymbol{x}$ at a speed of c.

The solution of the second formula in Eq. (3.24) can be obtained by applying "Ray tube cross-section" method. Associate rays from $\boldsymbol{x}_0$ to $\boldsymbol{x}$ with a tube, which consists of all the rays passing through a small area $A(\boldsymbol{x}_0)$ with $\boldsymbol{x}_0$ as the center, and $A(\boldsymbol{x}_0)$ is perpendicular to the ray path. When the ray tube arrives at $\boldsymbol{x}$ point, its sectional area becomes $A(\boldsymbol{x})$ (as shown in Fig. 19). Do integral in the second formula of Eq. (3.24) over all the volume of ray tube section connecting point $\boldsymbol{x}_0$ and point $\boldsymbol{x}$, and apply Gaussian theorem to transfer the volume integral to surface integral. Because all the paths in the ray tube are along the direction $\nabla\tau = \boldsymbol{s}$, the surface integral over side surfaces of the tube identically vanishes to 0, only the two ends contribute to the integral, Therefore,

$$P^2(\boldsymbol{x}_0)A(\boldsymbol{x}_0)(\nabla\tau \cdot \boldsymbol{n})_{x_0} = P^2(\boldsymbol{x})A(\boldsymbol{x})(\nabla\tau \cdot \boldsymbol{n})_x$$

Where $\boldsymbol{n}$ is the unit vector $A(\boldsymbol{x}_0)$ along the direction of the ray. Since the medium is not in motion, it is also the normal unit vector of the wave front. Moreover, because $\nabla\tau \cdot \boldsymbol{n} = \boldsymbol{s} \cdot \boldsymbol{n}$ is here $1/c$ (as shown in Eq. (2.80)) and c is a constant in this case, the above equation can be simplified as

$$P(\boldsymbol{x}) = P(\boldsymbol{x}_0)\left[\frac{A(\boldsymbol{x}_0)}{A(\boldsymbol{x})}\right]^{1/2} \tag{3.25}$$

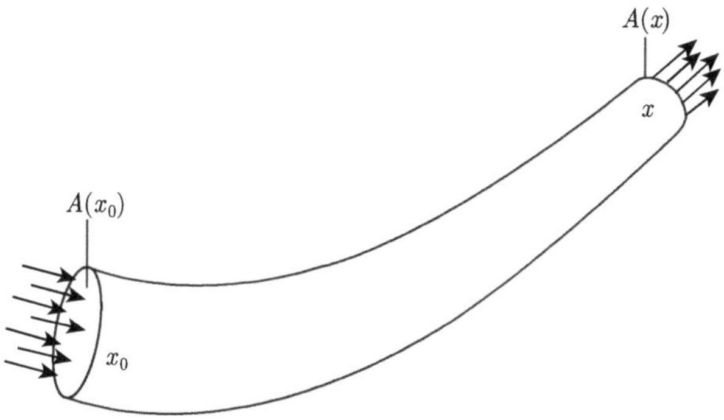

Fig. 19: Sketch diagram of ray tube.

Hence, it is clear that the change of wave amplitude along the ray is inversely proportional to the square root of the cross-section area of the ray tube. If the area shrinks (focus), the amplitude will increase.

The integral of $\nabla^2\tau$ over all of the volume of the ray tube section can be obtained in a similar way as $(1/c)[A(\boldsymbol{x})-A(\boldsymbol{x}_0)]$. Thus, for any short tube with a length of dl (the corresponding volume is approximately to $A(\boldsymbol{x})\mathrm{d}l$), we have

$$\nabla^2\tau = \frac{1}{cA}\mathrm{d}_l A \tag{3.26}$$

Where *A(l)* is the cross-sectional area of the ray tube at a distance of l along the ray. Furthermore, choose the coordinate system so that z axis points to the direction of ray; both x and y axes are at the directions of principal curvatures; also choose the interested point as the origin, in the neighborhood of this point, Eq. (3.3) gives

$$\tau \approx \text{const} + \frac{1}{c}\left(z - \frac{x^2}{2r_1} - \frac{y^2}{2r_2}\right) \tag{3.27}$$

Where r_1 and r_2 are the two principal curvatures radii of the wave front at the origin (positive at concave surfaces, and negative at convex surfaces). Thus, the following equation is obtained

$$c\nabla^2\tau = -\left(\frac{1}{r_1} + \frac{1}{r_2}\right) \tag{3.28}$$

Moreover, because $r_1 = r_1^0 - l$, and $r_2 = r_2^0 - l$ (obtained from Eq. (3.5), substituting $-1/r_1$ with $\mathrm{d}_l(\ln r_1)$, also doing the similar substitution to $-1/r_2$, and using $\mathrm{d}_l(\ln A)$ instead of $c\nabla^2\tau$ according to Eq. (3.26), we make an integral over Eq. (3.28). The following conclusion can be drawn: (A/r_1r_2) has nothing to do with l, meaning the ratio of the cross-sectional areas of ray tubes in Eq. (3.25) is the same as the ratio $r_1^0r_2^0/r_1r_2$. Therefore, the amplitude along the ray is

$$P(\boldsymbol{x}) = \left[\frac{r_1^0r_2^0}{(r_1^0 - l)(r_2^0 - l)}\right]^{1/2} P(\boldsymbol{x}_0) \tag{3.29}$$

and is inversely proportional to the geometric mean of the two principal curvatures of the wave fronts.

The above discussion can be extended to the case of the superposition of waves with different frequencies or the case of instantaneous waveforms. Because ray paths and travel time have nothing to do with frequency, and the amplitudes at different points along the ray path are uncorrelated with frequency either, the solution of wave equations with geometric acoustics approximation is

$$p = \hat{p}(l, \xi) f(t - \tau, \xi)$$

where the parameter ξ (strictly speaking, it represents a pair of parameters (ξ_1, ξ_2)) is used to distinguish different rays. The wave form $f(t - \tau, \xi)$ remains unchanged along any given ray, while the amplitude factor $\hat{p}(l, \xi)$ changes with distance along the ray. If $f(t - \tau, \xi)$ is chosen so that it takes a value of $p(\boldsymbol{x}, t)$ at the starting point of the ray $(l = 0)$, then $\hat{p}$ is the coefficient of $P(\boldsymbol{x}_0)$ in Eq. (3.29), and τ is $\tau_0 + l/c$.

3.4.2 Energy conservation along rays: extension to slowly-varying media

As for media where both $c(x)$ and $\rho(x)$ are slowly-varying functions of position (however, the media are still supposed to be quiescent), the above derivation can be done in a similar way. Nevertheless, we change another derivation based on the acoustic energy conservation which may be more enlightening.

First, let's assume that

$$p(\boldsymbol{x}, t) = \hat{p}(\boldsymbol{x}) f(t - \tau, \xi) \tag{3.30}$$

where τ is the solution of eikonal equation, while ξ is a constant along any given ray. If we use the above equation to express the plane wave propagating in any local region, the fluid velocity caused by it must be considered being $(\boldsymbol{n}/\rho c)p$ or $(\hat{p}/\rho)\nabla\tau f$ (because $n = c\nabla\tau$). As a result the energy density and acoustic intensity associated with this wave perturbation can be determined by Eq. (2.17) and (2.18)

$$E = \frac{\hat{p}^2}{\rho c^2} f^2(t - \tau, \xi), \quad \boldsymbol{I} = \boldsymbol{n} cE \tag{3.31}$$

Where eikonal equation (the first formula in Eq. (3.24)) is utilized, the law of acoustic energy conservation shown in (2.19) gives

$$2\frac{p^2}{\rho c^2} f \partial_t f + f^2 \nabla \cdot \left(\frac{\hat{p}^2}{\rho} \nabla\tau \right) + 2\frac{\hat{p}^2}{\rho} (\nabla\tau \cdot \nabla f) f = 0 \tag{3.32}$$

Omitting the weak dependence of f on position by way of $\xi(x)$, then we have $\nabla f = -\partial_t f \nabla\tau$. The first and third terms in Eq. (3.32) offset each other (reusing eikonal equation), and only the middle term remains

$$\nabla \cdot \left(\frac{\hat{p}^2}{\rho} \nabla\tau \right) = 0 \tag{3.33}$$

Integrating Eq. (3.33) over the whole ray tube, we can draw a conclusion by the similar means as in deriving Eq. (3.25): the quantity $(\hat{p}^2/\rho c)A$ is a constant along any ray tube (A is the cross-sectional area of ray tube). Therefore, if x_0 and x are any two points along the same ray,

$$\hat{p}(x) = \left[\frac{(A/\rho c)_{x_0}}{(A/\rho c)_x}\right]^{1/2} \hat{p}(\boldsymbol{x}_0) \tag{3.34}$$

The equation gives the general law for the variation of the acoustic pressure amplitude along a ray in quiescent and non-uniform media. For a wave with constant frequency, the relation may be interpreted as the following requirement that specifies that the time average of the energy flowing through a ray tube in unit time should have nothing to do with the distance its passing through the ray. This law is sometimes called the Green law for acoustic waves.[6]

3.5 Amplitude variations in moving media

3.5.1 Wave equation in moving media

In order to determine the effect on wave amplitudes of stable and inhomogeneous movements of media under geometrical acoustics approximation, we must first derive the acoustic wave equation for this case. Therefore, we will still start from the basic equations of fluid dynamics. While what is different from Sec. 1.1 of Chap. II is that, except v_0 not being 0 again, the state equation with the form of Eq. (1.12) (here we have the entropy $s_0 \neq$ const) is invalid because of the heterogeneity of media. We must use the general state equation $p = p(\rho, s)$ and the continuity equation (1.35a) of entropy. The remaining two equations are still Euler equation (1.30a) and continuity equation (1.31′).

Substituting the acoustic wave perturbation Eq. (2.1) into these equations, and only keeping the first-order terms of the acoustic perturbation quantities (linear processing), we obtain

$$\left.\begin{aligned} &D_t \boldsymbol{v}_1 + (\boldsymbol{v}_1 \cdot \nabla) v_0 + \frac{1}{\rho_0}\nabla p_1 - \frac{\rho_1}{\rho_0^2}\nabla p_0 = 0 \\ &D_t \rho_1 + \boldsymbol{v}_1 \cdot \nabla \rho_0 + \rho_1 \nabla \cdot \boldsymbol{v}_0 + \rho_0 \nabla \cdot \boldsymbol{v}_1 = 0 \\ &D_t s_1 + \boldsymbol{v}_1 \cdot \nabla s_0 = 0 \\ &p_1 = c^2 \rho_1 + (\partial_s p)_0 s_1 \end{aligned}\right\} \tag{3.35}$$

Now both sound velocity c and thermodynamic coefficient $(\partial_s p)_0$ are functions of position.

6 This law was found by G. Green in shallow water waves, you may refer to the original papers: *On the motion of waves in a canal of variable depth and width.* Trans. Camb. Phil. Soc. 1837, or N. M. Ferrers (ed.). *Mathematical Papers of the Late George Green.* London: Macmillan. 1871: 225∼230.

Similar to what we did in Sect. 1.1 of Chap. II, doing substitution among the coupling equations in Eq. (3.55). Eliminate ρ_1 from the first and second equations by using the fourth equation. The substitution gives the first and third terms in the second equation

$$c^2(D_t\rho_1 + \rho_1\nabla\cdot\boldsymbol{v}_0) = D_t p_1 - (\partial_s p)_0 D_t s_1 + c^2 p_1 \nabla\cdot\frac{\boldsymbol{v}_0}{c^2} - c^2 s_1 \nabla\cdot\left[\frac{\boldsymbol{v}_0}{c^2}(\partial_s p)_0\right]$$

The fourth expression in Eq. (3.35) describes the relationship of p_1, ρ_1 and s_1, which is also the relationship of ∇p_0, $\nabla\rho_0$ and ∇s_0. Therefore, from the third equation, we obtain

$$-(\partial_s p)_0 D_t s_1 = \boldsymbol{v}_1\cdot\nabla p_0 - c^2\boldsymbol{v}_1\cdot\nabla\rho_0$$

And, the second equation is transformed to

$$D_t p_1 + \boldsymbol{v}_1\cdot\nabla p_0 + c^2 p_1\nabla\cdot\frac{\boldsymbol{v}_0}{c^2} + \rho_0 c^2\nabla\cdot\boldsymbol{v}_1 - c^2 s_1\nabla\cdot\left[\frac{\boldsymbol{v}_0}{c^2}(\partial_s p)_0\right] = 0 \tag{3.36}$$

Because the medium is assumed slowly-varying, the spatial derivatives of the parameters describing its state all are very small. Thus, all the second-order terms can be omitted. For the acoustic wave in a homogeneous medium, s_1 should vanish. But now it is not equal to 0 anymore due to the spatial change of the parameters describing the medium state. Therefore, s_1 is also a first-order quantity. Consequently the last term in Eq. (3.36) is asecond-order quantity and can be neglected. Finally we obtain

$$\begin{aligned} D_t\boldsymbol{v}_1 + (\boldsymbol{v}_1\cdot\nabla)\boldsymbol{v}_0 + \frac{1}{\rho_0}\nabla p_1 - \frac{p_1}{(\rho_0 c)^2}\nabla p_0 = 0 \\ D_t p_1 + \boldsymbol{v}_1\cdot\nabla p_0 + c^2 p_1\nabla\cdot\frac{\boldsymbol{v}_0}{c^2} + \rho_0 c^2\nabla\cdot\boldsymbol{v}_1 = 0 \end{aligned} \tag{3.37}$$

3.5.2 Conservation of wave action quantities

Making some further approximations to the above equations, we can derive the relation of conservation of acoustic energy similar to that in a homogeneous medium.

Performing a dot product of the first formula in Eq. (3.37) with $\rho_0 v_1$, and adding it to the product of the second formula with $p_1/\rho_0 c^2$, we obtain

$$\begin{aligned} (\partial_t + \boldsymbol{v}_0\cdot\nabla)E - v_1^2(\boldsymbol{v}_0\cdot\nabla)\frac{\rho_0}{2} - p_1^2(\boldsymbol{v}_0\cdot\nabla)(2\rho_0 c^2)^{-1} \\ +\nabla\cdot\boldsymbol{I} + \rho\boldsymbol{v}_1\cdot\left[(\boldsymbol{v}_1\cdot\nabla)\boldsymbol{v}_0\right] + \frac{p_1^2}{\rho_0}\nabla\cdot\frac{\boldsymbol{v}_0}{c^2} = 0 \end{aligned} \tag{3.38}$$

Where E and $\boldsymbol{I}$ denote the acoustic energy density and sound intensity observed in the coordinate system moving with the medium, respectively. Consequently, they are given by Eq. (2.17) and (2.18) respectively. If we only care about the acoustic field which is locally similar to a plane traveling wave everywhere, then in the case which is consistent with the above approximation, we can let $v_1 = \boldsymbol{n}p_1/\rho_0 c$ in all those small terms which include the spatial derivatives of the parameters describing the medium state. Since $p_1^2 = \rho_0 c^2 E$ (the two relationships are strictly valid in the case of homogeneous medium, even if when $v_0 \neq 0$), the substitution gives

$$\begin{aligned}&\partial_t E + \boldsymbol{v}\cdot\nabla E - E\left[\frac{1}{\rho}(\boldsymbol{v}\cdot\nabla)\frac{\rho}{2} + \rho c^2(\boldsymbol{v}\cdot\nabla))(2\rho_0 c^2)^{-1}\right]\\&+\nabla\cdot\boldsymbol{I} + E\boldsymbol{n}\cdot\left[(\boldsymbol{n}\cdot\nabla)\boldsymbol{v}\right] + c^2 E\nabla\cdot\frac{\boldsymbol{v}}{c^2} = 0\end{aligned} \tag{3.39}$$

where the subscript 0 of ρ and $\boldsymbol{v}$ are omitted again, because without these subscripts-they will not be messed up with other parameters.

As for the next to the last term in Eq. (3.39), unit vector $\boldsymbol{n}$ can be replaced by $(c/\Omega)\boldsymbol{s}$ according to the second expression in Eq. (2.80), and $\boldsymbol{s} = \nabla\tau$. Applying vector analysis formula, we have

$$\boldsymbol{s}\cdot\left[(\boldsymbol{s}\cdot\nabla)\boldsymbol{v}\right] = (\boldsymbol{s}\cdot\nabla)(\boldsymbol{s}\cdot\boldsymbol{v}) - \boldsymbol{v}\cdot\left[(\boldsymbol{s}\cdot\nabla)\boldsymbol{s}\right]$$

The first term $\boldsymbol{s}\cdot\boldsymbol{v}$ on the right side can be replaced by $1-\Omega$, and $(\boldsymbol{s}\cdot\nabla)\boldsymbol{s}$ in the second term by $\frac{1}{2}\nabla(\Omega^2/c^2)$ in Eq. (2.84). Therefore, we obtain

$$\boldsymbol{n}\cdot[(\boldsymbol{n}\cdot\nabla)\boldsymbol{v}] = \Omega c\boldsymbol{n}\cdot\nabla\frac{1}{\Omega} + \frac{\Omega}{c}\boldsymbol{v}\cdot\nabla\frac{c}{\Omega} \tag{3.40}$$

With the same degree of approximation in the derivation of Eq. (3.39), we may let $\boldsymbol{n}E = \boldsymbol{I}/c$ in such terms as $E\boldsymbol{n}\cdot\nabla(1/\Omega)$ (in homogeneous media this term vanishes), hence, after the substitution just mentioned, Eq. (3.39) becomes

$$\begin{aligned}&\partial_t E + \boldsymbol{v}\cdot\nabla E + E\boldsymbol{v}\cdot\nabla\left[\ln\rho^{-1/2} + \ln(\rho c^2)^{1/2} + \ln\frac{c}{\Omega}\right]\\&+\nabla\cdot\boldsymbol{I} + \Omega\boldsymbol{I}\cdot\nabla\frac{1}{\Omega} + c^2 E\nabla\cdot\frac{\boldsymbol{v}}{c^2} = 0\end{aligned}$$

After further manipulation, the above equation gives

$$\partial_t\left(\frac{E}{\Omega}\right) + \nabla\cdot\left(\frac{\boldsymbol{I}+E\boldsymbol{v}}{\Omega}\right) = 0 \tag{3.41}$$

It is easily seen that, when $\boldsymbol{v}=0$, the above equation is reduced to acoustic energy conservation law for homogeneous medium[7] (see Eq. (2.19)). Although each term of

---

7 Eq. (3.41) is derived by introducing some approximations, and it only holds under geometrical acoustics approximation, but the deduction of the strict acoustic energy relation (whose form is still a sum of a time derivative and a spatial divergence) actually exits for the case of inhomogeneous medium in steady movements: (1) The final expression includes "Clebsch potential" (not the local characteristics of acoustic field), see W. Mohring, Z. Angew. Math. Mech., 1960, 50: T196~198; J. Sound Vib., 1971,18: 101~9 and 1973, 20: 93~101; (2) About the simple deduction applicable for movements with potential and equivalent entropy, you can refer to L.A.Tselnov, Journal of Theoretical Physics (in Russian), 1946, 16: 733~6; W. Cantren and R. W. Hart. J. Acoust. Soc. Am. 1964, 36: 697~706; (3) As for other energy expression for the fluid in motion, you can refer to: O. S. Ryshov & G. M. Shefter. J. Appl. Math. Mech. (USSR), 1962, 26: 1293~1309 and C. L. Morfey. J. Sound Vib., 1971, 14: 159~170.

the equation has an additional factor, Eq. (3.14) is still a conservation law, because it is a sum of a time derivative and a spatial divergence. To illustrate which physical parameter is conserved, let's consider the case of a constant frequency, and multiply both sides of the equation by $1/\omega$, making the obtained equation similar to Eq. (3.41), only with Ω being replaced by $\omega\Omega$. The quantity $\omega\Omega$ or $\omega - \omega v \cdot \nabla\tau \equiv \omega^*$ can be regarded as the observed frequency from the coordinate system attached to a moving fluid (intrinsic frequency), because of applying the operator D_t to $\exp[-i\omega(t-\tau)]$ is equivalent to multiplied by $-i\omega^*$. Replacing the exponential index t with $t-\tau(x)$ describes the dominant spatial dependence of a perturbation with fixed frequency under geometric acoustics approximation; where the form of $p(x,t)$ should take the real part of $\hat{p}(x)\exp(-i\omega(t-\tau))$, here, $\hat{p}(x)$ is a slowly-varying function of positions.

In analogy to mechanical systems described by Hamilton operator, the variable E/ω^* is similar to the acting variable in unit volume, which can be called the wave action quantity density (wave action quantity per unit volume, thus $(\boldsymbol{I}+E\boldsymbol{v})/\omega^*$ being the wave acting quantity flux). And the conservation relation Eq. (3.41) (substituting Ω with ω^*) can be regarded as the law of wave action quantity conservation. Although it is derived for the case of steady flows here, it is also applicable for the wave packet with nearly constant frequency propagating in slowly-varying media, whose property is a slowly-varying function of location and time [74]. When wave packets move, the frequency observed in quiescent coordinate system will change due to the double factors of sound speed depending on time and the motion of media. However, if E and $\boldsymbol{I}$ are defined as the previous discussion, and frequency ω^* is taken as the observed frequency in a moving coordinate system (which also depends on time), then the law of wave action quantity conservation still holds. This plausible assertion should be reasonable obviously, because the flows being homogeneous and time-independent for a stationary observerwill vary with time for the observer moving with media. Because E, $\boldsymbol{I}$ and ω^* are all invariants under the transformation of coordinate system, so is Eq. (3.41) by the substitution of Ω with ω^*.

3.5.2.1 The Blokhintzev (Blohincev) invariant

If we choose a coordinate system making the properties of its surrounding medium be independent of time, the advantages of the law of wave action quantity conservation is that such as Eq. (3.41) is also applicable for instantaneous disturbance. Therefore, assuming

$$p_1 = P(\boldsymbol{x})f(t-\tau(\boldsymbol{x}),\xi),\quad \boldsymbol{v}_1 = \frac{p_1}{\rho c}\boldsymbol{n} \tag{3.42}$$

where f is an arbitrary function (but mainly composed of high frequency), and ξ a constant along any given ray. Substituting Eq. (3.42) into (3.41), we obtain an equation about $P(\boldsymbol{x})$. According to these steps, and omitting each term which includes $\nabla\xi$, we obtain

$$E = \left(\frac{P^2}{\rho c^2}\right) f^2, \quad \boldsymbol{I} + E\boldsymbol{v} = E\boldsymbol{v}_{\text{ray}}$$

$$\nabla \cdot \left(\frac{\boldsymbol{I} + E\boldsymbol{v}}{\Omega}\right) = f^2 \nabla \cdot \left(\frac{P^2 \boldsymbol{v}_{\text{ray}}}{\rho c^2 \Omega}\right) - 2\left(\frac{P^2 \boldsymbol{v}_{\text{ray}}}{\rho c^2 \Omega}\right) \cdot \nabla\tau f \partial_t f \tag{3.43}$$

where $\boldsymbol{v}_{\text{ray}}$ is defined by Eq. (2.78), and from Eq. (2.79), we know $\boldsymbol{v}_{\text{ray}} \cdot \nabla\tau = 1$. Thus, the second term of the right side of Eq. (3.43) is $-\partial_t(E/\Omega)$, and Eq. (3.41) gives

$$\nabla \cdot \left(\frac{P^2 \boldsymbol{v}_{\text{ray}}}{\rho c^2 \Omega}\right) = 0 \tag{3.44}$$

This is one of the basic equations of geometrical acoustics. If the medium velocity is 0, the above equation is reduced to Eq. (3.33).

Volume Integrating Eq. (3.44) throughout the ray segments and repeating the steps done in Sect. 4.1, we can make the following conclusions: along any given infinitely small ray tube with variable sectional area, and **Blohincev** invariant [75] is a constant.

$$\frac{P^2 \boldsymbol{v}_{\text{ray}} A}{(1 - \boldsymbol{v} \cdot \nabla\tau)\rho c^2} = \text{const} \tag{3.45}$$

Eq. (3.45) can be used to explain some common phenomena. For example, it is able to partly explain the "abnormal" distribution of noise acoustic field produced by the jet of a nozzle: the distant place forming a small angle with the jet axis is the "relatively quiet" region [76, 77]. In analogy, the more familiar phenomena can be explained: the sound coming from a source near the ground always sounds louder along the direction of wind than against the direction of wind [3, 78]. Because wind speed increases with height, all the emitted sound rays close to the horizontal ground along downwind direction are bent downward due to refraction; thus the reduction of sound intensity with distance would be less than that specified by spherical spreading according to "inverse square law", while the situation is on the contrary when sound traveling direction is against the wind.

3.6 Sound wave reflection from the interface between two media

All the discussion before this section is about the sound propagation in unbounded atmosphere where the effect of ground is not considered. Actually the effect of the ground is very important, especially in low-layer atmosphere. Hence we begin to investigate the acoustic field generated by an isotropic point source (see Fig. 20) at a certain height, z_0, above the interface of two media.

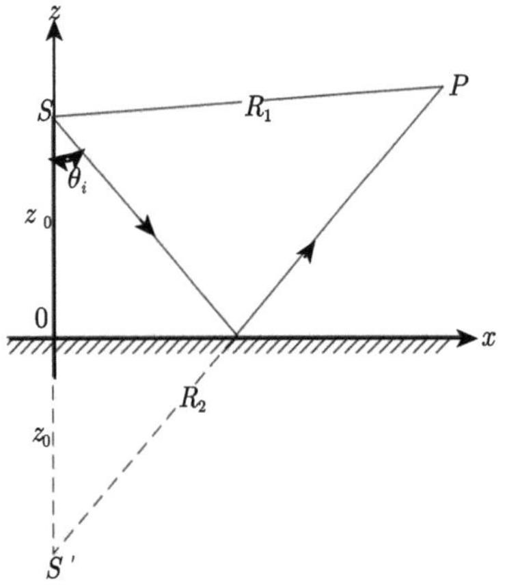

Fig. 20: Sketch of a point sound source above a plane interface.

3.6.1 Reflection of plane waves from rigid boundaries

As shown in Fig. 20, a plane wave is

$$p_i = f\left(t - \frac{1}{c}\boldsymbol{e}\cdot\boldsymbol{x}\right), \quad \boldsymbol{v}_i = \frac{p_i}{\rho c}\boldsymbol{e}_i \tag{3.46}$$

which is incident onto a rigid plane $z = 0$, where $\boldsymbol{e}_i$ denotes the unit vector along the incident direction, because $\boldsymbol{e}_i$ can be regarded as having no y component, thus we obtain

$$\boldsymbol{e}_i = \boldsymbol{e}_x \sin\theta_i - \boldsymbol{e}_z \cos\theta_i \tag{3.47}$$

where θ_i (incident angle) is the angle between $\boldsymbol{e}_i$ and the normal vector $\boldsymbol{e}_z$, pointing to the interior of the plane surface.

If Eq. (3.46) is the solution of wave equation (2.6) when the boundary at $z = 0$ does not exit, the solution when the plane does exit can be written as $p_i + p_r$ and $\boldsymbol{v}_i + \boldsymbol{v}_r$, where p_r, $\boldsymbol{v}_r$ must also be the solution of Eq. (2.6). Moreover, the boundary condition at $z = 0$ being $\boldsymbol{v}\cdot\boldsymbol{e}_z = 0$ requires $(\boldsymbol{v}_i + \boldsymbol{v}_r)\cdot\boldsymbol{e}_z = 0$ at $z = 0$. In this particular case, the solution of reflected wave can be easily obtained from another boundary condition: $\partial_z p = 0$ at $z = 0$. This condition can be satisfied when[8]

$$p_r(x, y, z; t) = p_i(x, y, -z; t) \tag{3.48}$$

Under this circumstance, the sum $p_i + p_r$ is equivalently relating to z. Hence, its derivative with respect to z at $z = 0$ is 0. Since $p_i(x, y, z; t)$ is given by Eq. (3.46), p_r can be expressed as $f\left(t - \frac{1}{c}\boldsymbol{e}_r\cdot\boldsymbol{x}\right)$ accordingly. The difference between $\boldsymbol{e}_r$ and $\boldsymbol{e}_i$ is

8 This is an example of the mirror image method, which was first put forward by Euler between 1759 and 1767.

that their normal components have opposite signs, i.e., $\boldsymbol{e}_r = \boldsymbol{e}_x \sin\theta_i + \boldsymbol{e}_z \cos\theta_i$. The angle between $\boldsymbol{e}_r$ and $\boldsymbol{e}_z$ is also θ_i, and the reflection angle is equal to incident angle (Mirror law).

Similar to Eq. (3.46), the fluid particle velocity caused by plane wave $f\left(t-\frac{1}{c}\boldsymbol{e}_r\cdot\boldsymbol{x}\right)$ propagating in $\boldsymbol{e}_r$ direction is

$$\boldsymbol{v}_r = \frac{\boldsymbol{e}_r}{\rho c} f\left(t - \frac{1}{c}\boldsymbol{e}_r \cdot \boldsymbol{x}\right) = \frac{\boldsymbol{e}_r}{\rho c} p_r \tag{3.49}$$

that satisfies the boundary condition $(\boldsymbol{v}_r + \boldsymbol{v}_i) \cdot \boldsymbol{e}_z = 0$ at $z = 0$.

It can be deduced from the above solutions that at $z = 0$ plane both the pressure and the tangential component of particle velocity caused by the total wave disturbance are twice the corresponding quantities when only the incident wave exists. If the frequency of the incident wave is constant, then

$$p_i + p_r = Re\left[\hat{p}\mathrm{e}^{-\mathrm{i}\omega t}\mathrm{e}^{\mathrm{i}k_x x}(\mathrm{e}^{-\mathrm{i}k_z z} + \mathrm{e}^{\mathrm{i}k_z z})\right] = 2\cos(kz\cos\theta_i)f\left(t - \frac{1}{c}x\sin\theta_i\right) \tag{3.50}$$

where $k = \omega/c$, $k_x = k\sin\theta_i$, $k_z = k\cos\theta_i$. Hence, once $kz\cos\theta_i$ is an odd multiple of $\pi/2$, the incident and reflected waves cancel each other out.

3.6.2 Reflection of plane waves at planes with finite specific acoustic impedances

Now let's extend the above discussion to the case when the specific acoustic impedances of an interface are finite (possibly depending on incident angles). We still start from Eq. (3.46) and (3.47). The total acoustic field still consists of incident and reflected waves. Nevertheless, the reflected wave now has the form of $g\left(t - \frac{1}{c}\boldsymbol{e}_r \cdot \boldsymbol{x}\right)$, which is not always having the same form as the incident wave $f(t)$.

If $f(t)$ is regarded as a kind of superposition, such as a Fourier series consisting of constant frequency components, each component has the form of $Re\,\hat{f}\mathrm{e}^{-\mathrm{i}\omega t}$. Defining the reflection coefficient of a sound pressure amplitude as $\Re(\theta_i, \omega)$, which makes $Re\Re(\theta_i, \omega)\hat{f}\mathrm{e}^{-\mathrm{i}\omega t}$ be the corresponding component of $g(t)$. Thus $\hat{g} = \Re(\theta_i, \omega)\hat{f}$ is obtained. Or, if both $f(t)$ and $g(t)$ are transient waveforms, $\Re(\theta_i, \omega)$ is the ratio of the Fourier transforms of $g(t)$ to $f(t)$. In above two cases, sound pressure and particle velocity can be expressed as

$$\hat{p} = \hat{f}\mathrm{e}^{\mathrm{i}k_x x}\left[\mathrm{e}^{-\mathrm{i}k_z z} + \Re(\theta_i, \omega)\mathrm{e}^{\mathrm{i}k_z z}\right]$$

$$\hat{v}_z = \frac{\cos\theta_i}{\rho c}\hat{f}\mathrm{e}^{\mathrm{i}k_x x}\left[-\mathrm{e}^{-\mathrm{i}k_z z} + \Re(\theta_i, \omega)\mathrm{e}^{\mathrm{i}k_z z}\right] \tag{3.51}$$

The boundary condition at $z = 0$ is $\hat{p}/(-\hat{v}_z) = Z(\omega)$ being the specific acoustic impedance (acoustic impedance in unit area) of the interface, which has the form of

$$\frac{Z(\omega)\cos\theta_i}{\rho c} = \frac{1 + \Re(\theta_i, \omega)}{1 - \Re(\theta_i, \omega)}, \quad \Re(\theta_i, \omega)\frac{\xi(\omega)\cos\theta_i - 1}{\xi(\omega)\cos\theta_i + 1} \tag{3.52}$$

where $\xi(\omega) \equiv Z(\omega)/\rho c$ denotes the ratio of specific acoustic impedances over medium characteristic impedance. It is also named as normalized characteristic impedance.

When and only when the real part of Z is positive, the magnitude of $\Re$ is less than 1, and any interface with this property will absorb acoustic energy. The time-averaged acoustic power per unit area flowing into the boundary (for a single frequency component) is equal to[9] $\frac{1}{2}Re\hat{p}(-\hat{v}^{+}) = \frac{1}{2}\left|-\hat{v}\right|^2 ReZ(\omega)$ (where the plus sign denotes complex conjugate). Substituting Eq. (3.51) into it, we obtain

$$[p(-v_z)]_{\mathrm{av}} = \frac{1}{2}\frac{\cos\theta_i}{\rho c}\left|\hat{f}\right|^2(1-|\Re|^2) \tag{3.53}$$

(Notice that the real part of $(1+\Re)(1-\Re^{+})$ is $1-|\Re|^2$). If $ReZ(\omega) > 0$, or equivalently, $|\Re| < 1$, the interface is regarded absorbing acoustic energy.

Expression $\frac{1}{2}\left|\hat{f}\right|^2(\cos\theta_i)/\rho c$ specifies the energy per unit area carried into the boundary by incident wave in unit time. Multiplying this quantity by $|\Re|^2$, we obtain the energy per unit area carried out of the boundary by the reflection wave in unit time. Therefore, Eq. (3.53) demonstrates the following principle: in time average viewpoint, the incident acoustic energy is the sum of the reflected energy and the absorbed energy, and the portion absorbed is expressed by absorption coefficient $\alpha(\theta_i,\omega)$, whose value is $[p(-v_z)]_{\mathrm{av}}$ divided by $\frac{1}{2}\left|\hat{f}\right|^2(\cos\theta_i)/\rho c$, or equivalently, is $1-|\Re|^2$, where $|\Re|^2$ is energy reflection coefficient.

If the reflection coefficient of sound pressure amplitude is unity ($\Re = 1$), then the reflected wave given above will make $\hat{v}_z = 0$ at $z = 0$, which is just the case of reflection at rigid boundary. Because $\Re = 1$ corresponds to $|Z| \to \infty$, specific acoustic impedance being infinity corresponds to a rigid interface. Another extreme case is $Z \to 0$, which causes $\Re = -1$, and $\hat{p}$ is always required to be 0 at $z = 0$ whatever $-\hat{v}_z$ is. The interface in this extreme case is referred as the pressure-release surface (see the discussion of Sec. 2.3)

3.6.3 Locally-reacting surfaces

As mentioned in section 3.1 of Chapter II, except for the previously mentioned two ideal extreme cases (rigid interface corresponding to $Z \to \infty$ and pressure-release boundary to $Z = 0$), the reflection coefficient, $\Re$, of pressure amplitudes always changes with incident angles. However, in some cases, Z, the ratio of specific acoustic impedances is nearly independent of incident angles, For example, the surfaces of certain thick or thin typical porous materials and those of typical porous materials

9 If $X = Re\hat{X}e^{-i\omega t}$, $Y = Re\hat{Y}e^{-i\omega t}$, the time average value of the product of X and $Y (XY)_{\mathrm{av}} = \frac{1}{2}Re\hat{X}\hat{Y}^{+}$.

with air cushion (with or without impenetrable hard layers, with or without support structures), etc. Its precondition is that the $Z(\omega)$ calculated by Eq. (3.52) for a given θ_i and a practical $\Re(\theta_i, \omega)$ should be nearly unrelated to θ_i. The Z calculated by $\Re(0, \omega)$ when $\theta_i = 0$ is named as the normal incidence interface impedance. Therefore, $\Re(\theta_i, \omega)$ at any θ_i can be determined by Eq. (3.52). A conclusion is that, if Z is limited, $\Re(\theta_i, \omega) \to -1$ in the extreme case of $\theta_i \to \frac{\pi}{2}$ (grazing angle), which is just like the case in pressure-release surface.

The requirement that Z is independent with θ_i is in accordance with the following assumption: the value of v_x at a given point on the interface only depends on the acoustic pressure at the same point. That is to say, the pressure imposed at a point doesn't make other points on the interface move. Therefore, we can imagine a locally-reacting surface, on which $\hat{p} = Z(-\hat{v}_z)$ is valid for each point where $Z(\omega)$ is fixed, whatever the properties of the acoustic field outside the interface are. This model does not exclude the possibility of curve surfaces. It can even deal with Z varying from point to point on the surface.

Locally-reacting surface model approximately counts the boundary vibration caused by outside sound pressures, as well as the change of normal velocity of particles on the surface resulted from the fluid medium pushed into or out of small holes on the boundary by outside pressure fluctuations. Although this model neglects the possible effect of the pressure at one point on the velocity at another point, it has apparent advantages in simplifying boundary reflection issue, because only one rather than two boundary conditions is needed as in general cases. This advantage is more obvious in discussing the reflection and diffraction of spherical waves.

The opposite of locally-reacting surface is that the behavior at one point on the surface is related with that of its neighbor points, such that its reaction to acoustic wave depends on incident waves. This surface is called extended reacting surface.

3.6.4 Sound field above reflecting surfaces

Based on the above discussion, we can calculate the sound field above reflecting surfaces.[10] Sound field at any point can be treated as the combination of direct arrival wave and reflected wave (see Fig. 20), and the latter can be regarded as the emission from an virtual source at point $(0, 0, -z_0)$. Notice we ignore the tangential displacement of

---

10 Obviously, in atmospheric acoustics only sound fields above the boundary are interested. However, in some occasions, people will be interested in the sound fields under the boundary (for example, in hydroacoustics).

reflected rays along the interface here.[11] Our other assumption is that the change of the amplitude and phase for the reflection of spherical waves is the same as those for plane waves with the same incident angle. If the ranges from sound source and virtual source are denoted as R_1 and R_2 respectively, the sectional areas of the incident and reflected ray tubes are also directly proportional to R_1^2 and R_2^2. For non-perfect reflection boundaries, the only correction is multiplying the complex amplitude of each frequency component of reflected wave with $\Re(\theta_i, \omega)$. According to Eq. (3.52), we have

$$\Re(\theta_i, \omega) = \frac{Z(\theta_i, \omega) - \rho c/\cos\theta_i}{Z(\theta_i, \omega) + \rho c/\cos\theta_i} \tag{3.52'}$$

For the wave with a constant frequency $f(t) = Re\hat{f}e^{-i\omega t}$, based on the previous discussion, the solution of complex pressure amplitude $\hat{p}$ under the approximation of geometric acoustics is given by the following equation

$$\hat{p} = \frac{\hat{f}}{R_1}e^{i(\omega/c)R_1} + \frac{\hat{f}}{R_2}\Re(\theta_i, \omega)e^{i(\omega/c)R_2} \tag{3.54}$$

When the estimated value of $\hat{p}$ is very small using the above equation, the validity of the equation is doubtable because any correction made based on the full wave analysis (for geometrical acoustical approximation) will be a significant portion of the amplitude of the total sound pressure. An example of this is the sound field near surface $z = 0$ when $\Re(\theta, \omega)$ is approaching -1(hence $R_1 \approx R_2$). For example, this scenario will happen fora reflection from locally-reacting surface when $\cos\theta_i \ll \rho c/|Z|$.

Take $\hat{f}$ and $\hat{p}$ as the Fourier transformations of $f(t)$ and $p(x, t)$, the corresponding transient solution can be obtained. After applying the Fourier integral theorem, we have

$$p = \frac{1}{R_1}f\left(t - \frac{R_1}{c}\right) + \frac{1}{R_2}g\left(t - \frac{R_2}{c}, \theta_i\right) \tag{3.54'}$$

where the waveform corresponding to the reflected wave $g(t, \theta_i)$ is the inverse Fourier transform of the product of the Fourier transforms of $\Re(\theta_i, \omega)$ and $f(t)$.

3.7 Effects of ground surfaces

Let's go further to discuss the effect of ground surfaces in detail. This issue has been received a widespread attention since the middle of the 1970s, and breakthroughs were

11 When narrow sound beamsare obliquely incident on an interface, this kind of tangential displacements can be introduced indeed, and the energy distribution in the cross section of sound beam will also change. For more information, refer to the paper written by A. Schoch. in Acustica, 1952, 2: 18~22 and the paper by M. A. Breazeale et al. in J. Acoust. Soc. Am., 1974, 56: 866~872. However, for a very wide sound beam or a spherical wave, the displacement effect for reflection is very small.

made in mid 80's. The main contribution of whichlies in two aspects: one is accurate mathematic description of the acoustic field of point sources near complex impedance interfaces [79], and the other is the "microscopic" description of the interaction between air-borne sounds and the ground medium [80]. Owing to the two progresses, a deep understanding was achieved on the physical processes of sound propagation above uniform flat earth surfaces.

3.7.1 Expressions of sound fields above porous half-space media

Referring to Fig. 20, we assumed a porous medium filling the halfspace beneath the boundary. The total sound field at receiving point P can be expressed by the following form [81, 82]

$$\phi_{\text{total}} = \frac{e^{ikR_1}}{R_1} + \frac{e^{ik_1R_2}}{R_2}\left\{\Re(\theta_i) + B[1 - \Re(\theta_i)]F(\omega)\right\} \tag{3.55}$$

where R_1 and R_2 are the distances between observation point P and sound source S as well as virtual source S' respectively. The sound field can be thought of consisting of direct wave, reflected wave and an additional term. The additional term is created by the mathematical requirement of matching boundary conditions, which considers the curvature change of wave front with range [83]. The expression inside curly braces in Eq. (3.55) denotes the "modified" reflection coefficient of a spherical wave on a flat interface with complex impedance, which consists the ordinary reflected coefficient of plane waves $\Re(\theta_i)$((3.52) or (3.52$'$)) and a term relating with B and $F(\omega)$, where factor B mainly counts the phase change caused by the reflection of the interface with complex impedance; while $F(\omega)$ is introduced by the need of matching curve surface wave front with the plane interface. ω, named as numerical distance, can be used to measure the propagation distance with impedance. With the progress of ω, wave fronts become closer and closer to the plane. However, $F(\omega)$ decreases.

With the introduction of refractive index $n = k_2/k_1$ and normal characteristic admittance $\beta = mn = 1/\zeta$, Eq. (3.52′) can be rewritten as

$$\Re(\theta_i) = \frac{\cos\theta_i - m(n^2 - \sin^2\theta_i)^{1/2}}{\cos\theta_i + m(n^2 - \sin^2\theta_i)^{1/2}} \tag{3.52''}$$

For more generally extended reacting interfaces, we have [79]

$$B = \frac{\left[\cos\theta_i + \beta(1 - \sin^2\theta_i/n)^{1/2}\right](1 - n^{-2})^{1/2}}{\left[\cos\theta_i + \beta(1 - n^{-2})^{1/2}/(1 - m^2)^{1/2}\right](1 - \sin^2\theta_i/n^2)^{1/2}} \times \frac{\left[(1 - m^2)^{1/2} + \beta(1 - n^{-2})^{1/2}\cos\theta_i + \sin\theta_i(1 - \beta^2)^{1/2}\right]^{1/2}}{(1 - m^2)^{3/2}(2\sin\theta_i)^{1/2}(1 - \beta^2)^{1/2}} \tag{3.56}$$

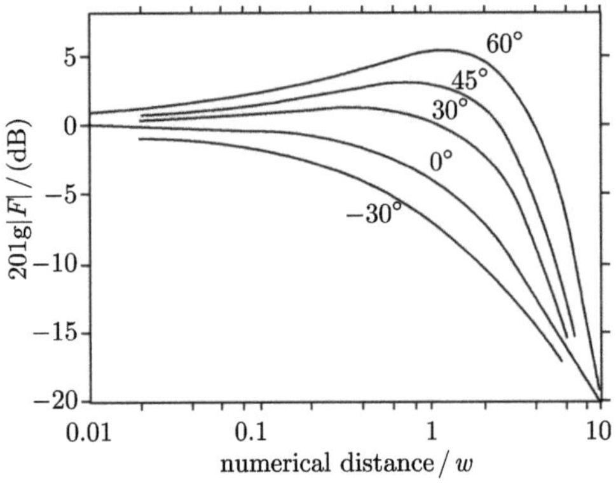

Fig. 21: $F(\omega)$ at different phase angles of interface impedance ϕ.

For locally reacting surfaces where $R \approx 1$ with a good approximation, the "spherical" factor $F(\omega)$ can be expressed by

$$F(\omega) = 1 + \mathrm{i}\pi^{1/2}\omega \mathrm{e}^{-\omega^2}\mathrm{erfc}(-\mathrm{i}\omega) \tag{3.57}$$

where erfc denotes complementary error function [84]

$$\mathrm{erfc}(\eta) = 1 - \mathrm{erf}(\eta) = \frac{2}{\pi^{1/2}}\int_{\eta}^{\infty} \mathrm{e}^{-u^2}\mathrm{d}u$$

And the numerical distance ω can be obtained by

$$\omega^2 = \mathrm{i}k_1R_2\left\{1 + \left[\beta\cos\theta_i(1-n^{-2})^{1/2} - \sin\theta_i(1-\beta^2)^{1/2}\right]/(1-m^2)^{1/2}\right\} \tag{3.58}$$

All the square root terms in the above equations should have their real parts be not negative. For different phase angles, $\phi = \arctan(X/R)$, of normal interface impedance, $\zeta(= R+\mathrm{i}X)$, function $F(\omega)$ is shown in Fig. 21.

3.7.2 Ground wave and surface wave

In order to understand the physical meaning of the third term in Eq. (3.55), we can consider the specified case, in which both sound source and receive spot are located on ground surfaces ($\theta_i = 90°$). According to Eq. (3.52), we know $\Re(\theta_i) = -1$ in Eq. (3.55), and because$R_1 = R_2$, the first two terms cancel out completely forming a shadow zone. Thus, only the third term can penetrate the shadow zone. For pure resistance interface $\phi = 0$, this term is called ground wave (this name is in analogy to the nomenclature in electromagnetic wave propagation). The curve corresponding to $\phi = 0$ in Fig. 21 demonstrates that, for short ranges, ground wave does not go through excess attenuation beyond the spherical spreading, while for long ranges, it has additional 6dB attenuation when range is doubled.

Tab. II: Propagation distance and existing height calculated based on the impedance of the ground surface where grass is cut

f/Hz	d_ω/m	$z_{s\omega}$/m
50	4000	58
100	1500	23
200	270	7.3
500	24	0.9
1000	3.6	0.5
2000	1	0.15

It will be shown in the next section that, for actual ground surfaces, ϕ varies between 30° and 60°. The corresponding value of F in Fig. 21 grows a lot when $\omega \geqslant 1$ comparing with that when $\phi = 0$. This rise is due to the contribution of surface waves propagating in the air, whose amplitude attenuates exponentially with the height above the boundary.[12] When $\omega < 1$, the contribution of surface wave is less than that of ground wave, because the amplitude of surface wave only reduces 3dB when distance is doubled. When $\omega > 1$, because of the viscosity loss in the small holes of porous interfaces, the amplitude of surface waves is less than the ground wave again.

As an example, the "propagation distance" d_ω (for $\omega = 1$) of ground wave, and the existing height $z_{s\omega}$ (the height at which the amplitude of surface wave is reduced to $1/e$ of the value on the interface) of surface wave are listed in Table II, which are calculated according to the impedances of the grass field when grass is cut. The relative contributions of various waves are given in Fig. 22: the excess attenuation is relative to the attenuation of the point source above an ideal rigid boundary.

3.7.3 Four-parameter semi-empirical expression for calculating ground impedances

The values of ω, ϕ and $\Re(\theta_i)$ depend on the normal complex characteristic impedance of ground surfaces. The semi-empirical expression for calculating ground impedances has several models. The best of which is the so-called "Four-parameter model" [85], which describes the structure of ground surfaces using four parameters: flow resistance, porosity (the void rate), particle shape factor, and void shape factor ratio.

12 Those waves with such attenuation characteristics are called by a joint name "non-uniform wave" (the direction of attenuation is vertical to the propagating direction).

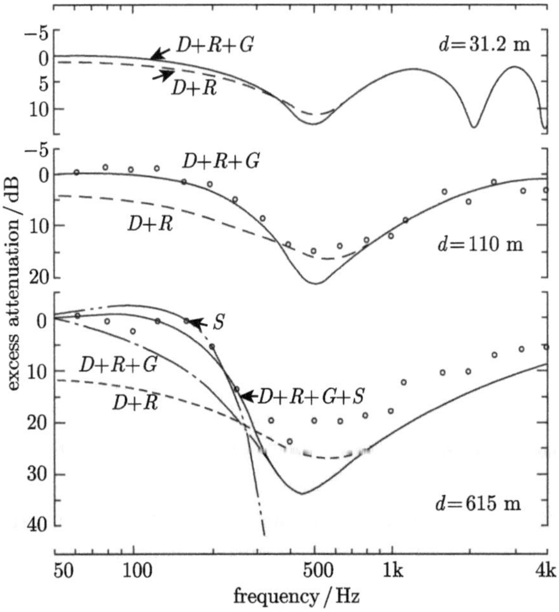

Fig. 22: Excess attenuation for sound propagation of point sources above a grass-cut ground surface. Source height $z_0 = 1.8$ m, receiver height $z = 1.5$ m. Theoretically calculated curve shows that the contributions are from each of the following waves: direct wave D, reflected wave R, ground wave G, and surface wave S. The circle denotes measured results using jet source at comparable distances [84].

For familiar ground surface structures, the normal complex characteristic impedance can be expressed approximately by

$$\zeta = \left(\frac{4q^2}{3\Omega} + \frac{\mathrm{i} S_p^2 \sigma}{\omega \rho_0} \right) \Big/ k_b \tag{3.59}$$

where Ω is the void rate,[13] $q^2 = \Omega^{-S_g}$ (S_g is particle shape factor, relating to the contortion of q with Ω); S_p is the void shape factor rate; σ is fluid impedance (i.e., pressure drop needed by unit flow air passing through ground media, in unit of $\mathrm{kgs^{-1}m^{-3}}$); ω is the angular frequency ($\mathrm{rads^{-1}}$); ρ_0 is air density ($\mathrm{kgm^{-3}}$), and k_b is the so-called normal wave number ($\mathrm{m^{-1}}$), which can be calculated by

$$k_b \approx (\gamma\Omega)^{1/2} \left[\left(\frac{4}{3} - \frac{\gamma - 1}{\gamma} N_{\mathrm{pr}} \right) \frac{q^2}{\Omega} + \frac{\mathrm{i} S_p^2 \sigma}{\omega \rho_0} \right]^{1/2} \tag{3.60}$$

where γ is specific heat ratio ($C_p/C_v \approx 1.4$), and N_{pr} is Prandtl number ($\mu C_p/\kappa \approx 0.72$).

13 Assuming the small holes in the medium beneath ground surface connect each other in a random and isotropic manner to make the air permeate easily in all directions. Therefore, the volume is not occupied by solid stuff and can be occupied by air completely. This volume portion is void ratio Ω.

Tab. III: Values of parameters required by four-parameter ground surface model (S_g is always 0.750, S_p is always 0.875)

Description of ground surface	Ω	σ/cgs
0.1 m new snow on top of 0.4 m-thick old snow	0.850	23
Sugar granulated snow	0.825	48
Pine forest and Chinese fir forest	0.825	60
Grass land, rough pasture, airport, public buildings	0.675	330
Roadside loose soil, small stones difficult to determine (screening out from 0.1m meshes)	0.575	960
Sandy silt compressed by vehicle	0.475	3470
Thick layer of pure limestone (screening out from 0.01∼0.025 m meshes)	0.425	6470
Old dirt road whose gaps are filled with gravels (screening out from 0.05 m meshes)	0.400	7500
Bare land filled with rain	0.350	17100
Substantial quarry dust being pressed tightly	0.300	41700
Dusty and rarely used bituminous pavement surface	0.250	120000

Several other variables S_p, Ω and S_g can be adjusted and corrected independently, until the theoretical expected value and actual measured value match under various boundaries. Repeated comparison results are that S_g is always 0.75, and S_p equals 0.875. As for Ω and σ, certain recommended values for several different boundaries are listed in Table III.

In the medium to high frequency range, the following semi-empirical expressions can be adopted [86, 87]

$$\zeta = (R + \mathrm{i}X)/[1 - (k/k_a)]^{1/2} \tag{3.61}$$

where

$$\begin{aligned} R &= \left[1 + 9.08(f/\sigma_e)^{0.75}\right] \\ X &= 11.9(f/\sigma_e)^{-0.73} \\ k_a &= (\omega/c_0)\left[1 + 10.8(f/\sigma_e)^{-0.70} - \mathrm{i}10.3(f/\sigma_e)^{-0.59}\right] \end{aligned} \tag{3.62}$$

These expressions give the expected values which agree with the experimental results very well. In these expressions, the unit of frequency f is Hz, and the unit of the "equivalent flow resistance" $\sigma_e \approx S_p^2\sigma/\Omega$ is cgs Rayl. For very large σ_e, the four-parameter model is simplified as

$$\zeta \approx 0.218(1 + \mathrm{i})(\sigma_e/f)^{1/2} \tag{3.63}$$

Some ground surface has a stratified structure due to the progressive settlement of the substances on soil base. For a layer with a thickness of d and an impedance of ζ_1 on top of a semi-infinite layer with a impedance of ζ_2, its equivalent impedance $\zeta(d)$ is given by

$$\zeta(d) = \frac{\zeta_2 - \mathrm{i}\zeta_1 \tan(k_b d)}{\zeta_1 - \mathrm{i}\zeta_2 \tan(k_b d)}\zeta_1 \tag{3.64}$$

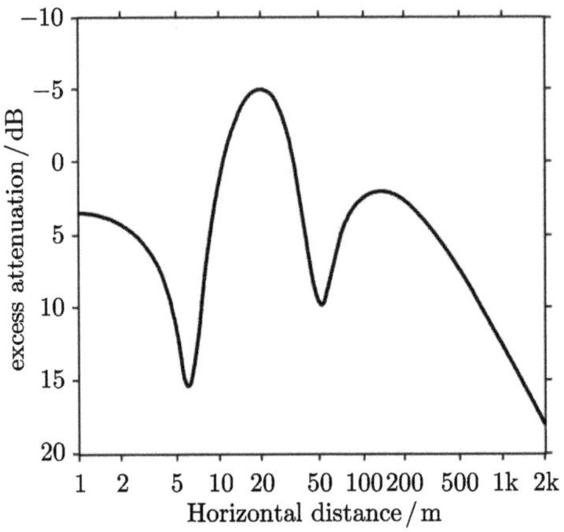

Fig. 23: Excess attenuation due to ground surface.

where k_b is the normal wave number ("volume propagation constant") of the upper layer; k_b, ζ_1 and ζ_2 can be obtained by Eq. (3.60) and (3.60) or (3.63).

3.7.4 Excess attenuation due to the ground surfaces

As discussed previously, the effect of ground surface will ultimately be demonstrated by the excess attenuation of sound wave propagating on it.

Fig. 23 shows the calculated curves for the case. When specific impedance is 9.34~8.75, sound source height 15 m, receiver height 0.55 m, and frequency range is 500Hz. The excess attenuation is relative to the attenuation in free space (excess attenuation greater than 0 means lower sound level).

As shown in Fig. 23, there are several notches: the first one is the so-called ground effect notch, caused by the interference between direct wave and reflected wave. With the increase of frequency, sound level rises due to the constructive interference, followed by another interference minimum. At a very large range, the path of direct wave is approximately equal to the reflected wave, which results in a shadow zone caused by the interface. The acoustic field in the shadow zone will be discussed in the next Chapter.

3.7.5 Effects of topography

In discussing acoustic fields near the ground, generally it is self-evident to assume the ground is flat. However, in many actual cases, small hills, trees and buildings often alter the amplitudes of the acoustic field at receiving points to some extent. In the following two typical scenarios are to be discussed.

(a) Semi-infinite barrier[14]

For the acoustic field behind a semi-infinite rigid barrier, it can be calculated applying Fresnel integral equation in the absence of the ground [88]. In this simple processing, the upper edge of the barrier can be treated as a line source diffracting acoustic energy to the rear of the barrier. Nevertheless, the problem is more difficult and complex actually.

Considering an infinite long barrier [89], the bilateral ground situations are different. A series of sound propagation paths are shown in the leftmost column in Fig. 24, where S and R denote the source and the receiver, respectively; T and X are their images under the ground. The second column shows the equivalent paths used to calculate the diffraction field Ψ (making some modification for reflection and the possible presence of ground wave). The third column lists the computational formula for the corresponding components of the acoustic field (see Eq. (3.65)). The first line is for the case that no barrier exists, which gives the acoustic field Φ_g, necessary for depicting acoustic field loss caused by the insertion of barrier.

The calculating formula of Φ_g and each component of Φ come from Eq. (3.55), where P is the reflection coefficient (subscript C, A, and D respectively correspond to the reflection coefficients of ground in the following cases: non-barrier, barrier and sound source on the same side, barrier and receiver on the same side, and P_B is the reflection coefficient when barrier faces the sound source); ω is the numerical distance for corresponding cases. Therefore, the total acoustic field at receiver R is the superposition of all the contributions

$$\Phi = \Phi_{SR} + \Phi_{TR} + \Phi_{SX} + \Phi_{TX} \tag{3.65}$$

where Φ_{SR} is the direct diffraction field of sound source S at receiving point R, Φ_{TR} the diffraction field of image T at R, Φ_{SX} the diffraction field of S at image receiving point X, Φ_{TX} is the diffraction field of image source T at X. More strictly speaking, we should not ignore the contribution of image source U produced by the barrier (i.e., the image of S caused by barrier). Therefore, the right side of Eq. (3.65) should add two items: $\Phi_{UR} + \Phi_{UX}$, being the diffraction field of the image source U at R and X respectively. Furthermore, if R can be illuminated by S or T directly (meaning both S and T can directly connect R by a straight line, but not by two lines as discussed above), then the acoustic field caused by the geometrical spreading of S and T should be added into the total field.

With the estimation of each term above, we can obtain the attenuation spectra shown in Fig. 25. From the Figure, barriers can provide only very small shielding effect as expected in low frequencies.

---

14 Its practical value exists in the application of sound barriers unilaterally (or bilaterally) in highways, which becomes common in Japan, America and other countries.

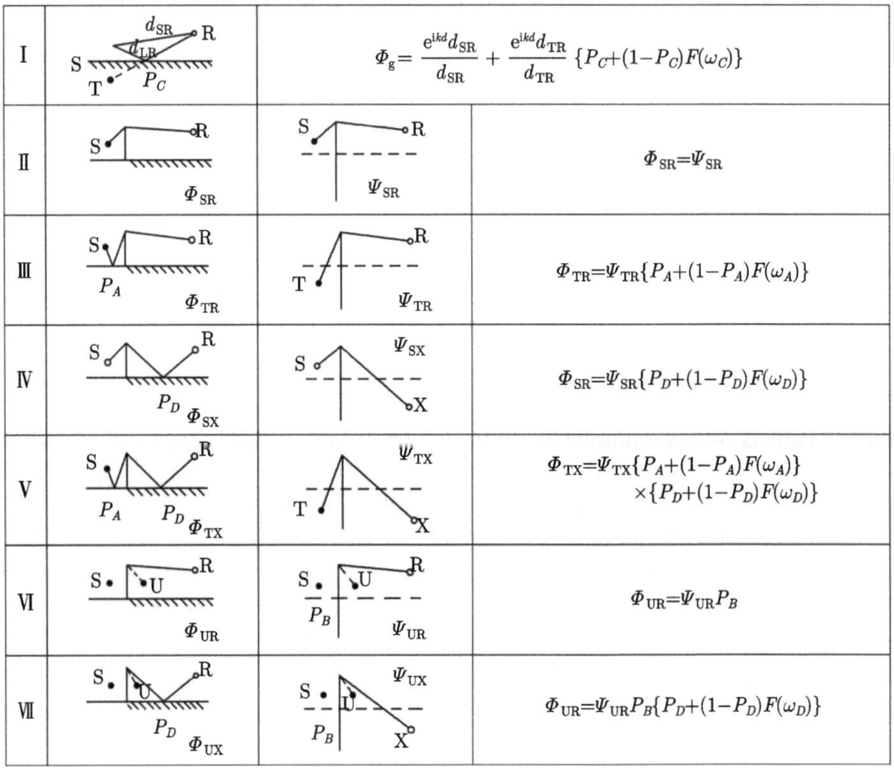

Fig. 24: All kinds of possible sound propagation paths (constituting each component of the total field) when source (S) and receiver (R) locate in different sides of the barrier

Natural barriers, such as hill, can be simulated by a wedge-shaped barrier or a wedge with a flat top (wide barrier). It can be observed again that the largest insertion loss happens in high frequencies. In actual applications, insertion loss near 5 kHz in Fig. 25 is very difficult to achieve, because refraction sound and scattering sound often fill in the notch caused by the interference and make sound loss decrease. Fig. 26 shows a typical scenario [90]: outside the asphalt road surface of a highway, we erected a 2.44 m high barrier, and the receiver is positioned 1 m above the ground on the other side of the barrier. Notice that at high frequencies where the largest insertion loss should happen, predicted by diffraction theory, the measured loss is much higher than the predicted one, illustrating that scattering reduces the shielding effect of barriers.

(b) Trees

Using plants (trees, bushes) to reduce noise is often desirable [91]. For sound with frequencies lower than 800 Hz, grown plants along the propagating paths are modified surface impedance formed by sedimentary organic matter. In this range, surface flow resistance usually decreases to lower than 100 cgs Rayl. Moreover, organic layer

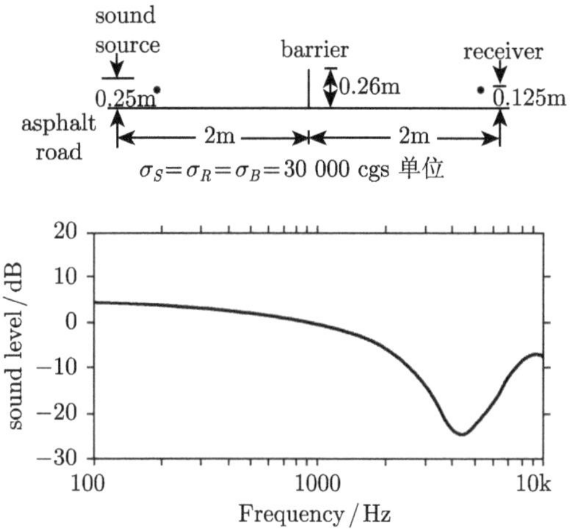

Fig. 25: Attenuated spectra predicted by diffraction theory for symmetrical structures of sound source, receiver, barrier and hard ground shown in the top graph.

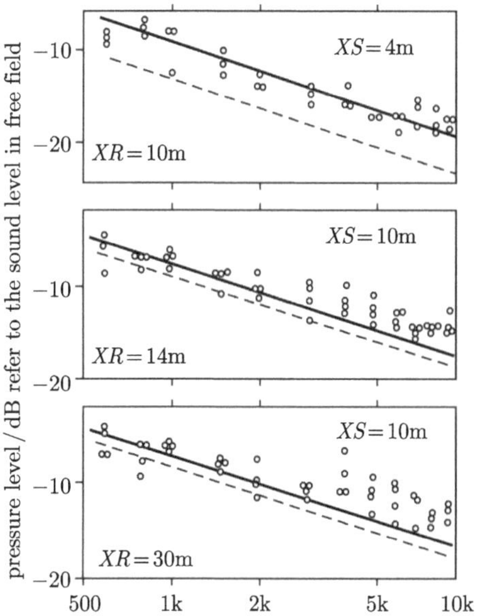

Fig. 26: Comparison of barrier "insertion loss" between theoretical (dotted line) and measured values *XS* and *XR* are the horizontal distances from sound source (highway noise) and receiver to the barrier, respectively.

requires using the two-layer model of surface impedances. In high frequency range, dense trees can create very strong scattering. However, even in high frequencies, the insertion loss produced by trees is usually very small. Fig. 27 shows the results

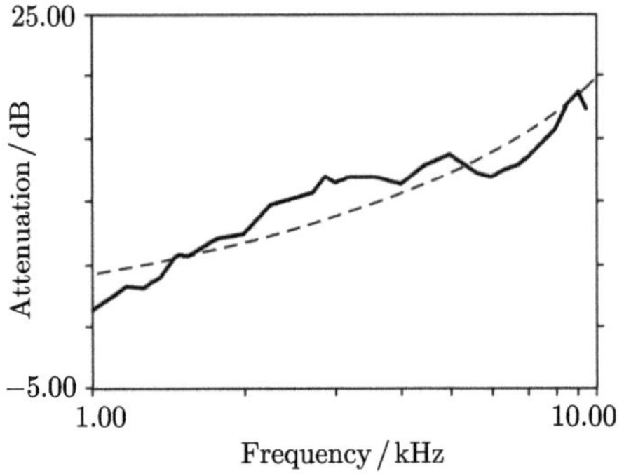

Fig. 27: Average attenuation measured in a coniferous forest (dB/24 m); Dotted line denotes theoretically calculated results considering the following two groups of scattering cylinder column arrays, (a) rigid, radius: 0.066 m, density 0.0724 m^{-2}; (b) non-rigid, radius: 0.001 m, density: 100 m^{-2}, surface flow resistance: $\sigma = 860\ g^{-1}cm^{-3}$.

obtained from a dense coniferous forest, where the measured and theoretical results are given in the meantime. The theoretical results are obtained by modeling the trees as two groups of equivalent, vertical and infinite long scattering cylindrical columns, and multi-scatterings are allowed. Two groups of layout are adopted to try to indicate that scattering is produced by both tree trunks and leaves.

Chapter 4
Sound Scattering and Diffraction in Atmosphere

Random non-homogeneity (non-uniformity) ranks second in the atmosphere non-homogeneity mentioned in Sec. 2.2 and Sec. 2.3 in Chap. 1. It mainly refers to two types of non-homogeneity: the non-homogeneity of atmospheric composition (such as rain, snow, fog, dust, etc.) and that of atmospheric motion (e.g. atmospheric turbulence). No matter which kind of non-homogeneity causes sound to scatter, it is obvious that scattering generated by turbulence is more frequent, and therefore more important.

The occurrence of turbulence is closely related to the ground, an objective existence, and turbulence exists only in the thermal-viscous boundary layer near the ground. Big vortices are formed due to instabilities within the layer, and furthermore, instabilities create bursts of vortices that occur at increasingly smaller scales, until energy dissipates into the viscous medium. Eventually very tiny vortices are formed. Therefore, a statistical distribution of vortices is always present in the atmosphere. The intensity of the turbulence depends on meteorological conditions. Generally, it is the strongest in summer afternoons and weakest at night with temperature inversion distributions.

There are two specific sources of turbulence: those arising from thermal sources and those related to the wind gradient. The former is caused by the heating and cooling of the earth's surface due to changes in solar radiation, which is called thermal force-driven turbulence. The latter type is caused by wind gradients created by the earth surface, which results in shear and the formation of turbulence vortices; this type is called wind force-driven turbulence. In most cases, wind-gradient caused turbulence is dominant.

The theoretical properties of the medium, such as the energy spectrum and coherence length of the turbulence, must be taken into account when considering sound propagation in a fluctuating (non-stable) medium. Both the amplitude and phase of the sound wave are fluctuating quantities associated with the medium's statistical properties. The turbulence can interact with other phenomena depending on its coherence (such as interference). In addition to causing a "new" phenomena such as scattering, for example, it can also reduce the acuity and depth of the interference minimum, and in unique cases, it can make them disappear completely.

The effect of atmospheric turbulence should be certainly become more distinct with an increase in propagating distance, and in general, it will be more significant with an increase in frequency.

4.1 Basic concepts of scattering

When a sound wave encounters an obstacle or an inhomogeneity in the medium along its propagation path, part of the wave will deviate from its original path and produce secondary waves in different directions, which is called scattering.[1]

The most distinct feature of many scattering phenomena (except cases that excite resonances) is that the scattering of high-frequency waves is much stronger than that of low-frequency waves. Tyndall and Rayleigh explained the color of the sky just based on this point: the light in the sky is scattered, and the scattering of blue light is much stronger than that of red light because the former is in a higher frequency range, while the latter is in a lower frequency range. Thus, the sky is blue. The concept of high-frequency and low-frequency depends on the relative relationship between the corresponding wavelength and the size of scatterer (including the obstacle and/or the non-homogeneity of the medium). When the size of the scatterer is much larger than the wavelength (common in the optical case, but rare in the acoustic case), half of the scattered waves spread evenly more or less along all the directions of the scatterer (called a reflected wave), while the other half are focused to the rear of the scatterer, generating a shadow with a clear edge that is similar to counteractive interference with the original undisturbed waves (interference waves). When the scatterer is much smaller than the wavelength (common in the acoustic case), all of the scattered waves spread in all directions without a clear-edge shadow. For the intermediate case, where the scatterer has the same size as the wavelength, a variety of wonderful interference phenomena can happen. Apparently, we are only interested in the latter two cases.

4.1.1 Scattering of fixed rigid object

The model of Rayleigh scattering (for low frequencies) describes the scattering produced by a plane wave with a constant frequency that is projected on a fixed, rigid object. Let the center of gravity of the object be located at the origin (see Fig. 28), and the unit vector of the plane wave along the marching direction be $\boldsymbol{e}_k$ (wave vector $\boldsymbol{k} = k\boldsymbol{e}_k$). The total sound pressure at a point in space can then be written as

$$\hat{p} = \hat{p}_{\mathrm{i}} \mathrm{e}^{\mathrm{i}\boldsymbol{k}\cdot\boldsymbol{x}} + \hat{p}_{\mathrm{sc}}(\boldsymbol{x}) \tag{4.1}$$

where $\hat{p}_{\mathrm{i}}$ is the amplitude of the incident wave p_{i}, and $\hat{p}_{\mathrm{sc}}$ is the complex amplitude of the scattered wave. The latter should satisfy the Helmholtz equation and the Sommerfeld radiation condition. The requirement of a rigid surface $\nabla\hat{p}\cdot\boldsymbol{n} = 0$ specifies that at the object's surface S (unit normal vector $\boldsymbol{n}$ points away from the object)

1 Sometimes the phenomenon that the propagation direction is disarranged by a rough surface is also called scattering. However, its more appropriate title should be diffuse reflection.

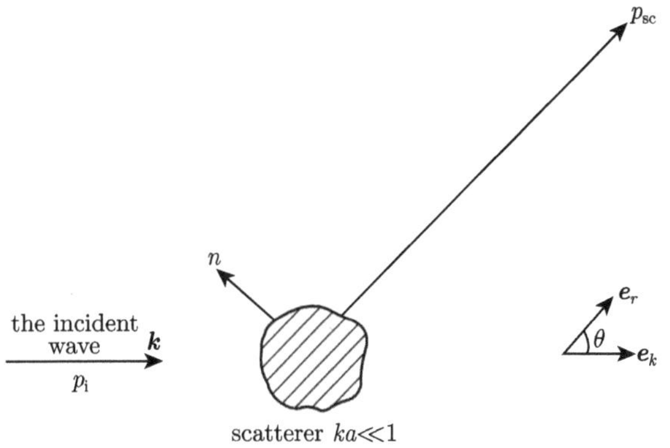

Fig. 28: A plane sound wave is scattered by an immovable rigid object whose size is much shorter than the wavelength (the incident wave; scatterer).

$$\nabla \hat{p}_{sc} \cdot \boldsymbol{n} = -i\hat{p}_i e^{i\boldsymbol{k}\cdot\boldsymbol{x}} \boldsymbol{k} \cdot \boldsymbol{n} \tag{4.2}$$

The expansion of the exponential function in the above equation accurate to the first order of k is

$$\hat{v}_{sc} \cdot \boldsymbol{n} = -\frac{\hat{p}_i}{\rho c} \boldsymbol{e}_k \cdot \boldsymbol{n} - i\frac{\hat{p}_i}{\rho c} (\boldsymbol{k} \cdot \boldsymbol{x}) \boldsymbol{e}_k \cdot \boldsymbol{n} \tag{4.3}$$

The first term corresponds to the back-and-forth motion of the rigid object that has a velocity amplitude of $-\hat{p}_i/\rho c$ along the direction parallel to $\boldsymbol{e}_k$, which creates dipole radiation independently. The second term creates monopole radiation. Although it is one order of ka smaller than the first term, the effect of two in the far field is of the same order of magnitude, because the radiation efficiency of the monopole is higher than that of the dipole.

According to the relevant radiation theory [56], the complex amplitudes of the monopole and dipole can be calculated independently, and then accordingly added to get the amplitude of the scattered wave

$$\hat{p}_{sc} = -\frac{k^2 \hat{p}_i}{4\pi} \left[V - \boldsymbol{e}_r \cdot \boldsymbol{M} \cdot \boldsymbol{e}_k \left(1 + \frac{i}{kr} \right) \right] \frac{e^{ikr}}{r} \tag{4.4}$$

where V is the total volume of the scatterer; $\boldsymbol{e}_r$ is the unit vector along the direction of radius vector r from the origin to the observation point in the range of R; $\boldsymbol{M}$ is the so-called magnetic polarization tensor [92], whose matrix expression depends on the geometric shape of the scatterer. For example, for a sphere and a thin disk with radius a, with the face being perpendicular to z axis, $\boldsymbol{M}$ is given respectively by the following

$$\frac{3}{2} V \begin{bmatrix} 1 & 0 & 0 \\ 0 & 1 & 0 \\ 0 & 0 & 1 \end{bmatrix} \text{ or } \frac{8}{3} a^3 \begin{bmatrix} 0 & 0 & 0 \\ 0 & 0 & 0 \\ 0 & 0 & 1 \end{bmatrix}$$

Hence, the amplitude of the corresponding scattered wave can be obtained from Eq. (4.4)

$$\hat{p}_{\mathrm{sc,sphere}} = -\frac{k^2\hat{p}_{\mathrm{i}}}{4\pi}\left(\frac{4}{3}\pi a^3\right)\left[1-\frac{3}{2}\cos\theta\left(1+\frac{\mathrm{i}}{kr}\right)\right]\frac{\mathrm{e}^{\mathrm{i}kr}}{r} \tag{4.4a}$$

$$\hat{p}_{\mathrm{sc,disk}} = \frac{k^2\hat{p}_{\mathrm{i}}}{4\pi}\frac{8}{3}a^3\cos\theta\left(1+\frac{\mathrm{i}}{kr}\right)\frac{\mathrm{e}^{\mathrm{i}kr}}{r} \tag{4.4b}$$

where $\cos\theta = \boldsymbol{e}_k \cdot \boldsymbol{e}_r$. Thus θ is the angle of scattering and incident direction. No monopole term exists in Eq. (4.4b), because the volume of the thin disk vanishes. It can be seen from Eq. (4.4) and the above two expressions that the amplitude of the far-field scattering is directly proportional to $k^2 a^3/r$

4.1.2 Scattering cross section

At a large distance r, the time-averaged intensity of the scattered wave, I_{sc}, is an asymptotic value of $\frac{1}{2}\left|\hat{p}_{\mathrm{sc}}\right|^2/\rho c$, which is directly proportional to the time-averaged incident intensity I_{i}, which decreases with r according to $1/r^2$, and usually depends on the direction from the scatterer to the position where the scattering pressure is measured. The specific value $r^2 I_{\mathrm{sc}}/I_{\mathrm{i}}$ denotes the scattered power per unit solid angle of unit incident intensity, which is called the differential cross section $\mathrm{d}\sigma/\mathrm{d}\Omega$. Integrating it over the entire solid angle yields the scattering cross section σ. It is known from the definition that the scattering cross section is the quotient of the power of incident wave scattered to the intensity of the incident wave (power per unit area), which actually is the part of the cross-sectional area of the incident wave being moved away by the scatterer. The scattering cross section pointing backwards to the sound source is called the back scattering cross section $\sigma_{\mathrm{back}} = 4\pi\mathrm{d}\sigma/\mathrm{d}\Omega$. For isotropic scatterers, $\mathrm{d}\sigma/\mathrm{d}\Omega$ is independent of direction, and it is equal to $\sigma/4\pi$. Therefore, the back scattering cross section σ_{back} is equal to the scattering cross section σ.

The target strength is closely related to the back scattering cross sectio, which is defined as

$$\mathrm{TS} = 10\lg\frac{\sigma_{\mathrm{back}}}{4\pi R_{\mathrm{ref}}^2}\,(\mathrm{dB}) \tag{4.5}$$

where the reference length R_{ref} was taken as 1 yard ≈ 0.914 m in earlier works, and is now taken as 1 m in current literature [93]. The ratio of the logarithmic argument can also be regarded as the differential cross section in the back scattering direction being divided by the reference differential section 1 m^2/solid radian. If the incident sound pressure level is L_{i}, and the sound pressure level of the back scattered wave at a distance of R_0 from the scatterer is $L_{\mathrm{back}}(R_0)$, the definition of target strength is

$$\mathrm{TS} = L_{\mathrm{back}}(R_0) + 10\lg\frac{R_0^2}{R_{\mathrm{ref}}^2} - L_{\mathrm{i}} \tag{4.5'}$$

Here the scattered wave is supposed to decrease with distance just like spherical spreading.

The expression of the differential cross section of the low-frequency scattering produced by a fixed rigid object can be derived from Eq. (4.4)

$$\frac{d\sigma}{d\Omega} = \frac{k^4}{4\pi}|V - \boldsymbol{e}_r \cdot \boldsymbol{M} \cdot \boldsymbol{e}_k|^2 \tag{4.6}$$

For the back scattering cross section, let $\boldsymbol{e}_r = -\boldsymbol{e}_k$, and further multiply it by 4π, we obtain

$$\sigma_{\text{back}} = \frac{k^4}{4\pi}|V + \boldsymbol{e}_r \cdot \boldsymbol{M} \cdot \boldsymbol{e}_k|^2 \tag{4.7}$$

The frequency dependence shown in the above two equations ($\sim f^4$) is also valid for the scattering cross section σ. The required integral of $d\sigma/d\Omega$ over the solid angle turns out to be simpler when taking the z axis along the direction of $\boldsymbol{M} \cdot \boldsymbol{e}_k$ (here $\boldsymbol{e}_r \cdot \boldsymbol{M} \cdot \boldsymbol{e}_k = |\boldsymbol{M} \cdot \boldsymbol{e}_k| \cos\theta$), because $\cos\theta$ is very small in the vicinity of $\theta = \pi/2$, making the integral of the cross term vanish. Therefore, the scattering sound power associated with the monopole and dipole can be superimposed. The remaining two integrals can be given by a simple expression. Because the average of $\cos^2\theta$ on the surface of a sphere is 1/3, and the total solid angle at a point is 4π, the final result is

$$\sigma = \frac{k^4}{4\pi}[V^2 + \frac{1}{3}(\boldsymbol{M} \cdot \boldsymbol{e}_k)^2] \tag{4.8}$$

4.2 Scattering due to non-homogeneity

4.2.1 Differential equation for scattering

Assume that the scatterer has a region R with acoustic properties being different to its neighbor medium. Also assume that the non-homogeneity is caused by a difference in density ρ and compression ratio κ. The non-homogeneity can then be described by the following two parameters (referring to Section 8.1.5 of [31]):

$$\gamma_\rho = \begin{cases} 1 - \dfrac{\rho_0}{\rho_e} \\ 0 \end{cases}; \gamma_\kappa = \begin{cases} \dfrac{\rho_0 c_0^2}{\rho_e c_e^2} - 1, & \text{within} \quad R \\ 0, & \text{beyond} \quad R \end{cases} \tag{4.9}$$

where subscript e denotes the quantity within the region of R, and subscript 0 denotes a quantity beyond the region of R. By only considering a sound wave with a constant frequency ω, the Helmholtz equation (2.14) can be transformed into

$$\nabla^2\hat{p} + k^2\hat{p} = -k^2\gamma_k\hat{p} - \nabla \cdot (\gamma_\rho \nabla\hat{p}) \tag{4.10}$$

where $k = \omega/c_0$. The two terms in the right-hand side are associated with monopole scattering and dipole scattering, respectively. The spatial scale characterizing the domain of the non-homogeneity, a, satisfies the conditions of $kac_0/c \gg 1$ and

$(ka)^2\rho_0/\rho \gg 1$ in the range of interest. As before, the incident sound pressure has a complex amplitude $\hat{p}_i e^{ik\cdot x}$, and hence $\hat{p} - \hat{p}_i e^{ik\cdot x}$ satisfies the Sommerfeld radiation conditions.

4.2.2 Integral equation for scattering

The same as when dealing with the radiation problem, application of the integral equation of the pressure wave is more effective than the use of differential equations with boundary conditions. The integral equation can be obtained by using the Green function method (refer to Sec. 7.1.7 of [31]). Regarding the right-hand side of Eq. (4.10) as the source function f, we first obtain an expression for $\hat{p}$ that is the sum of the volume integral of fG, and the surface integral of $\hat{p}$ and $G(x|x_s)$. For unbounded media, the Green function is $g = e^{ikr'}/4\pi r'$. The surface integral over the surface of an infinite sphere yields the incident wave p_i that comes from infinity and is scattered by the region R. Therefore, when the incident wave p_i is scattered by a single region R that has a different density and compression rate to its surrounding medium, the integral equation that corresponds to Eq. (4.10) becomes

$$\begin{aligned}\hat{p}(x) &= \hat{p}_i e^{ik\cdot x} + \iiint_R \left\{k^2\gamma_k(\boldsymbol{x}_s)\hat{p}(\boldsymbol{x}_s) + \nabla\cdot[\gamma_\rho(\boldsymbol{x}_s)\nabla_s\hat{p}(\boldsymbol{x}_s)]\right\}g(\boldsymbol{x}|\boldsymbol{x}_s)\mathrm{d}V_s \\ &= \hat{p}_i e^{ik\cdot x} + \iiint_R (k^2\gamma_\kappa\hat{p}g + \gamma_\rho\nabla_s\hat{p}\cdot\nabla_s g)\mathrm{d}V_s \end{aligned} \tag{4.11}$$

where

$$g(\boldsymbol{x}|\boldsymbol{x}_s) = \frac{1}{4\pi|\boldsymbol{x}-\boldsymbol{x}_s|}\exp(ik|\boldsymbol{x}-\boldsymbol{x}_s|)$$

Subscript s indicates quantities relating to the sound source. Hence, $\boldsymbol{x}_s$ denotes the position of the sound source ($\boldsymbol{x}$ denotes the position of the observing points (field point)), and ∇_s indicates the gradient over the coordinates of the sound source. The first term of the equation comes from the part of the non-homogeneity where the compression rate κ in R is different to the surrounding medium, which results in a monopole distribution with a source density of $(ik/\rho c)\gamma_\kappa(\boldsymbol{x})\hat{p}(\boldsymbol{x})$. The second term introduces dipole wavelets that are scattered by each $\mathrm{d}V_s$. Because the densities are different, region R does not respond to the force $\nabla\hat{p}$ in order to move at the same velocity with its surrounding medium. Thus it creates a dipole with a density of $(1/ik\rho c)\gamma_\rho(x)\nabla\hat{p}(\boldsymbol{x})$. Notice that both sources depend only on the real sound pressure field $\hat{p}$ in R, and not the incident sound pressure $\hat{p}_i$.

4.2.3 Asymptotic expression for scattered waves

Usually a scattered wave must be measured at positions far away from the scatterer in order to separate the scattered waves from the incident waves. Hence, it has special meaning to obtain the asymptotic expressions of scattered waves at positions far away from the scattered region $R(x \gg a$, a is the radius of the sphere surrounding $R)$. Here the Green function has the simple form

$$g(\boldsymbol{x}|\boldsymbol{x}_s) \to \frac{1}{4\pi r}\exp(\mathrm{i}kr - \mathrm{i}\boldsymbol{k}\cdot\boldsymbol{x}_s) \tag{4.12}$$

where $\boldsymbol{k}_s = k\boldsymbol{e}_r$. The asymptotic expression of the scattered wave can be obtained from Eq. (4.11)

$$\begin{aligned}\hat{p}_{sc} &\to \hat{p}_i\left(\frac{\mathrm{e}^{\mathrm{i}kr}}{r}\right)\Phi_s(\boldsymbol{k}_s)\\ \Phi_s(\boldsymbol{k}_s) &= \frac{k^2}{4\pi\hat{p}_\mathrm{i}}\iiint\limits_R\left[\gamma_\kappa(\boldsymbol{x}_s)\hat{p}(\boldsymbol{x}_s) - \mathrm{i}\gamma_\rho(\boldsymbol{x}_s)\boldsymbol{e}_r\cdot\boldsymbol{\nabla}_s\hat{p}(\boldsymbol{x}_s)\right]\mathrm{e}^{-\mathrm{i}\boldsymbol{k}_s\cdot\boldsymbol{x}_s}\mathrm{d}V\end{aligned} \tag{4.13}$$

Therefore, if the accurate form of the acoustic field inside R, $\hat{p}(\boldsymbol{x}_s)$, is known, the accurate form of the far-field scattered wave can be obtained according to the above equation. The amplitude depends on the wave vector $\boldsymbol{k}_s$, which points to the observing point.

Function $\Phi_s(\boldsymbol{k}_s)$ is named as the angular distribution factor of the scattered wave, or the scattered amplitude. Therefore, the intensity of the scattered wave at $r \gg a$ is $|\Phi_s|^2/\rho c r^2$. What is important is that Φ_s is the 3-D Fourier transformation of the following function for the special value of $\boldsymbol{K} = \boldsymbol{k}_s$.

$$\begin{aligned}\varphi_s(\boldsymbol{x}) &= \frac{2\pi^2k^2}{\hat{p}_\mathrm{i}}[\gamma_\kappa(\boldsymbol{x})\hat{p}(\boldsymbol{x}) - \mathrm{i}\gamma_\rho(\boldsymbol{x})\boldsymbol{e}_r\cdot\nabla\hat{p}(\boldsymbol{x})] = \iiint\Phi_s(\boldsymbol{K})\mathrm{e}^{\mathrm{i}\boldsymbol{K}\cdot x}\mathrm{d}V_K\\ \Phi_s(\boldsymbol{K}) &= \frac{1}{8\pi^3}\iiint\varphi_s(x)\mathrm{e}^{-\mathrm{i}\boldsymbol{K}\cdot\boldsymbol{x}}\mathrm{d}V_x\\ \mathrm{d}V_{\boldsymbol{K}} &= \mathrm{d}\boldsymbol{K}_x\mathrm{d}\boldsymbol{K}_y\mathrm{d}\boldsymbol{K}_z,\ \mathrm{d}V_x = \mathrm{d}x\mathrm{d}y\mathrm{d}z\end{aligned} \tag{4.14}$$

This property of the scattered amplitude is very useful when analyzing scattered waves.

4.2.4 Born approximation

It has been seen from the above discussion that, in the scattering region R, once the exact form of the pressure distribution is determined, all of the properties of the acoustic field outside of R, including the asymptotic behavior of the scattered wave, can be calculated. In fact, this is the same assumption what makes the Born approximation[2]

2 The approximation method stems from similar scattering problems in quantum mechanics. Refer to M. Born, *Quantum Mechanics of Collision Processes.* Z. Phys., 1926, 38, 803~827; J. Mathews and R. L. Walker. Mathematical Methods of Physics, New York: Benjamin. 1965: 289.

valid: the acoustic waves scattered by scattering region R have nothing to do with the waves scattered by other scatterers. The integral equation (4.11) can be solved exactly for some relatively simple-shaped scatters (such as a sphere, cylinder, disk, ellipsoid, etc.). Thus the exact formula of the sound scattering can be obtained. In other more complex cases, the integral equation can be solved approximately using the variation method or a successive approaching method. In the following we will discuss the latter method in detail.

In the integral of Eq. (4.11), replace the unknown $\hat{p}$ with the known function, $\hat{p}_i$ (incident wave) to obtain the first-order approximation

$$p(\boldsymbol{x}) \approx p_{\mathrm{i}}(\boldsymbol{x}) + p_1(\boldsymbol{x})$$
$$p_1(\boldsymbol{x}) = \iiint_R [k^2 \gamma_k(\boldsymbol{x}_s) p_{\mathrm{i}}(\boldsymbol{x}_s) g(\boldsymbol{x}|\boldsymbol{x}_s) + \gamma_\rho(\boldsymbol{x}_s) \nabla_s p_i(\boldsymbol{x}_s) \cdot \nabla_s g(\boldsymbol{x}|\boldsymbol{x}_s)] \mathrm{d}V \tag{4.15}$$

In the case when $p_1(\boldsymbol{x})$ is much less than $p_{\mathrm{i}}(\boldsymbol{x})$ in the whole region of R, the above equation yields an expression the satisfactorily describes the behavior of scattered waves. If the aforementioned condition does not hold, high-order terms should be added in the approximate series

$$p_1(\boldsymbol{x}) = p_{\mathrm{i}}(\boldsymbol{x}) + p_1(\boldsymbol{x}) + p_2(\boldsymbol{x}) + p_3(\boldsymbol{x}) + \cdots$$
$$p_{n+1} = \iiint_R [k^2 \gamma_\kappa(\boldsymbol{x}_s) p_n(\boldsymbol{x}_s) g(\boldsymbol{x}|\boldsymbol{x}_s) + \gamma_\rho(\boldsymbol{x}_s) \nabla_s p_n(\boldsymbol{x}_s) \cdot \nabla_s g(\boldsymbol{x}|\boldsymbol{x}_s)] \mathrm{d}V_s \tag{4.16}$$

Although the above series always converges, it usually converges slowly unless both γ_κ and γ_ρ are relative small (which means that ρ_e is relatively close to ρ_0, and c_e also close to c_0). At this time, the p_1 approximation is enough. When the Born approximation does not work, the variation method can be used. Of course, exact solutions can be obtained for specific possible scenarios.

When the two γs are small, and the Born approximation is effective, the calculation of scattered waves is very simple. If the incident wave is a plane wave $p_{\mathrm{i}}(\boldsymbol{x}) = \hat{p}_{\mathrm{i}} \mathrm{e}^{\mathrm{i}\boldsymbol{k}\cdot\boldsymbol{x}}$, the angular distribution function is approximated to

$$\begin{aligned}\Phi_b(\boldsymbol{\mu}) &= \frac{k^2}{4\pi} \iiint_R \left[\gamma_\kappa(\boldsymbol{x}_s) + \gamma_\rho(\boldsymbol{x}_s)\cos\theta\right] \mathrm{e}^{\mathrm{i}\boldsymbol{\mu}\cdot\boldsymbol{x}_s} \mathrm{d}V_s \\ &= 2\pi^2 k^2 \left[\Gamma_\kappa(\boldsymbol{\mu}) + \Gamma_\rho(\boldsymbol{\mu})\cos\theta\right]\end{aligned} \tag{4.17}$$

where $\boldsymbol{\mu} = \boldsymbol{k}_s - \boldsymbol{k}_i$, $\cos\theta = \boldsymbol{k}_i \cdot \boldsymbol{k}_s / k^2$ is the cosine of the cross angle between $\boldsymbol{k}_s$ and $\boldsymbol{k}_i$ (i.e., the angle between the incident wave and the radial vector $\boldsymbol{x}$ pointing to the observation point), and each Γ is the spatial Fourier transformation of each γ.

If R is a sphere with radius a, and the two γs are only related to the radial vector $\boldsymbol{x}_s$, the integral of $\mathrm{d}V_s$ with respect to the angular coordinates can be calculated. Therefore,

$$\Phi_b(\mu) = \frac{k^2}{\mu}\int_0^a \left[\gamma_\kappa(\boldsymbol{x}_s) + \gamma_\rho(\boldsymbol{x}_s)\cos\theta\right]\sin(\mu \boldsymbol{x}_s)\mathrm{d}\boldsymbol{x}_s$$

$$\to \frac{1}{3}k^2a^3\left(\left\langle\frac{\rho_0 c_0^2}{\rho_e c_e^2} - 1\right\rangle + \left\langle 1 - \frac{\rho_0}{\rho_e}\right\rangle\cos\theta\right), \quad \mu a \ll 1 \qquad (4.18)$$

where $\mu = |\boldsymbol{k}_s - \boldsymbol{k}_i| = 2k\sin(\theta/2)$. Hence, in the limit of long waves, the dependence of the dipole wave on ka has the same order of magnitude with the dependence of the monople on ka, where both are proportional to the average $\langle\gamma_\rho\rangle$ in R.

4.3 Interactions between atmospheric turbulences and acoustic waves

The non-homogeneity of the movement of the atmosphere itself (or turbulence) and acoustic waves are different forms of fluid movement. The former is a kind of random velocity field, while the latter is an ordered wave. When the two movements exist in the atmosphere simultaneously, complex interactions may occur. The description of this interaction depends on two approximations: firstly, the changes of densities due to acoustic waves is a reversible adiabatic process. That is, the entropy of each air particle (not molecular) or air mass remains constant in time. This adiabatic condition will hold as long as the sound wave frequencies are not nearly as high as the molecular collision frequency, or not as low as the infrasonic frequency when radiation absorption starts to act. Secondly, turbulence is an incompressible random flow. That is, the density of each particle is kept constant. In typical atmospheric conditions, turbulence approximately behaves as an incompressible fluid. However, because turbulence itself will produce random aerodynamic noise, this approximation is valid on the premise that either noise level to a certain minimum level is ignored or the turbulence intensity is restricted.

4.3.1 Separating acoustic waves from turbulence

To make a quantitative description of the interaction is a complicated process. Therefore, first of the above two approximations must be incorporated into the equations of motion, meaning that we must find a definite way to decompose particle velocities into the turbulence-associated part w and the acoustic wave-associated part u. For example, we can "freeze" the dynamic and thermodynamic states of the fluid at successive points in the time domain [94, 95], then impose a temporary additional pressure and "body force" (which make atmospheric particles or air masses be compressed or expanded adiabatically) at the points on the density $\rho_0(\boldsymbol{X})$ when acoustic waves and turbulence do not exist. Notice that the "entity coordinate" $\boldsymbol{X}$ is used to denote each of the atmospheric particles in order to differentiate it from the geometric coordinate of space point $\boldsymbol{x}$.

For every time t, the particle velocity after an adiabatic process is denoted as $\boldsymbol{w}(\boldsymbol{X})$, and kept unchanged, i.e., $\boldsymbol{w}(\boldsymbol{X}) = \boldsymbol{v}(\boldsymbol{X})$. Because the particles must move back and forth during the adiabatic process to obtain the expected density adjustments, there must be $\boldsymbol{w}(\boldsymbol{x}) \neq \boldsymbol{v}(\boldsymbol{x})$. This means that a new velocity field associated with an incompressible turbulence $\boldsymbol{w}(\boldsymbol{x})$ is defined. However, this definition is not precise – to eliminate this uncertainty, a third assumption is introduced. $\boldsymbol{w}(\boldsymbol{x}) = \boldsymbol{v}(\boldsymbol{X})$, however, particle displacement in an adiabatic processes requires $\nabla \boldsymbol{w}(\boldsymbol{X}) = \nabla \boldsymbol{v}(\boldsymbol{x})$. Except for an acoustic wave near the interface and infrasound waves, the vorticity associated with acoustic waves can be ignored. On the other hand, being a 3-D random velocity field, the turbulence must have a non-zero vorticity. Hence a third assumption is introduced: $\nabla \times \boldsymbol{w}(\boldsymbol{X}) = \nabla \times \boldsymbol{v}(\boldsymbol{X})$, which results in a single-value turbulence velocity field and a single-value acoustic particle velocity field that has $\boldsymbol{u}(\boldsymbol{x}) = \boldsymbol{v}(\boldsymbol{x}) - \boldsymbol{w}(\boldsymbol{x})$. Also notice that the previous deduction requires $\partial_t \boldsymbol{w}(\boldsymbol{X}, t) = \partial_t \boldsymbol{v}(\boldsymbol{X}, t)$, while $\partial_t \boldsymbol{w}(\boldsymbol{x}, t) \neq \partial_t \boldsymbol{v}(\boldsymbol{x}, t)$. Even with the new assumption, our deduction still has its domain of validity: the average density gradient in the atmosphere is non-zero, which means that as frequency drops to infrasound bands, the particle path should vary from line to ellipse gradually, which further means that the vorticity does not vanish. Even in audible-sound frequency bands, without a certain vorticity, it is generally not satisfied that the tangential velocity component of particle velocity is null on the solid surface.

4.3.2 Wave equation in turbulent atmosphere

The three assumptions applied previously are that the acoustic wave is adiabatic and irrational; the turbulence is incompressible; and the acoustic wave equation in a turbulent atmosphere can be derived. In order to simplify the study, some linearization assumptions should be made. Let A denote any normalized acoustic variables, and T any normalized turbulence variables. All terms with the same order as, or higher than, A^2 and AT^2 will be ignored. The separation of the acoustic wave from the turbulence-caused wave discussed in Sec. 4.3.1 means that turbulence variables with the same type as T and T^2 that satisfy their own equations can be removed from the acoustic wave equation. Nevertheless, the Mach number of the average wind $\bar{m} = \bar{\omega}/c_0$ needs special treatment because the value of $\bar{m}$ can approach 0.1 (maybe higher during a hurricane or a Chinook wind). Although the root-mean-square value of the Mach number of a turbulence velocity fluctuation $m'_{\mathrm{rms}} = \langle \omega'^2 \rangle^{1/2} / c_0$ rarely reaches 0.01, the major effects of an average wind can be considered using the "average substantial derivative" without any restriction to the results. That is, if the total velocity is written as $V_i = v_i + \bar{\omega}_i$, and the operator in (1.29) denotes the ordinary substantial derivative, the average substantial derivative is $\bar{\mathrm{d}}_t = \partial_t + \bar{\omega}_i \partial_i$. An exception is the appearance of a term related to an acoustic perturbation, a turbulence perturbation with an order of magnitude of m'_{rms}, and the Mach number of the average wind $\bar{m}$. However, the term can be neglected since it is still much smaller than the other terms.

Based on the above method to separate the turbulence and acoustic velocities, we can write v_i as $v_i = \omega'_i + u_i$, where u_i is the acoustic particle velocity. Similarly, the pressure can be expressed as $p = p_\omega + p_a$, where the subscript ω denotes that turbulence is a random velocity field (ω), and a denotes the acoustic field. Because of $p_\omega = \bar{p}_\omega + \overline{p'_\omega} = p_0 + p'_\omega$, and p'_ω is the second-order term of ω'_{rms}, we can make $p_\omega \approx p_0$ in any terms that include acoustic wave parameters. Similarly, since turbulence is assumed to be incompressible, for any turbulence density under uniform situations, we have $\rho_\omega = \rho_0$. Nevertheless, if the turbulence mixes up air sections with different temperatures, the density is not constant any more, nor it is uniform. If such an air mass is denoted using the sign σ ($\sigma = s/c_V$ represents the ratio of entropy densities s, and the specific heat at a constant volume c_V), the mixed turbulence density can be written as $\rho_\sigma = \rho_0 + \rho'_\sigma$, here ρ'_σ is a non-zero random turbulence variable. Besides the adiabatic condition $\mathrm{d}_t\sigma = 0$, if ρ_T is a "passive (no-source) additive term", that is, the turbulence density remains unchanged during its motion (which also assumes that the shift of air mass is not too large such that gravity creates an obvious change), we obtain an additional equation $\mathrm{d}_t\rho_\sigma = 0$. Because the characteristic time of turbulent movement is much shorter than that associated with heat flux, we can assume that the pressure of air masses with different densities ρ_σ will reached the pressure value of the turbulence field. Therefore, in the ideal gas scenario, any displacement d over time or space can be written as $\mathrm{d}\ln\rho_\sigma = -\mathrm{d}\ln c^2$. Here c is the random sound velocity associated with the turbulence field.

Based on the above symbols and assumptions, the equation of continuity (conservation of mass), of motion (momentum conservation equation), of adiabatic condition, as well as the restriction that ρ_σ (or c^2) is a passive addictive item, are taken as the following forms, respectively (comparison can be referred to the discussion in the fourth section of the first chapter)

$$\bar{\mathrm{d}}_t\upsilon + \partial_j(\rho v_j) = 0 \tag{4.19}$$

$$\bar{\mathrm{d}}_t(\rho v_i) + \partial_j(\rho v_i v_j) = -\partial_i p \tag{4.20}$$

$$\bar{\mathrm{d}}_t\sigma + v_j\partial_j\sigma = 0 \tag{4.21}$$

$$\bar{\mathrm{d}}_t\rho_\sigma + v_j\partial_j\rho_\sigma = 0 \tag{4.22}$$

Expanding Eq. (4.20) using the differential law of product and substituting $\bar{\mathrm{d}}_t\rho$ of Eq. (4.19) into it, Euler's equation can be obtained as per the following

$$\bar{\mathrm{d}}_t v_i + v_j\partial_j v_i = -\rho^{-1}\partial_i p \tag{4.23}$$

On the other hand, by taking the divergence of Eq. (4.20), and substituting $\partial_j(\rho v_j)$ into Eq. (4.19), we obtain

$$\partial_i\partial_j(\rho v_i v_j) + \nabla^2 p - \bar{\mathrm{d}}_t^2\rho = 0 \tag{4.24}$$

The difference between the turbulence and acoustic fields determines ω'_i and u_i. Nevertheless, the mechanism for determining ρ_σ is different. That is, ρ_σ is generated by the adiabatic motion among different positions in a gradient atmosphere. Or we can say that ρ_σ is a passive addictive terms for both turbulence motion represented by ω'_i and acoustic motion represented by u_i. Therefore, for a given gradient, ρ_σ, the time derivative $\bar{\mathrm{d}}_t\rho_\sigma$ required by Eq. (4.22) is split into two parts: $(\bar{\mathrm{d}}_t\rho_\sigma) = -\omega'_j\partial_j\rho_\sigma$ for the rate of change caused by the advection of ρ_σ introduced by the turbulence, and $(\bar{\mathrm{d}}_t\rho_\sigma)_a = -u_j\partial_j\rho_\sigma$ for the rate of change caused by the shaking back and forth when an acoustic wave passes by. As a result, the pure turbulence field equations that correspond to Eq. (4.23) and Eq. (4.24) are reduced to

$$\bar{\mathrm{d}}_t\omega'_i + \omega'_j\partial_j\omega'_i = -\rho_\sigma^{-1}\partial_i p_\omega \tag{4.23'}$$

$$\partial_i\partial_j(\rho_\sigma\omega'_i\omega'_j) + \nabla^2 p_\omega - \partial_i(\omega'_i\omega'_j\partial_j\rho_\sigma) + (\partial_i\rho_\sigma)\bar{\mathbf{d}}_t\omega'_i = 0 \tag{4.24'}$$

Because of the shaking effect, the acoustic wave equation is much simpler when using the ratio $\xi = \rho_a/\rho_\sigma$ as the argument than using ρ_a as the argument. Therefore, let $\rho = \rho_0(1+\xi)$, and expand the time differential quotient in Eq. (4.24) as

$$\bar{\mathrm{d}}_t^2\rho = (1+\xi)\bar{\mathrm{d}}_t^2\rho_\sigma + 2\bar{\mathrm{d}}_t\rho_\sigma\bar{\mathrm{d}}_t\xi + \rho_\sigma\bar{\mathrm{d}}_t^2\xi \tag{4.25}$$

We have $\bar{\mathrm{d}}_t^2\rho_\sigma = -\bar{\mathrm{d}}_t(u_i\partial_i\rho_\sigma) - \bar{\mathrm{d}}_t(\omega'\partial_i\rho_\sigma)$ from Eq. (4.22). When using the assumptions of A^2 and AT^2 mentioned above, the first term of Eq. (4.25) reduces to

$$(1+\xi)\bar{\mathrm{d}}_t^2\rho_\sigma = -\bar{\mathrm{d}}_t(u_i\partial_i\rho_\sigma) - \bar{\mathrm{d}}_t(\omega'_i\partial_i\rho_\sigma) \tag{4.26}$$

Based on the same assumption and Eq. (4.22), $\bar{\mathrm{d}}_t\rho_\sigma = -u_i\partial_j\rho_\sigma - \omega'_j\partial_j\rho_\sigma$, the second term on the right side reduces to 0. Furthermore, the first term on the right of Eq. (4.26) reduces to $-\bar{\mathrm{d}}_t(u_i\partial_i\rho_\sigma) = -(\bar{\mathrm{d}}_t u_i)(\partial_i\rho_\sigma)$, and the second term reduces to a term consisting of pure turbulence variables: $-\bar{\mathrm{d}}_t(\omega'_i\partial_i\rho_\sigma) = -(\bar{\mathrm{d}}_t\omega'_i)\partial_i\rho_\sigma + \omega'_i\partial_i(\omega'_i\partial_j\rho_\sigma)$. Substitute these results into Eq. (4.24), subtracting all of the pure turbulence quantities from Eq. (4.23′) and Eq. (4.24′), and reusing the assumptions of A^2 and AT^2, we obtain

$$2\rho_0\partial_i\partial_j(\omega'_i u_j) + \nabla^2 p_a + (\bar{\mathrm{d}}_t u_i)(\partial_i\rho_\sigma) - \rho_\sigma\bar{\mathrm{d}}_t^2\xi = 0 \tag{4.27}$$

Notice that the third term is already a term of AT type. Therefore, we can use Euler's equation (4.23) and thus all of those terms about T and $\boldsymbol{A}^2$ are canceled out to replace $\bar{\mathrm{d}}_t u_i = -\rho_0^{-1}\partial_i p_a$. Next, when considering to the magnitudes, we have $\rho_0^{-1}\partial_i\rho_\sigma = \partial_i \ln\rho_\sigma = -\partial_i \ln c^2$. Thus the third term of Eq. (4.27) becomes $(\partial_i \ln c^2)(\partial_i p_a)$.

Next, apply the adiabatic condition (4.21) to get rid of the hat-bar denoting average, when then becomes $\mathrm{d}_t\sigma = 0$. According to thermodynamics relations satisfied by ideal gases $T\mathrm{d}s = c_V\mathrm{d}T - (p/\rho^2)\mathrm{d}\rho$ and the definition of $\sigma = s/c_V$, $\mathrm{d}_t\sigma = \mathrm{d}_t\ln(p/p_0) - \gamma\mathrm{d}_t\ln(\rho/\rho_0)$ can be derived. Then, reuse $\rho = \rho_\sigma(1+\xi)$, $p = p_0 + p_a$ and Eq. (4.22), with hat-bar dropped, i.e, $\mathrm{d}_t\rho_\sigma = 0$. We obtain $\mathrm{d}_t\ln(1 + p_a/p_0) = \gamma\mathrm{d}_t\ln(1+\xi)$. Or after considering the assumption about $\boldsymbol{A}^2$, then we have $\mathrm{d}_t(p_a/p_0) = \gamma\mathrm{d}_t\xi$. Because we can always assume that the acoustic field does not exist at an earlier moment (it can be set at $t = -\infty$ if necessary), if we integrate the above equation along every air mass path, in principle, we obtain

$$p_a/p_0 = \gamma\xi = \gamma\rho_a/\rho_\sigma \tag{4.28}$$

By applying the above relation, we can write out the acoustic wave equation with either ξ or p_a. Substituting p_a in Eq. (4.28) into (4.27) (the third term has become $(\partial_i \ln c^2)(\partial_i p_a)$), and reusing the relation $c^2 = \gamma p_0/\rho_\sigma$, we obtain

$$2\partial_i\partial_j(\omega_i' u_j) + \partial_i(c^2\partial_i\xi) - \bar{\text{d}}_t^2\xi = 0 \tag{4.29}$$

Noticing that the first term is also a type of AT. Therefore, Eq. (4.23) can be used again in which all terms about A^2 and T then cancel out. According to Eq. (4.28), we can write the reduced form of (4.23) as $\bar{\text{d}}_t u_i = -c_0^2\partial_i\xi$. However, before using this equation, all of the parameters need to be transformed into a time derivative by applying a Fourier time transformation F_t, meaning that the same transformation is needed for Eq. (4.29). For example, $P = P(k) = \text{F}_t\xi$ (having only a constant difference with $F_t p_a$), $U_i = F_t u_i$, $N' = F_t n'$ (here, $n' = n - 1 = (c/c_0) - 1$), $M_i' = F_t\omega_t'/c_0$, $\bar{M}_i = \bar{m}_i\delta(k)$ and $c_0^{-1}F_t\bar{d}_t = -\text{i}k + \bar{m}_i\partial_i$, then the reduced form of Eq. (4.23) becomes

$$U_i/c_0 = -\text{i}k^{-1}\partial_i P - \text{i}k^{-1}\bar{M}_j\partial_j U_i/c_0 \tag{4.30}$$

We can further eliminate U_i from the right of the above equation by using the iterative method. Nevertheless, notice that the hidden in the first term on the right-hand side of Eq. (4.30) are all small factors $\bar{M}$, M' and P, and should be given in the results after the substitution of the Fourier transformation of the first term in (4.29). As mentioned at the end of the first paragraph of this section, these terms are still one order of magnitude smaller than the other terms. Hence, they can be ignored. Because in general ω_i', u_i, c^2 and ξ all are functions of time, we apply a Fourier transform to Eq. (4.29) to transform all of the products of the functions related to time into a convolution of k. The one-dimensional convolution denoted by "$\star$" is defined as

$$\begin{aligned} F_t\left\{f(t)g(t)\right\} &= F(k) \star G(k) = \int_{-\infty}^{\infty} \text{d}k' F(k-k')G(k') \\ F_k^{\dagger}\left\{F(k)G(k)\right\} &= f(t) \star g(t) = \int \text{d}t' f(t-t')g(t') \end{aligned} \tag{4.31}$$

And the 3-D convolution is denoted with a "O", and defined as

$$\begin{aligned} F_x^{\dagger}\left\{f(\mathbf{x})g(\mathbf{x})\right\} &= F(\boldsymbol{\kappa}) \otimes G(\boldsymbol{\kappa}) = \int_{-\infty}^{\infty} \text{d}^3\boldsymbol{\kappa}' F(\boldsymbol{\kappa}-\mathbf{k}')G(\mathbf{k}') \\ F_x^{\dagger}\left\{F(\boldsymbol{\kappa})G(\boldsymbol{\kappa})\right\} &= f(\mathbf{x}) \otimes g(\mathbf{x}) = \int_{-\infty}^{\infty} \text{d}^3\mathbf{x}' f(\mathbf{x}-\mathbf{x}')g(\mathbf{x}') \end{aligned} \tag{4.32}$$

Furthermore, the assumption of AT^2 means that N^{-2} in the second term should be approximated by $1 - 2N'$. Applying all of the above transformations into Eq. (4.29) and arranging the term orders, the acoustic wave equation in a turbulent atmosphere is finally created

$$\nabla^2 P + (k + \mathrm{i}M_j\partial_j)^2 P = 2\partial_j(N' \star \partial_j P) + 2\mathrm{i}\partial_j\partial_h[M' \star (k^{-1}\partial_h P)] \tag{4.33}$$

Because this equation plays a particularly important role in acoustic remote sensing in the atmosphere (refer to Chapter 8), it is also called the "acoustic remote sensing equation".

On the assumption of the so-called "frozen turbulence", where the average wind velocity is 0 and the change rate of turbulence is much smaller than the acoustic wave frequency, Eq. (4.33) is simplified to

$$\nabla^2 P + k^2 P = 2\partial_j(N' \star \partial_j P) + 2ik^{-1}\partial_j\partial_h(M_j'\partial_h P) \tag{4.33a}$$

We denote the acoustic field when turbulence does not exist as P_0. On the right-hand side of the above equation, we replaced P with P_0 which is in terms of the refractive index fluctuation N' and the velocity fluctuation M_i'. Then, the normal unit vector $\boldsymbol{e} = \boldsymbol{e}_i$ of initial wave P_0 is not random, and we can express this "effective refractive index fluctuation" in the form of

$$N_e' = N' - \boldsymbol{e} \cdot \boldsymbol{M}' \tag{4.34}$$

Finally, the acoustic wave equation under the Born approximation is given

$$\nabla^2 P_1 + k^2 P_1 = 2\partial_i(N_e'\partial_i P_0) \tag{4.33a'}$$

where the acoustic field is decomposed into an initial field and a scattered field $P = P_0 + P_1$. By applying the angle brackets to denote the total average, we then have $P_1 = \langle P_1 \rangle + P_1'$ and $\langle P_1' \rangle = 0$. It must be emphasized that, only under the Born assumption does $\langle P_1 \rangle = 0$, and therefore $P_1 = P'$.

4.3.3 Interaction mechanisms between turbulence and acoustic waves

The terms "propagation" and "scattering" have various meanings. For example, the special case of a wave propagating along a direction with a smooth refractive index gradient is called "coherent forward scattering". However, a stricter definition is used for scattering (in fact a more proper name should be "simple scattering"). Assume that there is an initial wave P_0, and choose a direction where P_0 vanishes, or time and space is limited in the propagating direction, and an acoustic wave P_1, if found propagating in this region, is named as the "scattered wave" generated by the interaction between P_0 and scattering volume V_s of the scatter. The much more complicated process that P_0 and P_1 propagate in the same region and interact with each other is simply called "propagation". Because the scattering spreads a part of the original energy flow along one direction into many directions, the intensity of the scattered wave at a scattering angle θ_S is always much smaller than the original intensity. Therefore, we can often deal with scattering by using the Born approximation or the "single scattering" approximation, which assumes that, if part of a wave is scattered by turbulence, it will not be scattered for most future times, and thus it arrives at the receiver directly. This approximation does not mean that there is only one scatterer or there is only a single interaction center.

The right-hand side of Eq. (4.33) clearly shows that the physical process of the interaction between an acoustic wave and turbulence depends on the fluctuation of the refractive index N', which in turn depends on the fluctuation of the temperature T', vapor pressure fluctuation e', and the fluctuation of the turbulence Mach vector $M' = \omega'/c_0$. Firstly, let us only consider the temperature fluctuation T'. For homogeneous turbulence, the interaction with a temperature fluctuation can be ignored because the temperature fluctuation T' is proportional to the fluctuation of the turbulence pressure, where the latter is a second-order quantity of M'. However, a common situation occurring in the atmosphere is that the existence of an average temperature gradient in space results in a quite different conclusion. If turbulence and a special temperature gradient exist in the same time, the turbulence will mix particles with different positions and temperatures together. Thus, a turbulence mixture with a normalized velocity fluctuation can be the same magnitude as the normalized temperature fluctuation. Note that what is required here is a positional temperature gradient rather than a temperature gradient, and the same must be noticed that, only a small turbulence is completely capable of having a position-temperature gradient. On the other hand, even if the average position-temperature is very small, there may still be turbulence caused by wind shear.

When an acoustic wave passes through a random fluctuation zone of a refractive index, the different point systems on the wave front will march forward with different random velocities. Applying Huygens principle to the resultant irregular wave front shows that, part of the energy flows will deviate from their original directions to many random and forward directions. Whereas, when the refractive index gradient is big enough, backwards propagation and reflection can occur. For an actual planar discontinuity ΔN, the classical Fresnel equation gives a power reflection coefficient $\Gamma^2 = \Gamma_0^2 = |\Delta N/2|^2$. For a smooth refractive index profile, the power reflection coefficient can be obtained from the integral of the profile. For example, for a profile layer with thickness l, Γ^2 is about 70% of Γ_0^2 when the ratio of the thickness and wavelength $l/\lambda < 0.1$, while Γ^2 drops to 1% of Γ_0^2 directly when $l/\lambda > 0.5$. Therefore, if a turbulence has a large enough random refractive index gradient, the acoustic wave will be scattered not only in the forward scattering angle, but also in the backward hemisphere. However, a feature of the interaction between sound and turbulence is that, whether arising from the fluctuation of N or the fluctuation of M, scattering never occurs at $\theta_s = 90°$. This feature results from the appearance of N' and M' themselves and their differential quotients in the acoustic wave equation (4.33).

At the beginning of this chapter, when we discussed the separation of turbulent motion from acoustic wave motion, the incompressible property of the air applies only for the slowly pushing of air, whereas, there will be a sound generation for rapid pushing, such that a push produced by a transducer or an explosion source. This property means that slow motion cannot couple with rapid motion, which can only "move" them. Thus, turbulence motion generates a random turbulence velocity fluctuating vector superposed on the wave phase velocity. When the wave goes into the turbulence far enough, so that the random wave vector will have in the forward hemisphere, the component of the wave vector of the plane vertical to the initial propagating direction is capable of having the same magnitude as the turbulence velocity fluctuation, and will thus cause scattering in the backwards direction. But there is an exception, because velocity is always merely the vector sum of the wave and turbulence velocities, the only component along the purely backwards scattering direction ($\theta_s = 180°$) is the turbulence velocity fluctuations, which do not consist of propagating waves whose average oscillation frequency is nonzero. Therefore, there is no back-scattering wave from the velocity fluctuation. Actually, only fluctuations in the refractive index can generate scattering waves.

From the following comparison, a further insight can be obtained regarding the difference between scattering caused by a refractive index fluctuation and that by a velocity fluctuation. As mentioned above, a suitably strong uniform refractive index gradient can generate an incomplete reflection, such as a back-scattering wave, like a incomplete mirror. Assuming that we can introduce a uniform laminar flow, and that there is a velocity gradient in the flow direction, there is not any change in the refractive index. Hence, when the acoustic wave propagates along the flow direction, only the wave front's acceleration or deceleration due to the passage of an average flow can be expected. However, the generation of back-scattering waves or reflected waves cannot be expected, because a Galilean transformation can be performed at any point to obtain a local coordination system moving in a average flow velocity at each point. Naturally, when we change a uniform gradient into a random gradient, the involved physical process remains unchanged (although the diffractive effect should be added).

Unlike a mirror, it is impossible for the atmosphere to maintain an actual discontinuity of N' and M'. The two types of fluctuations keep a smooth variation and correlate with each other at a considerable distance. As a rule, we use l_0 (inner scale) to represent the minimum distance at which turbulence keeps inertial-zone characteristics, and L_0 (outer scale) to denote the maximum distance at which the correlation of isotropic turbulence cannot be ignored. Measurements show that l_0 varies from several millimeters to about 1 centimeter, and that L_0 from about 10 m to dozens of meters. However, the correlation can be extended to a few hundred meters in anisotropy turbulence. Therefore, although turbulence is random, the nonzero correlation distance means that the interaction of scattering, or the scattering generated by a point scattering field, is completely different to a random distribution. As long as the wavelength λ is not longer than L_0, then the scattered waves from adjacent areas will be able to not only interfere constructively, but also destructively, to generate a part of coherent scattered waves. Thus, the interaction itself can be described as a "coherent scattered process", which is caused by the integral effect of many small scatterings in a nonzero spatial area.

Coherent wave and coherent scattering processes are different, which implies that for a typical acoustic wave wavelength, the intensity of coherent scattering is much larger than that of incoherent scattering (such as Rayleigh scattering), and that for back-scattering, the fluctuation of refraction index can be treated as a group of parallel partially reflecting mirrors with a spacing of $\lambda/2$. Or speaking more generally, for any scattering angle, the turbulence can be considered as a regular lattice with a pitch that satisfies the Bragg condition $l = (\lambda/2)/\sin(\theta_s/2)$, as long as $(\lambda/2) \leqslant L_0$. We can apply a Fourier space transformation to decompose the random turbulence field, and obtain the Fourier components that satisfy the Bragg condition and generates coherent scatterings. Owing to this, a Fourier transformation and the fluctuated 3-D spatial frequency spectrum play very important roles in the theoretical analysis of the interaction between an acoustic wave and turbulence.

By definition, sound power is the acoustic energy flow rate per unit cross-sectional area $\boldsymbol{S}(\boldsymbol{x})$. A power spectrum or power spectrum density $\boldsymbol{I}(\boldsymbol{x}, k)$ is the energy flow quantity per unit area or per unit frequency pitch. Since $p(\boldsymbol{x}, t)$ is the force per unit area, and $\boldsymbol{u}(\boldsymbol{x}, t)$ is the displacement per unit time, so $p(\boldsymbol{x}, t)\boldsymbol{u}(\boldsymbol{x}, t)$ is the instantaneous energy flow per unit area in the direction of $\boldsymbol{u}$. The energy flow density we are concerned with is the time domain average of $p(\boldsymbol{x}, t)\boldsymbol{u}(\boldsymbol{x}, t)$ over one cycle for the propagation of a continuous wave with a single frequency through a nonrandom medium. However, this definition is invalid for multi-frequency acoustic waves, and for acoustic waves interacting with a random velocity fluctuation in a turbulent atmosphere. Therefore, as a replacement, we consider the propagation and scattering of sound pulses with a length of l_p, and one of the two "gating" types of signal receiving is required to be applied.

Regarding propagation, we are interested in the "time gating" which is an on/off switch receiver with a period of l_p/c_0 around the moment T, when pulse passes through. Regarding scattering, we are also only interested in similar measurements. For example, when we consider backwards scattering, the scattered signal of a pulse with a length of l_p and a space range of R, comes from a area having a thickness of $l_p/2$. Hence, $l_p/2$ is the minimum spatial resolution, and the corresponding area generates a scattered signal with a length of l_p. Therefore, for scattering, we are interested in the "distance gating", which means the average of the received signal over the period l_p/c_0 in the vicinity of moment $2R/c_0$. This average gives all of the information able to be obtained from the atmospheric layer that is a distance R away and has a thickness of $l_p/2$. Therefore, for propagation and scattering scenarios, the sound power we are interested is the time-domain average of the instantaneous power over the pulse length (the average of total pulse) l_p at moment T or $2R/c_0$. The acoustic energy flow density is

$$\mathbf{S}(\boldsymbol{x}) = \frac{c_0}{l_p} \int_{-l_p/2}^{l_p/2} \mathrm{d}t \left\langle p(x,t) u(x,t) \right\rangle \tag{4.35}$$

Where $\langle\rangle$ denotes the overall average.

Expressing $p(t)$ and $\boldsymbol{u}(t)$ by their Fourier transforms $P(k)$ and $\boldsymbol{U}(k) = \mathrm{i}k^{-1}\nabla P(k)$, and assuming that there are many wavelengths in a pulse length (actually at least 10), the above equation can be written as

$$\mathbf{S}(\boldsymbol{x}) = (2\pi/l_p) \int_{-\infty}^{\infty} \mathrm{d}k \left\langle P^{\dagger}(k) \left[-\mathrm{i}k^{-1} \nabla P(k) \right] \right\rangle \tag{4.36}$$

Moreover, if $P(k)$ is not assumed as being too wide, the following approximation can be made: $-\mathrm{i}k^{-1}\nabla P(k) \approx \boldsymbol{e}P(k)$, where $\boldsymbol{e}$ is the unit vector along the $\boldsymbol{S}$ direction. Then

$$\mathbf{S}(\boldsymbol{x}) = \boldsymbol{e}(2\pi/l_p) \int_{-\infty}^{\infty} \mathrm{d}k \left\langle P(k) P^{\dagger}(k) \right\rangle \tag{4.36a}$$

Or, by defining $\mathbf{S}(\boldsymbol{x}, \theta_s) = \boldsymbol{e} \cdot \mathbf{S}(\boldsymbol{x})$, where θ_s denotes the direction of §, the scalar form can be written as

$$\mathrm{S}(\boldsymbol{x}, \theta_s) = (2\pi/l_p) \int_{-\infty}^{\infty} \mathrm{d}k \left\langle P(k) P^{\dagger}(k) \right\rangle \tag{4.36a'}$$

In statistical theory, the "power spectrum density" of a function is defined as the Fourier transformation of the auto covariance of the function. This definition can be fully modified to illustrate the "non-ergodic process", which may be an unstable process.[3] This method can be applied to a pulse series. In essence, it means that the meaningless interval time between pulses can be ignored [96]. This method includes two average steps, the first of which is taking the average of the product of two acoustic pressures with a time delay τ over each pulse time, which generates a random function ensemble only about τ. The starting time of each pulse is reduced to a label of each unit in the ensemble; then, we perform a Fourier transform to this two-fold integral from τ to k, and thus obtain the power spectrum density. Applying the "intermittent" auto covariance of the sound pressure, we can get the statistical power spectrum density of the sound pressure

$$I(k) = (2\pi/l_p) \left\langle P(k) P^{\dagger}(k) \right\rangle \tag{4.37}$$

Comparing the above equation with Eq. (4.36a′), it can be seen that the total power (per unit area) is just the integral of $I(k)$ over all of the wavenumbers

3 Refer to Jenkins G. M, Watls D. G. *Spectral Analysis and its Applications*. Holden-Day, 1968; or Koopmans L. H., *The Spectral Analysis of Time Series*. Academic, 1974.

$$S(\boldsymbol{x}) = \int_{-\infty}^{\infty} \mathrm{d}kI(k) \tag{4.38}$$

The above expression demonstrates the relation between the frequency spectrum and energy flow, and it also shows that $I(k)$ is the density of $S(\boldsymbol{x})$ in the frequency domain. In fact, the statistical power spectrum density of the sound pressure is the energy flow rate per unit frequency spacing and per unit area, i.e., the acoustic power spectrum density. For a continuous wave with a wavenumber of k_0, and a non-widened frequency spectrum, Eq. (4.37) is reduced to $I(\boldsymbol{x}, k) = S(\boldsymbol{x})\delta(k - k_0)$, where δ denotes the Dirac function.

4.4 Sound scattering in turbulent atmosphere

4.4.1 Scattering cross section

The acoustic wave scattering caused by a fluctuation of N and M is an important physical basis for atmospheric remote-sensing technology (refer to Chapter 8). A measurement of the characteristics of the received signal and the comparison of it with the known characteristics of the emitted signal provides all of the information able to be obtained from the acoustic method inside the scattering volume V_S at a distance R. Such a comparison must be made using the following steps: firstly, the scattered energy flow density should be compared with that of the incident wave on V_S; secondly, the variation of the flow density needs to be determined. The variation is caused by the diffraction expansion which results from the waves travelling in and out of the transmitting and receiving antennae, the molecular absorption, and the average motion effects. Finally, the relation between the acoustic signal and the electric signal needs to be resolved. Let Ω be the solid angle in the direction θ_S where the energy flow is scattered, and l_S is the volume thickness of the scatterer at the incidence direction, so we obtain $V_S = \Omega R^2 l_S$. Hence, the scattering cross section is defined as the ratio of the scattered energy flow density per unit scatterer volume thickness in unit solid angle $S(\boldsymbol{x}, \theta_S)$ over the incident energy flow density $S_0(\boldsymbol{x})$. Thus, we can compare the incident and scattering radiation in the scattering volume. Using the above expression about V_S, the scattering cross section is expressed as

$$\sigma_S(\theta_S) = R^2 S(\theta_S)/(V_S S_0) \tag{4.39}$$

Note that, for a plane wave $S_0(\boldsymbol{x}) = (2\rho_0 c_0)^{-1} A_0^2(\boldsymbol{x})$. Here, A_0 is the amplitude of P_0.

With the application of the incompressible condition into the relation between wave scattering and the 3-D frequency spectrum of the turbulence, and by applying the Born approximation to the acoustic wave equation (4.33a′), we can obtain a strict solution to $\sigma_S(\theta_S)$. The Green function G in the wave equation satisfies the fundamental equation

$$\nabla^2 G + k_0^2 G = \delta_{(\boldsymbol{x})} \tag{4.40}$$

where $\delta(\boldsymbol{x})$ is the space Dirac function. The well-known solution of the above equation is

$$G(\boldsymbol{x}, k_0) = (-\exp ik_0R)/(4\pi R) \tag{4.41}$$

where $R = |\boldsymbol{x}|$. The Born solution for the scenario of "frozen turbulence" (4.33a′) expressed by G is

$$P_1(\boldsymbol{x}, k_0) = 2G \odot \nabla \cdot (N_e' \nabla P_0) \tag{4.42}$$

where the 3-D spatial convolution of $\boldsymbol{x}$ is defined by Eq. (4.32).

Applying the above equation to (4.36a′) and (4.39), the scattering cross section for inertial zone is given by

$$\begin{aligned}\sigma_s(\theta_s) = 1.52k_0^{1/3}\cos^2\theta_s\{0.13C_n^2 \\ + \cos^2(\theta_s/2)C_v^2/(4c_0^2)\}[2\sin(\theta_s/2)]^{-11/3}\end{aligned} \tag{4.43}$$

where C_n^2 depends on the fluctuating intensity of the temperature and water vapor described by Eq. (1.6). It can be seen that the scattering cross section is 0 when $\theta_s = 90°$, and only the fluctuation of N contributes when $\theta_s = 180°$. We notice that, far-field (the region is defined by $R \geqslant 2l^2/\lambda$) scattering is never the case in the vicinity of $\theta_s \to 0$, because the size of the turbulence vortices contribute to the scattering length l, which satisfies $l = (\lambda/2)/\sin(\theta_s/2)$ and $l \to \infty$ when $\theta_s \to 0$. Especially, Eq. (4.43) cannot be used in propagation any more. When the average wind is no longer 0, $\sigma_s(\theta_s)$ is expected to have some change, even if the frozen turbulence condition still holds in the coordinate system moving with the wind. Its main effect is the position where the scattering cross section being 0 shifts from being a small angle to $\theta_s = 90°$ [97], which comes from the rotation of the incident and scattered wave vectors. The typical example of this effect is given in Fig. 29, which shows that very big fluctuations may occur in the region near $\theta_s = 90$. The existence of an average wind only makes σ_s change a little for a back-scatter or a near back-scatter.

4.4.2 Power ratio

The ratio of the received power over the transmitted power is a very important parameter. Analogeous to the "Radar Equation" of electromagnetic waves, it can be expressed as the so-called "SODAR (or echo sound detector) Equation".[4]

For a transmitted wave with a very narrow angular width, the diffraction loss of the incident acoustic energy flux density in the whole emission beam path length R_0 can be very small. Nevertheless, scattered sound always suffers a loss, which contributes

4 The derivation of this equation (and other related ones), as well as further discussion regarding acoustic scattering will be carried out with specific acoustic remote sensing systems in the atmosphere in the second section of Chapter 8.

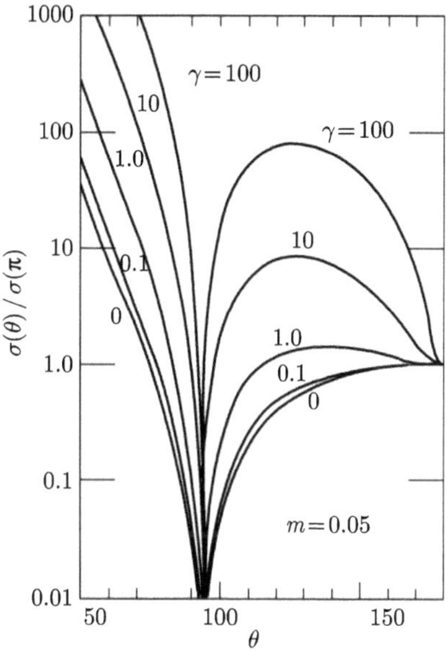

Fig. 29: The typical variation of the scattering cross section caused by an average wind with a velocity of 15 m/s, $\gamma = C_V^2 T_0^2 / C_T^2 c_0^2$.

to the factor R_S^{-2} in the far field. Of course, both the transmitted and the received beams should be attenuated due to molecular absorption in their paths. When the scattering volume deviates from the transmitter, the received energy rate of this volume will decrease; inversely, when the volume moves towards the emitter, the received energy rate will increase [98]. There exists a similar effect between the scattering volume and receiver. Denoting the unit vector along the direction of transmitted wave as $\boldsymbol{e}_0$, the initial energy flux density will be the original value multiplied by a factor $(1 + \boldsymbol{e}_0 \cdot \bar{\boldsymbol{M}})^{-1}$ when the wave moves towards the scattering volume. Expressing the unit vector along the scattering direction as $\boldsymbol{e}_S$, the scattering energy flux density will be the original value multiplied by a factor $(1 - \boldsymbol{e}_S \cdot \bar{\boldsymbol{M}})^{-1}$, as the wave moves towards the receiver. In the meantime, the pulse length elongates owing to the average wind, which makes the scattering volume related to the pulse length and beam width on the ground be also changed to its original value by being multiplied by $(1 + \boldsymbol{e}_0 \cdot \bar{\boldsymbol{M}})^{-1}$. Therefore, when comparing the transmitted power with the received power, only a factor of $(1 - \boldsymbol{e}_S \cdot \bar{\boldsymbol{M}})^{-1} \approx 1 + \boldsymbol{e}_S \cdot \bar{\boldsymbol{M}}$ is left. Although the energy flux rate changes, no energy loss or increase occurs. Because only linear sound propagation is considered, the total energy only expands according to the time-varying distance between the ground and the scattering volume. For the energy reflection on moving rigid mirror surfaces, the change of the energy flux rate should be two times of that of the case of scattering on V_S, which strictly obeys the Doppler frequency shift law. Here we noticed that when $\boldsymbol{e}_0 \cdot \bar{\boldsymbol{M}}$ is positive, the expansion of V_S introduced by winds compensates the decrease of transmitted power caused by the same wind, and only has the effect of the reduction of the radiation flux scattered by the receiver.

When considerating the specific equipment used in atmospheric probing via scattering,[5] let $\mathcal{P}_R$ denote the received electrical power, ε_R is the acoustic electric conversion efficiency, $\mathcal{P}_T$ is the transmitting power, ε_T is the electric acoustic conversion efficiency, $\hat{a}$ is the effective attenuation coefficient of sound intensity ($\mathrm{N_p/m}$), A_R/R_S^2 is the solid angle of the antenna aperture A_R at a distance of R_S, A_T is the transmitter area, A_c is the common area shared by the transmitting and receiving beams, and g is the directivity gain factor of the antenna. We can express the equation of the acoustic detector as follows[6]

$$\mathcal{P}_R/\varepsilon_R = \varepsilon_T\mathcal{P}_T \exp[-\hat{a}(R_0 + R_s)](1 + \boldsymbol{e}_s \cdot \bar{\boldsymbol{M}})A(\theta_s)(l_p/2)(gA_R/R_s^2)\sigma(\theta_s) \quad (4.44)$$

The "condition factor" of the unit of $A(\theta_s)$ will be discussed in detail with the involvement of Eq. (8.30) in Chapter 8. It is not difficult to verify directly that Eq. (4.44) is only another form of the definition of σ_s in Eq. (4.39), which can be proved by substituting in the expression showing the relationship between the energy flow fluxes at transmitter, receiver and scattering volume V_s (Notice that, $\boldsymbol{e}_s = -\boldsymbol{e}_0$ for backwards backscattering). Eq. (4.44) specifies the relationship between theory and experiment, as well as the relationship between the detector's electric power measurements and the values of C_n^2 and C_v^2 (which appear in the theoretical relation between wave scattering and turbulence).

4.4.3 Power spectra

Early literatures studied the relation between the scattering angle θ_s and the power spectrum $I(k)$ generated only by the scattering of a velocity fluctuation [99]. The basic assumption is that, because scattering depends on vortices with magnitudes of $\lambda/2$, such as $\lambda \ll L_0$ (L_0 is a certain objective scale), an approximate separation can be performed, which separates large vortices that shift and push small vertices into small scattering vortices. This condition is called the "local freezing" hypothesis. By making a Galilean transformation to the coordinate system by translating the velocity locally, the spectrum of turbulence in a locally isotropic, uniform inertial region can be applied. For example, if $\boldsymbol{K}$ is the wave scattering vector given by $K(\boldsymbol{e}_0 - \boldsymbol{e}_s)(K = 2k_0 \sin(\theta_s/2)$, Brag condition), $\boldsymbol{M}_L$ is the Mach vector of large-scale vortices, and τ is

---

5 To be discussed in Chapter 8.

6 This equation and Eq. (8.30) in Chapter 8 are only different in forms as long as we notice that $\mathcal{P}_R/g\varepsilon_R A_R = I_{\mathrm{sac,ap}}$, $\varepsilon_T\mathcal{P}_T = (4\pi r^2 I_i)_0\Delta\Omega_{\mathrm{tr}}/4\pi$, $R_s = \bar{R}$, $\sigma_s = \eta/4\pi$, $l_p = c\tau$. It should also be noticed that the background wind is not considered in those relations, nor is the attenuation along the ray (from the transmitter to the scattering volume and then from the scattering volume to the receiver). Forms such as Eq. (4.44) are convenient in the discussion here, and it is beneficial to get familiar with the two different forms of the equations.

the time delay of the correlation introduced by the large scale motion, the turbulence spectrum that varies with time is related to the local fixed spectrum by the following "local freezing turbulence spectrum"

$$\Phi_{ij}(\boldsymbol{K}, \tau) = \langle \exp(\mathrm{i}\boldsymbol{K} \cdot \boldsymbol{M}_L c_0 \tau)\rangle_V \Phi_{ij}(\boldsymbol{K}) \tag{4.45}$$

Where $\langle\rangle$ with subscript V denotes the volume average of large scale motion.

$I_{(k)}$, the scattering spectrum due to velocity fluctuation, can be obtained by using Eq. (4.45). When using the Born approximation assumption and an effective refractive index N_e', the result can be generalized to the situation in the existence of both a velocity and a refractive index fluctuation simultaneously. Hence, $I(k)$ can be written in the following form [94]

$$I(k - k_0) = S(\boldsymbol{x}, k)\mathcal{F}_T\langle \exp(\mathrm{i}\boldsymbol{K} \cdot \boldsymbol{M}_L c_0 \tau)\rangle_V \tag{4.46}$$

Assuming that the component of $\boldsymbol{M}_L$ has a Gaussian distribution, the equation $\langle \exp x\rangle = \exp x \exp(1/2\langle x'^2\rangle)$ can be used. Here, $\boldsymbol{x} = \bar{\boldsymbol{x}} + \boldsymbol{x}'$, and $\bar{\boldsymbol{x}} = \langle \boldsymbol{x}\rangle$. Therefore, Eq. (4.46) becomes

$$I(k) = S(\boldsymbol{x})[2\pi\langle(\boldsymbol{K} \cdot \boldsymbol{M}_L')^2\rangle]^{-1/2} \exp[-(k - k_0 + \boldsymbol{K} \cdot \bar{\boldsymbol{M}}_L)^2/2\langle(\boldsymbol{K} \cdot \boldsymbol{M}_L')^2\rangle] \tag{4.47}$$

When a sound wave's wave front passes through or is scattered by turbulence, the wave fronts have irregularities caused by random impacts, and when those irregularities drift along the path, their phases will change with time, and the instantaneous frequency $\mathrm{d}_t\varphi$ will fluctuate. Therefore, the observed signal frequency spectrum will be widened. The widen range depends on the size of the averaged transverse wind. For example, if we let $\boldsymbol{M}_L' = 0$ in Eq. (4.45) or (4.47), $I(k)$ will not degenerate into the widened frequency spectrum caused only by average wind. This step makes the Gaussian function in Eq. (4.47) be reduced to the Dirac function. So we can assume $I(k) = S(\boldsymbol{x})\delta(k - k_0 + \boldsymbol{K}\cdot\bar{\boldsymbol{M}})$, i.e., $I(k)$ is a non-broadened frequency spectrum, while only wind causes Doppler frequency drift. This contradiction is because of our omitting the phase fluctuation in local frozen processing in Eq. (4.45).

A more cautious process should count both the phase fluctuation and use the local frozen hypothesis to determine the effects on the scattering frequency spectrum. In the literature [100], only the phase fluctuation and frozen turbulence hypothesis (Taylor hypothesis) are combined to determine $I(k)$, where a finite scattering volume V_s was considered in the study, which was determined via a finite pulse length l_p and a half-beam angle width of the limited receiver ψ. Because of the limited beam width, not all of the scattering elements are at $\theta_s = 180°$. Therefore, both the refractive index fluctuation and the velocity fluctuation contribute to the frequency spectrum. A pulse with a limited length has a nonzero spectral width, and the finiteness of V_s also widens the frequency spectrum (but only to a small degree). Therefore, this study considered all broadening mechanisms rather than just the pushing and moving of the larger vortices to scattering vortices.

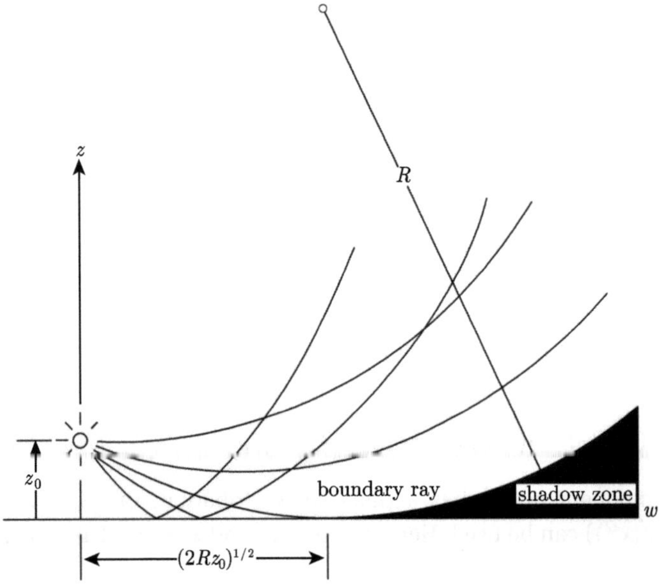

Fig. 30: The shadow zone formed in an atmosphere with a velocity profile that decreases linearly with height.

4.5 Sound diffraction in quiescent atmosphere

In general terms, “from scattering to diffraction” is a process from a quantitative to a qualitative change: when the size of a scatterer is much larger than the acoustic wavelength, we usually say that the sound is reflected and diffracted, rather than being scattered. However, even though the actual mechanisms and effects of scattering and diffraction are the same, there exist great differences in the relative magnitudes. Thus, it seems a qualitative difference exists between them. However, in atmospheric acoustics, diffraction mainly refers to the phenomenon that acoustic waves penetrate into the shadow zone.

In some profiles of the atmosphere, acoustic rays starting from an acoustic source (direct, reflected, refracted) cannot reach a certain area, which forms a shadow zone. An example is shown in Fig. 30, where the amplitude of the sound pressure is so small that it nearly disappears in the shadow zone. However, the amplitude is not decreased to zero discontinuously from the “bright zone (the area that acoustic rays can reach)” to the shadow region. It oscillates with increasing amplitude in the vicinity of the value, reaches its maximum near the front of the shadow edge, then decreases monotonically, and tends to approach 0 at quite a deep distance into the shadow zone. Next we will study the essence of this phenomenon by using a special scenario.

Assume that the sound velocity $c_{(z)}$ decreases linearly with height, and the sound source is located at height z_0 near the ground (Fig. 30). In Section 3.1 of this Chapter, for such cases that the refraction rays were found to be circular arcs that curve upward with radii of curvature $R = c/|c'(z)$ (as usual, here and later the prime denotes the derivative with respect to the variable z). Among those rays there is a "boundary ray", which is tangent to the ground and forms the edge of the shadow zone. It is easy to see from a simple geometrical relation that, if the given horizontal distance w is much smaller than R, then the shadow zone is composed of points that satisfy $w > (2Rz_0)^{1/2} + (2R_z)^{1/2}$. In the following we will calculate the acoustic field in the shadow zone for many circumstances with w being much larger than the wavelength. Because geometric acoustics are no longer valid, we will start with the wave method directly.

4.5.1 Point source above a locally-reacting surface

Assume that the point source is a monopole with an amplitude $\hat{S}$, which radiates an acoustic wave with angular frequency ω. With the omission of a density variation, the complex amplitude of the sound pressure satisfies the active Helmholtz equation – the source term $-4\pi\hat{S}\delta(x)\delta(y)\delta(z-z_0)$ is added to the right end of Eq. (2.11) (where δ is the Dirac function) and $\omega^2/c^2(z)$ is substituted for k^2.

To obtain the solutions, we use the following expressions for the dual Fourier transformation for x and y as the starting point

$$\hat{p} = \frac{\hat{S}}{\pi} \lim_{\varepsilon \to 0} \int_{-\infty}^{\infty} e^{-\varepsilon^2(\alpha^2+\beta^2)} e^{i\alpha x} e^{i\beta y} Z(z, \alpha, \beta) d\alpha d\beta \tag{4.48}$$

If function Z satisfies the following equation, Eq. (4.48) will satisfy the inhomogeneous Helmholtz equation (active).

$$Z'' + \left[\frac{\omega^2}{c^2(z)} - k^2\right] Z = \delta(z - z_0) \tag{4.49}$$

where $k^2 = \alpha^2 + \beta^2$.

Because the acoustic field is cylindrically symmetric, it will not change the equation when we substitute 0 for y in Eq. (4.48) and let $x = w$. Replacing the integral variable of k by $\theta(\alpha = k\cos\theta, \beta = k\sin\theta)$, we can complete the integration with respect to θ. According to Eq. (4.49), it is believed that Z is independent of θ. The integral of the exponential expression $\exp(ikw\cos\theta)$ with respect to θ from 0 to 2π happens to be the zeroth-order Bessel function[7]

$$\int_0^{2\pi} e^{ikw\cos\theta} d\theta = 2\pi J_0(kw)$$

7 Refer to G. N. Watson, A Treatise on the Thery of Bessel Functions, 2nd ed, Cambridge Univ., London, Press, 1966: 24~25; 328~338.

Therefore, we obtain

$$\hat{p} = -\hat{S} \lim_{\varepsilon \to 0} \int_0^\infty e^{-\varepsilon^2 k^2} 2J_0(kw)Z(z,k)k\mathrm{d}k \tag{4.50}$$

As was pointed out at the beginning of this section, we are only interested in large values of kw. Hence, we can use the corresponding asymptotic expressions of the Bessel function

$$J_0(\eta) \approx \left(\frac{1}{2\pi}\right)^{1/2} e^{-i\pi/4} \left[\frac{1}{\eta^{1/2}} e^{i\eta} - \frac{1}{(-\eta)^{1/2}} e^{-i\eta}\right] \tag{4.51}$$

For $\eta > 0$, $(-\eta)^{1/2}$ can be expressed as $e^{i\pi/2}\eta^{1/2}$. Therefore, by considering $Z(z, k)$ as an even function of k, Eq. (4.50) can be rewritten as

$$\hat{p} \approx -\left(\frac{2}{\pi\omega}\right)^{1/2} \hat{S} e^{-i\pi/4} \int_{-\infty}^{\infty} k^{1/2} e^{ikw} Z(z,k)\mathrm{d}k \tag{4.50'}$$

where for $k < 0$, $k^{1/2}$ equals $e^{i\pi/2}|k|^{1/2}$. This integral can now be regarded as a loop integral, where $k^{1/2} = |k|^{1/2} \exp(i\varphi_k/2)$ and the value of the phase of φ^k is restricted between $-\pi/2$ and $3\pi/2$. The convergence factor $\exp(-\varepsilon^2 k^2)$ in (4.50) now can be omitted, because even if the integral converges very slowly, we can always change the integral loop to be away from the real axis, which makes e^{ikw} always approach 0 exponentially when $|k|$ goes to ∞ from each end of the loop.

Function $Z(z, k)$, which satisfies Eq. (4.49), can be regarded as the two solutions, $\Psi(z, k)$ and $\Phi(z, k)$, of the homogeneous equations in the case that k is positive. The two solutions satisfy the upper boundary conditions, are consistent with the Sommerfeld radiation condition, as well as the lower boundary condition at $z = 0$, respectively. The upper boundary condition that must be specified is that Ψ must disappear exponentially, or else appear as an oblique upward propagating wave for real k. The lower boundary condition that corresponds to a local reaction interface with a specific acoustic impedance of Z_s, i.e.[8]

$$\Phi' + i\frac{k_0 \rho c}{Z_s}\Phi = 0, \quad z = 0 \tag{4.52}$$

where $k_0 = \omega/c(0)$. For complex k, both Ψ and Φ should be understood analytically except on the branch line.

The solution $Z(z, k)$ of the non-homogeneous equation (4.49) is $A\Psi(z, k)$ for $z > z_0$ and $B\Phi(z, k)$ for $z < z_0$. Here, the constants A and B make Z continuous at z_0. But its slope has a discontinuity with a value being 1. Therefore, we have

$$Z(z,k) = \frac{\Psi(z>,k)\Phi(z<,k)}{[\Psi'\Phi - \Phi'\Psi]_{z_0}} \tag{4.53}$$

8 As pointed out in section 3.1 of Chapter 2, the corresponding boundary condition of a rigid interface is $\Phi' = 0$ (because $Z_s \to \infty$), and of a pressure release interface, is $\Phi = 0$.

where $z<$ and $z>$ represent the smaller and the larger values of z_0 and z, respectively. Because both Ψ and Φ satisfy the homogeneous differential equation (4.49) (the right end of the equation vanishes), the denominator of Eq. (4.53) (the Wronski determinant about Ψ and Φ) is thus independent of z_0. So Eq. (4.52) can be rearranged as

$$(\Psi'\Phi - \Phi'\Psi)_{z_0} = \left(\Psi' + \frac{ik_0\rho c}{Z_s}\Psi\right)_0 \Phi(0, k) \tag{4.54}$$

In the confined range of an acoustic field near surfaces (low level), it can be assumed that $c(z)$ decreases with height infinitely. This idealized assumption may make it possible to predict whether the undetermined function, Ψ, satisfies the boundary conditions from the behavior at appropriately small values of Z. The differential equation can be transformed approximately by replacing $1/c^2(z)$ with $[1/c^2(0)](1 + 2z/R)$, where $R = C(0)/|c'(0)$ is the curvature radius of a ray emitted from the sound source horizontally. In this approximation, the homogeneous equation becomes

$$\Psi'' + \left(k_0^2 + k^2 + \frac{2k_0^2 z}{R}\right)\Psi = 0 \tag{4.55}$$

4.5.2 Sound field expressions in the shadow zone

It is not hard to see that Eq. (4.55) belongs to the same type as Eq. (2.51) being studied. Thus, it can be also be solved by applying the Airy function. We first introduce the following parameters

$$\tilde{\tau} = (k^2 + k_0^2)l^2, \quad y = \frac{z}{l}, \quad l = \left(\frac{R}{2k_0^2}\right)^{1/3} \tag{4.56}$$

Then, by rearranging Eq. (4.55), we know that one of the possible solutions is $Ai(\tilde{\tau} - y)$, while $Ai((\tilde{\tau} - y)e^{i2\pi/3})$ and $Ai((\tilde{\tau} - y)e^{-i2\pi/3})$ are the other solutions. Notice that there are only two linearly independent solutions, and combinations of the two solutions when multiplied by any constant are also solutions. After studying various solutions of similar problems, the following two solutions are recommended[9]

$$v(\eta) = -\pi^{1/2} Ai(\eta), \quad \omega_1(\eta) = 2\pi^{1/2} e^{i\pi/6} Ai(\eta^{i2\pi/3}) \tag{4.57}$$

here we have $\eta = \tilde{\tau} - y$.

The reason for choosing $\omega_1(\eta)$ since it has an asymptotic behavior

$$y \to \infty : \omega_1(\tilde{\tau} - y) \to \frac{e^{i\pi/4}}{y^{1/4}} e^{i(2/3)y^{3/2}} e^{-i\tilde{\tau}y^{1/2}} \tag{4.58}$$

9 Refer to Fock V. A. *Electromagnetic Diffraction and Propagation Problem*. London, Pergamon. 1965: 237, 379~381.

This property can be derived from Eq. (2.53a) and Eq. (2.53b), which represents an oblique, upward propagating wave. Therefore $\omega_1(\tilde{\tau} - y)$ is an appropriate solution of $\psi(z, k)$.

As the function $\Phi(z, k)$ satisfies Eq. (4.52), it can take the following form

$$\Phi(z, k) = v(\tilde{\tau} - y) - \frac{v'(\tilde{\tau}) - qv(\tilde{\tau})}{w_1'(\tilde{\tau}) - qw_1(\tilde{\tau})} w_1(\tilde{\tau} - y) \tag{4.59}$$

where $q = ik_0 l\rho c/Z_s$ and the prime denotes the derivative of variables. Substituting these results into Eq. (4.53), we find

$$Z(z, k) = \frac{w_1(\tau' - y >)\Phi(z <, k)l}{v'(\tilde{\tau})w(\tilde{\tau}) - w_1'(\tilde{\tau})v(\tilde{\tau})} = -w_1(\tau' - y >)\Phi(z <, k)l \tag{4.60}$$

A property of the Airy differential equation was used in the second step: the Wronski determinant about the two solutions $v'w_1 - w_1'v$ is a constant, whose value can be -1 by substituting the asymptotic expressions into the determinant.

For receiving points close to surfaces (low height), the main contribution to the integral of (4.50′) comes from those values of k not vastly different to k_0 in the shadow zone and nearby. This is just what is expected, because the integral can be regarded as the superposition of a plane wave and an evanescent wave, and waves propagating with a horizontal phase velocity of the order of $c(0) = \omega/k_0$ dominate at larger horizontal distances near to the ground surface. Therefore, we initially allowed the approximation be consistent with this expectation; and the derived results support this assumption in the end. In particular, we replaced the multiple factor of integrand $k^{1/2}$ with $k_0^{1/2}$, and of $k^2 - k_0^2$ in the expression of $\tilde{\tau}$ in (4.56) with $2k_0(k - k_0)$. Hence, we have $\tilde{\tau} \to \tau = (2k_0 l)^2(k - k_0)$.

Transform the integral variables in (4.50′) into τ, the amplitude of the complex pressure can be reduced as per the following standard expression

$$\hat{p} = \frac{\hat{S}}{w} e^{ik_0 w} V(\xi, y_0, y, q) \tag{4.61}$$

where

$$V(\xi, y_0, y, q) = e^{i\pi/4} \left(\frac{\xi}{\pi}\right)^{1/2} \int_{-\infty}^{\infty} e^{i\xi\tau} w_1(\tau - y >) \cdot \left[v(\tau - y <) - \frac{v'(\tau) - qv(\tau)}{w_1'(\tau) - qw_1(\tau)} w_1(\tau - y <)\right] d\tau \tag{4.62}$$

Which is the Van der Pol-Bremmer diffraction formula in the form of Φ_{0K}, where $\xi \equiv w/2k_0 l^2 = (k_0 R/2)^{1/3} w/R$, $y_0 \equiv z_0/l(2k_0^2 R^2)^{1/3}(z_0 R)$.

4.5.3 Series expansion of diffraction formula

A conclusion can be drawn from the definition in Eq. (4.57) and from the asymptotic equations (2.53a) and (2.53b): when τ goes to ∞ in the upper half plane

$\mathrm{Im}\tau > 0$, the integrand in Eq. (4.62) goes to 0. If $\xi - y_0^{1/2} - y^{1/2} > 0$ (equivalent to $w > (2Rz_0)^{1/2} + (2Rz)^{1/2}$, meaning that the observer is in the shadow zone – refer to Fig. 30). In this case the integral can be done by using the method of loop deformation. That is, the integral is equal to the sum of the residues of all of the poles in the upper half plane multiplied by $2\pi\mathrm{i}$. These poles are just the zeros of the denominator of the integrand $w_1'(\tau) - qw_1(\tau)$, and $\tau_n(n = 1, 2, \cdots)$. The denominator can be approximately expressed as $[w''_1(\tau_n) - qw'_1(\tau_n)](\tau - \tau_n)$ in the vicinity of $\tau = \tau_n$, or $(\tau_n - q^2) \cdot w_1(\tau_n)(\tau - \tau_n)$, as $w'_1(\tau_n) = qw_1(\tau_n)$, and $w''_1(\tau) = \tau w_1(\tau)$ are derived by Eq. (2.51). Moreover, the Wronski being -1 and the definition of $w'_1(\tau_n)$ used above gives $v'(\tau_n) - qv(\tau_n) = -1/w(\tau_n)$. Therefore, the series expression of the residues of V becomes

$$V = (\xi, y_0, y, q) = (4\pi\xi)^{1/2} \mathrm{e}^{\mathrm{i}\pi/4} \sum_n \frac{\mathrm{e}^{\mathrm{i}\tau_n \xi} w_1(\tau_n - y_0) w_1(\tau_n - y)}{(\tau_n - q^2)[w_1(\tau_n)]^2} \tag{4.63}$$

Substituting w_1 with w'_1/q, the denominator of the above equation can be rewritten as $[(\tau_n/q^2) - 1][w'_1(\tau_n)]^2$. The former form is suitable for the restricted case of $q \to 0$, $Z_s \to \infty$, which corresponds to rigid ground. The latter form is suitable for the restricted case of $q \to \infty$, $Z_s \to 0$, which corresponds to the pressure release interface.

Because most of the interesting practical situations can be approximated by one of these two cases, and because the zeroes of the Airy function $Ai(\eta)$ and its derivative $Ai'(\eta)$ all are real numbers, we replace τ_n with $b_n \mathrm{e}^{-\mathrm{i}2\pi/3}$.[10] For rigid interfaces, b_n is just a'_n, and the root of $Ai'(a'_n) = 0$. For pressure release interfaces, b_n is just a_n, and the root of $Ai(a_n) = 0$. These conclusions can be verified by the requirement that Eq. (4.57) and τ_n should satisfy $w'_1(\tau_n) - qw_1(\tau_n) = 0$. Because a'_n and a_n are all negative, each of the corresponding τ_n should be located on the line with the phase of τ being $\pi/3$ in the first quadrant of τ-complex plane. Therefore, the imaginary part of each τ_n gradually increases with n in succession. If ξ is large enough, and after y and y_0 are given, the summation formula (4.63) can be approximated by the first term. Based on this, the following equations are given for rigid and pressure release interfaces respectively.

$$V(\xi, y_0, y, 0) \approx (4\pi\xi)^{1/2} \mathrm{e}^{-\mathrm{i}\pi/12} \exp(\mathrm{i}a'_1 \xi \mathrm{e}^{-\mathrm{i}2\pi/3}) \frac{f_1(y_0) f_1(y)}{(-a'_1)} \tag{4.64a}$$

$$V(\xi, y_0, y, \infty) \approx (4\pi\xi)^{1/2} \mathrm{e}^{\mathrm{i}\pi/12} \exp(\mathrm{i}a_1 \xi \mathrm{e}^{-\mathrm{i}2\pi/3}) g_1(y_0) g_1(y) \tag{4.64b}$$

---

10 Since b_n is the root of $Ai'(b) + \mathrm{i}\mathrm{e}^{\mathrm{i}\pi/3}(\rho c/Z_s) k_0 l Ai(b) = 0$. So for $|\rho c/Z_s| k_0 l \ll 1$ (close to rigid interfaces), $b_n \approx a'_n + \mathrm{e}^{-\mathrm{i}\pi/6}(\rho c/Z_s) k_0 l / a'_n$, while for $|\rho c/Z_s| k_0 l \gg 1$ (close to soft interfaces), $b_n \approx a_n + \mathrm{e}^{\mathrm{i}\pi/6} Z_s/\rho c k_0 l$. As $k_0 l = (k_0 R/2)^{1/3}$ increases with frequency, so any interface with limited impedance will be close to soft interfaces when the frequency is enough high. For example, when $\nu = 1000$ Hz, $c = 340$ m/s, and the ground impedance is $Z_s = 5\rho c(1 + \mathrm{i})$, $|\rho c/Z_s| k_0 l$ will be $0.30 R^{1/3}$ (R is the curvature radius in the units of meters). Therefore, it is proper to treat the boundary of atmospheric profiles with values of R larger than 10 km as a pressure released boundary. The following paper can be referred to: R. Onyeonwu. *Diffraction of sonic boom past the nominal edge of the corridor*, J. Acoust. Soc. Am., 1975, 58: 326~330.

where an apostrophe is adopted,

$$f_1(y) = \frac{Ai(a_1' - ye^{i2\pi/3})}{Ai(a_1')} = \frac{w_1(a_1'e^{-i2\pi/3} - y)}{2\pi^{1/2}e^{i\pi/6}Ai(a_1')} \tag{4.65a}$$

$$g_1(y) = \frac{Ai(a_1 - ye^{i2\pi/3})}{Ai'(a_1)} = \frac{w_1(a_1e^{-i2\pi/3} - y)}{2\pi^{1/2}e^{i\pi/6}Ai'(a_1)} \tag{4.65b}$$

Where the specific numerical values involved are: $e^{i2\pi/3} = (-1 + i\sqrt{3})/2$, $a_1' = -1.0188$, $Ai(a_1') = 0.5357$, $a_1 = -2.3381$, $Ai(a_1) = 0.7012$. If $\xi - y_0^{1/2} - y^{1/2}$ is slightly larger than 1, it can be proved in both cases that the "shortened" series of taking only the first term is a reasonable approximation.[11]

If y and y_0 are of medium magnitudes, the function $w_1(b_1e^{-i2\pi/3} - y)$ and $w_1(b_1e^{-i2\pi/3} - y_0)$ can be replaced by their asymptotic expression, such as the form shown in Eq. (4.58). Thus, the first term of Eq. (4.63) can be reduced as

$$V \approx \frac{e^{i\pi/12}\xi^{1/2}e^{i(2/3)y^{3/2}}e^{i(2/3)y_0^{3/2}}}{K_1(q)y_0^{1/4}y^{1/4}}\exp[e^{-i\pi/6}b_1(\xi - y_0^{1/2} - y^{1/2})] \tag{4.66}$$

$$K_1(q) = (4\pi)^{1/2}(-b_1 + q^2e^{i2\pi/3})[Ai(b_1)]^2 \tag{4.67}$$

$$= (4\pi)^{1/2}\left(1 - \frac{b_1}{q^2}e^{-i2\pi/3}\right)[Ai'(b_1)]^2 \tag{4.67'}$$

The two forms were applied to the case of $q \to 0$ (rigid interface; while $\boldsymbol{K}_1(0) = 1.036$) and the case of $q \to \infty$ (pressure release interface, $\boldsymbol{K}_1(\infty) = 1.743$).

4.5.4 Creeping wave

The connotation implied in Eq. (4.61) and (4.63) is that, in the shadow zone and on the boundary sound pressure amplitude or the amplitudes of any other acoustic field related parameters should decrease asymptotically along the interface w according to $w^{-1/2}e^{-aw}$. Here the attenuation coefficient (Np/m) is given by the following equations

$$a = \text{Re}(-e^{i\pi/6}b_1)\left(\frac{k_0}{2R^2}\right)^{1/3} \tag{4.68}$$

$$= \frac{n}{2c}f^{1/3}(-d_zc)_0^{2/3} \tag{4.68'}$$

11 The criterion is revealed by the comparison of Eq. (4.63) and (4.66): $|\exp[e^{-i\pi/6}(b_2 - b_1)(\xi - y_0^{1/2} - y^{1/2})]| \ll 1$. This condition is approximately satisfied when $\xi - y_0^{1/2} > 2/\{\text{Re}[(-b_2 + b_1)e^{-i\pi/6}]\}$. The numerical values for the rigid interface and pressure release interface are 1.034 and 1.3198, respectively.

where $n = 2\pi^{1/3}\mathrm{Re}(-\mathrm{e}^{-\mathrm{i}\pi/6}b_1)$; f is the frequency (Hz). For a rigid interface, $n = 2.58$, and for a pressure release interface, $n = 5.93$. The corresponding rate of the curves with constant phase (phase velocity) along the interface can be derived similarly

$$v_{\mathrm{ph}} = \frac{c(0)}{1 + \mathrm{Im}(\mathrm{e}^{-\mathrm{i}\pi/6}b_1)/(2k_0^2R^2)^{1/3}} \tag{4.69}$$

Which is always less than the sound velocity $c(0)$.

Weak attenuation and slightly slow phase velocities are the two obvious characteristics of a creeping wave.[12] Those waves move along the ray perpendicular to the constant phase surfaces (tangential to the interface in wind-less conditions) everywhere on the interface (Fig. 31). In the scenarios considered here, creeping wave rays are horizontal straight lines extending from the sound source, while in other circumstances rays will bend along the interface. Besides very weak exponential attenuation along the propagation distance, the amplitude also varies along the interface inversely proportional to the square root of the vertical distance between the two adjacent rays propagating to the interface (Ray band width). In this situation, the ray band width is directly proportional to w. Therefore, a factor of $w^{-1/2}$ will appear when substituting Eq. (4.66) into (4.61).

By the way, for propagation along a curved interface in a homogeneous medium [101], the requirements that the creeping wave ray should be perpendicular to the constant phase surfaces, and that it should move with a rate close to the sound velocity, lead to the conclusions that the rays should be geodesic lines and that the path connecting two points on the interface should be the shortest of all possible paths. This property is similar to Fermat's minimum-time principle. For the two most meaningful ideal cases, "spherical and cylindrical", these paths are large circles and spiral lines, respectively. In these cases, the coordinate system and the origin can be adjusted locally so that the interface equation can be given by $z = -x^2/2R_1 - y^2/2R_2$, where R_1 and R_2 represent the two principal radii of the curvature of the interface. The perturbation near the origin takes the form $\mathrm{e}^{\mathrm{i}k_x\xi}\mathrm{e}^{\mathrm{i}k_y\eta}F(\zeta)$, and here $\xi \approx x - xz/R_1$, $\eta \approx y - yz/R_2$, and $\xi \approx z + x^2/2R_1 + y^2/2R_2$. Almost the same differential equation is given below by applying a similar approximation method as derived in Eq. (4.55) (whose typical form is (2.51))

12 This very vivid name was firstly introduced by W. Franz and K. Depperman. Its original German phrasing is Kriechwelle (the English translation is creeping waves, usually being translated as "creeping waves"). See the articles, published in Ann Phys. 1952, (6) 10: 361~373. However, the predictions for this kind of wave can be traced back to the paper of G. N. Watson in 1919 "*The diffraction of electric waves by the earth*", Proc. R. Soc. London, A95: 83~99. Nevertheless, the concept that the wave penetrating into the shadow zone forms above a flat interface in a stratified medium be able to be regarded as creeping wave was proposed by G. D Malyuzhinets. See his paper published to commemorate the 130 anniversary of the death of T. Young, "Development in our concepts of diffraction phenomena", Sov. Phys. Usp. 1959, 69: 749.

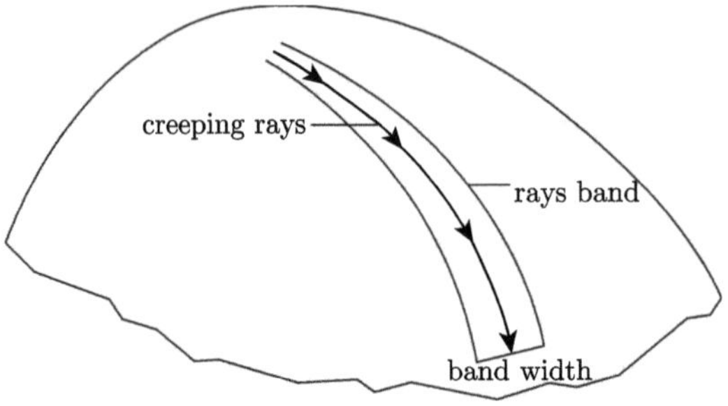

Fig. 31: The concept of a creeping wave propagating along an interface. If the velocity is constant, the creeping wave ray is geodesic, and the amplitude at the interface decreases inversely with the square root of the radiation bandwidth, as well as decreasing exponentially with distance along the ray.

$$F'' + \left(k_0^2 - k^2 + \frac{2k^2\zeta}{R_{\text{eff}}}\right)F = 0 \tag{4.55'}$$

where

$$k_0^2 = \frac{\omega^2}{c^2}, \quad k^2 = k_x^2 + k_y^2, \quad R_{\text{eff}}^{-1} = R_1^{-1}\cos^2\theta_k + R_2^{-1}\sin^2\theta_k \tag{4.70}$$

Here θ_k is in the direction of (k_x, k_y) relative to the x axis. If $\zeta \ll R_{\text{eff}}$, k^2 in the last term of (4.55′) can be approximated by k_0^2. Thus we retrieve Eq. (4.55), with only R owning a new implication; ζ may be understood as the distance across the interface. The choice of the boundary condition and minimum attenuation wave leads to the Airy function with the form of $w_1(\tau_1 - \zeta/l_{\text{eff}})$, which is just the first term of Eq. (4.63), with only z/l being replaced by ζ/l_{eff}, here $l_{\text{eff}} = (R_{\text{eff}}/2k_0^2)^{1/3}$.[13]

4.5.5 Geometric-acoustical interpretation of creeping waves

Although the concept of the creeping wave comes purely from the wave method, it can also be interpreted by the ray concept of geometric acoustics. The connotation implied by (4.66) is that the perturbation propagates along ordinary geometric acoustics rays in a deep shadow zone rather than near the interface (z is slightly greater than l). However, the origin of these rays is a creeping wave and not a sound source (see Fig. 32). This can be verified by substituting V of Eq. (4.66) into (4.61)

$$p = \frac{e^{i\pi/12}(R^2/4k_0)^{1/6}\hat{S}e^{-a\Delta w}\exp[iw\tau_{\text{TR}}(z_0) + i(w/v_{\text{ph}})\Delta w + iw\tau_{\text{TR}}(z)]}{w^{1/2}[K_1(q)/2^{1/2}][(2Rz_0)(2Rz)]^{1/4}} \tag{4.71}$$

13 This approximation comes from the similar theory of radio waves propagating along the earth surface (taking into account the curvature of the earth itself, which should be done sometimes in atmospheric acoustics), and is famous for the "flattening of the earth" approximation.

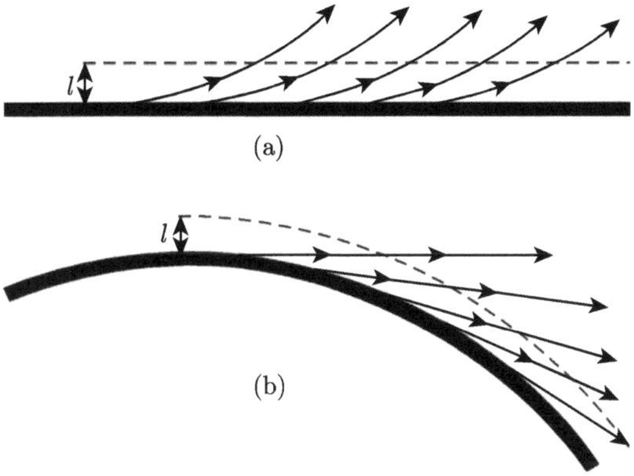

Fig. 32: Rays emitted by creeping waves. (a) Atmospheric profiles for a sound velocity that linearly increases with height over the flat ground; (b) a homogeneous atmosphere over spherical ground, where the sound velocity is constant.

where

$$\Delta w = w - (2Rz_0)^{1/2} - (2Rz)^{1/2} \tag{4.72}$$

$$c_0\tau_{TR}(z) = (2Rz)^{1/2} + \frac{2}{3}\left(\frac{2z^3}{R}\right)^{1/2} \tag{4.73}$$

Here, $(2Rz_0)^{1/2}$ is the horizontal distance from the sound source to the rim of the shadow zone (see Fig. 30), $(2Rz)^{1/2}$ is the horizontal distance from the interface to the observer along a ray. The ray leaves the interface at $(w_0, 0)(w_0 = w - (2Rz)^{1/2}$ at a grazing angle, and subsequently passes through the observer's position. Parameter, $\tau_{TR}(z)$, is the traveling time along the ray segments, which can be derived from Eq. (3.16). The equation predicts $d\tau_{TR}/dw$ to be c_0/c^2, because $s_w = 1/c_0$. For rays that are tangential to the interface from the beginning, c_0/c^2 is approximately equal to $(1+2z/R)/c_0$, and becomes $(w-w_0)^2/2R$ along the ray. Therefore, after the integral of $d\tau_{TR}/dw$, we obtain $c_0\tau_{TR} = (w - w_0) + \frac{1}{3}(w - w_0)^2/R^2$. Replacing $w - w_0$ with $(2Rz)^{1/2}$, we obtain Eq. (4.73).

Similarly, $\tau_{TR}(z_0)$ corresponds to the travelling time of the ray from the sound source to the edge of the shadow zone at the interface (horizontal distance $(2Rz_0)^{1/2}$). Therefore, the phase change $w\tau_{TR}(z_0)+(w/v_{ph})\Delta w+w\tau_{TR}(z)$ corresponds to a "broken" ray: it reaches the ground from the sound source with the sound velocity, and then extends for a distance Δw with the phase velocity along the ground, and finally arrives at the observer at the sound velocity.

The above conclusion implies the following interpretation: the sound arriving at the observer at (w, z) is emitted out by the creeping wave at $(w_0, 0)$. This viewpoint is further supported by the attenuation factor $e^{-a\Delta w}$. The disturbance at (w, z) is carried by a creeping wave only in the interval between $[(2Rz_0)^{1/2}, 0]$ and $[w_0, 0]$ (the net distance is Δw).

The factor $w^{1/2}$ and $(2Rz)^{1/4}$ in the denominator of Eq. (4.71) can also be interpreted similarly in terms of geometric acoustics. Their product is directly proportional to the square root of the area of the tube associated with the rays passing through the position of the observer. The vertical spacing between two successive rays emitted from w_D and $w_D + \delta w_D$ is approximated as $\delta z \approx -\delta[(w - w_O)^2/2R]$ or $(w - w_O)(\delta w_O/R)$ after the rays cross a distance of $(w - w_O)$. Therefore, the change of cross-sectional area of a ray tube with z is the same as $w - w_O$ or $(2Rz)^{1/2}$. The other factor in the expression of the ray tube section w comes from the cylindrical expansion (from the sound source).

4.6 Sound diffraction in moving atmosphere

Much less that the case where the sound source is a pulse point source, the existence of wind makes the situation more complex. We will adopt the Friedlander method in processing pulse diffraction [102], and enjoy the normal mode expansion method. The latter plays an important role in wave acoustics through this process.

4.6.1 Fundamental equations and formal solutions

If not taking the irreversible processes of acoustic waves into account, and we instead only limit ourselves to the application of the linear approximation, the acoustic potential Φ for a sound in an isentropic medium with a steady non-rotation motion will satisfy the following equation [75]

$$\mathrm{d}_t^2\Phi = c^2\nabla^2\Phi + \frac{\nabla p}{\rho}\cdot\nabla\Phi + (v\cdot\nabla\ln c^2)\mathrm{d}_t\Phi \tag{4.74}$$

where c is sound velocity, p, ρ and v are the atmosphere pressure, density and moving velocity (wind speed), respectively, when the atmosphere is not disturbed.

Assuming that the Mach number of the wind speed $M \equiv v/c \ll 1$, and that all atmospheric parameters are only slowly-varying functions of height z, as well as taking the Cartesian coordinates and making the wind speed $\boldsymbol{v}$ point in the positive direction of the x axis, Eq. (4.74) can be reduced to

$$L(\Phi) \equiv (1 - M^2) + \partial_x^2\Phi + \partial_y^2\Phi + \partial_z^2\Phi - \frac{1}{c^2}\partial_t^2\Phi - 2\frac{M}{c}\partial_{x,t}^2\Phi - \frac{\rho'}{\rho}\partial_x\Phi = 0 \tag{4.74a}$$

The "prime" denotes the derivative of z.

Now, let us assume that there is a pulse point source at $(0, 0, z_0)$, whose behavior can be described by the δ function. Hence, Φ at the $z > 0$ half-space satisfies the following equation

$$L(\Phi) = -4\pi\delta(x)\delta(y)\delta(z - z_0)\delta(t) \tag{4.75}$$

The initial condition is satisfied in the half-space, except at the origin

$$\begin{array}{l} \Phi|_{t=0} = 0 \\ \partial_t\Phi|_{t=0} = 0 \end{array} \quad x^2 + y^2 + (z - z_0)^2 \neq 0 \tag{4.76}$$

One of the boundary conditions that is satisfied at the interface $z = 0$ (i.e., (2.40) or (2.41))

$$\Phi|_{z_0} = 0 \text{or} \partial_z\Phi|_{z=0} = 0 \tag{4.77}$$

The radiation condition should be satisfied at infinity.

The Laplace transformation of Φ gives

$$\bar{\Phi}(x, y, z; s) = \int_{-\infty}^{\infty} \Phi(x, y, z; t)\mathrm{e}^{-st}\mathrm{d}t \tag{4.78}$$

Its two-fold Fourier transformation is

$$\tilde{\Phi}(z; s, \kappa_x, \kappa_y) = \iint_{-\infty}^{\infty} \bar{\Phi}(x, y, z; s)\mathrm{e}^{-\mathrm{i}(\kappa_x x+\kappa_y y)}\mathrm{d}x\mathrm{d}y \tag{4.79}$$

which satisfies the corresponding boundary and radiation conditions, and the equation

$$\tilde{\Phi}'' - \frac{\rho'}{\rho}\tilde{\Phi}' + q\tilde{\Phi} = -4\pi\delta(z - z_0) \tag{4.80}$$

where

$$q \equiv \left(\frac{\mathrm{i}s}{c} - M_{\kappa_x}\right)^2 - \kappa^2, \quad \kappa^2 \equiv \kappa_x^2 + \kappa_y^2 \tag{4.81}$$

Let $\tilde{\Phi}_1$ and $\tilde{\Phi}_2$ be two linearly independent solutions of the homogeneous equation that corresponds to (4.80), then the solution of the equation that satifies both the boundary and the radiation conditions has the following form[14]

$$\tilde{\Phi}(z) = -\frac{4\pi}{W}\begin{cases} \tilde{\Phi}_2(z_0)[\tilde{\Phi}_1(z) - w\tilde{\Phi}_2(z)], & z \leqslant z_0 \\ [\tilde{\Phi}_1(z_0) - w\tilde{\Phi}_2(z_0)]\tilde{\Phi}_2(z), & z \geqslant z_0 \end{cases} \tag{4.82}$$

where the Wronski determinant is

$$W(\tilde{\Phi}_1, \tilde{\Phi}_2) \equiv \tilde{\Phi}_1, \tilde{\Phi}_2', -\tilde{\Phi}_2\tilde{\Phi}_1' \tag{4.83}$$

It should take the value at $z = z_0$, and w has a different definition with different boundary conditions

$$w \equiv \begin{cases} \tilde{\Phi}_1(0)/\tilde{\Phi}_2, \text{ when } \tilde{\Phi}(0) = 0 \\ \tilde{\Phi}_1'(0)/\tilde{\Phi}_2', \text{ when } \tilde{\Phi}'(0) = 0 \end{cases} \tag{4.84}$$

14 For brevity, we often write only a key argument z of such multivariate functions as $\tilde{\Phi}$ hereafter.

4.6.2 Normal mode expansions

Let us discuss the asymptotic behavior of $\bar{\Phi}$ when $s \to \infty$. Because the asymptotic behavior can determine the initial behavior of Φ, it has special meaning in pulse diffraction theory. The acoustic field can be expressed as the superposition of all of the normal modes at this time.

According to Langer,[15] the homogeneous equation that corresponds to Eq. (4.80) has two linearly independent asymptotic solutions when s is large

$$\begin{aligned}\tilde{\Phi}_1 &\sim PQ^{1/2}q^{-1/4}H^{(1)}_{1/3}(Q)\\ \tilde{\Phi}_2 &\sim PQ^{1/2}q^{-1/4}H^{(2)}_{1/3}(Q)\end{aligned} \tag{4.85}$$

where

$$P \equiv \left[\frac{\rho(z)}{\rho(z_1)}\right]^{1/2}, \quad Q \equiv \int_{z_1}^{z} q^{1/2}(z')\mathrm{d}z' \tag{4.86}$$

z_1 is the first-order zero of q, H is the Hankel function.

In order to calculate $\bar{\Phi}$, the inverse Fourier transform of $\tilde{\Phi}$, we shall use polar coordinates, $x = r\cos\varphi$, $y = r\sin\varphi$; $\kappa_x = \kappa\cos\theta$ and $\kappa_y = \kappa\sin\theta$. Therefore,

$$\bar{\Phi}(r,\varphi,z;s) = \frac{1}{(2\pi)^2}\int_0^{2\pi}\mathrm{d}\theta\int_0^{\infty}\tilde{\Phi}(z;s,\kappa,\theta)\mathrm{e}^{\mathrm{i}\kappa r\cos(\theta-\varphi)}\kappa\mathrm{d}\kappa \tag{4.87}$$

When $kr \gg 1$, the integral with respect to θ can be carried out by the method of stationary phase,[16] which gives the following results

$$\begin{aligned}\bar{\Phi}(r,\varphi,z;s) = &\frac{1}{(2\pi)^{3/2}r^{1/2}}\left[\int_0^{\infty}\tilde{\Phi}(\varphi,z;s,\kappa)\mathrm{e}^{\mathrm{i}\left(\kappa r-\frac{\pi}{4}\right)}\kappa^{1/2}\mathrm{d}\kappa\right.\\ &\left.+\int_0^{\infty}\tilde{\Phi}(\varphi+\pi,z;s,\kappa)\mathrm{e}^{-\mathrm{i}\left(\kappa r-\frac{\pi}{4}\right)}\kappa^{1/2}\mathrm{d}\kappa\right] + O(\kappa r)^{-1}\end{aligned} \tag{4.88}$$

The integral path of the first integral consists of the positive real axis of κ, the circle arc with an infinite radius sweeping across the first quadrant in the anticlockwise direction and a reverse positive imaginary semi axis. The integral path of the second integral consists of a circle arc with an infinite radius sweeping across the fourth quadrant in a clockwise direction, and a straight-forward negative imaginary semi

15 Refer to Langer R. E. Trans. Amer. Math. Soc.; 1931, 33, 23~64; or Lighthill, M. J., Quart. J. Mech. Appl. Math., 1950, 3: 311.

16 Refer to Jeffereys H., B. S., Methods of Mathematical Physics, 3rd ed., Cambridge, 1956.

axis. Taking advantage of Eqs. (4.82), (4.84), (4.85), the asymptotic expression of the Hankel function is

$$H_{1/3}^{(1),(2)}(x) \sim \left(\frac{2}{\pi x}\right)^{1/2} e^{\pm i\left(x-\frac{5}{12}\pi\right)} \tag{4.89}$$

It can be proved that $\lim\limits_{|\kappa|\to\infty} \tilde{\Phi}_{\kappa}^{1/2} \to 0$. Hence, according to the Jordan lemma, two integrals along a quadrant circle with an infinite radius vanish, respectively. Furthermore, it can be seen through direct substitution that the two integrands have equal values and opposite signs on the positive and minus imaginary semi axes, respectively, and hence they cancel out each other. Finally, the integral along the edge of the cut of the integrand branch point usually gives a side-wave whose amplitude is inversely proportional to κr^2, and is negligible due to the approximation level applied here.

Therefore, the two integrals along the real semi axes of κ in Eq. (45.88) can be reduced to two loop integrals to be carried out in the first and fourth quadrants, which can be completed using the residue theorem. The physical condition requires that the integrand should have no poles in the fourth quadrant. Hence, the value of the second integral of Eq. (4.88) should be 0; otherwise, it will introduce a series of waves converging at the source, which violates the physical essence of the problem. We will prove this in the next section. In the meantime, we can see that the integrand has an infinite number of first-order poles $\kappa = \kappa_n (n = 1, 2, 3, \cdots)$ located in the first quadrant, or on the real axis. If we choose the integral path just under the real axis but parallel to it, the second integral is always 0 when the loop is built, and $\bar{\Phi}$ is only given by the value of the first integral. According to the famous law of acquiring residues, we obtain

$$\begin{aligned}\bar{\Phi}(r,\varphi,z;s) &= \sum_{n=1}^{\infty} \bar{\Phi}_n(r,\varphi,z;s)\\ &= i2(2\pi)^{1/2} \sum_{n=1}^{\infty} \left(\frac{\kappa_n}{r}\right)^{1/2} \frac{\chi_n}{W} \tilde{\Phi}_2(\varphi, z_0; s, \kappa_n)\\ &\cdot \tilde{\Phi}_2(\varphi, z; s, \kappa_n) e^{i\left(\kappa_n r-\frac{\pi}{4}\right)} + O(kr)^{-1}\end{aligned} \tag{4.90}$$

where

$$\chi_n \equiv \begin{cases} \dfrac{\tilde{\Phi}_1(\varphi, 0; s, \kappa_n)}{\partial_\kappa \tilde{\Phi}_2(\varphi, 0; s, \kappa_n)|_{\kappa=\kappa_n}}, & \text{when } \tilde{\Phi}(0) = 0\\[2ex] \dfrac{\tilde{\Phi}_1'(\varphi, 0; s, \kappa_n)}{\partial_\kappa \tilde{\Phi}_2'(\varphi, 0; s, \kappa_n)|_{\kappa=\kappa_n}}, & \text{when } \tilde{\Phi}'(0) = 0 \end{cases} \tag{4.91}$$

Eq. (4.90) is valid for regions $z \ll z_0$ and $z \gg z_0$.

4.6.3 Asymptotic expressions for the eigen-values

It can be seen from Eqs. (4.82) and (4.84) that, for soft interfaces, the poles of $\tilde{\Phi}$ are the zeroes of $\tilde{\Phi}_2(0)$, and according to Eq. (4.85), the latter is the approximate zeroes of

$H^{(2)}_{1/3}(Q_0)$, here

$$Q_0 \equiv Q(0) = \int_{z_1}^{0} q^{1/2} \mathrm{d}z = \int_{0}^{q_0} \frac{q^{1/2}}{q'} \mathrm{d}q \tag{4.92}$$

Hereafter, we will use subscript 0 to denote the value at $z = 0$. From the familiar relation

$$H^{(2)}_{1/3}(x\mathrm{e}^{\mathrm{i}\pi}) = \frac{2}{3^{1/2}} \mathrm{e}^{\mathrm{i}\pi/6}[J_{1/3}(x) + J_{-1/3}(x)] \tag{4.93}$$

the zeroes of $H^{(2)}_{1/3}(Q_0)$ can be written as

$$Q_{0n} = \beta_n \mathrm{e}^{\mathrm{i}\pi} \tag{4.94}$$

where β_n is a positive real number that satisfies the following relation

$$J_{1/3}(\beta_n) + J_{-1/3}(\beta_n) = 0 \tag{4.95}$$

Expanding $q(z)$ and $q'(z)$ into the Maclaurin series, respectively, eliminating z from it, substituting the obtained expression of q' into (4.92), and completing the integral, we have [103]

$$Q_0 = \frac{2}{3} \frac{q_0^{3/2}}{q_0'} \left[1 + \frac{2}{5} \frac{q_0 q_0''}{q_0'^2} + \cdots \right] \tag{4.96}$$

In the usual case, $M_0 = v_0/c_0 = 0$ is always valid. Thus, from Eq. (4.81), we have

$$q_0 = -\frac{s^2}{c_0^2} - \kappa^2 \tag{4.97}$$

$$q_0' = 2\frac{s}{c_0} \left[\frac{s}{c_0} \frac{c_0'}{c_0} - \mathrm{i}\kappa M_0' \cos\varphi \right] \tag{4.98}$$

It is not difficult to prove that the magnitudes of the second term in Eq. (4.95) and all the following terms are no more than $O(s^{-2/3})$, therefore, when $s \to \infty$, we obtain the following relation from Eqs. (4.94), (4.96) and (4.97)

$$\kappa_n = \frac{\mathrm{i}s}{c_0} + \frac{\mathrm{i}}{2} \left(\frac{3}{2} \beta_n \mathrm{e}^{\mathrm{i}\pi} q_0' \right)^{2/3} \left(\frac{s}{c_0} \right)^{-1} + O(S^{-1/3})$$

From the zeroth-order approximation, $\kappa_n \sim \frac{\mathrm{i}s}{c_0}$, we substitute it into Eq. (4.98), and introduce a function in the meantime,

$$f(z) \equiv (c + v\cos\varphi)^{-2} \tag{4.99}$$

We have

$$q_0' \sim -s^2 f_0'$$

Hence

$$\kappa_n = \mathrm{i} f_0^{1/2} s + \frac{\mathrm{i}}{2} \alpha_n f_0^{-1/2} f_0'^{2/3} s^{1/3} + O(s^{-1/3}) \tag{4.100}$$

where

$$\alpha_n \equiv \left(\frac{3}{2}\beta_n\right)^{2/3} \tag{4.101}$$

Because Re$s > 0$, and $f > 0$ is always valid, it is obvious that according to (4.100) $\kappa = \kappa_n$ cannot be maintained in the lower half-space, and therefore there are no poles of $\tilde{\Phi}$ in the third and fourth quadrants. Moreover, because the integral of (4.88) is carried out on the right half of the plane, it is enough to just consider the poles of $\tilde{\Phi}$ in the first quadrant and on the real semi-axis. It proves that the analysis made in last section is right.

The meaning of Eq. (4.101) can be demonstrated by the following consideration. Let us compare the Airy function and the Bessel function [104]

$$Ai(-\alpha) = \frac{1}{3}\alpha^{1/2}[J_{1/3}(\beta) + J_{-1/3}(\beta)]$$

With the help of Eq. (4.95), we can immediately see that, β_n can be replaced with the zeroes of $Ai(-\alpha)$, α_n. Eq. (4.100) is for soft interfaces. For hard interfaces, κ_n is approximately determined by the zeroes of $\dot{H}^{(2)}_{1/3}(Q_0)$ (the upper dot denotes the total derivative of the function with respect to its arguments) (formally written as $\dot{Q}_{0n} = \beta_n e^{i\pi}$). Therefore, α_n should be replaced with the zeroes of $\dot{A}i(-\alpha)$, $\dot{\alpha}_n$.

4.6.4 Asymptotic expressions of the eigen-functions

Rewrite Eq. (4.81) as

$$q = s^2(1 - M^2\cos^2\varphi)(\mu - f^{1/2})\left(\mu + \frac{1}{c - v\cos\varphi}\right) \tag{4.102}$$

where

$$\mu = \frac{\kappa}{is} = f_0^{1/2} + \frac{1}{2}\alpha f_0^{-1/2} f_0'^{2/3} s^{-2/3} + O(s^{-4/3}) \tag{4.103}$$

By taking the common situation in practice $c_0' + v_0'\cos\varphi < 0$,[17] then

$$f_0' > 0$$

By definition, Φ given by Eq. (4.90) is the Laplacian transformation of a certain function Φ when $t < 0$. Thus, it should be an analytical function in the half-plane of Re$s > C$ (C is a certain positive constant). The value of Φ determines its inverse transformation uniquely when s is a positive real number. Therefore, for the sake of simplicity, we take s to be a positive real number. Therefore $\mu \geqslant f_0^{1/2} = 1/c_0$.

17 It is not difficult to see from the second section of Chapter 3 that this is just the condition for forming an acoustic shadow zone.

Because f is a positive real number, the condition of forming a shadow zone demonstrates that f is a function that monotonously increases with z, and that q has a unique first-order zero $z = z_1$. Because both the first and third brackets in (4.102) cannot be 0, the only possibility is

$$\mu = f^{1/2}(z_1) \tag{4.104}$$

When $z < z_1$, $\mu > f^{\frac{1}{2}}$, and when $z > z_1$, $\mu < f^{1/2}$. Therefore, the real function of z is

$$\begin{aligned} \frac{2}{3}(-\zeta)^{3/2} &= \int_{z}^{z_1} \left[(1 - M^2\cos^2\varphi)(\mu - f^{1/2})\left(\mu + \frac{1}{c - v\cos\varphi}\right)\right]^{1/2} dz', z \leqslant z_1 \\ \frac{2}{3}\zeta^{3/2} &= \int_{z_1}^{z} \left[(1 - M^2\cos^2\varphi)(f^{1/2} - \mu)\left(\mu + \frac{1}{c - v\cos\varphi}\right)\right]^{1/2} dz', z \geqslant z_1 \end{aligned} \tag{4.105}$$

Comparing Eq. (4.105) and the second formula of Eq. (4.86), it can be seen that[18] (where q is replaced by (4.102))

$$Q = -\mathrm{i}\frac{2}{3}s\zeta^{3/2} \tag{4.106}$$

According to (4.93), we have

$$H_{1/3}^{(2)}(Q) = \frac{3}{3^{1/2}}\mathrm{e}^{\mathrm{i}2\pi/3}\left[I_{-1/3}\left(\frac{2}{3}s\zeta^{3/2}\right) - I_{1/3}\left(\frac{2}{3}s\zeta^{3/2}\right)\right]$$

Considering the relationship of

$$Ai(\alpha) = \frac{1}{3}\alpha^{1/2}[I_{-1/3}(\beta) - I_{1/3}(\beta)]$$

The expression $\tilde{\Phi}_2$ in Eq. (4.85) can be reduced to

$$\tilde{\Phi}_2 \sim 2^{3/2}\mathrm{e}^{\mathrm{i}\pi/6}\zeta^{-1/3}Ai(s^{2/3}\zeta)P \tag{4.107}$$

Expansion of the Airy function [105] gives

$$Ai(\alpha) = \frac{1}{2}\pi^{-1/2}\alpha^{-1/4}\mathrm{e}^{-\beta}\left[1 - \frac{3\cdot 5}{1!216}\beta^{-1} - \frac{5\cdot 7\cdot 9\cdot 11}{2!216^2}\beta^{-2} - \cdots\right] \tag{4.108}$$

we can see that $Ai(s^{2/3}\zeta_0)$ tends to 0 when $s \to \infty$, according to the rule of $s^{-1/6}\mathrm{e}^{-s}$. Therefore, when s is very large, we should take $s^{2/3}\zeta_0$ as the value nearby one of the zeroes of the Airy function $\alpha = -\alpha_n$

$$\zeta_0 \sim -\alpha_n s^{-2/3} \tag{4.109}$$

18 The minus sign must be taken from two possible signs, because if we take the positive sign, we will obtain $\tilde{\Phi} \sim q^{-1/4}\exp\left[\frac{2}{3}s\zeta^{3/2}\right]$ (when $z \to \infty$) from Eqs. (4.82) and (4.85), and this form of $\tilde{\Phi}$ does not satisfy the radiation condition.

Accordingly, we may find out that

$$\zeta_0' \sim f_0^{1/3}, \quad \zeta_0'' \sim \frac{v' \cos\varphi}{2c_0} f_0^{1/3} \tag{4.110}$$

Since all are finite quantities, (4.107) gives[19]

$$\tilde{\Phi}_2(0) \sim 2^{3/2} e^{i\pi/6} f_0^{-1/6} s^{-1/3} Ai(-\alpha_n) P_0 \tag{4.111}$$

At the same time,

$$\tilde{\Phi}_2'(0) \sim 2^{3/2} e^{i\pi/6} f_0^{1/6} s^{1/3} \dot{A}i(-\dot{\alpha}_n) P_0 \tag{4.112}$$

From the above two equations, we have

$$\partial_\kappa \tilde{\Phi}_2(0)|_{\kappa=\kappa_n} \sim 2^{5/2} e^{i2\pi/3} f_0^{1/2} f_0'^{-5/6} s^{-2/3} \dot{A}i(-\alpha_n) P_0 \tag{4.113}$$

$$\partial_\kappa \tilde{\Phi}_2'(0)|_{\kappa=\kappa_n} \sim 2^{5/2} e^{i\pi/3} f_0^{1/2} f_0'^{-1/2} \dot{\alpha}_n Ai(-\dot{\alpha}_n) P_0 \tag{4.114}$$

As for $\tilde{\Phi}_1$, we still take the form shown in Eq. (4.58). Substituting (4.102) and (4.106) into it, we obtain

$$\tilde{\Phi}_1 \sim \left(\frac{2}{3}\right)^{1/2} \zeta^{1/2} \zeta'^{-1/2} H_{1/3}^{(1)} \left(-i\frac{2}{3} s \zeta^{3/2}\right) P \tag{4.115}$$

Considering (4.109), (4.110) and (4.111), we have

$$\tilde{\Phi}_1 \sim \left(\frac{2}{3}\right)^{1/2} e^{i\pi/2} \alpha_n^{1/2} f_0^{-1/6} s^{-1/3} H_{1/3}^{(1)}(-\beta_n) P_0 \tag{4.116}$$

Taking the derivative of Eq. (4.115) with respect to z, substituting the relevant values, and ignoring terms of the order of $O(s^{-1/3})$, we can obtain

$$\tilde{\Phi}_1'(0) \sim \left(\frac{2}{3}\right)^{1/2} e^{i\pi/2} \dot{\alpha}_n f_0^{1/6} s^{1/3} H_{-2/3}^{(1)}(-\dot{\beta}_n) P_n \tag{4.117}$$

By substituting Eqs.(4.113), (4.116) and (4.114), (4.117) into Eq. (4.91), we will obtain

$$\chi_n \sim \frac{1}{4 \cdot 3^{1/2}} f_0^{-1/2} f_0^{2/3} s^{1/3} \begin{cases} e^{i\pi/6} \varepsilon_n, \text{ when } \tilde{\Phi}(0) = 0 \\ e^{i\pi/6} \dot{\varepsilon}_n, \text{ when } \tilde{\Phi}'(0) = 0 \end{cases} \tag{4.118}$$

where

$$\varepsilon_n \equiv \alpha_n^{1/2} \frac{H_{1/3}^{(1)} - (\beta)}{\dot{A}i(-\alpha_n)}, \quad \dot{\varepsilon}_n \equiv \frac{H_{-2/3}^{(1)}(-\dot{\beta}_n)}{Ai(-\dot{\alpha}_n)} \tag{4.119}$$

19 The terms of the order of $O(s^{-1/3})$ are ignored. This means that the corresponding boundary condition is $\tilde{\Phi}(0) = 0$ when taking the derivative of $\tilde{\Phi}_2$ or $\tilde{\Phi}_1$ with respect to z to obtain the value of $\tilde{\Phi}_2'(0)$ or $\tilde{\Phi}_1'(0)$. As pointed out before, substitution of $\beta_n \to \dot{\beta}_n$, $\alpha_n \to \dot{\alpha}_n$ should be made.

We must perform this transformation to the expression of Φ_2 (4.107). Expanding (4.104) to the power series of z_1 and comparing it with Eq. (4.103), we find $z_1 = O(s^{-2/3})$. Therefore, for each fixed $z(\neq 0)$ and in s being very large case, we can regard ζ in Eq. (4.107) as defined by the second expression in (4.105). Take the integral lower limit as 0, expand the integrand, and substitute the relevant values, we obtain

$$\frac{2}{3}\zeta^{3/2} = |F(z)| - \frac{1}{2}\alpha_n f_0^{2/3} s^{-2/3} G(z) + O(s^{-4/3}) \tag{4.120}$$

where

$$F(z) \equiv \frac{1}{c_0} \int_0^z \frac{[(c_0 - v\cos\varphi)^2 - c^2]^{1/2}}{c} \mathrm{d}z' \tag{4.121}$$

$$G(z) = c_0 \int_0^z \frac{c^2 + c_0 v \cos\varphi - v^2 \cos^2\varphi}{c[(c_0 - v\cos\varphi)^2 - c^2]^{1/2}} \mathrm{d}z' \tag{4.122}$$

When s is large, $s^{2/3}\zeta$ is also large. The Airy function in Eq. (4.107) can be replaced by the first term of its expansion (4.108). Thus

$$\tilde{\Phi}_2 \sim \left(\frac{2}{\pi}\frac{1}{sF'}\right)^{1/2} P \exp\left[-sF + \frac{1}{2}\alpha_n f_0'^{2/3} s^{1/3} G + \mathrm{i}\frac{\pi}{6} + O(s^{-1/3})\right] \tag{4.123}$$

Taking advantage of Eq. (4.100), we can directly get the following formula

$$\left(\frac{\kappa_n}{r}\right)^{1/2} \mathrm{e}^{\mathrm{i}\left(\kappa_n r - \frac{\pi}{4}\right)} = f_0^{1/4} s^{1/2} r^{-1/2} \exp\left[-f_0^{1/2} s r - \frac{1}{2}\alpha_n f_0^{-1/2} f_0'^{2/3} s^{1/3} r + O(s^{-1/3})\right] \tag{4.124}$$

Finally, by substituting all of the relevant equations into Eq. (4.83), and by considering the relationship

$$\boldsymbol{W}(\boldsymbol{H}_\nu^{(1)}, \boldsymbol{H}_\nu^{(2)}) \equiv \boldsymbol{H}_\nu^{(1)}(x)\dot{\boldsymbol{H}}_\nu^{(2)}(x) - \dot{\boldsymbol{H}}_\nu^{(1)}(x)\dot{\boldsymbol{H}}_\nu^{(2)}(x) = -\frac{4\mathrm{i}}{\pi x}$$

It is not difficult to obtain

$$\boldsymbol{W}(\tilde{\Phi}_1, \tilde{\Phi}_2) = -\frac{4\mathrm{i}}{\pi} P^2 \tag{4.125}$$

4.6.5 Approximated expressions for the diffraction field

Using the result in last section, the asymptotic expression of $\bar{\Phi}$ can be obtained immediately. Substituting the expressions given by Eqs. (4.118), (4.123)~(4.125) into the expression of the n-th order normal modes, we obtain

$$\bar{\Phi}_n = \frac{1}{2}\left(\frac{\pi}{6}\right)^{1/2} f_0^{2/3} s^{-1/6} \left[\frac{c_0\rho(z)}{\rho(z_0)F'(z_0)F'(z)r}\right]^{1/2} [1 + O(s^{-1/3})]$$

$$\times \begin{pmatrix} \varepsilon_n \\ \dot{\varepsilon}_n \end{pmatrix} \exp\left\{-s\tau - \frac{1}{2} f_0^{2/3} s^{1/3} \delta \begin{pmatrix} \alpha_n \\ \dot{\alpha}_n \end{pmatrix} - \mathrm{i}\begin{pmatrix} 5\pi/6 \\ \pi/2 \end{pmatrix}\right\} \tag{4.126}$$

where the upper line corresponds to soft interfaces, and the lower line corresponds to hard interfaces. For the sake of simplicity, we introduce a couple of symbols

$$\tau \equiv \frac{r}{c_0} + |F(z_0)| + |F(Z)| \tag{4.127}$$

$$\delta \equiv c_0 r - G(z_0) - G(z) \tag{4.128}$$

It can be proved that τ is the time that the pulse arrives at the observation point, and δ is the distance of the observing point entering the shadow zone. (4.126) is valid not only for positive real s, but also for all complex s when $\mathrm{Re}s > 0$.

The expansion of $\bar{\Phi}$(4.90) gives its inverse Laplacian transformation Φ only when it is convergent because τ is always positive. So if $\delta > 0$, (4.126) gives $\bar{\Phi}_n \to 0$ when $\mathrm{Re}s_n \to \infty$. From this we can see that the series in the shadow zone (4.90) is convergent. It is not difficult to see from Eq. (4.126) that the rate of the main term of the n-th order normal mode to the correction term tends to approach 0 when $|\delta| \to \infty$. In other words, the correction term of the first-order normal mode is much larger than that of the main terms of any high-order normal mode. Therefore when substituting (4.126) into (4.90), we can only keep the term $n = 1$. So we can get

$$\bar{\Phi} = \frac{1}{2}\left(\frac{\pi}{6}\right)^{1/2} f_0'^{2/3} s^{-1/6} \left[\frac{c_0\rho(z)}{\rho(z_0)F'(z_0)F'(z)r}\right]^{1/2} [1 + O(s^{-1/3})]$$

$$\times \begin{pmatrix} \varepsilon_1 \\ \dot{\varepsilon}_1 \end{pmatrix} \exp\left\{-s\tau - \frac{1}{2} f_0^{2/3} s^{1/3} \delta \begin{pmatrix} \alpha_1 \\ \dot{\alpha}_1 \end{pmatrix}\right.$$

$$\left. - \mathrm{i}\begin{pmatrix} 5\pi/6 \\ \pi/2 \end{pmatrix}\right\}, \quad z_0 > 0, z > 0 \tag{4.129}$$

Actually, the above equation is only useful for the cases of $z_0 > 0$ and valid for $z > 0$. Because if one of z_0 and z is 0, an infinite factor of $F_0'^{-1/2}$ will appear according to Eq. (4.129), and at this time Eq. (4.129) must be corrected.

At first, let us discuss the case of hard interfaces. Assuming $z_0 > 0$ and $z = 0$, in order to obtain $\bar{\Phi}$, we should replace the expression of $\bar{\Phi}_2(z)$ (4.123) with $\bar{\Phi}_2(0)$ (4.111) (as already pointed out, at this time α_1 should be replaced by $\dot{\alpha}_1$). Therefore, when replacing Eq. (4.129), we have

$$\bar{\Phi}_0 = \frac{\pi}{6^{1/2}} f_0'^{1/2} \left[\frac{c_0 \rho_0}{\rho(z_0) F'(z_0) r} \right]^{1/2} \boldsymbol{H}^{(1)}_{-2/3}(-\dot{\beta}_1)[1 + O(s^{-1/3})]$$
$$\times \exp\left\{ -s\tau_0 - \frac{1}{2} f_0'^{2/3} s^{1/3} \delta_0 \dot{\alpha}_1 - \mathrm{i}\frac{\pi}{2} \right\}, \quad z_0 > 0, z = 0 \tag{4.129a}$$

For the inverse case: $z_0 = 0$ and $z > 0$, the only substitution that needs to be made in the above expression is $z_0 \to z$, $\rho_0/\rho(z_0) \to \rho(z)/\rho_0$. If z_0 and z are 0 in the meantime, the only additional substitution that is needed in the above equation corresponds to Eq. (4.123) → Eq. (4.111). At this time, we have

$$\bar{\Phi}_{00'} = \pi \left(\frac{2\pi}{3} \right) f_0'^{1/3} s^{1/6} \left(\frac{c_0}{r} \right)^{1/2} Ai(-\dot{\alpha}_1) \boldsymbol{H}^{(1)}_{-2/3}(-\dot{\beta}_1)[1 + O(s^{-1/3})]$$
$$\times \exp\left\{ -s\tau_{00'} - \frac{1}{2} f_0'^{2/3} s^{1/3} \delta_{00'} \dot{\alpha}_1 - \mathrm{i}\frac{\pi}{2} \right\}, \quad z_0 = z = 0 \tag{4.129b}$$

where the first subscript 0 corresponds to $z = 0$ as before, and the second subscript 0′ corresponds to $z_0 = 0$.

Thus, for soft interfaces, if $z_0 > 0$ and $z = 0$, $\bar{\Phi}_0 = 0$ is obtained basing on the boundary conditions. If $z_0 = 0$ and $z > 0$, $\Phi_{(z)} = 0$ is obtained, because the substitution of (4.123) → (4.111) gives a factor of $Ai(-\alpha_n)$, which is the same for each $\bar{\Phi}_n$.

Now we can find the inverse Laplacian transformation of Φ, $\bar{\Phi}$

$$\bar{\Phi}(v, \varphi, z; t) = \frac{1}{2\pi \mathrm{i}} \int_{c-\mathrm{i}\infty}^{c+\mathrm{i}\infty} \bar{\Phi}(v, \varphi, z; s) \mathrm{e}^{st} \mathrm{d}s, \, c > 0$$

By substituting various expressions for $\bar{\Phi}$ given in (4.129)∼(4.129b), we obtain the integral needed for the calculation, which has the following form

$$I_k(\Upsilon, T) = \frac{1}{2\pi \mathrm{i}} \int_{c-\mathrm{i}\infty}^{c+\mathrm{i}\infty} s^{-k} \exp\{T_s - \Upsilon_s^{1/3}\} \mathrm{d}s \tag{4.130}$$

where k takes values of $1/6$, 0, $-1/6$ for different cases, whereas

$$\Upsilon \equiv \frac{1}{2} f_0'^{2/3} \delta \begin{pmatrix} \alpha_1 \\ \dot{\alpha}_1 \end{pmatrix} > 0 \tag{4.131}$$

$$T \equiv t - \tau \tag{4.132}$$

When $T < 0$, integral (4.130) can be shown to be 0; when $T \geqslant 0$, we can use the crossing method to estimate the value I_k, which gives the following results

$$I_k = \frac{3^{(6k-1)/4}}{2\pi^{1/2}} \Upsilon^{(3-6k)/4} T^{(6k-5)/4} U(T) \left[1 + O\left(\frac{T^{1/2}}{\Upsilon^{3/2}} \right) \right] + \exp\left\{ -\frac{2}{3^{3/2}} \frac{\Upsilon^{3/2}}{T^{1/2}} \right\} \tag{4.133}$$

where $U(T)$ denotes the Heaviside unit function.

Using Eq. (4.133) and taking into account the values α_1, $\dot{\alpha}_1$ and $\dot{A}i(-\alpha_1)$, $Ai(-\dot{\alpha}_1)$, and after obtaining the values of β_1, $\dot{\beta}_1$, and $\boldsymbol{H}^{(1)}_{1/3}(-\beta_1)$, $\boldsymbol{H}^{(1)}_{-2/3}(-\dot{\beta}_1)$, as well as introducing the following parameter D,

$$D \equiv \left(\frac{\rho}{F'r}\right)^{1/2} \tag{4.134}$$

we can obtain the approximation expressions for a diffraction field for large enough δ and small enough T,

(1) For the case of hard interfaces

$$\Phi \sim \mathrm{i}0.0797 f_0' \delta^{1/2} \frac{(c_0 r)^{1/2}}{\rho(z_0)} D(z_0) D(z) \frac{U(T)}{T} \exp\left\{-0.140 f_0' \frac{\delta^{3/2}}{T^{1/2}}\right\}, z_0 > 0, z > 0 \tag{4.135}$$

$$\Phi_0 \sim \mathrm{i}0.0967 f_0' \delta^{3/4} \frac{(\rho_0 c_0)^{1/2}}{\rho(z_0)} D(z_0) \frac{U(T)}{T_0^{5/4}} \exp\left\{-0.140 f_0' \frac{\delta^{3/2}}{T_0^{1/2}}\right\}, z_0 > 0, z = 0 \tag{4.135a}$$

$$\Phi_{0'} \sim \mathrm{i}0.0967 f_0' \delta_{0'}^{3/4} \left(\frac{c_0}{\rho_0}\right)^{1/2} D(z) \frac{U(T_{0'})}{T_{0'}^{5/4}} \exp\left\{-0.140 f_0' \frac{\delta_{0'}^{3/2}}{T_{0'}^{1/2}}\right\}, z_0 = 0, z > 0 \tag{4.135b}$$

$$\Phi_{00'} \sim \mathrm{i}0.1180 f_0' \delta_{00'} \left(\frac{c_0}{r}\right)^{1/2} \frac{U(T_{00'})}{T_{00'}^{3/2}} \exp\left\{-0.140 f_0' \frac{\delta_{00'}^{3/2}}{T_{00'}^{1/2}}\right\}, z_0 = 0, z = 0 \tag{4.135c}$$

(2) For soft interfaces

$$\Phi \sim \mathrm{i}0.1237 f_0' \delta^{1/2} \frac{(c_0 r)^{1/2}}{\rho(z_0)} D(z_0) D(z) \frac{U(T)}{T} \exp\left\{-0.486 f_0' \frac{\delta^{3/2}}{T_0^{1/2}}\right\}, z_0 > 0, z > 0 \tag{4.136}$$

$$\Phi_0 = \Phi_{0'} = \Phi_{00'} = 0, \quad z_0 = 0 \begin{array}{c}\text{or}\\ \text{and}\end{array} z = 0 \tag{4.136a}$$

4.6.6 Analyses and conclusions

It is not difficult to prove from the above diffraction field expressions that, on the wave fronts of a diffracted wave of $T = 0$, Φ and its time derivatives are all of order 0, which means that the increase of the field with time is slower than the powers of any order at the moment the pulse arrives. Because all of the above expressions of the field are valid only for the case of enough large δ, the above conclusion is only correct in this scope. In the vicinity of the shadow zone's edges, the field establishing process should be rapid. However, we still lack an appropriate processing method up to now for the field in this region.

It can be easily seen from (4.135) that the field is 0 for both $T = 0$ and $T \to \infty$, and that the field reaches the maximum $\Phi^{\max}$ when $T = T^{\max}$. The "normalized" field can be obtained by using this method

$$\frac{\Phi}{\Phi^{max}} = \left(\frac{T^{max}}{T}\right) \exp\left\{2\left[1 - \left(\frac{T^{max}}{T}\right)^{1/2}\right]\right\} \tag{4.137}$$

where

$$T^{max} \equiv 0.0049 f'^2_0 \delta^3 \tag{4.138}$$

Apparently T^{max} is directly proportional to the third power of δ. Therefore, the establishing process is slowed rapidly with the increase of the distance inside the shadow zone.

Similarly, the other expressions of the field can be written as

$$\frac{\Phi_0}{\Phi_0^{max}} = \left(\frac{T_0^{max}}{T_0}\right)^{5/4} \exp\left\{\frac{5}{2}\left[1 - \left(\frac{T_0^{max}}{T_0}\right)^{1/2}\right]\right\} \tag{4.137a}$$

$$\frac{\Phi_{00'}}{\Phi_{00'}^{max}} = \left(\frac{T_{00'}^{max}}{T_{00'}}\right)^{3/2} \exp\left\{3\left[1 - \left(\frac{T_{00'}^{max}}{T_{00'}}\right)^{1/2}\right]\right\} \tag{4.137b}$$

Where T_0^{max} and $T_{00'}^{max}$ denote the value of T when Φ_0(or $\Phi_{0'}$) and $\Phi_{00'}$ reach the maxima respectively.

$$\frac{T_0^{max}}{T^{max}} = \left(\frac{4}{5}\right)^2 = 0.640$$

$$\left(\frac{T_{00'}^{max}}{T^{max}}\right) = \left(\frac{2}{3}\right)^2 \sim 0.444$$

If we rewrite Eq. (4.136) in the form of Eq. (4.137), then T^{max} should be replaced by T_{soft}^{max} (for the comparison, T^{max} in (4.137) can be denoted as T_{hard}^{max} here), and we have

$$\frac{T_{soft}^{max}}{T_{hard}^{max}} \approx 12.05 \tag{4.138a}$$

We can clearly see from the above discussion that in case of hard interfaces, the establishing process of Φ_0 (or $\Phi_{0'}$) is faster than that of Φ, and $\Phi_{00'}$ is faster still. On the other hand, the establishing process of a field in case of soft interfaces is much slower than in the case of hard interfaces. Fig. 33 shows the dependence of Φ/Φ^{max} on T/T^{max} in all kinds of cases. Because T/T^{max} is a function of coordinates, and depends on the details of the atmosphere, the actual pictures of the field are not the same at all space points, which just demonstrates that distortion is created during the propagation of pulses. But the "relative" establishing process shown in Fig. 33 does not vary from place to place.

Now let us study the spatial distribution of Φ^{max}. It can be proved that, $\frac{1}{c_0}G(z_0) \equiv r_1$ is the distance between the edges of the shadow zone and the

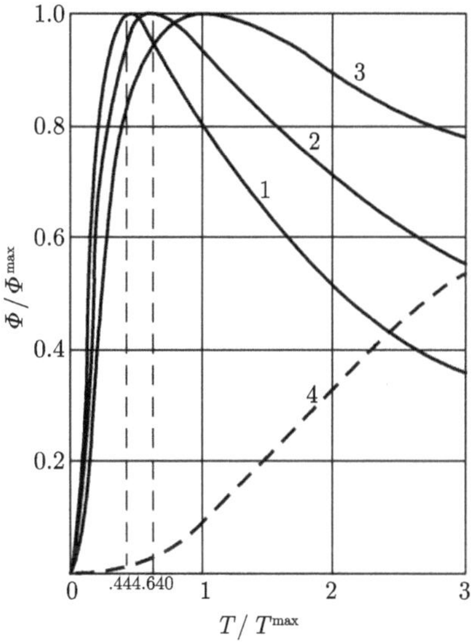

Fig. 33: The establishing process of a pulse diffraction field in the atmospheric shadow zone with wind. Hard interface scenario: $z_0 = 0$ and $z = 0$ (Curve 1); $z_0 = 0$ or $z = 0$ (Curve 2); $z_0 > 0$ and $z > 0$ (Curve 3); Soft interface scenario: $z_0 > 0$ and $z > 0$ (Curve 4).

coordinate origin, and $\frac{1}{c_0}G(z_0) \equiv \lambda$ is the horizontal distance between the observation point and the foot of the diffracted ray. Hence, according to Eq. (4.128)

$$\delta = c_0(r - r_1 - \lambda) \tag{4.128′}$$

By substituting (4.128′) and (4.138) into (4.135), we obtain

$$\Phi^{\max} = \mathrm{i}\boldsymbol{K}^{1/2}\left(\frac{r}{r_1}\right)^{-1/2}\left(\frac{r}{r_1} - 1 - \frac{\lambda}{r_1}\right)^{-5/2} \tag{4.139}$$

where

$$\boldsymbol{K}^{1/2} \equiv \frac{2.2}{c_0^2 f_0'}\left[\frac{\rho(z)}{\rho(z_0)F'(z_0)F'(z)}\right]^{1/2} r_1^{-3} \tag{4.140}$$

Similarly, we can obtain

$$\Phi_0^{\max} = \mathrm{i}\boldsymbol{K}_0^{1/2}\left(\frac{r}{r_1}\right)^{-1/2}\left(\frac{r}{r_1} - 1\right)^{-3} \tag{4.139a}$$

$$\boldsymbol{K}_0^{1/2} \equiv \frac{10.7}{c_0^{5/2} f_0^{3/2}}\left[\frac{1}{\rho(z_0)F'(z_0)}\right]^{1/2} r_1^{-7/2} \tag{4.140a}$$

$$\Phi_{0'}^{\max} = \mathrm{i}\boldsymbol{K}_0^{1/2}[\rho(z_0)\rho(z)]^{1/2} r^{-1/2}(r - \lambda)^{-3} \tag{4.139b}$$

$$\Phi_{00'}^{\max} = \mathrm{i}\frac{57.8}{c_0^3 f_0'^2} r^{-4} \tag{4.139c}$$

In this way, we are able to separate out the dependence of the field on r, and make it possible to study the generic law of the decrease of sound intensity with horizontal

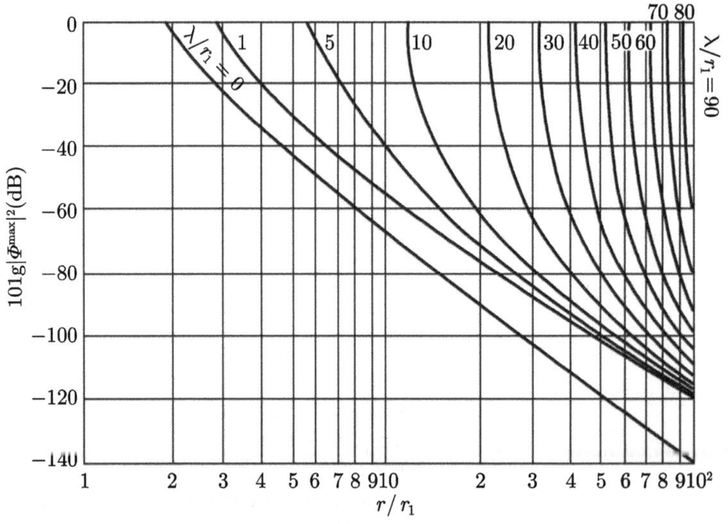

Fig. 34: The dropping law of diffracted sound intensity with horizontal distance in shadow zones.

distance for certain values of φ, z and z_0. As an example, the graphic representation of this rule when $z_0 \neq 0$ is given in Fig. 34 where the vertical axis denotes $10\lg|\Phi^{max}|^2$, i.e., the dB value of sound intensity. The zero dB sound level in the dB calculation is chosen randomly, and corresponds to disregard the constant multiplier when determining Φ^{max}. The dimensionless parameter r/r_1 for the horizontal axis provides a proper measure of the distance entering the shadow zone from the observing point. Parameter λ/r_1 reflects the effect of observing point height in the horizontal attenuation law. We observe that the sound intensity drops rapidly upon deep entrance to the shadow zone. Generally speaking, the larger the value of λ/r_1 is, the more quickly it drops.

Using Eqs. (4.138) and (4.138a) to derive the expression of Φ^{max} from (4.136) for the soft interface scenario, we can precisely obtain the same formula as (4.139), only with K being replaced by K_{soft}, and $K_{soft} \approx 0.0166K_{hard}$.

Thus the decline law of the soft interface case is completely identical to the hard interface case, though the absolute value of the former is only 1/60 of that of the latter, i.e., a difference of 17.8 dB. The dependence of sound intensity on φ and z depends on the details of atmosphere. Analysis for a simple example shows that this relation is weaker than that on r.

Chapter 5
Sound Absorption in Atmosphere

The propagation of acoustic waves with certain energies in the atmosphere certainly cannot last forever. The waves will gradually decay with the propagation distance, which acts to weaken the sound's intensity. The decay is caused firstly by a spherical wave spreading that results in a decrease of energy flow per unit area, and secondly by a change of the propagation path due to reflection, refraction and scattering, which causes the energy to deviate from its expected route. The above two kinds of attenuation are collectively referred to as "geometric attenuation", both of which involve only the redistribution of the acoustic energy, rather than the conversion of energy as well. Yet another type of attenuation involves the absorption of acoustic energy by the medium, in which the acoustic energy associated with ordered molecular motions is converted into thermal energy associated with random thermal agitations.

The acoustic absorption process by the atmosphere mainly includes two mechanisms, one of which is "classical absorption"[1] caused by atmospheric viscosity and thermal conductivity. Acoustic waves induce vibrations of atmospheric molecules, and viscous forces makes the vibration energy partly dissipate via internal friction among the molecules. In the meantime, an acoustic wave can also make some part of the atmosphere be compressed and hot, and make other part become cold due to expansion. The thermal conductivity of the atmosphere reduces temperature differences between regions, equivalent to transforming acoustic energy into thermal energy.

A second loss mechanism results from the dual atomic structure of atmospheric molecules. It is an irreversible process for the energy stored in the lumped movement of molecules (the average translation of molecules) to be transferred to that stored in the internal degrees of freedom of the molecules (rotation and vibration). Because sound pressure depends only on translation of the molecule, this transfer of energy will result in the attenuation of acoustic energy. This absorption mechanism is called "molecular absorption" or "relaxation absorption" because the energy conversion requires a certain amount of time, and is thus a type of relaxation process.

A comprehensive calculation of sound absorption should include not only the contribution of the two separate mechanisms themselves, but also their mutual

1 The origin of this name is from the mid-nineteenth century, the "golden age" of classical physics. Initially the absorption of sound by a medium was regarded as being caused only by shear viscosity and thermal conductivity. However, in his rougher measurement at the end of the nineteenth century, Duff found that the actual absorption is far greater than the calculated values based on the classic formula introduced by Stokes (1845) and Kirchhoff (1868) (Refer to Phys. Rev. 1900, 11: 65 – this is the first ever published paper about atmospheric acousticabsorption experiment).

dependencies. For example, the influence of molecular relaxation processes on the classical absorption mechanisms. Fortunately, both theory (the approximate solution of the Boltzmann equation [106]) and experiment (measured data at low pressure [107]) have confirmed that classical absorption and molecular absorption can be added together at frequencies below 10 MHz.

The absorption effect can be expressed with a sound absorption coefficient α through which the relationship between sound pressure amplitude and distance is shown as $p(r) = p^0 e^{-\alpha r}$. The unit of α can be dB/m, but more commonly Np/m. To sum up, α consists of three added parts $\alpha = \alpha_{\text{cl}} + \alpha_{\text{rot}} + \alpha_{\text{vib}}$. Here, α_{cl}, α_{rot} and α_{vib} represent the classical, rotation, and vibration absorption coefficient, respectively. α_{vib} is the superimposition of the contributions of molecular vibration of each component of atmosphere, where experimental results demonstrate that the contribution of O_2 and N_2 are the most dominant. Therefore

$$\alpha = \alpha_{\text{cl}} + \alpha_{\text{rot}} + \alpha_{\text{vib,O}} + \alpha_{\text{vib,N}} \tag{5.1}$$

Although the following discussion is mainly aimed at the frequency range of 100 Hz~10 MHz, actually, all mechanisms are guaranteed to be applicable[2] for infrasonic frequencies up to 10 Hz. As for infrasound below 10 Hz, contributions from radiation absorption may become more important. However, this aspect still lacks systematic research. We refer the reader to Chapter VI for some specific cases.

5.1 Classical absorption

5.1.1 Equation of motion for viscous fluid – Navier-Stokes equation

When considering the viscosity and heat conduction of a medium, the basic equations of fluid dynamics are given in Eqs. (1.30)~(1.32). Only the continuity equation still maintains its original form, while other two should be modified.

In the Euler equation shown in the form of (1.30b), the momentum flow density tensor defined in Eq. (1.33) only describes the reversible "macroscopic" transfer of momentum. That is, it is only related to the dynamic movement of the fluid from one place to another as well as the pressure acting in the fluid. The notion of fluid viscosity reflects the idea that there exists an irreversible "microscopic" transfer of fluid motion, i.e. momentum transfer from a place with greater speed to a place with smaller speed.

Therefore, the equation of motion of a viscous fluid can be easily obtained by only adding a term corresponding to the irreversible "viscous" momentum flow on top of the "ideal" momentum flow expression (1.33). Then the momentum flow density tensor in viscous fluid is written as

---

2 The major author of the main reference of this chapter in [110], H. E. Bass, emphasized this point in a private communication with the author of this book.

$$\Pi_{ik} = p\delta_{ik} + \rho v_i v_k - \sigma'_{ik} = -\sigma_{ik} + \rho v_i v_k \tag{5.2}$$

The tensor here

$$\sigma_{ik} = p\delta_{ik} + \sigma'_{ik}$$

is called the fluid stress tensor (comparing with the illustration of Eq. (1.33)) and σ'_{ik} is called the viscous stress tensor.

Only when each part of the fluid moves at different speeds relatively can the viscidity (internal friction) be regarded as significant, thus, σ'_{ik} depends on the spatial derivative of the speed. If the velocity gradient is not very large, it can be regarded as only depending on the first-order derivative, $\partial_k v_i$, and this dependence is treated as being linear. σ'_{ik} is 0 not only when v = const, but also when the whole fluid rotates at a constant angular velocity $\boldsymbol{\Omega}$ (i.e. when $\mathbf{v} = \boldsymbol{\Omega} \times \mathbf{r}$) (because there is no shear strain at this time, there is no internal friction). The linear combination that satisfies this requirement is called the symmetric composition of $\partial_k v_i + \partial_i v_k$. Notice that in this combination, some items, such as $\partial_l v_l \equiv \nabla \cdot \mathbf{v}$, should be excluded, because they only represent the output flow rate of the fluid per unit volume rather than the pure shear(s) of the fluid. Removing these items in a symmetric way, furthermore, by introducing corresponding coefficients σ'_{ik}, can be written in the following form[3]

$$\sigma'_{ik} = \mu(\partial_k v_i + \partial_i v_k) - \frac{2}{3}\mu\delta_{ik}\partial_l v_l \tag{5.3}$$

where μ is the shear viscosity coefficient, which is directly proportional to the velocity gradient.

Substituting Eqs. (5.2) and (5.3) into the right-hand side of Eq. (1.33), and using the continuity equation, motion equations for viscous fluids in the most common form can be obtained. Generally speaking, as η is a function of pressure and temperature, we can treat it as being constant, but in most cases where this change is not significant then we can obtain the Navier-Stokes Equation

$$\begin{aligned} \rho(\partial_t v_i + v_k \partial_k v_i) &= -\partial_i p + \partial_k \sigma'_{ik} \\ &= -\partial_i p + \mu\partial_k\left(\partial_k v_i + \partial_i v_k - \frac{2}{3}\delta_{ik}\partial_l v_l\right) \end{aligned} \tag{5.4}$$

or write it in vector form

$$\rho[\partial_t \mathbf{v} + (\mathbf{v}\cdot\nabla)\mathbf{v}] = -\nabla p + \mu\nabla^2\mathbf{v} + \frac{\mu}{3}\nabla(\nabla\cdot\mathbf{v}) \tag{5.4'}$$

3 In older references, items taking account of the volume viscosity coefficient η, $\eta\delta_{ik}\partial_l v_l$ were also included. Because η is only connected with the molecular rotation energy, it is properly considered in rotational absorption. However, according to the mechanism used by us, the concept of volume viscosity is not introduced even in rotational absorption.

By the way, we should point out that if the fluid can be regarded as non-compressible, and since $\nabla \cdot \boldsymbol{v} = 0$, then the above equation can be reduced as

$$\partial_t \boldsymbol{v} + (\boldsymbol{v} \cdot \nabla)\boldsymbol{v} = -\frac{1}{\rho}\nabla p + \nu \nabla^2 \boldsymbol{v}$$

here $\nu \equiv \mu/\rho$ is the kinematic viscosity coefficient. This parameter is more suitable than η for describing the movement characteristics of a fluid.

5.1.2 Equation of heat-conduction

The law of the conservation of energy for a ideal fluid, Eq. (1.36), must be modified for a viscous fluid. In addition to items related to simple mass transfer (the expression in the curly braces on the right-hand side of Eq. (1.36)), the energy flow density needs to include items that are related to the internal friction process and items caused by thermal conductivity.

By definition, the viscous stress tensor σ'_{ik} is just the momentum flow caused by viscosity (friction acting on a unit area), whose scalar product with the speed gives the energy flow. Therefore, the energy flow density related to viscosity is $v_i \sigma'_{ik}$.

The energy flow caused by heat conduction is a kind of energy transfer among molecules, which is independent of the fluid's macroscopic motion. That is, molecules with high speeds in a high-temperature region spread to the lower temperature region, which eventually makes a thermal balance of the whole region.

For any form of diffusion process, the flux density is directly proportional to the rate of change of the "concentration". Therefore, the heat flux density is directly proportional to the temperature gradient

$$\boldsymbol{j}_h = -\kappa \nabla T \tag{5.5}$$

where $\kappa(> 0)$ is the thermal conductivity coefficient, which, in general, is a function of both temperature and pressure. The minus sign indicates that the heat flux direction is opposite to the temperature gradient – i.e. heat flows from high-temperature regions to low-temperature regions.

In conclusion, the reference to the discussion presented when we derived Eq. (1.36), the energy conservation equation that is adapted to a viscous, heat-conducting fluid should be (i.e. the Fourier-Kirchhoff-Neumann energy equation)

$$\partial_t \left(\frac{\rho v^2}{2} + \rho \varepsilon \right) = -\nabla \cdot \left[\rho \boldsymbol{v} \left(\frac{v^2}{2} + w \right) - v_i \sigma'_{ik} - \kappa \nabla T \right] \tag{5.6}$$

The equation can be transformed into a more convenient form for application purposes. Firstly, using operator (1.29) as the definition of enthalpy w, and the relevant formula for vector analysis, we rewrite the above equation into

$$D_t \left(\frac{\rho v^2}{2} + \rho \varepsilon \right) + \left[\frac{\rho v^2}{2} + \rho \varepsilon + p \right] \nabla \cdot \boldsymbol{v} = -\boldsymbol{v} \cdot \nabla p + \partial_k (v_i \sigma'_{ik}) + \nabla \cdot (\kappa \nabla T) \tag{5.7}$$

Then expanding the left-hand side and applying the continuity equation (1.31) and the second law of thermodynamics, we obtain

$$\rho D_t\left(\frac{v^2}{2}\right)+\rho T D_t s=-\boldsymbol{v}\cdot\boldsymbol{\nabla}p+\partial_k(v_i\sigma'_{ik})+\boldsymbol{\nabla}\cdot(\kappa\boldsymbol{\nabla}T)$$

Take the scalar product of $\boldsymbol{v}$ and D in Eq. (5.5), We obtain

$$\rho D_t\left(\frac{v^2}{2}\right)=-\boldsymbol{v}\cdot\boldsymbol{\nabla}p+v_i\partial_k\sigma'_{ik}=-\boldsymbol{v}\cdot\boldsymbol{\nabla}p+\partial_k(v_i\sigma'_{ik})-\sigma'_{ik}\partial_k v_i$$

By subtracting the last two equations, we can obtain the general equation (i.e. the Kirchhoff-Fourier Equation)

$$\rho T(\partial_t s+\boldsymbol{v}\cdot\boldsymbol{\nabla}\boldsymbol{s})=\sigma_{ik}\partial_k v_i+\boldsymbol{\nabla}\cdot(\kappa\boldsymbol{\nabla}T) \tag{5.8}$$

When viscosity and thermal conductivity do not exist, the equation degenerates into an adiabatic equation for an ideal fluid, Eq. (1.35a). Eq. (5.8) has a very obvious physical meaning: the left-hand side denotes the heat obtained by the unit volume fluid in unit time, while the first term of the right-hand side denotes the part of the energy due to heat transformed from viscosity, and the second term expresses the heat directly brought to the volume via heat conduction.

5.1.3 Energy relationships of acoustic waves in viscous and heat-conducting fluids

So far we have obtained three equations that describe the motion of viscous heat-conducting fluids, which are the continuity equation (1.31a), the Navier-Stokes equation (5.4) and the equation of heat conduction (5.8). For small amplitude waves, (1.31a′) can be reduced to Eq. (2.3). The other two equations can also be linearized as before

$$\rho^0\partial_t\boldsymbol{v}=-\boldsymbol{\nabla}p_1+\partial_k\sigma'_{ik} \tag{5.4a}$$

$$\rho^0 T^0\partial_t s_1=\kappa\boldsymbol{\nabla}^2 T_1 \tag{5.8a}$$

What is different from before is that the superscript 0 is used to express the corresponding undisturbed value, as well to denote a certain reference value, while subscript 1 (omitted in $\boldsymbol{v}$, as usual) still corresponds to the parameter values after being disturbed by an acoustic wave. The difference from the ideal fluid case is that, because two new variables s_1 and T_1 are introduced in Eq. (5.8a), two equations must be added. The first is the equation of state (though different from Eq. (1.12) because the temperature disturbance must be considered), while the other describes the relationship between the entropy disturbance and the pressure disturbance & temperature perturbation. Both equations can be expressed in the form of partial derivatives. Taking advantage of the definition of the expansion coefficient

$$\beta=-\frac{1}{\rho}(\partial_T\rho)_p \tag{5.9}$$

and using the related thermodynamic relations, we have

$$\rho_1 = (\partial_{p^0}\rho^0)_{s^0}p_1 + (\partial_{s^0}\rho^0)_{p^0}s_1 = \frac{1}{c^2}p_1 - \left(\frac{\beta}{c_p}\rho^0 T^0\right)s_1 \tag{5.10}$$

$$T_1 = (\partial_{p^0}T^0)_{s^0}p_1 + (\partial_{s^0}T^0)_{p^0}s_1 = \frac{\beta T^0}{c_p\rho^0}p_1 + \frac{T^0}{c_p}s_1 \tag{5.11}$$

Firstly, let us consider the energy relationships of acoustic waves in a lossy medium. Dot multiplying Eq. (5.4a) by $\boldsymbol{v}$, multiplying Eq. (2.3) by p_1/ρ^0, and multiplying Eq. (5.8a) by T_1/T^0, and then adding them together, we obtain

$$\partial_t\left(\frac{1}{2}\rho^0 v^2\right) + \frac{p_1}{\rho^0}\partial_t\rho_1 + \rho^0 T_1\partial_t s_1 = -\nabla\cdot(p_1\boldsymbol{v}) + \partial_k v_i\sigma'_{ik} - \sigma'_{ik}\partial_k v_i + \frac{\kappa}{T^0}\nabla\cdot(T_1\nabla T_1) - \frac{\kappa}{T^0}(\nabla T_1)^2 \tag{5.12}$$

Using (5.10) and (5.11), we can write the sum of the last two terms in the left-hand side of the above equation as $\partial_t\left[\frac{1}{2}\frac{p_1^2}{\rho^0 c^2} + \frac{1}{2}\frac{\rho^0 T^0}{c_p}s_1^2\right]$, and the third item in the right end can be transformed into

$$\sigma'_{ik}\partial_k v_i = \mu\partial_k v_i\left(\partial_k v_i + \partial_i v_k - \frac{2}{3}\delta_{ik}\partial_l v_l\right) = \frac{\mu}{2}\left(\partial_k v_i + \partial_i v_k - \frac{2}{3}\delta_{ik}\partial_l v_l\right)^2 \tag{5.13}$$

Thus, it is not difficult to see that the left end of Eq. (5.12) denotes the rate of change of time of the acoustic wave energy $\partial_t E$, here

$$E \equiv \frac{1}{2}\rho^0 v^2 + \frac{1}{2}\frac{p_1^2}{\rho^0 c^2} + \frac{1}{2}\left(\frac{\rho^0 T^0}{c_p}\right)s_1^2 \tag{2.17'}$$

And E can be thought as the acoustic energy density in the discussed case. When the irreversible process caused by viscidity and heat conductivity is not considered, the entropy disturbance is $s_1 = 0$. Therefore, Eq. (2.17′) is reduced to the acoustic energy density equation for an ideal fluid (2.17). We further introduce the acoustic energy flux density vector

$$\boldsymbol{I} = p_1\boldsymbol{v} - v_i\sigma'_{ik} - \frac{\kappa}{T^0}T_1\nabla T_1 \tag{5.14}$$

where the second and third items correspond to the acoustic energy flow densities caused by viscidity and by heat conductivity, respectively. They are of course in the opposite direction of $p\boldsymbol{v}$ and lead to a reduction of sound intensity. When these two items are not considered, it is reduced to the common expression for the acoustic energy flow density vector in an ideal fluid, i.e. Eq. (2.18).

Therefore, Eq. (5.12) can be finally written in the form of

$$\partial_t E + \nabla\cdot\boldsymbol{I} = -\mathcal{D} \tag{5.15}$$

where

$$\mathcal{D} \equiv \frac{\mu}{2}\left(\partial_k v_i + \partial_i v_k - \frac{2}{3}\delta_{ik}\partial_l v_l\right)^2 + \frac{\kappa}{T_0}(\nabla T_1)^2 \tag{5.16}$$

Which denotes the energy loss per unit volume per unit time. Eq. (5.15) is actually the generalized law of the conservation of acoustic energy in a viscous heat-conducting medium (the right-hand side is 0 in the form of an ideal fluid). Its physical meaning is quite obvious: the first term on the left end is for the rate of change of time of the acoustic energy per unit volume, which must be equal to the sum of the net flux of the acoustic energy flowing through the surface of the volume (the second item is moved to the right end) and energy dissipated via internal friction (the first item on the right end of (5.16)) and heat-conduction (the second item).

5.1.4 Sound absorption coefficient in viscous and heat-conducting fluids

Applying the above results to a plane wave, the particle velocity can be written as

$$v_x = \hat{v}\cos(kx - \omega t), \quad v_y = v_z = 0$$

The first item of Eq. (5.16) becomes

$$\frac{\mu}{2}\left\{\left[\left(1 + 1 - \frac{2}{3}\right)\partial_x v_x\right]^2 + \left[-\frac{2}{3}\partial_x v_x\right]^2 + \left[-\frac{2}{3}\partial_x v_x\right]^2\right\}$$
$$= \frac{4}{3}\mu(\partial_x v_x)^2 = \frac{4}{3}\mu k^2 \hat{v}^2 \sin^2(kx - \omega t)$$

Now let us focus on its time-averaged value, which is

$$\frac{1}{2}\left(\frac{4}{3}\mu\right)k^2\hat{\mathbf{v}}^2$$

Before going further, we should first calculate the relationship between a temperature disturbance T_1 and the particle velocity $\mathbf{v}$ of the plane wave. It can be written as (under a linear approximation)

$$T_1 \approx (\partial_p T)_s p_1 = -\frac{(\partial_p s)_T}{(\partial_T s)_p}$$

According to the second law of thermodynamics, the following equation can be obtained directly

$$\mathrm{d}\left(\varepsilon - Ts + \frac{p}{\rho}\right) = -s\mathrm{d}T + \frac{1}{\rho}\mathrm{d}p$$

which means that is a Maxwellian relationship.

$$(\partial_p s)_T = -\left(\partial_T \frac{1}{\rho}\right)_p = \rho^{-2}(\partial_T \rho)_p$$

Therefore,

$$(\partial_p T)_s = \frac{-\rho^{-1}(\partial_T \rho)_p}{\rho T^{-1}[T(\partial_T s)_p]} = \frac{\beta T}{\rho c_p}$$

where β is defined in Eq. (5.9). By applying the relationship of the ratio of pressure and the characteristic impedance $\rho^0 c$ of the particle velocity, we obtain

$$T_1 = \left(\frac{\beta T^0}{\rho^0 c_p}\right) p' = \frac{\beta c T^0}{c_p} v_x \tag{5.17}$$

And, by taking the average over time, we obtain

$$\frac{\kappa c^2 T^0 \beta^2}{2c_p^2} k^2 \hat{v}^2 = \frac{(\gamma - 1)}{2c_p} k^2 \hat{v}^2$$

where the thermodynamic relation $\gamma - 1 = \dfrac{T^0 \beta^2 c^2}{c_p}$ is used. By substituting the above relationship into Eq. (5.16), we obtain the time-averaged energy dissipation of a plane wave

$$\bar{\mathfrak{D}} = \frac{1}{2}\left[\frac{4}{3}\mu + \frac{\gamma - 1}{c_p}\kappa\right] k^2 \hat{v}^2$$

For the energy density formula of the lossless ideal gas case, (i.e. the third term of Eq. (5.13) is equal to 0), by replacing the pressure $p_1 (= \rho c v)$ with the plane-wave particle velocity v, we can obtain the time-averaged energy density of a plane wave:

$$\bar{E} = \frac{1}{2}\rho^0 \hat{v}^2$$

For a plane wave propagating in an ideal gas with an intensity $\boldsymbol{I} = c\bar{E}$, the energy will reduce to a share of $2_{\alpha_{\text{cl}}} = \bar{\mathfrak{D}}/\boldsymbol{I}$ whenever it passes through a unit distance. Hence we abtain the classical acoustic absorption coefficient (i.e. the Stokes-Kirchhoff formula)

$$\alpha_{\text{cl}} = \frac{\omega^2}{2\rho^0 c^3}\left(\frac{4}{3}\mu + \frac{\gamma - 1}{c_p}\kappa\right) \equiv \frac{\omega^2}{2\rho^0 c^3}\delta(\text{N}_\text{p}/\text{m}) \tag{5.18}$$

Next, by introducing the characteristic length that describes the viscous and heat-conducting processes, l_v and l_h, the total length is given as

$$l_{\text{cl}} \equiv \frac{1}{\rho^0 c}\left[\frac{4}{3}\mu + (\gamma - 1)\frac{\kappa}{c_p}\right] \equiv l_\text{v} + (\gamma - 1) l_\text{h} \tag{5.19}$$

Then, (5.18) can be written as

$$\alpha_{\text{cl}} = \frac{\omega^2}{2c^2} l_{\text{cl}}$$

The magnitude of l_{cl} is the same as the molecular mean free path, about 10^{-5} cm in normal temperature and atmospheric pressure.

5.1.5 Practical classical sound absorption coefficient

The coefficients discussed can be reduced to a more practical form. Firstly let us connect κ and μ by using the Eucken expression [108]

$$\kappa = (15R_\mu/4)[4c_v/(15R) + 3/5] \tag{5.20}$$

For dry air the ratio c_v/R in the temperature range of 255~370 K(−48~97° C) is no more than 5/2 ± 0.4%, while for wet air in the same temperature range, the ratio slightly exceeds the ideal gas value with an increase of water vapor concentration. Therefore, it is inaccurate to only 1% when we take $c_v/R = 5/2$ in (5.20) is less than 1%. Hence, we obtain

$$\kappa = 19R\mu/4$$

Substituting it into Eq. (5.18) yields

$$\alpha_{\mathrm{cl}} = [2\pi^2 f^2/(\gamma p_c)](1.88\mu) \tag{5.18'}$$

Because the sound velocity c and viscous coefficient μ change with temperature, α_{cl} given by the above equation is also temperature dependent. According to Eq. (1.17), $c \sim T^{1/2}$, and the change of μ with temperature is given by (for dry air, the Sutherland equation [108])

$$\mu = \frac{\beta T^{1/2}}{1 + (S/T)} \tag{5.21}$$

where both β and S are empirical parameters, whose standard values are $\beta = 1.458 \times 10^{-6}$ kg m^{-1} s^{-1} K$^{-1/2}$, $S = 110.4$ K respectively. Using the standard values of γ, R, and M, as well as the reference temperature T_0 being 293.15 K, the dependence of c on temperature is[4]

$$c = 343.23(T/T_0)^{1/2}\mathrm{m/s} \tag{1.17'}$$

$$\mu = 7.318 \times 10^{-3}\frac{(T/T_0)^{3/2}}{T + 110.4}(\mathrm{kg/m}) \tag{5.21'}$$

Therefore, (5.18′) becomes

$$\alpha_{\mathrm{cl}} = 5.578 \times 10^{-9}\frac{T/T_0}{T + 110.4}f^2/(p^0/p_0) \quad (\mathrm{N_p/m}) \tag{5.18'a}$$

where p_0 is the reference pressure, equal to 1 at atmospheric pressure, or 1.011325×10^5 N/m^2.

5.1.6 Wave modes in viscous and heat-conducting media

In principle, using similar steps as taken in the first section of Chapter II, any four variables can be eliminated from the five basic equations in Section 1.1~1.3 of this

4 Here the sound velocity variation caused by absorption processes is ignored. To a fairly good approximation, the sound velocity can be written as [109] $c = c_{\mathrm{re}}(1 + \alpha^2 c_{\mathrm{re}}/\omega^2)$, in which c_{re} should be measured in the absence of an absorption process, which is in fact the sound speed calculated in Eq. (1.17). For the values of α in the temperature range of 273, 15~313.15 K (0~40° C), and frequency range of 50~10 MHz, the assumption of $c = c_{\mathrm{re}}$ means $\alpha^2 c_{\mathrm{re}}^2/\omega^2 \ll 1$, and the introduced calculation error is not more than 0.3%.

chapter to obtain the "wave equation" about the last (5th) variable. However, this is a rather complicated process. Furthermore, the final result is not in a convenient form for demonstrating its physical meaning, and for practical application. A more convenient approach is to use several equations respectively to describe the different motion modes rather than using a "uniform" equation. These modes are independent of each other under a linear approximation, and therefore they can be superimposed as they only couple at the boundaries.

In the following, we first obtain the relationship, the so-called dispersion relation, between the wavenumber k and frequency ω in various modes.

Substituting ρ_1 shown in Eq. (5.10) into the linearized continuity Eq. (2.3), and substituting T_1 shown in Eq. (5.11) into the linearized heat-conducting equation (5.8a), the equation of motion shown in Eq. (5.4′) obtains the following form after linearization

$$\rho^0\partial_t \boldsymbol{v} = -\nabla p_1 + \mu\nabla^2\boldsymbol{v} + \frac{\mu}{3}\nabla(\nabla\cdot\boldsymbol{v}) \tag{5.4′a}$$

Let the plane-wave solution of this basic equations be

$$\psi(x,t) = \hat{\Psi}\mathrm{e}^{-\mathrm{i}\omega t}\mathrm{e}^{\mathrm{i}\boldsymbol{k}\cdot\boldsymbol{x}} \tag{5.22}$$

Here, ψ represents any one of the five variables: ρ_1, $p_1\boldsymbol{v}$, s_1, or T_1. Substituting them into the basic equations, we obtain

$$\omega\left(\frac{1}{c^2}\hat{p}_1 - \frac{\beta\rho^0T^0}{c_p}\hat{s}_1\right) = \rho^0\boldsymbol{k}\cdot\hat{\boldsymbol{v}} \tag{5.23}$$

$$\mathrm{i}\omega\rho^0\hat{\boldsymbol{v}} = \mathrm{i}\boldsymbol{k}\hat{p}_1 + \mu k^2\hat{\boldsymbol{v}} + \frac{\mu}{3}\boldsymbol{k}(\boldsymbol{k}\cdot\hat{\boldsymbol{v}}) \tag{5.24}$$

$$\mathrm{i}\omega\rho^0c_p\hat{s}_1 = \kappa k^2\left(\frac{\beta}{\rho^0}\hat{p}_1 + \hat{s}_1\right) \tag{5.25}$$

Cross and dot multiplying Eq. (5.26) with $\boldsymbol{k}$, respectively, we have

$$(\mathrm{i}\omega\rho^0 - \mu k^2)(\boldsymbol{k}\times\hat{\boldsymbol{v}}) = 0 \tag{5.26}$$

$$\left[\omega\rho^0 + \mathrm{i}\frac{4}{3}\mu k^2\right](\boldsymbol{k}\cdot\hat{\boldsymbol{v}}) = k^2\hat{p}_1 \tag{5.27}$$

From solving Eq. (5.26) we obtain

$$k^2 = \frac{\mathrm{i}\omega\rho^0}{\mu} \tag{5.28}$$

As well as we can get $\boldsymbol{k}\times\hat{\boldsymbol{v}} = 0$, the latter possible solution corresponds to the longitudinal wave whose $\boldsymbol{k}$ is parallel to $\hat{\boldsymbol{v}}$. Eliminating $\boldsymbol{k}\cdot\hat{\boldsymbol{v}}$ in Eqs. (5.23) and (5.27) first, then and coupling with Eq. (5.25), we obtain

$$(\gamma-1)\left(1+\mathrm{i}\frac{c}{\omega}l_v k^2\right)\hat{s}_1 - \left(1+\mathrm{i}\frac{c}{\omega}l_v k^2 - \frac{\omega^2}{c^2}k^2\right)\left(\frac{\beta}{\rho^0}\hat{p}_1\right) = 0$$

$$\left(1+\mathrm{i}\frac{c}{\omega}l_h k^2\right)\hat{s}_1 + \mathrm{i}\frac{c}{\omega}l_h k^2\left(\frac{\beta}{\rho^0}\hat{p}_1\right) = 0 \tag{5.29}$$

Where the thermodynamic relation $\beta^2 T^0 c^2/c_p = \gamma - 1$ is used; l_v and l_h are defined in Eq. (5.19).

The premise that Eq. (5.29) has no invalid solutions is that its characteristic determinant is 0. Thus the quadratic equation about k^2 (the Kirchhoff dispersion relation) is obtained as follows

$$l_h\left(l_v + i\frac{c}{\omega}\right)k^4 + \left[1 - i\frac{c}{\omega}(l_v + \gamma l_h)\right]k^2 - \left(\frac{\omega}{c}\right)^2 = 0 \tag{5.30}$$

As long as the frequency is not very high (such as below 10 MHz), and both $\frac{\omega}{c}l_v$ and $\frac{\omega}{c}l_h$ are much less than 1. Then the approximate solutions shown by Eqs. (5.31) and (5.32) are valid,

$$k^2 \approx \left(\frac{\omega}{c}\right)^2 + il_{cl}\left(\frac{\omega}{c}\right)^3 \tag{5.31}$$

$$k^2 \approx i\frac{\omega}{cl_h} + (\gamma - 1)\left(1 - \frac{l_v}{l_h}\right)\left(\frac{\omega}{c}\right)^2 \tag{5.32}$$

Therefore, the three dispersion relations, which describe three different modes given by Eqs. (5.28), (5.31), and (5.32) can be unified in the following form

$$k^2 + f(i\omega) = 0$$

But in the discussed cases, k^2 is equivalent to the operator $-\nabla^2$, while $i\omega$ is equivalent to $-\partial_t$. Hence, for plane waves whose wave vector satisfies the dispersion relation given by Eq. (5.22), we have

$$[\nabla^2 - f(-\partial_t)]\psi(\boldsymbol{x}, t) = 0 \tag{5.33}$$

where ψ can be chosen as the most suitable one of the five parameters, ρ_1, $p_1\boldsymbol{v}$, s_1 or T_1, based on different situations.

For example, taking $\psi = v_{vor}$ (the subscript 'vor' corresponds to the vortex) in Eq. (5.28), we obtain

$$\nabla^2\boldsymbol{v}_{vor} = \frac{\rho^0}{\mu}\partial_t\boldsymbol{v}_{vor} \tag{5.34}$$

Substituting (5.28) into (5.27), (5.23) and (5.25), and solving the coupling equations simultaneously, we obtain the result that $\boldsymbol{k}\cdot\hat{\boldsymbol{v}}_{vor}$, $\hat{p}_{vor}$ and $\hat{s}_{vor}$ all are 0. Thus, it is obvious that $\boldsymbol{k}$ is vertical to $\boldsymbol{v}_{vor}$, meaning that this mode is a type of shear wave (transverse wave), and all of the disturbances of pressure, density, temperature and entropy are 0. In fact, this is a "vortex" movement that corresponds to a non-compressible flow. Additionally, equation Eq. (5.34) is a diffusion equation rather than a wave equation. Because $\boldsymbol{k}\cdot\boldsymbol{v}_{vor} = -i\nabla\cdot\boldsymbol{v}_{vor} = 0$, the equation can also be written as

$$\nabla\times\nabla\times\boldsymbol{v}_{vor} = -\frac{\rho^0}{\mu}\partial_t\boldsymbol{v}_{vor} \tag{5.34'}$$

In the same way, Eq. (5.35) in the following can be obtained from (5.31) (the subscript "ac" is for "acoustic")

$$\nabla^2 p_{\mathrm{ac}} - \frac{1}{c^2}\partial_t^2 p_{\mathrm{ac}} + \frac{l_{\mathrm{cl}}}{c^3}\partial_t^3 p_{\mathrm{ac}} = 0 \tag{5.35}$$

This equation can be regarded as an acoustic perturbation wave equation that is slightly modified for viscosity and heat-conducting processes. Here we have $\boldsymbol{k} \times \boldsymbol{v} = -\mathrm{i}\nabla \times \boldsymbol{v} = 0$, thus the acoustic wave is still non-rotational. Taking into account $\frac{\omega}{c} l_{\mathrm{h}} \ll 1$, and substituting Eq. (5.31) into the first formula of (5.29), we obtain the entropy disturbance in these types of modes (or propagating modes)

$$s_{\mathrm{ac}} \approx \frac{\beta l_{\mathrm{h}}}{\rho^0 c}\partial_t p_{\mathrm{ac}} \tag{5.36}$$

The disturbance is a first or second small quantity. To the same degree of approximation, we can obtain the following from Eqs. (5.23) and (5.24)

$$\rho^0 \partial_t \boldsymbol{v}_{\mathrm{ac}} \approx -\left[1 + \frac{l_{\mathrm{v}}}{c}\partial_t\right] \nabla p_{\mathrm{ac}} \tag{5.37}$$

From Eqs. (5.10) and (5.11), and also taking into account Eq. (5.36), we have

$$\rho_{\mathrm{ac}} \approx \left[1 - \frac{\gamma - 1}{c} l_{\mathrm{h}} \partial_t\right] \frac{p_{\mathrm{ac}}}{c^2} \tag{5.38}$$

$$T_{\mathrm{ac}} \approx \left[1 + \frac{l_{\mathrm{h}}}{c}\partial_t\right] \left(\frac{T^0 \beta}{\rho^0 c_p}\right) p_{\mathrm{ac}} \tag{5.39}$$

Comparing these results (the so-called polarization relations) with the corresponding non-modified relations (ignoring the second term in the square brackets), we can see that in the present case, in addition to the acoustic wave being attenuated, $\boldsymbol{v}_{\mathrm{ac}}$, ρ_{ac} and T_{ac} all have a tiny phase shift relative to p_{ac}. The disturbance to the entropy s_{ac} is very small, and has a phase difference of $\pi/2$ with p_{ac}. As long as $\frac{\omega}{c} l_{\mathrm{cl}} \ll 1$, the propagatino speed is still close to the adiabatic sound speed.

$$\nabla \times \boldsymbol{v}_{\mathrm{ent}} = 0$$

Finally, Eq. (5.32) (ignoring the second term) gives (the subscript "ent" corresponds to "entropy")

$$\nabla^2 s_{\mathrm{ent}} \approx \frac{1}{c l_{\mathrm{h}}}\partial_t s_{\mathrm{ent}} = \frac{\rho^0 c_p}{\kappa}\partial_t s_{\mathrm{ent}} \tag{5.40}$$

The form is exactly the same as (5.34), the well-known heat diffusion equation (Fourier Equation), and the corresponding mode is called a thermal mode (or entropy mode). In this mode, contrary to acoustic modes, the entropy disturbance is the principal disturbance, and the pressure disturbance ≈ 0. Because we also have $\nabla \times \boldsymbol{v}_{\mathrm{ent}} = 0$, this kind of modes is also non-rotational. Substituting Eq. (5.32) into (5.25) (here the second term should be taken into account, otherwise the result is 0), we can obtain the pressure disturbance at this time

$$p_{\mathrm{ent}} = -\frac{\gamma - 1}{\beta c}\rho^0(l_{\mathrm{h}} - l_{\mathrm{v}})\partial_t s_{\mathrm{ent}} \tag{5.41}$$

It is a much smaller quantity than the small quantity of the 2nd level (because of the existence of a factor of $l_{\mathrm{h}} - l_{\mathrm{v}}$). Substituting Eq. (5.41) into (5.10) and (5.11), we can, respectively, obtain

$$\rho_{\mathrm{ent}} \approx -\rho^0\left(\frac{\gamma - 1}{\beta c^2}\right)\left[1 + \frac{l_{\mathrm{h}} - l_{\mathrm{v}}}{c}\partial_t\right] s_{\mathrm{ent}} \tag{5.42}$$

$$T_{\mathrm{ent}} \approx \frac{T^0}{c_p}\left[1 - (\gamma - 1)\frac{l_{\mathrm{h}} - l_{\mathrm{v}}}{c}\partial_t\right] s_{\mathrm{ent}} \tag{5.43}$$

Obtaining the formula for $\boldsymbol{v}_{\mathrm{ent}}$ is slightly more difficult. Firstly, we eliminate $\boldsymbol{k} \cdot \hat{\boldsymbol{v}}$ in (5.24) using (5.27), and, ignoring the small terms, we have $\boldsymbol{k}\hat{p}_1 \approx \left[\omega\rho^0 + \mathrm{i}\frac{4}{3}\mu k^2\right]\hat{\boldsymbol{v}}$. Then we substitute Eq. (5.41) and Eq. (5.32) with the second level items being ignored into it. After some simplification, we obtain

$$\boldsymbol{v}_{\mathrm{ent}} \approx \frac{\gamma - 1}{\beta c} l_{\mathrm{h}} \nabla s_{\mathrm{ent}} \tag{5.44}$$

5.2 Molecular rotational relaxation absorption[5]

5.2.1 Absorption mechanism for modes of the internal degrees of freedom

The relaxation process of internal degrees of freedom can also cause energy losses. When the local temperature rises, the energy of the internal degrees of freedom mode also increases. If the temperature changes very slowly, the change of internal energy is equal to $c_{\mathrm{int}}\Delta T$, here c_{int} is the specific heat capacity of internal mode, and ΔT is the local difference between the temperatures with and without the existence of a sound wave. For low frequencies, when the local temperature at a maximum, the energy entering into the inner modes will immediately decrease as the local temperature begins to fall. However, the energy transferred back and forth between the translational and internal energy of the gas only happens when gas molecules collide with each other, which in turn and depend on the inner modes involved with each collision which all have certain possibilities for energy transfer. Therefore, with an increase in frequency, a time delay will be introduced into the mutual transfer of energy between the internal

5 Regarding sound absorption in atmosphere, especially molecular absorption, [110] gives very good comments, and the sections in this chapter are mainly based on this reference.

and translational modes. The delay is between the time needed for the fall of local temperature and that for the internal energy to be released into translational modes. This time delay caused the loss of acoustic energy is called attenuation. For a given energy mode, absorption will not happen at very low frequencies, but it will increase as frequency rises to a certain value f_r, i.e. the relaxation frequency of the mode. Then, because the change of local temperature is so rapid that it causes the internal modes to always lag behind in energy transfer, relaxation absorption becomes a constant. For any specific relaxation process, absorption can be expressed by the following expression [108, 111]

$$\alpha = \frac{\pi s}{c} \frac{f^2/f_r}{1 + (f/f_r)^2} \tag{5.45}$$

Where s is the relaxation intensity (in units of Np). It depends on the specific heat capacity of the relaxation modes. And $f = \omega/2\pi$ is the frequency of acoustic wave.

In the next section we will show that for a relaxation with only one degree of freedom, the above equation can be written as

$$\alpha = -\frac{\delta K_S/K_S^\infty}{2c} \frac{\omega^2 \tau'_{VS}}{1 + (\omega^2 \tau'_{VS})^2} \tag{5.45'}$$

where K_S^∞ is the fluid's instantaneous adiabatic compression rate ($= pc_p^\infty/c_V^\infty$, with c_p^∞ and c_V^∞ is the specific heat at a constant pressure and constant volume, respectively, when the frequency is much higher than f_r). δK_S is the relaxation compressional rate, while τ'_{VS} is the relaxation time under the partial pressure of the reaction in the mixture under the adiabatic and constant volume condition. For a single relaxation degree of freedom, $\delta K_S/K_S^\infty = -Rc'/[c_p^\infty(c_V^\infty + c')]$, where c' is the specific heat of the relaxation mode.

5.2.2 Rotational relaxation contributions

The main composition of atmosphere is O_2, N_2 and CO_2, where all have two rotational degrees of freedom. Thus, the specific heat capacity owing to rotational relaxation is $c' = R$. H_2O has three rotational degrees of freedom, and its rotational specific heat capacity is $3R/2$. However, because the mole fraction of water vapor in air in the temperature has a range of 273~313 K (0~40° C) is 8% at most, the contribution of water vapor to the air rotating specific heat capacity can be omitted, as a fairly accurate approximation.

The relaxation time of each rotational energy level is not the same, which causes the absorption function of the frequency to be quite complex [112, 113]. Nevertheless, when comparing with the average heat energy, the rotational energy levels of the main components of the air are closer. Therefore, the air rotational relaxation process behaves as if the rotational energy levels are continuous, which can thus be described by a single constant-volume relaxation time under adiabatic conditions $\tau'_{VS,\mathrm{rot}}$. For all frequencies less than 10 MHz, and all $\tau'_{vs,\mathrm{rot}} \ll \omega^{-1}$ [114], as well as at the above discussed value of $\delta K_S/K_S^\infty$, with the consideration of $c_V = c_V^\infty + c' = c_V^\infty + R$ and $R/c_V = \gamma - 1$, Eq. (5.45) becomes

$$\alpha_{\mathrm{rot}} = \{[\pi R(\gamma - 1)]/(cc_p^\infty)\}(f^2/f_{\mathrm{r,rot}}) \tag{5.46}$$

where,

$$f_{\mathrm{r,rot}} = 1/(2\pi\tau'_{vs,\mathrm{rot}}) \tag{5.47}$$

Which is the rotational relaxation frequency.

The relaxation frequency is needed to calculate the contribution of the relaxation to the absorption. Its value can be determined via the normalized reaction rate of dual components (between two components) occuring under normal atmospheric pressures.

5.2.3 Collision reaction rate

When a multi-atomic gas is compressed, the total compression power will not be immediately divided by each degree of freedom, because the rotation mode and vibration mode can share energy transfered only through dual-component collisions. The energy transfer reaction can be expressed by the reaction expression of two types of chemical compounds A and B in the form of

$$A^* + B \underset{k'_b}{\overset{k'}{\rightleftharpoons}} A + B \tag{5.48}$$

where the asterisk (*) indicates an excited molecule. k' and k'_b are, respectively, the forwards and backwards rates of the reaction in the partial pressures of the two components in the mixture (in units of s^{-1}).

When the rate constants k' and k'_b are known, the specific reaction rate equation that determines the ratio of the change of vibrant and rotational energy can be solved, and the effective specific heat capacity of the corresponding degrees of freedom is then given immediately. The meaning of the so-called effective specific heat capacity is the specific heat capacity that sound waves involve.

The dual component reaction rate k' can be written as the product of the collision frequency ν' and the energy transfer probability Π for a certain collision

$$k' = \nu'\Pi \tag{5.49}$$

The collision frequency[6] under standard atmospheric pressure p_0 is $\nu = [\sigma n\bar{v}]_{p_0}$, here, σ is the collisional cross section of the molecules in the reaction, n is the number

6 The quantity with prime corresponds to conditions under a practical gas pressure, while the quantity without prime corresponds to the conditions under standard atmospheric pressures.

of molecules per unit volume, and $\bar{v}$ is the mean speed of a molecule. Because for a given temperature, n is directly proportional to the pressure, v' can be denoted using v, $v' = [\sigma n\bar{v}]_{p_0}[p/p_0] = vp/p_0$ or $k' = v\Pi p/p_0 = kp/p_0$, where k is the reaction rate under a standard pressure. Using the assumption that all molecules are rigid balls, the collision frequency is related to the fluid viscosity coefficient μ by [108]: $v' = 1.25p/\mu$. If only one energy transfer reaction occurs, the relaxation time τ'_{vT} under the constant-volume and isotherm condition can be connected with the difference between k' and k'_b

$$1/\tau'_{vT} = k' - k'_b \tag{5.50}$$

For a single energy transfer reaction [115], we have Eq. (5.51), which is obtained according to "the principle of detailed balance" $k'_b = k' \exp[-\Delta E/(k_B T)]$. Here ΔE is the forward transfer energy per moore in the reaction, and k_B is the Boltzmann constant,

$$\tau'_{vs} = (c_p^\infty/c_p^0)\tau'_{vT} \tag{5.51}$$

In which c_p^0 is the specific heat capacity at a constant pressure when $f \ll f_r$ ($= 1/2\pi\tau'_{vs}$). Because both k' and k'_b are directly proportional to pressure, τ'_{vs} is then inversely proportional to the pressure. The relaxation frequency is also directly proportional to the pressure since it is inversely proportional to τ'_{vs}. Traditionally, the energy transfer rate is designated as a constant, and the relaxation time is specified to be the value under a standard pressure.

5.2.4 Absorption coefficient due to rotational relaxation

Substituting Eq. (5.49) into (5.50), then substituting the results into (5.51), and assuming that $\exp[-\Delta E/(k_B T)]$ is very small, we obtain $\tau'_{vs,rot} = c_p^\infty/(c_p^0 v'\Pi_{rot})$. Or, by applying Eq. (5.47), we obtain

$$f_{r,rot} = c_p^0 v'\Pi_{rot}/(2\pi c_p^\infty) \tag{5.52}$$

Where Π_{rot} is the average probability of multiple rotational energy modes. This probability is customarily written as $\Pi_{rot} = 1/Z_{rot}$, where Z_{rot} is the collision number generally needed to establish rotating balance after a change of temperature. Combining Eqs. (5.46) and (5.52), applying the substitution $v' = 1.25p/\mu$, and finally arranging the results in the form of Eq. (5.18′), we have

$$\alpha_{rot} = [2\pi^2 f^2/(\gamma pc)]\mu\{\gamma(\gamma-1)R/(1.25c_p^0)\}Z_{rot} \tag{5.53}$$

When we substitute $\gamma = 1.4$ and $c_p^0/R = 7/2$ into the above equation, the item in the curly bracket is found to be 0.128, which is independent of temperature, because the rotational specific heat capacity does not change with temperature. Comparing Eq. (5.18′) with (5.53), we can write

$$\alpha_{rot}/\alpha_{cl} = 0.128Z_{rot}/1.88 = 0.0681Z_{rot} \tag{5.54}$$

For dry air, the collision number of rotation Z_{rot} was measured at near-room temperature [114, 116] and higher temperatures [117]. Summarizing these experimental results [118], we can express the collision number in the temperature range of 293~690 K (20~417° C) as

$$Z_{rot} = 61.1\exp(-16.8T^{1/3}) \tag{5.55}$$

If we include the existence of water vapor, but ignoring the relaxation process of specific heat capacity for very small water vapor rotation, the collision number of rotation can be re-written as

$$\begin{aligned}Z_{rot} = \{&[X(N_2+O_2)/Z_{rot}(N_2+O_2)] \\ &+ [X(H_2O)/Z_{rot}(N_2+O_2+H_2O)]\}^{-1}\end{aligned} \tag{5.56}$$

Where $X(N_2+O_2)$ is the sum of the Moore fraction of nitrogen and oxygen, $Z_{rot}(N_2+O_2)$ is the rotation collision number of dry air, $X(H_2O)$ is the Moore fraction of water, and $Z_{rot}(N_2+O_2+H_2O)$ is the collision number of H_2O molecules required by building a rotary balance of N_2 and O_2. The last quantity can take the value from ∞ to 1 (corresponding to the probability being 0~1). The final Z_{rot} mainly depends on $X(H_2O)$ (under the condition that $Z_{rot}(N_2+O_2+H_2O) = 1$). Because when $X(H_2O) \lesssim 0.02$, $Z_{rot}(N_2+O_2) \approx 5$ [119, 120], the change of the mixture's rotation number should not be more than 2%, meaning that $Z_{rot} \approx Z_{rot}(N_2+O_2)$. Thus we have proved that the effect of vapor on the collision number of rotation can be ignored.

Combining Eqs. (5.45) and (5.55), the absorption coefficient caused by the combination of classical absorption factors and rotational relaxation is given by

$$\alpha_{cr} = 5.578\times10^{-9}\frac{T/T_0}{T+110.4}\frac{[1+4.16\exp(-16.8T^{-1/3})]}{p/p_0}f^2 \tag{5.57}$$

The estimation of the above equation for different temperatures shows that in the range of temperature 213~373 K (−60~100° C) the following simplified empirical formula can be used.

$$\alpha_{cr} = 1.83\times10^{-11}\frac{(T/T_0)^{1/2}}{p/p_0}f^2 \tag{5.57a}$$

The difference between the results from (5.57a) and those from (5.57) is less than 2%.

5.3 Molecular vibrational relaxation absorption

5.3.1 The exchange rate in mole numbers for vibration excited molecules

In the discussion about the rotational relaxation absorption in the previous section, the mathematical expression of the relaxation absorption can be significantly simplified when we consider the assumption that all of the rotational modes exchange energy with translational modes in a single relaxation time. However, though the assumption is valid for rotational modes, it is not valid for vibrational modes, because of the very different rates of energy exchange between vibrational and translational modes. The situation becomes very complicated, and indeed to quite a formidable degree, due to vibration energy exchange among different vibrational modes of molecules in collisions. When there is no energy being transferred among vibrational modes, the differential equation describing the number of vibrationally excited molecules of the mode depends only on the energy transfer rate constant and the pressure contribution of each part undergoing collisions. However, energy transfer processes among the vibrational modes couple to all of the differential equations because of the dependence of the vibrational relaxation rates of Type i on the excited state of other types, such as Type j.

The following energy transfer system is considered in the comprehensive absorption theory of multi-component relaxation [115, 121–123],

$$M_i + M_j \underset{k_{qb}}{\overset{k_q}{\rightleftharpoons}} M_k + M_l + \Delta E_q \tag{5.58}$$

where each M represents the component compositions (mol) of an atmosphere at a vibration energy level. ΔE_q represents the energy of the atmospheric composition per mole transferred from vibrational modes in the reaction (where a negative value indicates that the energy is transferred into the translational modes), k_q represents the rate of reaction $q(\mathrm{s}^{-1})$ under a standard atmospheric pressure, and k_{qb} represents the reverse reaction rate. Notice the difference of reaction rates between (5.58) and (5.48). The former is for a standard pressure case while the latter is for a partition-pressure case.

The net rate of the energy transfer in the reaction is $[\mathrm{d}_t\xi_q]\Delta E_q$, ($\xi_q$ is the net number of moles of the simulated atmospheric components by a given vibration generated in the energy transfer reaction q in a unit of kg·mol. $\mathrm{d}_t\xi_q$ is measured by the number of moles of M_k or M_l generated per unit time, and is given by the following formula

$$\mathrm{d}_t\xi_q = m(p/p_0)(k_q X_i X_j - k_{qb} X_k X_l) \tag{5.59}$$

where m is the total number of moles, p/p_0 is the ratio of the pressure of each component over the standard atmospheric pressure, and each X is the mole fraction of each corresponding component.

Let $r_q = k_q^0 X_i^0 X_j^0$, (identical to $r_{qb} = k_{qb}^0 X_k^0 X_l^0$ in the steady state), expand the above equation near the equilibrium state specified by subscript 0, and apply

$$k_q/k_{qb} = \exp[-\Delta E_q/(RT)]$$

we obtain

$$\mathrm{d}_t\xi_q = m\left(\frac{p}{p^0}\right)^0 \gamma_q \left[\sum_g \frac{\partial \ln[X_i X_j/(X_k X_l)]}{\partial m_g}(m_g - m_g^0) + \frac{\Delta E_q}{RT^2}(T - T^0)\right] \tag{5.60}$$

where $m_g = mX_g$ is the number of moles of the g th kind of molecules, and

$$\frac{\partial \ln[X_i X_j/(X_k X_l)]}{\partial m_g} = (mX_g^0)^{-1} \times \begin{cases} 2, & \text{when } g = i = j \\ 1, & \text{when } g = i \text{ or } j \\ 0, & \text{when } g \neq i, j, k \text{ or } l \\ -1, & \text{when } g = k \text{ or } l \\ -2, & \text{when } g = k = l \end{cases}$$

Each reaction q has a reaction rate equation of the form shown in Eq. (5.60).

Matrix notation should be adopted when dealing with a large number of reactions of potential importance in a polyatomic gas mixture, where all $d_t\xi_q$ are included in matrix ζ. All $r_q = r_{qb}$ located in the diagonal of diagonal matrix $\boldsymbol{k}$, and all $(m_g - m_g^0)$ are incorporated in the matrix $\boldsymbol{m}$. The diagonal matrix $\boldsymbol{X}$ contains the mole fractions of X_g^0 in its diagonal, while all other components in the rest positions are zero. Each row in the non-square matrices corresponds to different energy levels, and each column corresponds to different reactions. Therefore, N_{iq} with values of 1, 2, −1, or −2 demonstrate, respectively, the energy levels, once or twice, participating in reaction q, as well as the reaction directions, from the right/left ends. If the energy levels are not involved with either reaction part, or if they are involved with both parts with an equal number of times, then we have $N_{iq} = 0$. We thus denote ΔE_q as $\tilde{\boldsymbol{N}}\boldsymbol{u}$, with $\boldsymbol{u}$ being a matrix whose elements are energy levels per mole, and $\tilde{\boldsymbol{N}}$ is the transpose of matrix $\boldsymbol{N}$. Then, the change of rate of mole numbers of vibrationally excited molecules can be written as

$$d_t\boldsymbol{m} = -(p/p_0)\boldsymbol{N}\boldsymbol{k}\tilde{\boldsymbol{N}}[\boldsymbol{X}^{-1}\boldsymbol{m} - (m/RT^2)\boldsymbol{u}(T - T^0)] \qquad (5.61)$$

Because acoustic experiments are usually conducted in adiabatic conditions, Eq. (5.61) can be rewritten as

$$T - T^0 = [-\tilde{\boldsymbol{u}}\boldsymbol{m} - pm(v - v^0)]/(mc_v^\infty) \qquad (5.62)$$

where v is the specific volume (the reciprocal of density), c_v^∞ is the non-relaxation specific heat capacity (i.e. the specific heat capacity caused by the modes of translation and rotation when vibrational relaxation is considered). Combining Eq. (5.61) and Eq. (5.62), letting $\boldsymbol{I}$ be a unit matrix, replacing $v - v^0$ with $-(v/\rho)(\rho - \rho^0)$, and taking into account that $d_t = i\omega$ is a sine wave, the following equation can be obtained

$$\boldsymbol{m} = \left[\frac{i\omega\boldsymbol{I}}{p/p_0} + \boldsymbol{N}\boldsymbol{k}\tilde{\boldsymbol{N}}\left(\boldsymbol{X}^{-1} + \frac{\boldsymbol{u}\tilde{\boldsymbol{u}}}{Rc_v^\infty T^2}\right)\right]^{-1}\boldsymbol{N}\boldsymbol{k}\tilde{\boldsymbol{N}}\frac{\boldsymbol{u}pmv}{\rho Rc_v^\infty T^2}(\rho - \rho^0) \qquad (5.63)$$

Under adiabatic and isometric conditions, the temperature change $T - T^0$ in Eq. (5.62) is equal to $-\tilde{\boldsymbol{u}}\boldsymbol{m}/(mc_v^\infty)$. This is because an increase in the number of excited molecules leads to a decrease of translational energy, resulting in energy being transferring from vibrational modes to translational modes. For adiabatic cases, Eq. (1.10) works. Applying ideal gas law, $T - T^0$ equals to $(T/p)(p - p^0)$ therefore,

$$(p - p^0)/(\rho - \rho^0) = (p/T)(-\tilde{\boldsymbol{u}}\boldsymbol{m}/(mc_v^\infty))/(\rho - \rho^0) \qquad (5.64)$$

5.3.2 Dynamic adiabatic compression modulus

The attenuation of an acoustic wave is demonstrated by its wavenumber being complex, shown as

$$\omega/k = c[1 - \mathrm{i}(\alpha\lambda/2\pi)]^{-1} = (K_s^{\mathrm{dyn}}/\rho^0)^{1/2} \tag{5.65}$$

where K_s^{dyn} is the frequency-dependent (or dynamic) complex adiabatic compressional coefficient in unit of Np/m^2, and which can be written as

$$K_s^{\mathrm{dyn}} = K_{s,\mathrm{int}}^{\mathrm{dyn}} + K_s^{\infty} \tag{5.66}$$

where K_s^{∞} is the instantaneous contribution, while $K_{s,\mathrm{int}}^{\mathrm{dyn}}$ is the contribution of relaxation internal energy modes. Generally speaking, the adiabatic compressional coefficient K_S can be determined by the change of pressure and density in the form of $\rho^0(\Delta p/\Delta\rho)_S$.

The instantaneous contribution K_s^{∞} can be calculated from Eq. (1.10). Taking its derivative with respect to ρ, we obtain $K_s^{\infty} = \rho(\partial_\rho p)_S = \gamma p$. Because of the assumption of adiabatic condition, γ should be replaced by the instantaneous specific heat capacity ratio. Therefore we obtain $K_s^{\infty} = pc_p^{\infty}/c_V^{\infty}$. $K_{s,\mathrm{int}}^{\mathrm{dyn}}$ can be calculated using Eq. (5.64). Substituting $\boldsymbol{m}$ given in Eq. (5.63) into the formula, and making the approximation of $pv/(RT) \approx 1$, and remembering that $\rho^0/\rho \approx 1$, the following equation is obtained

$$K_{s,\mathrm{int}}^{\mathrm{dyn}} = \left(\frac{p}{c_V^{\infty}T^2}\right)\tilde{\boldsymbol{u}}\left[\frac{\mathrm{i}\omega I}{p/p_0} + \boldsymbol{Nk\tilde{N}}\left(X^{-1} + \frac{\boldsymbol{u\tilde{u}}}{Rc_V^{\infty}T^2}\right)\right]^{-1}\boldsymbol{Nk\tilde{N}u} \tag{5.67}$$

By expressing the second matrix in the square brackets by its eigen-values and eigenvectors (let it be $\boldsymbol{Nk\tilde{N}}(\mathcal{L})$), the above equation can be simplified. Solving the following equations can yield the right-handed eigenvector $\boldsymbol{r}_j$ and left-handed eigenvector $\boldsymbol{I}_j$.

$$\boldsymbol{R\Lambda} = \boldsymbol{Nk\tilde{N}}(\mathcal{L})^{-1}\boldsymbol{R} \tag{5.68}$$

$$\boldsymbol{\Lambda\tilde{L}} = \boldsymbol{\tilde{L}Nk\tilde{N}}(\mathcal{L})^{-1} \tag{5.69}$$

Where $\boldsymbol{R}$ is a matrix with the right-handed eigenvectors, $\boldsymbol{r}_j$, as its columns, and $\boldsymbol{\tilde{L}}$ is a matrix with left-handed eigenvectors as its row. $\boldsymbol{\Lambda}$ is a matrix with eigen-values as the diagonal elements, and where are all of the other elements are zeroes. Comparing the transposed form of Eq. (5.68) with Eq. (5.69), the following equation can be obtained,

$$\boldsymbol{\tilde{L}} = \boldsymbol{\tilde{R}}(\mathcal{L})^{-1} \tag{5.70}$$

Because the left-handed and right-handed eigenvectors are orthogonal and normalized to 1, we have $\boldsymbol{R\tilde{L}} = I$. Multiplying Eq. (5.70) by $\boldsymbol{R}$, and substituting it into Eq. (5.69) which is also multiplied by $\boldsymbol{R}$, we have

$$\boldsymbol{R\Lambda\tilde{R}} = \boldsymbol{NK\tilde{N}} \tag{5.71}$$

Thus, Eq. (5.67) can be written as

$$K_{s,\text{int}}^{\text{dyn}} = -\left(\frac{p}{c_V^\infty T^2}\right) \tilde{\boldsymbol{u}}\boldsymbol{R}\left[\frac{\mathrm{i}\omega \boldsymbol{I}}{p/p_0} + \boldsymbol{\Lambda}\right]^{-1} \boldsymbol{\Lambda}\tilde{\boldsymbol{R}}\boldsymbol{u} \tag{5.67a}$$

Now the above equation is in a pure quadratic form of $\tilde{\boldsymbol{R}}\boldsymbol{u}$, which can be expressed in a single summation form

$$K_{s,\text{int}}^{\text{dyn}} = -\left(\frac{p}{c_V^\infty T^2}\right) \sum_j \frac{\tilde{\boldsymbol{u}}\boldsymbol{r}_j^2}{1 + \mathrm{i}\omega\tau_{vs,j}} \tag{5.67b}$$

where $1/\tau_{vs,j}$ is the jth diagonal element of matrix $\boldsymbol{\Lambda}$. Denote $-p(\tilde{\boldsymbol{u}}\boldsymbol{r}_j)^2/(c_V^{\infty 2} T^2)$ as $\delta_j K_s$, and substituting it into Eq. (5.66), we have

$$K_S^{\text{dyn}} = K_S^\infty \left[1 + \sum_j \frac{\delta_j K_s / K_S^\infty}{1 + \mathrm{i}\omega\tau_{vs,j}}\right] \tag{5.72}$$

Let the square of Eq. (5.65) be equal to Eq. (5.72) divided by ρ^0, and then by separating the real and imaginary parts, the following is obtained

$$\left(\frac{c}{c^\infty}\right)^2 = 1 + \sum_j \frac{\delta_j K_s / K_S^\infty}{1 + (\omega\tau_{vs,j})^2} \tag{5.73}$$

$$\alpha\lambda\left(\frac{c}{c^\infty}\right)^2 = -\pi \sum_j \frac{\delta_j K_s / K_S^\infty}{1 + (\omega\tau_{vs,j})^2}\omega\tau_{vs,j} \tag{5.74}$$

where $c^\infty = (K_S^\infty/\rho^0)^{1/2}$ is the sound speed at frequencies above the relaxation frequency.

Eq. (5.74) represents the sum of a group of relaxation absorption curves, where each case corresponds to a different relaxation time $\tau_{vs,j}$ and different relaxation strength $\delta_j K_s / K_S^\infty$. Generally speaking, the number of relaxation times and relaxation strengths is the same as the number of energy modes present in the gas. However, as most relaxation strengths of these processes are very small, their contribution to the sound absorption can be neglected.

5.3.3 Vibration relaxation sound absorption coefficient

To determine the magnitude of the sound absorption coefficients, the energy transfer process that happens in the air must be identified, where the reaction rate can then be determined experimentally for dual-molecular mixtures of each reaction.

It is very complex to use a computer program to calculate vibrationally caused relaxation absorption by applying Eq. (5.74). Besides, each rate constant is not known very precise, which leads to an uncertainty of the sound absorption of order ±10%.

A verification of the complete calculation shows that only three terms are significant in the summation given in Eq. (5.74), of which the contribution of one of the three terms is less than 1% [123]. Because the uncertainty of basic reaction rate limits the calculation precision to ±5%, the 1% contribution of the third term can be completely neglected. Moreover, the difference between the ratio of $(c/c^{\infty})^2$ and 1 is only 0.1% at frequencies below 10 MHz, and under one atmospheric pressure. Therefore Eq. (5.74) can be written as

$$\alpha\lambda = -\pi\frac{\delta_1 K_S/K_S^{\infty}}{1+\omega^2\tau_{vs,1}^2}\omega\tau_{vs,1} - \pi\frac{\delta_2 K_S/K_S^{\infty}}{1+\omega^2\tau_{vs,2}^2}\omega\tau_{vs,2} \tag{5.75}$$

Where subscripts 1 and 2 represent the relaxation time and the relaxation strength associated with the two maximum values of $\delta_j K_S/K_S^{\infty}$, respectively. Careful examination of the magnitudes of $\delta_1 K_S/K_S^{\infty}$ and $\delta_2 K_S/K_S^{\infty}$ calculated using general theory [123] shows that both of them are almost independent of frequency. Thus, they can be further written as

$$\begin{aligned} \delta_1 K_S K_S^{\infty} &\approx -c'(O_2)R/\{c_p^{\infty}[c_V^{\infty} + c'(O_2)]\} \\ \delta_2 K_S K_S^{\infty} &\approx -c'(N_2)R/\{c_p^{\infty}[c_V^{\infty} + c'(N_2)]\} \end{aligned} \tag{5.76}$$

where $c'(O_2)$ and $c'(N_2)$ are the vibration-related specific heat capacities of O_2 and N_2, respectively.

The relaxation strengths of the two important processes in air are close to the expectation value of the single relaxation of each molecule. Thus, it can be concluded that (at least approximately) process 1 corresponds to O_2 molecular relaxation, while process 2 corresponds to N_2 molecular relaxation. Hence, subscript 1 and 2 can be replaced with H and U, respectively. By substituting $\lambda = c/f$ and $\omega = 2\pi f$, Eq. (5.75) can be written as

$$a_{\text{vib},j} = \frac{\pi s_j}{c}\frac{f^2/f_{r,j}}{1+(f/f_{r,j})^2} \quad (\text{Np/m}) \tag{5.77}$$

where j = O or N, $s_j = c'_j R/[c_p^{\infty}(c_V^{\infty}+c'_j)]$, and $f_{r,j} = 1/(2\pi\tau_{vs,j})$. The relaxation strength s_j can be related to the specific atmospheric compositions and temperature via the Planck-Einstein relationship [124]

$$c'_j/R = \frac{X_j(\theta_j/T)^2 e^{-\theta_j/T}}{(1-e^{-\theta_j/T})^2} \tag{5.78}$$

where X_j is the molar fraction of the components considered – 0.20948 for oxygen and 0.78084 for nitrogen [125]. θ_j is the characteristic vibration temperature, being 2239.1 K for oxygen, and 3350.0 K for nitrogen [126]. In the temperature range 0~40° C, c'_j is small relative to c_p^{∞}, making $c_p^{\infty} \approx c_p$, but $c_V^{\infty}+c'_j = c_V$, so $s_j \approx (c'_j R)(R^2/c_p c_V)$. Applying the approximation made when Eq. (5.18′) was derived, $(R^2/c_p c_V)$ can be replaced with $(\gamma-1)R/c_p$, whose value is 4/35. Therefore, Eq. (5.77) can be further written as

$$a_{\text{vib},j} = \frac{4\pi X_j}{35c}\frac{(\theta_j/T)^2 e^{-\theta_j/T}}{(1-e^{-\theta_j/T})^2}\frac{f^2/f_{r,j}}{1+(f/f_{r,j})^2} \tag{5.77′}$$

If the relaxation frequencies of oxygen and nitrogen $f_{r,j}$ are known, the acoustic absorption caused by vibrational relaxation can be calculated according to the above equation.

5.3.4 Vibration relaxation frequencies for oxygen and nitrogen

The highest absorption frequencies for the vibrational relaxation of oxygen and nitrogen can be calculated using general theory [120], where it is seen that the calculation demonstrates the dependence of frequency $f_{r,j}$ on water vapor concentration. However, there is no simple relationship between the various energy transfer rates and the highest absorption frequency. The latter is a function of water vapor content, $f_{r,j}$ and can be calculated for different water vapor concentrations. Then, a certain type of numerical expansion can be applied for $f_{r,j}$ as a function of water vapor content, in which the constants can be found by using the known reaction rates from general theory. The approximate solution of function $f_{r,O}$ is adopted in the following [127, 128].

$$f_{r,O} = (p/p_O)\{24 + 4.41 \times 10^4 h[(0.05 + h)/(0.391 + h)]\} \tag{5.79}$$

where h is the molar percentage of water vapor. All of the constants in the formula can be determined from general theory or by experimental measurements in the atmosphere. General theory predicts that the energy transfer rates of the modes are related to vibrational energy from O_2 to CO_2, where the translational modes decide the lower limit of $f_{r,O}$ for low humidity [120]. The existence of this low limit specifies the necessity of considering a general reaction scheme before adopting the approximate relation of $f_{r,O}$. However, Eq. (5.79) should be acceptable for calculating sound absorption in an atmosphere under usual meteorological conditions met in the low-layer atmosphere.

The molecular energy transfer rate constant, and the relaxation frequency, depend on temperature in several ways. Firstly, for a given pressure, an increase in temperature results in a decrease of density. Secondly, an increase of temperature causes an increase of the average molecular rate. Both effects lead to a net decrease of the collision frequency, which is inversely proportional to the change of viscosity coefficient μ. However, a decrease of the collision frequency is usually sufficiently offset by an increase in the probability of energy transfer Π in Eq. (5.49), due to an increase in temperature. The dependence of $f_{r,O}$ on temperature is difficult to accurately determined in experiments. Nevertheless one thing is certain, i.e., it can be predicted either from theory using known and assumed rates [120] or from experimental data [129]. In the temperature range of $0 \sim 40°$ C, $f_{r,O}$ is not sensitive to temperature. Thus, there is no apparent temperature factor in Eq. (5.79) and the effect of temperature only exists indirectly in water vapor molar percentage h.

The maximum absorption frequency of nitrogen is not as difficult to determine as that of oxygen, in theory, because it is dominated by direct “reduced excitation” (i.e. energy transferred from H_2O molecules to excited nitrogen molecules in a single step from vibration to translation ($V - T$) or from vibration to vibration ($V - V$)). Carbon dioxide provides another relaxation path when the water vapor concentration is low. The following expression [131] can be expected when the temperature dependence relation specified by theory [130] is applied.

$$f_{r,N} = (p/p_O)(T/T^0)^{-1/2}[9 + 350h \exp\{-6.142[(T/T_O)^{-1/3} - 1]\}] \tag{5.80}$$

Vibrational relaxation data for wet nitrogen [132] demonstrate that Eq. (5.80) may display a stronger dependence on humidity in wet air. These data are consistent with measurements [131, 133], but are different to the humidity dependence determined by some laboratories and field data [134, 135]. This difference needs to be further studied. According to the data given in reference [132], Eq. (5.80) can be replaced by the following formula

$$f_{r,N} = (p/p_O)(9 + 200h) \tag{5.80′}$$

5.3.5 Mole fraction (molecular concentration) of water vapor

If the molar percentage of water vapor is known, the contribution of the vibration relaxation process to atmospheric absorption can be determined using Eqs. (5.77′), (5.79) and (5.80), or (5.80′). According to Avogadro's law, the molar percentage is equal to the ratio of the partial pressure of water vapor p_W to the atmospheric pressure p of the wet air in sampling volumes. Therefore, $h = 100p_W/p$. We then introduce the saturated vapor pressure p_{sat} of pure water with liquid surface, to obtain

$$h = (100p_W/p_{sat})(p_{sat}/p) = h_r(p_{sat}/p) \tag{5.81}$$

where h_r is the relative humidity of a given wet air scenario under pressure p and temperature T. For convenience, we rewrite the above equation in standard atmospheric pressure p_O

$$h = \frac{h_r(p_{sat}/p_O)}{p/p_O} \tag{5.81′}$$

Because p_{sat} is usually an unknown in most acoustic measurements, another expression is needed for the calculation of (5.81′). For a given atmospheric pressure, the saturated vapor pressure is only a function of temperature [136]. The values of p_{sat} that correspond to different values of T can be found in the tables of various relevant manuals (such as in [126] and [137]). However, the following interpolation formula adopted by the World Meteorological Organization (WMO) is more convenient

$$\begin{aligned} \lg(p_{sat}/p_O) = {} & 10.79586[1 - (T_{01}/T)] - 5.02808 \log(T/T_{01}) \\ & + 1.50474 \times 10^{-4}(1 - 10^{-8.29692}[(T/T_{01}) - 1]) \\ & + 0.42873 \times 10^{-3}(10^{4.76955}[1 - (T_{01}/T] - 1) \\ & - 2.2195983 \end{aligned} \tag{5.82}$$

where, T_{01} is the isothermal temperature at the triple-phase point, whose exact value isrecognized internationally as being 273.16K [125]. The p_{sat}/p_O values provided by the above equation and by the table demonstrate that the difference between them is within ±0.06% [126] when the temperature is higher than 273K (0° C), and within ±0.04% [137] when temperature is between 233~273 K (−40~0° C).

For a specified T, the value of p_{sat}/p_0 can be obtained by taking the antilogarithm of the right-hand side of Eq. (5.82). The value of h can be determined by using (5.81′) and known values of h_r and p (or p/p_0).

A simplified form of (5.82) is

$$\log(p_{sat}/p_0) = 8.422 - 10.06(T_0/T) - 5.023\log(T/T_0) + 23 \times 10^{-4.44}(T_0/T) \tag{5.82a}$$

where $T_0 = 293.15$ K (20° C). The valid temperature range for the equation is 233~ 373 K (−40~100° C). A match of the result computed using Eq. (5.82a) and the published value is within ± 0.06% in the temperature range of 253~373 K, and within ± 0.4% in the range of 233~253 K. Therefore, with a moderate sacrifice of precision, the calculation can be simplified by replacing Eq. (5.82) with (5.82a).

Other forms of humidity information may also be provided, including ratios of mixing h_{mr}, specific humidity h_s and absolute humidity h_a. The relationship between these parameters and water vapor molecule concentration h is described below.

The mixing ratio h_{mr} is defined as the ratio of m_w, the water vapor mass of a given wet-air sample, over m_a, the mass of dry air combined with the water vapor mass in the sample. By definition, the molar fraction X_w is the ratio of the molar mass n_w of water vapor in a given wet-air sample over the total molar mass n_{ma} of the material contained in the wet-air sample $X_w = n_w/n_{ma}$. Since $n_{ma} = n_w + n_a$, (n_a, the amount of dry air substance), we have

$$X_w = n_w/(n_w + n_a)$$

The amount of material present in a given sample is the ratio of the mass of the substance over its relative molecular mass. Hence, for wet air containing only clear dry air and water vapor mixture

$$X_w = (m_w/M_w)/[(m_w/M_w) + (m_a/M_a)] = (m_w/m_a)/[(M_w/M_a) + (m_w/m_a)]$$

where M_w and M_a are the relative molecular mass of water and dry air, respectively. Plugging in their values of 18.016 and 28.9644 kg(mol)$^{-1}$ and $m_w/m_a = h_{mr}$ into the above equation, and noticing that $h = 100X_w$, we have

$$h = 100h_{mr}/(0.622 + h_{mr}) \tag{5.83}$$

The specific humidity h_s is the ratio of water vapor mass m_w contained within a given wet air sample over the total mass m_{ma} of the sample: $h_s = m_w/m_{ma}$.

As done in (5.83), h_s can be derived as

$$h_s = h_{mr}/(1 + h_{mr})$$

Or an equivalent relationship

$$h_{mr} = h_s/(1 - h_s) \tag{5.84}$$

Thus, if h_s is known, (5.84) and (5.83) can be used to acquire h.

The absolute humidity h_a is defined as the ratio of the water vapor mass m_w in a given wet air sample over the sample's total volume V under pressure p and temperature T: $h_a = m_w/V$. Assuming the sample follows the ideal gas law, the vapor component in the mixture can be written as

$$m_w/V = M_w p_w/RT = h_a$$

where p_w is the partial pressure of water vapor in a given sample. Introducing p_0 and p, the following equation is obtained

$$h_a = (M_w p_w/RT)(p_0/p_0)(p/p)$$

According to Dalton's partial pressure law $p_w/p = n_w/n_{ma} = X_w$ or $p_w/p = h/100$, the above equation can be transformed to

$$h_a = (M_w p_0/100R)(h/T)(p/p_0)$$

or

$$h = \frac{100R}{M_w p_0}\frac{Th_a}{p/p_0} \tag{5.85}$$

Substituting the values of the parameters cited above, we have

$$h = 4.5546 \times 10^{-1}\frac{Th_a}{p/p_0} \tag{5.85'}$$

where h_a is in kg/m^3.

5.4 Total absorption coefficient and additional absorption

5.4.1 Total absorption coefficient

The above description of atmospheric sound absorption is quite complicated and superfluous. It is therefore necessary to make a brief summary.

For stationary and homogeneous "pure" (but may contain water vapor) air, atmospheric sound absorption is caused by several mechanisms. First is the "classic" contribution caused by viscosity and thermal conductivity, which depends on temperature, pressure, and frequency (equation (5.18′a)). The second is the "rotational relaxation" contribution caused by the relaxation of rotationally excited molecules, which also depends on temperature, pressure, and frequency, and so can be expressed as a coefficient combined with classical contribution[7] (Eq. (5.57′)). The third is a lengthy

7 Because of this, the "volume viscosity coefficient" is introduced to describe the rotational relaxation contribution in earlier literatures. This historical perspective seems to "unify" some notions, but using relaxational process to introduce energy equation in this book is more in line with its physical nature.

and tedious part, which is a contribution of the loss associated with the relaxation processes of vibrating excited molecule. The vibrational absorption coefficient $\alpha_{\mathrm{vib},j}$ is very complex, which not only depends on temperature, pressure and frequency, but also on the specific compositions of the atmosphere (mainly O_2 and N_2, assuming their proportions are 20.9% and 78%, respectively) and the molar percentage of contained water vapor (Eq. (5.77′)). Thus, the total atmospheric absorption coefficient (5.1) can be written as

$$\alpha = f^2 \left\{ 1.83 \times 10^{-11} \left(\frac{p_0}{p}\right) \left(\frac{T}{T_0}\right)^{1/2} + \left(\frac{4\pi X_j}{35c}\right) \left(\frac{\theta_j}{T}\right)^2 \frac{\mathrm{e}^{-\theta_j/T}}{f_{\mathrm{r},j} + (f^2/f_{\mathrm{r},j})} \right\} \tag{5.86}$$

where we assume the factor $(1 - \mathrm{e}^{-\theta_j/T})^2)$ is equal to 1 in the calculation of the atmospheric absorption in a critical temperature range. Plugging the values of X_j and θ_j of oxygen and nitrogen given in Eq. (5.78) into the above equation, and using formula (1.17′) for c, the equation becomes

$$\alpha = f^2 \left\{ 1.83 \times 10^{-11} \left(\frac{p_0}{p}\right) \left(\frac{T}{T_0}\right)^{1/2} + \left(\frac{T_0}{T}\right)^{5/2} \left(1.278 \times 10^{-2} \frac{\mathrm{e}^{-2239.1/T}}{f_{\mathrm{r,O}} + (f^2/f_{\mathrm{r,O}})}\right) \right.$$
$$\left. + \left(1.069 \times 10^{-1} \frac{\mathrm{e}^{-3352/T}}{f_{\mathrm{r,N}} + (f^2/f_{\mathrm{r,N}})}\right) \right\} \tag{5.86′}$$

where $f_{\mathrm{r,O}}$ and $r_{\mathrm{r,N}}$ are given by equation (5.79) and (5.80), respectively. The molar percentage h is related to the relative humidity h_{r} by Eqs. (5.81′) and (5.82), or (5.82a), or links to the mixing ratio h_{mr}, the humidity ratio h_{s}, or the absolute humidity h_{a} respectively, as shown by Eqs. (5.83), (5.84), (5.83), and (5.85′).

The dependence of α on frequency, as a whole and as each component, is shown in Fig. 35. Molecular vibration loss of O_2 and N_2 is "catalyzed" by water vapor content in the air. Therefore, this part of the loss is very sensitive to the content of water vapor content in the air (humidity).

5.4.2 Additional sound absorption

Another sound absorption mechanism that needs to be considered is that gas mixtures of low-mass molecules with higher heating rates reach equilibrium faster than high-mass molecules do when a local gradient in pressure or temperature exists. The diffusion caused by pressure gradients accompanied by the "first" diffusion of the low-mass molecules due to a thermal gradient introduces another additional absorption [109]

$$\alpha_{\mathrm{ad}} = \frac{2\pi^2 f^2 \gamma X_1 X_2 p D_{12}}{pc^3} \left[\frac{M_2 - M_1}{M} + \frac{(\gamma - 1)k_T}{\gamma D_{12} X_1 X_2}\right]^2 \tag{5.87}$$

where X_1, X_2 and M_1, M_2 are the molar fractions and the relative molecular masses of gases 1 and 2, respectively. $M's$ is the mixed molecular mass of gas 1 and 2, while D_{12}

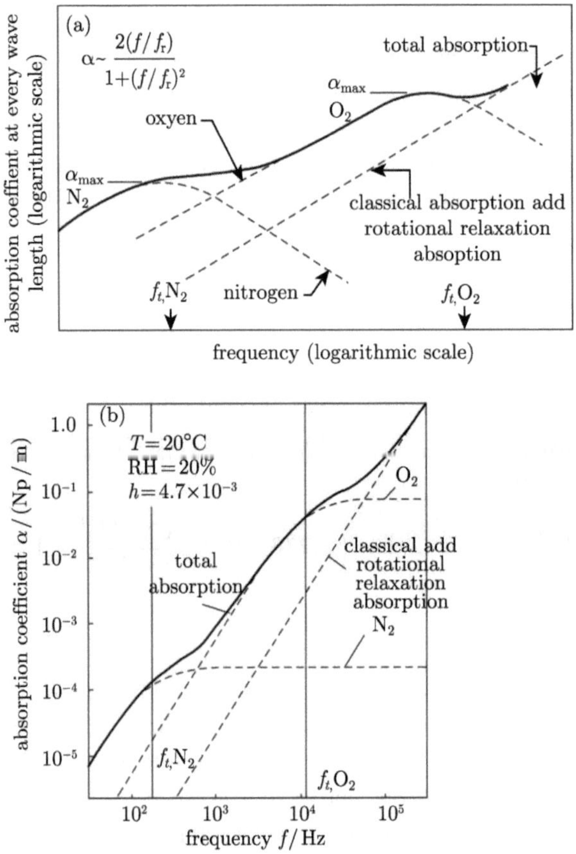

Fig. 35: Dependency of the absorption coefficient on frequency. (a) Losses per wave length; (b) Losses per unit distance.

is the concentration diffusion coefficient between gas 1 and 2, and k_T is the thermal diffusion coefficient.

To calculate α_{ad} in Eq. (5.87), the product pD_{12} and quotient k_T/D_{12} must be known. They can be calculated by applying molecular motion theory [108]. However, experimental data of this are rare. pD_{12} is at its maximum when considering the collision of O_2/H_2O; but in this case both X_1X_2 and $k_T/(X_1X_2)$ are very small. Collisions between O_2/N_2, X_1 and X_2 in air are relatively large, however the molecular mass difference $M_2 - M_1$ and k_T are both very small. Neglecting relaxation processes, it can be proved that for the air [106] 99.5% of the total classical absorption are counted by Eq. (5.18′). Thus, the additional absorption owing to diffusion can be neglected under the precision range specified by the equation.

By the way, there exists another additional absorption due to thermal radiation. When local gas temperature rises, the number of molecules in an excited vibrational state increases according to the Maxwell-Boltzmann Statistical Law. An example is an electric dipole moment stimulated by inner modes. The energy may be released by spontaneously emitting photons. The energy leaves the local high temperature areas and is reabsorbed in the low-temperature region or is dissipated in regions where acoustic waves reside. In air, molecules that possess a strong electric dipole moment are H_2O and CO_2. The theoretical study of sound absorption caused by thermal radiation is rare, and not detailed enough. For example, because of the rather unconvincing assumption of ignoring energy re-absorption, experimental results deviate essentially from practical cases. Experiments (summarized in [135]) do not find signs of thermal radiation. Thus, it seems that it is only important beyond the frequency ranges that experimental data covers, which is 500∼100 MHz.

5.5 Sound absorption in fog and suspended particles

Droplets and other particles suspended in an atmosphere may introduce additional sound absorption mechanisms, including evaporation and condensation around the particles; some of the particles being pushed away due to sound velocity fluctuation; the exchange of thermal energy between particles and the surrounding air; the stimulation of particle modality oscillations; as well as the temperature and velocity gradients accompanied by heat flow and viscosity effects. For a variety of reasons, experimental data regarding these effects (i.e. a real fog) are rare. Thus, understanding of these mechanisms relies mainly on theoretical analyses and the laboratory (artificial fog) data.

For actual atmospheric fogs, the range of droplet sizes and concentration (particle number per unit volume) is considerably large. Nevertheless, a droplet's radius is at the micron (μm) level (the average about 8 μm). The mass ratio between liquid water and air in a unit volume (notified as g_{wo} in the following chapters) is typically about 1.5×10^{-3}. "Submicron" droplets of the order of 0.01 μm can be obtained in laboratories. In theory, different sized droplets have essentially no differences [138, 139] except when droplet size decreases, the relaxation process shifts to higher frequencies and the maximum of acoustic absorption increases with the same factor.

5.5.1 Historical review

Acoustic wave propagation in a fog is the second historical controversial issue[8] after the "wind effect". As early as in 1708, Derham thought that fogs would introduce additional attenuation, and thus inhibit sound propagation. Field experiments carried out by Tyndall (1874) negated the above conclusion. Tyndall thought that

8 It caused controversy because of Rayleigh's conceptual error in this area. See footnotes on page 74.

temperature fluctuations were reduced in fogs, which explains the observed result why sound attenuation becomes smaller within them.[9] The dispute was finally resolved in the 1970s. The conclusion is that in the range of low audible frequencies and in infrasonic region, the following factors cause an increase in attenuation – the transform relaxation of heat and momentum, mass transfer, and heat release between droplets and vapor. However, when the frequency becomes higher than the relaxation frequency, sound attenuation in fogs becomes smaller than that related with moisture. The development of a new model makes this more satisfactory theory possible. The model treats the "sea" of small water droplets as a continuum medium, i.e. treats it as a third kind of gas besides air and water vapor mixtures. In a wide frequency range, three types of transfer relaxation processes (mass, momentum and heat) are introduced. Marble and others developed the equations of this mixture [140]. Cole, etc. obtained numerical predictions based on similar theories, though experimental results [141, 142] were not in good agreement with their predictions. Later, Davidson found that it was because of ignorance of the difference between air and vapor in the energy conservation equation, and because of the omission of some terms having the same order of the terms being kept in the equations. The re-introduction of these items saw that the modified predictions were consistent with observations [143]. According to the calculation, the maximum attenuation in fogs occurs below 1 Hz in frequency or between 6∼60 s in period.

Davidson further introduced the Burgers equation from nonlinear acoustics (see section 1.4 in Chapter 9), by replacing the complex problem described by nine partial differential equations with a single general equation that included all of the transfer process [144] in classical absorption and a "mono-disperse" fog (an ideal model where all fog droplets have the same size). He also tried to account for molecular absorption of wet air and the "poly-dispersion" of fogs (i.e. the droplet size obeys certain distribution rules) [145], and comprehensively discussed the theory of audible sound absorption in fog.

The study of the surface evaporation and condensation phenomenon of fog droplets when sound propagates in fog was proposed by Ostwald at the beginning of the twentieth Century. Ostwatitsch determined the attenuation and dispersion caused by this effect in 1941, and also calculated the maximum absorption at low frequencies (the absorption is 0 when frequencies are 0 and ∞).[10] Knudsen et al. studied the effect experimentally for the first time in 1948 [146].

9 It could also be thought that the attenuation decrease is not caused by the fog itself, but rather by "the atmosphere" in which the fog exists. In any case, both natural and artificial fogs are completely different, which also explains why so few field data exists, and why major studies focus on theory and lab data.

10 Refer to, Oswald W. Chem, Zbl, 1908, 1(1):1 104 and Oswatitsch K. L. Physik, Zeit., 1941, 42: 365 respectively.

Chinese scholar Wei Rongjue did a lot of work regarding sound absorption in a fog in the 1950s and 1980s with his students, and made important contributions to both theory and experiment. He modified Oswatitisch's theory, and got the result of absorption for similar molecules by introducing the relaxation time τ expressed by the fog droplet constant, and found the "bell" shape dependence of attenuation per unit wavelength on $\omega\tau$. When $\omega\tau = 1$, maximum absorption appears. However, when comparing with experiments, it only shows qualitative agreement. He thought it was because of the difficulty of determining a droplets' parameters both quickly and accurately. This issue has not yet been solved despite considerable effort.

5.5.2 Basic analyses: mass transfer process

Assume that the phase velocity of a sound wave in a fog is a complex number with limiting values being real at both low and high frequencies. Wet air can be treated as an ideal gas because the mass of water vapor with relative humidity lower than 100% is much smaller than that of the air, and interaction among fog droplets can be neglected. At low frequencies, the number of fog droplets per unit mass of air does not change with and without the existence of an acoustic field.

Consider the one-dimensional plane acoustic wave in foggy air, similar to the process applied in section 2.1 of Chapter 2. The "relative" fluctuation equations for pressure, density, and humidity in the foggy air can be obtained from the basic equations of fluid dynamics: To get the second equation about p' and T', the following relationship can be derived in low-frequency limit cases.[11]

$$\partial_t^2 p' - \frac{p_0}{\rho_0} \partial_x^2 p' = \partial_t^2 T' \tag{5.88}$$

(1) The fog droplet "Growth Equation"

$$\mathrm{d}_t M_r = 4\pi r D \rho_{v0} (\rho'_v - p'_{vr} + T'_r - T') \tag{5.89}$$

where M_r is the mass of a single fog droplet, and D is the diffusion rate of water vapor in air.

(2) The energy conservation law requires

$$L\mathrm{d}_t M_r = 4\pi r \kappa T_0 (T'_r - T') + M_r c_w T_0 \partial_t T'_r \tag{5.90}$$

11 Subscript "0" represents parameters undisturbed by an acoustic field, while "′" represents the relative fluctuation of the corresponding parameters caused by a sound field, such as $p' = (p - p_0)/p_0$ and so on. Subscript "a" corresponds to the parameters of dry air, subscript "v" corresponds to the parameters of water vapor, and subscript "m" corresponds to those of the mixture of air and vapor. Subscript "r" corresponds to the surface of the fog droplets (r is the radius of the droplets).

and

$$4\pi r\kappa T_0(T'_r - T')\frac{n_0}{\rho_0} = c_p T_0 \partial_t T' - (c_p - c_v) T_0 \partial_t p' \tag{5.91}$$

where L is the heat arising from the evaporation of fog droplets, c_w is the specific heat capacity of water, n_0 is the density of droplet number in the absence of sound.

(3) When water transfers from a liquid state to a gaseous state, the quantitative change relationship gives

$$\mathrm{d}_t M_r = \frac{\rho_0 m_v p_{v0}}{n_0 m p_0}(\partial_t p' - \partial_t p'_v) \tag{5.92}$$

and

$$\partial_t p'_{vr} = \frac{T_0}{p_{v0}}(\partial_T - \rho_v)_0 \partial_t T'_r \tag{5.93}$$

where m is the molecular mass. The second equation can be derived from Eqs. (5.88)~(5.93)

$$\frac{\gamma - 1}{\gamma}(1 + \lambda_s\lambda_D - \lambda_f\lambda_D)\partial_t^2 p' + \omega^*\left(\lambda_f + \frac{\gamma - 1}{\gamma}\right)\partial_t p'$$
$$= (1 + \lambda_s\lambda_D - \lambda_f\lambda_D)\partial_t^2 T' + \omega^*(1 - \lambda)\partial_t T' \tag{5.94}$$

where

$$\omega^* \equiv 4\pi D n_0 r \tag{5.95}$$

is the relaxation frequency that corresponds to the mass transfer process, whose reciprocal

$$\tau_c \equiv 1/\omega^* = \rho_{w0} r^2 / 3D\rho_{a0} g_{w0} \tag{5.96}$$

is the relaxation time, where g_{w0} is the ratio of water content and air mass in a unit volume. Each λ is defined as

$$\lambda_D = c_p \rho_0 (D/\kappa)$$

$$\lambda_f = (L/c_p T_0)(\rho_{v0}/\rho_0)$$

$$\lambda_s = \frac{L\rho_{v0}}{c_p p_{v0} \rho_0}(\partial_T p_w)_0 \tag{5.97}$$

λ varies more or less with p and T. From the condition that p' and T' have a common solution in the form of $e^{-i\omega(t - x/v_\infty)}$ that satisfies Eqs. (5.88) and (5.94), the phase velocity of a sound wave V in a fog can be expressed as λ, ω^* and other parameters. The value when ω^* goes to infinity can be shown as

$$V_0^2 = \lim_{\omega \to \infty} V^2 = \gamma \frac{p_0}{\rho_0} \frac{1 + \lambda}{1 + \gamma(\lambda_s - \lambda_f)}$$

$$V_\infty^2 = \lim_{\omega \to \infty} V^2 = \gamma \frac{p_0}{\rho_0} \tag{5.98}$$

The losses per unit distance, per unit time, and per wavelength are shown in the following, respectively

$$\begin{aligned}\alpha_F^d &= \frac{\omega}{\mathrm{Re}V}\tan\varphi\\ \alpha_F^t &= \mathrm{Re}V\alpha_F^d = \omega\tan\varphi\\ \alpha_F^\lambda &= 2\pi\tan\varphi\end{aligned}\tag{5.99}$$

Where Re V is the real part of V

$$\tan\varphi = \left\{\left(1-\frac{V_0}{V_\infty}\right)^2\left[\left(\frac{V_0}{V_\infty}\right)^2+(\omega\tau_c)^2\left(\frac{1+\lambda_D\lambda_s-\lambda_D\lambda_f}{1+\gamma(\lambda_s-\lambda_f)}\right)^2\right]^{-1}\right\}$$
$$\omega\tau_c\frac{1+\lambda_D\lambda_s-\lambda_D\lambda_f}{1+\gamma(\lambda_s-\lambda_f)}$$

It can also be written in a form similar to that of molecular absorption [such as Eq. (5.45′)]

$$\alpha_F^\lambda = 2\pi\frac{\omega\tau_c}{1+(\omega\tau_c)^2}J(p,T)\tag{5.100}$$

The maximum occurs when $\omega\tau_c \approx 1$. Under normal atmospheric pressure, J is a complex function that is strongly dependent on T. For example, $J \approx 0.15$ when the temperature is 0° C, while $J \approx 0.22$ when the temperature is 21° C.

5.5.3 Further analyses

In [141], three relaxation processes (apart from the transfer of mass, the transfer of momentum and heat are also considered) are studied using the same physical model. Their results are consistent with those obtained by Wei Rongjue who used the mass transfer model.

As mentioned in Section 1, Davidson [143] pointed out the incorrectness of certain terms in reference [141]. He instead wrote the energy equation as

$$\begin{aligned}&\rho_\mathrm{m}\partial_t e_\mathrm{m} + \rho_\mathrm{m} u_\mathrm{m}\partial_x e_\mathrm{m}\\ &= -p_\mathrm{m}\partial_x u_\mathrm{m} + 4\pi n\kappa r(T_r - T_\mathrm{m})\\ &\quad + 6\pi\eta n r(u_r-u_\mathrm{m})^2 - n\mathrm{d}_t M_r\left[\frac{1}{2}(u_r-u_\mathrm{m})^2 + h_r - e_\mathrm{v}\right]\end{aligned}\tag{5.101}$$

Where e is the special internal energy, η is the air viscous coefficient, u is the particle velocity, and h is the vapor-specific enthalpy. The meaning of the subscripts is the same as in the previous chapters. Ignoring the 2nd order terms, the following equation is obtained

$$\rho_\mathrm{m}\partial_t e_\mathrm{m} = -p_\mathrm{m}\partial_x u_\mathrm{m} + 4\pi n_0\kappa r(T_r - T_\mathrm{m}) - (c_{p,\mathrm{v}} - c_{v,\mathrm{v}})n_0\mathrm{d}_t M_r T\tag{5.101′}$$

By substituting the mass conservation equation

$$-n_0 \mathrm{d}_t M_r = \partial_t \rho_{\mathrm{m}} + \rho_{\mathrm{m}} \partial_x u_{\mathrm{m}} \tag{5.102}$$

into Eq. (5.101), we obtain

$$\rho_{\mathrm{m}} \partial_t e_{\mathrm{m}} = -p_{\mathrm{m}} \partial_x u_{\mathrm{m}} + 4\pi n_0 \kappa r (T_r - T_{\mathrm{m}}) + (\partial_t \rho_{\mathrm{m}} + \rho_{\mathrm{m}} \partial_x u_{\mathrm{m}}) R_{\mathrm{v}} T \tag{5.103}$$

where R_{v} is the gas constant per unit mass. The state functions for the mixture and for the water vapor are

$$p_{\mathrm{m}} = \rho_{\mathrm{m}} R_{\mathrm{m}} T, \quad p_{\mathrm{v}} = \rho_{\mathrm{v}} R_{\mathrm{v}} T$$

Therefore,

$$p_{\mathrm{v}} = p_{\mathrm{m}} \frac{\rho_{\mathrm{v}} R_{\mathrm{v}}}{\rho_{\mathrm{m}} R_{\mathrm{m}}} = p_m \frac{\rho_{\mathrm{v}} m_{\mathrm{m}}}{\rho_{\mathrm{m}} m_{\mathrm{v}}} \tag{5.104}$$

Taking the continuity equation $\partial_x u = \frac{1}{\rho_0} \partial_t \rho_{\mathrm{a}}$ into account, the energy equation (5.103) becomes

$$\rho_{\mathrm{m}} \partial_t e_{\mathrm{m}} = \left(1 - \frac{m_{\mathrm{m}}}{m_{\mathrm{v}}}\right) \frac{p_{\mathrm{m}}}{\rho_{\mathrm{a}0}} \partial_t \rho_{\mathrm{a}} + 4\pi r n_0 \kappa (T_r - T_m) + \frac{p_{\mathrm{m}} m_{\mathrm{m}}}{\rho_{\mathrm{m}} m_{\mathrm{v}}} \partial_t \rho_{\mathrm{m}} \tag{5.103$'$}$$

By neglecting the differences between the vapor-air mixture and the vapor, and approximating all parameters by their corresponding air parameters, Eq. (5.103′) becomes Eq. (5.91). Because $m_{\mathrm{m}} \approx 28.8$, $m_{\mathrm{v}} \approx 18$, the first and third terms on the right-hand side of Eq. (5.103′) are $-0.6(p_{\mathrm{m}}/\rho_{\mathrm{m}})\partial_t \rho_{\mathrm{m}}$ and $1.6(p_{\mathrm{m}}/\rho_{\mathrm{m}})\partial_t \rho_{\mathrm{m}}$, respectively. Neither can be ignored because they have the same order of magnitude.

Davidson introduced Burger's equation to solve the problem presented in [144] further. Firstly he omitted the 3rd and high-order terms in the three conservation equations, and then derived general equations of each transfer-relaxation process by applying commonly used methods from nonlinear acoustics, and finally obtained the total absorption coefficient per wavelength caused by the three types of transfer processes.

$$\frac{\alpha_F^\lambda}{2\pi} = \sigma B (\gamma A)^{-1} \cdot \frac{\omega \tau_c}{1 + (\omega \tau_c)^2} + g_{\mathrm{w}0} \left[\frac{\omega \tau_D}{1 + (\omega \tau_D)^2} + \frac{\gamma - 1}{\gamma} \frac{c_{\mathrm{w}}}{c_{V,\mathrm{m}}} \frac{\omega \tau_T}{A^2 + (\omega \tau_T)^2} \right] \tag{5.105}$$

Where σ is the ratio of vapor and air contents in a unit volume, and

$$A \equiv 1 + \sigma L (L - 1)(\gamma_{\mathrm{v}} - 1) \frac{c_{V,\mathrm{v}}}{c_{V,\mathrm{m}}}$$

$$B \equiv [(\gamma - 1)(L - 1) - 1] \left[(\gamma_{\mathrm{v}} - 1)(L - 1) \frac{c_{V,\mathrm{v}}}{c_{V,\mathrm{m}}} - \frac{m_{\mathrm{a}}}{m_{\mathrm{v}}} \right] \tag{5.106}$$

τ_c is defined in Eq. (5.96), and the two parameters are defined by the following equations

$$\tau_D = 2\rho_{\mathrm{w}0} r^2 / 9\eta, \quad \tau_T = \rho_{\mathrm{w}0} c_{\mathrm{w}} r^2 / 3\kappa \tag{5.107}$$

Which are the relaxation times of the transfers of mass, momentum, and energy, respectively.

In order to compare with Eq. (5.100), the first term (which corresponds to the mass-transfer process) on the right-hand side of Eq. (5.105) can be written as

$$\alpha_F^\lambda = 2\pi \frac{\omega\tau_c}{1+(\omega\tau_c)^2} K(p, T)$$

Under a normal atmospheric pressure, when the temperature is 0° C or 21° C, K is roughly equal to 0.11 and 0.14, respectively. These results are 30%~35% smaller than those given by Eq. (5.100), but more consistent with observations in general.

Fig. 36(a) shows the relationship of α_F^λ versus frequency based on Eq. (5.105). There are two peaks: the first one is at low-frequency bands and caused by mass transfer, which is significantly higher than the second one at high-frequency bands caused by momentum transfer. It can also be seen that the first peak is located in the infrasonic frequency band for larger fog droplets (r = 8 μm, curve A), while the peak is in the low audible frequency band for smaller droplets (r = 1 μm, curve C). The figure demonstrates that when the radius of a fog droplet decreases, the mass

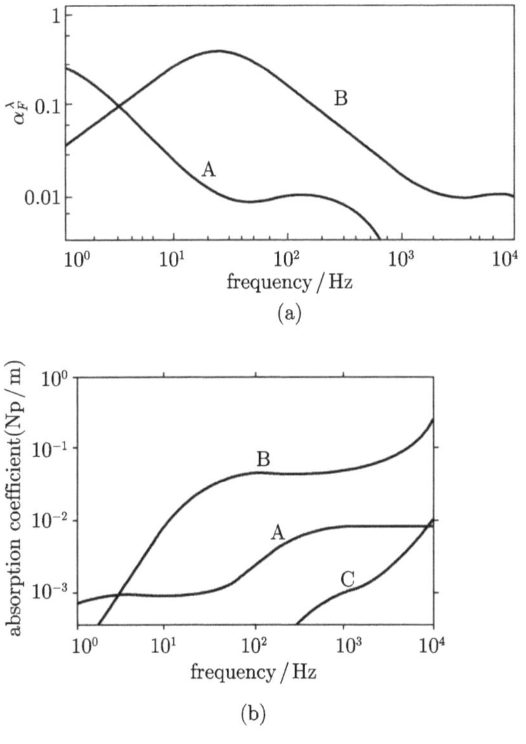

Fig. 36: Absorption coefficient versus frequency. (a) Absorption coefficient per wavelength; (b) Sound absorption coefficient per unit distance g_{w0} = 1.5×10^{-3}; r = 8 μm (curve A), 1 μm (curve B); and molecular absorption (curve C).

transfer process shifts to a higher frequency band and sound absorption in this band will increase significantly.

Fig. 36(b) shows the absorption coefficient per unit distance α_F^d. It can be seen that sound absorption increases inversely along with a decrease of wavelength of the same order of magnitude. This is an inevitable result of the above conclusion. The molecular absorption is also given (curve C) in the same figure for comparison.

Chapter 6
Effects from Gravity Field and Earth's Rotation

The atmospheric layer at the surface of the earth is affected by the earth's gravitational field and its rotation. These effects influence wave motions in the atmospheric layer, which can be regarded as a special form of fluid movements. In general, for sound waves in the audible range and infrasonic waves with higher frequencies (e.g. supersonic waves), the influence cannot be accounted for because in wave motion the acceleration is directly proportional to the frequency squared. Thus, in the above-mentioned frequency range, this acceleration is numerically far greater than acceleration due to gravity, g. However, when the frequency is below a certain value, for example, when it is g/c, these two accelerations will have the same order of magnitude.

In fact, the atmospheric density under the action of gravity will decrease exponentially with increasing altitude. For a sound wave with frequency $g/c \approx 0.029$ Hz and wavelength $c^2/g \approx 12$ km, the density will obviously change. When the steady atmosphere is disturbed by this sort of sound wave, a certain air mass will push down from the lighter upper layer to the heavier lower layer. Then, under the action of buoyancy the air mass will return to the original water level, and will even over it owing to inertia. As can be seen, in addition to the compression-restoring force of the medium in which the ordinary sound wave propagates, there is another restoring force – buoyancy (in fact that is the gravity) provided by the "stratified non-homogeneity" of the atmosphere's density. As mentioned in section 3.3 of Chapter 1, under the action of buoyancy, there will be an intrinsic vibrating frequency in the atmosphere called the buoyancy frequency or Väisälä-Brunt (V-B for short) frequency (Formula (1.28)). It is generally a function of altitude, and its value changes from 0.01 to 0.025 Hz.

For the above-mentioned wave, we must take gravity into account and the notion that the wave still maintains the effect of a resilient restoring force. The wave has the properties of a ordinary wave, as well as specific characteristics of a gravity wave (for example, gravity always points in only one direction, and is anisotropic); thus it is also called an acoustic gravity wave (AGW).

Along with a continuous decrease in frequency, the action of buoyancy will increase, and the action which originally dominated will become relatively weaker until it can be neglected. That is to say, under this sort of frequency (about 1/10 of the V-B value, namely about 0.003 Hz), the atmosphere can be regarded as "incompressible", and the corresponding wave is named an internal gravity wave (IGW). Its mechanism is completely similar to the (outer) gravity wave that usually exists at an incompressible fluid surface. It does not exist only at the boundary layer of two different media, but also in a medium that has a continuously stratified variation in density.

Except for gravity, the curvature of the earth itself and the earth's rotation will also influence a wave. When any substance (of course including vibrating air particles) on the earth's surface moves relative to the ground, as long as the direction of its relative velocity is not parallel to earth's axis, it will be affected by Coriolis acceleration $2\boldsymbol{\Omega}_{\rm E}\times\boldsymbol{v}$ (see the second term of formula (1.43)). The direction of the acceleration is perpendicular to the plane determined by $\boldsymbol{\Omega}_{\rm E}$ and $\boldsymbol{v}$, and the value $2\boldsymbol{\Omega}_{\rm E}\boldsymbol{v}\sin\theta$ will change with latitudinal angle θ. The Coriolis effect changes with latitude however, which generates another outer force field that creates a type of horizontal transverse wave with an extraordinary long wavelength of several kilometers. The vibrating direction of particles is perpendicular to the propagation direction in the same horizontal plane. This is a planetary wave, which was named after the discoverer as a Rossby wave. It establishes the wave-shape system that is distributed alternately with high and low pressure in the atmosphere.

Similar to an ocean, there are "atmospheric tides" in the atmosphere. These are different to oceanic tides caused primarily by the moon, where atmospheric tides are chiefly created by thermal effects induced by the sun. The attraction of the moon and sun are relatively smaller for the earth's atmosphere, where the period of the atmospheric tide is the divisor of the solar day, that is $24/m$ hour ($m = 1, 2, 3, \cdots$).

Strictly speaking, planetary waves and atmospheric tides are not in the acoustic category (clearly IGWs do not belong in this category). Although, as the important wave forms in a "generalized" atmosphere, both of them should have a niche in atmospheric acoustics. However, we stress that they are not the main objects of study here, although they have important meaning in meteorology.

In what follows, according to the system in [30], we discuss the atmospheric wave theory that considers earth's gravitational field and rotation.

6.1 Wave system in quiescent atmosphere

6.1.1 Fundamental equations and frequency dispersion equation

As in Chapter 2, we start with the three fundamental equations of fluid kinetics. To compute the effect of gravity, the gravity acceleration g should be added to the right-hand side of Euler's equation (1.30a). By then substituting the acoustic disturbances formula (2.1) into it, and taking the fluid statics equation (1.13) into account for linear processing, we get

$$\rho_0 \partial_t v = -\nabla p + \rho g \tag{6.1}$$

As before, a value with subscript 0 indicates that it is undisturbed, while a value without a subscript (in fact subscript 1 has been omitted) indicates that it is a disturbed quantity (i.e. after being perturbed). In the same way, respectively processing the continuous equation (1.31′) and the conditional equation (1.32) we get

$$\partial_t \rho + \boldsymbol{v} \cdot \nabla \rho_0 + \rho_0 \nabla \cdot \boldsymbol{v} = 0 \tag{6.2}$$

$$\partial_t p + \boldsymbol{v} \cdot \nabla p_0 = c^2 (\partial_t \rho + \boldsymbol{v} \cdot \nabla \rho_0) \tag{6.3}$$

Choosing the coordinate system to make the x axis point in the wave propagation direction and the z axis point in the vertical direction, according to the coordinate components the above equation can be written as follows

$$\left.\begin{aligned} \rho_0 \partial_t \mathbf{v}_x &= -\partial_x p \\ \rho_0 \partial_t v_z &= -\partial_z p - g\rho \\ \partial_t \rho - \frac{\rho_0}{H} v_z + \rho_0(\partial_x v_x + \partial_z v_z) &= 0 \\ \partial_t p - \rho_0 g v_z &= c^2 \left(\partial_t \rho - \frac{\rho_0}{H} v_z\right) \end{aligned}\right\} \tag{6.4}$$

In the previous equation formula (1.13) is used again, and in the last two equations the Halley law (1.15) was used, in which H is the scalar height defined by formula (1.16a).

In fact, the supposition of an isothermal atmosphere as mentioned above is not entirely correct, where, more strictly speaking, we should consider an atmosphere with a constant scale-level (i.e. H does not change with altitude), otherwise Halley's law will not be tenable. Now, we further suppose that all the disturbances change sinuously (plane wave solution), i.e. $\mathbf{v}$, p and ρ can be expressed as $\exp[\mathrm{i}(k_x x + k_z z - \omega t)]$, where k_x, k_z are complex wave numbers that are concerned with attenuation and growth. Thus Eq. (6.4) can be turned into four algebraic equations that include four variables v_x, v_z, ρ/ρ_0 and p/p_0, expressed in matrix form as follows

$$\begin{bmatrix} -\mathrm{i}\omega & 0 & 0 & \mathrm{i}k_x gH \\ 0 & -\mathrm{i}\omega & g & g + \mathrm{i}k_z gH \\ \mathrm{i}k_x & -\frac{1}{H} + \mathrm{i}k_z & -\mathrm{i}\omega & 0 \\ 0 & \frac{\gamma - 1}{H} & \mathrm{i}\omega\gamma & -\mathrm{i}\omega \end{bmatrix} \begin{bmatrix} v_x \\ v_z \\ \frac{\rho}{\rho_0} \\ \frac{p}{p_0} \end{bmatrix} = 0 \tag{6.5}$$

where using (1.17) and (1.16a) to express p_0/ρ_0 as $c^2/\gamma = gH$. In order for (6.5) to have a non-trivial solution, its characteristic determinant must be equal to 0, so we can obtain the important frequency dispersion equation

$$\omega^4 - \omega^2 c^2 (k_x^2 + k_z^2) + (\gamma - 1) g^2 k_x^2 + \mathrm{i}\omega^2 \gamma g k_z = 0 \tag{6.6}$$

Analyzing the above equation indicates that it is impossible for k_x and k_z to both be non-zero real numbers simultaneously, i.e. the attenuation or growth of amplitude will be generated in either the vertical or horizontal direction. Now in order to simplify the equation, we suppose that there is no change in amplitude in the horizontal direction, thus k_x should be purely real number. Thus (6.6) can be divided into real and imaginary parts

$$\omega^4 - \omega^2 c^2 [k_x^2 + (\mathrm{Re}k_z)^2 - (\mathrm{Im}k_z)^2] - \omega^2 \gamma g (\mathrm{Im}k_z) + (\gamma - 1) g^2 k_x^2 = 0 \tag{6.6a}$$

$$\omega^2 \gamma g (\mathrm{Re}k_z) - 2\omega^2 c^2 (\mathrm{Re}k_z)(\mathrm{Im}k_z) = 0 \tag{6.6b}$$

Therefore, under the condition of $\omega \neq 0$, or

$$\mathrm{Re}k_z = 0 \tag{6.7a}$$

The vertical wave number is a purely imaginary number, or

$$\mathrm{Im}k_z = \frac{\gamma g}{2c^2} = \frac{1}{2H} \tag{6.7b}$$

The first selection (6.7a) means there is no phase change in the vertical direction, and only the exponential growth or attenuation changes with altitude. They are just the characteristics of surface waves and evanescent waves: a surface wave is a type of wave whose energy is concentrated at the boundary layer or on the surface for which a certain parameter is not continuous, and the amplitude will change exponentially. Evanescent waves exist independent of the boundary layers, and can propagate in the massive fluid medium; these are sometimes called body waves. Both surface waves and evanescent waves are named "external waves", which are inversely different to "internal waves", and their most important characteristic is that they do not have a phase change in the vertical direction.

6.1.2 Internal waves

The dispersion Eq. (6.6) is a fourth-order equation of ω, when (6.7) is tenable,[1] the equation group (6.5) will have four independent solutions. Fig. 37(a)~(c) shows the relationship curve (dispersion diagram) between the other two quantities when k_x, k_z or ω is constant respectively. The last diagram (the relationship diagram between k_x and k_z when ω is given) is known as propagating as a curved surface diagram [148]. As there are three variables (ω, k_x, k_z) in (6.6), we can draw three diagrams to illustrate the interrelations among them.

There are two entirely different internal wave ranges in all dispersion diagrams. One includes the wave whose frequency is more than

$$\omega_\alpha = \frac{\gamma g}{2c} = \frac{c}{2H} \tag{6.8}$$

where ω is noted as "acoustic cut-off frequency", while the wave when $\omega > \omega_\alpha$ is confirmed as sound wave. The other one includes the low-frequency wave whose frequency takes the isothermal V-B frequency ω_g given by (1.28b) as its upper limit. Such waves with low frequencies and long periods are atmospheric gravity waves. The above-mentioned four solutions all correspond to the preceding two types of waves that propagate in opposite directions. Between those two ranges, Lamb waves propagate horizontally at the acoustic speed.

1 As mentioned above, this corresponds to an "internal wave". Figures to show their relations.

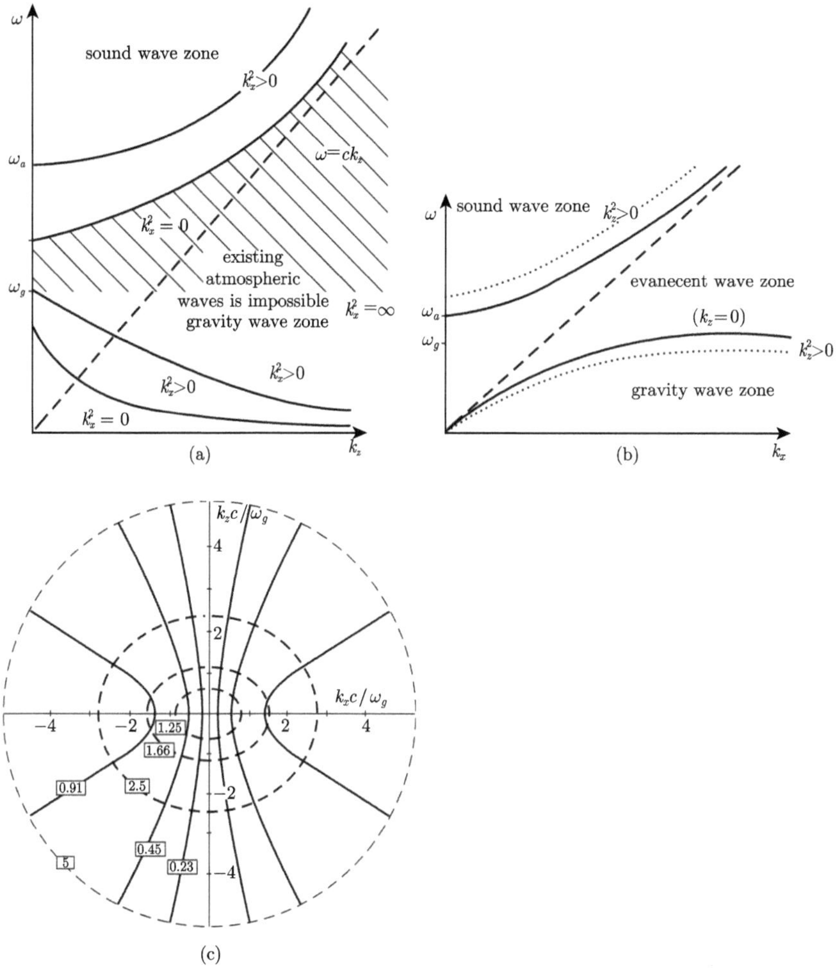

Fig. 37: Diagram of wave dispersion in isothermal atmosphere. (a) k_x = const; (b) k_z = const; (c) ω = const, the numerals in square frames are the ratio ω/ω_g; dotted lines indicate sound waves, solid lines indicate internal gravity waves.

In the case of internal waves (now (6.76) is tenable and Re $k_z \neq 0$), and the dispersion relation (6.6a) can be expressed by ω_α and ω_g as

$$k_z^2 = \left[\left(\frac{\omega_g}{\omega}\right)^2 - 1\right] k_x^2 + \frac{1}{c^2}(\omega^2 - \omega_\alpha^2) \tag{6.6a'}$$

or

$$k_x^2 = \frac{\omega^2}{c^2}\left[\frac{\omega^2 - \omega_\alpha^2 - c^2 k_x^2}{\omega^2 - \omega_g^2}\right] \tag{6.6a''}$$

where k_x and k_z indicate the real part of the corresponding wave number. If we suppose

$$l^2 = \left(\frac{\omega^2 - \omega_\alpha^2}{\omega^2 - \omega_g^2}\right)\frac{\omega^2}{c^2}, \qquad m^2 = \frac{\omega^2 - \omega_\alpha^2}{c^2}$$

then (6.6a′) can be transformed into the following conic curve equation

$$\frac{k_x^2}{l^2} + \frac{k_z^2}{m^2} = 1$$

In a real atmosphere, $\gamma < 2$, but when the altitude is less than 200 km, diatomic molecule gas N_2 and O_2 are dominant, so $\gamma = 1.4$. When the altitude is greater than 400 km, monatomic gas is dominant, so $\gamma = 1.67$. Therefore ω_α is always more than ω_g.

A propagating curved surface diagram is shown in Fig. 37(c), for the sound wave branch, $\omega > \omega_\alpha$, thus both l^2 and m^2 are more than zero. Therefore the curved surface is an ellipsoidal surface (the dotted line as shown in the diagram). The long axis of the ellipsoidal surface $l = \frac{\omega}{c}[(\omega^2 - \omega_\alpha^2)/(\omega^2 - \omega_g^2)]^{1/2}$ is in the horizontal direction, and its short axis $m = \frac{1}{c}[\omega^2 - \omega_\alpha^2]^{1/2}$ is in the vertical direction. Thereby its corresponding waves are anisotropic and are so called acoustic gravity waves, where their phase velocities are always more than the sound speed c. Under the ultimate condition when $\omega \to \infty$, (6.6a) can be simplified into $k_x^2 + k_z^2 = k^2 = \omega^2/c^2$, i.e. it is an isotropic ordinary wave.

For the internal gravity wave branch, $\omega < \omega_g < \omega_\alpha$, thus $l^2 > 0$ and $m^2 < c$, thereby the curved surface is a hyperbolic surface that rotates about the vertical axis (the solid line as shown in the diagram), which intersects with the horizontal axis at $\frac{\omega}{c}[(\omega_\alpha^2 - \omega^2)/(\omega_g^2 - \omega^2)]^{1/2}$. For a given gravity wave, k is always more than ω/c and its phase velocity is always less than the sound speed c.

6.1.3 Phase velocity and group velocity

In the above, the phase velocity of a sound wave and an internal gravity wave were mentioned. Now, by means of the wave dispersion diagram in Fig. 37(a) and (b) we will further analyze the phase velocity and group velocity in the vertical and horizontal directions. From Fig. 37(a) we can see, for gravity waves, that their phase velocity and group velocity are opposite in the vertical direction: if the vertical phase velocity is $V_{P_z} = \omega/k_z > 0$, the vertical group velocity is $V_{G_z} = \partial\omega/\partial k_z < 0$; on the other hand, if $V_{P_z} < 0$ then $V_{G_z} > 0$. This is an important characteristic of internal gravity waves. Note that for gravity waves traveling in the horizontal direction, they are in the same direction.

The group velocity represents the energy flow direction (except in high-dispersion media), and the phase velocity is the velocity of the observed wave crest–trough movement. Therefore if the energy that maintains the internal gravity wave propagates from up to down, the observed wave itself will represent the movement from up to down. It appears from this that we can explain the well-known downward drift of a traveling ionosphere disturbance (TID). The energy flow of an internal gravity wave emitted in the lower atmosphere transmits upward directly to the ionosphere. Because the atmospheric density will diminish with an increase in altitude and according to the continuity of energy flow, the amplitude is sure to increase with altitude. So at altitudes above 60 km, the wind profile will almost entirely be dominated by these wide-amplitude and long-period waves. This kind of wind can make the ionosphere redistribute, and form the basis of irregularities in the ionosphere that have been observed to move with the phase velocity of a wave. Radar sounding of irregularities indicates the downward drift of the ionosphere.

Using the dispersion relation (6.6), we can compute the phase velocity and group velocity, among which, the horizontal phase velocity can be obtained from (6.6a″)

$$V_{\mathrm{P}_x} = c\left[\frac{\omega^2 - \omega_g^2}{\omega^2 - \omega_\alpha^2 - c^2 k_z^2}\right]^{1/2} \tag{6.9}$$

while the horizontal group velocity is

$$V_{\mathrm{G}_x} = c\frac{(\omega^2 - \omega_g^2)^{3/2}(\omega^2 - \omega_\alpha^2 - c^2 k_z^2)^{1/2}}{\omega^4 + \omega_g^2(\omega_\alpha^2 + c^2 k_z^2 - 2\omega^2)} = \frac{c^2(\omega^2 - \omega_g^2)V_{\mathrm{P}_x}}{\omega^2 V_{\mathrm{P}_x}^2 - c^2\omega_g^2} \tag{6.10}$$

Similarly we can get the equation of phase velocity and group velocity in the vertical direction. The important conclusion we can get is: for gravity waves, as mentioned above, the direction of the phase velocity is opposite to the group velocity. For sound waves, the direction of both velocities are the same, but the group velocity will never exceed the phase velocity.

From (6.9) and (6.10) we can see that for gravity waves their horizontal phase velocity and group velocity are always less than the sound speed. When $\omega \to \infty$, both approach to the same limiting value $V_{\mathrm{P}_x}^m = V_{\mathrm{G}\partial x}^m = [\omega_g/(\omega_\alpha^2 + c^2 k_z^2)^{1/2}]c$. Therefore, in an isothermal atmosphere, gravity waves with exceedingly low frequencies move at the phase velocity, and the group velocity less than $0.9c$.

For a sound wave, when under the high-frequency limit, V_{P_x} and V_{G_x} both approach to sound speed as expected. But when $\omega \to \omega_\alpha$, both are related with k_z : $V_{\mathrm{P}_x} \to \infty$ but $V_{\mathrm{G}_x} \to 0$. The different velocity domains in the velocity–frequency plane are shown in Fig. 38 [29].

The real phase velocity V_P can be expressed by the track velocity V_{P_x} and V_{P_z} as follows

$$\frac{1}{V_\mathrm{P}^2} = \frac{1}{V_{\mathrm{P}_x}^2} + \frac{1}{V_{\mathrm{P}_z}^2} \tag{6.11}$$

As shown in Fig. 39, V_{P_x} and V_{P_z} are not true components of V_P. Owing to $k^2 = k_x^2 + k_z^2$, the corresponding wavelength relation is

$$\frac{1}{\lambda^2} = \frac{1}{\lambda_x^2} + \frac{1}{\lambda_z^2} \tag{6.11′}$$

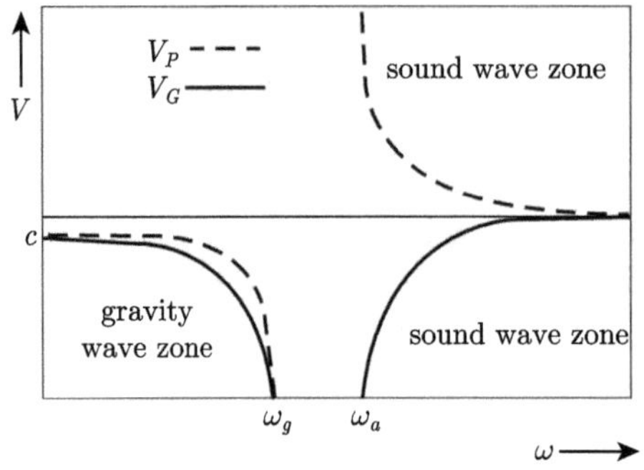

Fig. 38: Group velocity and phase velocity in the (V, ω) plane.

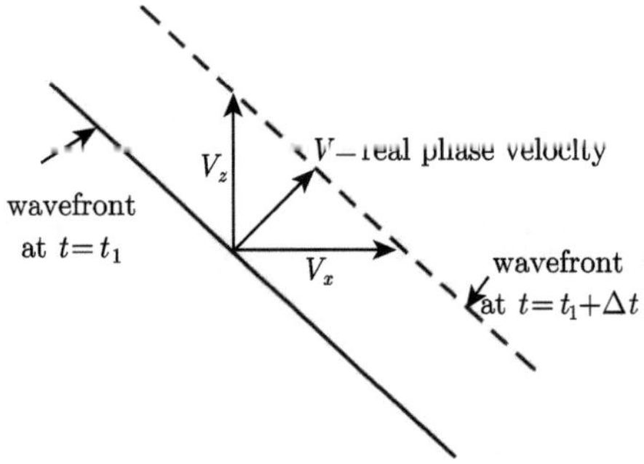

Fig. 39: The relationship between the trail velocity and the real phase velocity.

When analyzing the record of micro-pressure, relations (6.11) and (6.11′) are extremely important. Generally, when using the "three-point matrix" to receive atmospheric waves, we get the horizontal track velocity from the corresponding recorded time delay. To obtain V_P, we make use of relations (6.11) and (6.11′) to joint these two velocities together.

6.2 Waves in moving inhomogeneous atmosphere

6.2.1 Fundamental equations and the processing procedures

Now let us further account for the effects of wind, and simultaneously consider the stratified non-homogeneity of the atmosphere. To make it simple, we suppose that the wind is constant and moves in the horizontal direction, i.e. $|v_0| = v_{0x} = \text{const}$. To replace (6.4), we repeat what we have done in section 1.1 of this chapter and the fundamental equations are as follows

$$\left.\begin{aligned} \rho_0 D_t^0 v_x &= -\partial_x p \\ \rho_0 D_t^0 v_z &= -\partial_z p - g\rho \\ D_t^0 \rho + (\partial_z \rho_0) v_z + \rho_0(\partial_x v_x + \partial_z v_z) &= 0 \\ D_t^0 p + (\partial_z p_0) v_z &= c^2 [D_t^0 \rho + (\partial_z \rho_0) v_z] \end{aligned}\right\} \tag{6.12}$$

Apart from replacing the simple time differential quotient ∂_t in (6.4) with the operator $D_t^0 \equiv \partial_t + \boldsymbol{v}_0 \cdot \nabla = \partial_t + v_{0x}\partial_x$ (i.e. make $\boldsymbol{v} = \boldsymbol{v}_0$ in (1.29)), both (1.13) and (1.15) are not usable as we are not considering a isothermal atmosphere.

Using the ideal gas state equation (1.7), we get

$$\partial_z\left(\frac{p}{p_0}\right) = \frac{1}{p_0}\partial_z p + \frac{g\rho_0}{p_0^2}p = \frac{1}{p_0}\partial_z p + \frac{gM}{p_0RT}p$$

Supposing that M and R are kept constant and are stratified and non-homogeneous, we get

$$\frac{1}{\rho_0}\partial_z\rho_0 = -\frac{gM}{RT} - \frac{T'}{T}$$

Here and later the prime indicates the differential quotient respective to z. We can rewrite (6.12) as

$$\left.\begin{aligned} D_t^0 v_x &= -\frac{RT}{M}\partial_x\left(\frac{p}{p_0}\right) \\ D_t^0 v_z &= -\frac{RT}{M}\partial_z\left(\frac{p}{p_0}\right) - g\left(\frac{\rho}{\rho_0} - \frac{p}{p_0}\right) \\ D_t^0\left(\frac{\rho}{\rho_0}\right) + \partial_x v_x + \partial_z v_z - \left(\frac{gM}{RT} + \frac{T'}{T}\right) v_z &= 0 \\ D_t^0\left(\frac{p}{p_0}\right) + \gamma(\partial_x v_x + \partial_z v_z) - \frac{gM}{RT} v_z &= 0 \end{aligned}\right\} \tag{6.12'}$$

When deducing the fourth equation, we make use of the third.

Then, we cannot obtain linearized equations with corresponding matrices from equations (6,12′) by the procedures stated in section 1.1 (i.e. for an isothermal atmosphere) – if we insist on doing so, we will obtain contradictory results. On the one hand, the equation of the assumption that the scalar height H is a constant; on the other hand it also includes a non-zero T' term. So when solving Eq. (6.12′), we must pay special attention to maintaining the differential quotient of quantities that change with altitude. Thereupon we can get four two-order differential equations, or, express parameters with a group of coupled one-order differential equations.

As presented in section 1.1, suppose disturbances change sinuously. Taking v_z as an example, set

$$v_z = V(z)\exp[\mathrm{i}(k_x x - \omega t)] \tag{6.13}$$

then

$$D_t^0 = \mathrm{i}(v_{0x}k_x - \omega) \equiv -\mathrm{i}\omega^* \tag{6.14}$$

Where the defined Doppler frequency-shift $\omega^* \equiv \omega - v_{0x}k_x$ is also the natural frequency introduced in Section 5.2 Chapter 3. The wave amplitude variation in the horizontal direction is still to be omitted, so k_x can be treated as pure real number. $\partial_x = ik_x$ is a linear operator.

The procedure to get the differential equation of v_x is as follows: first, get p/p_0 from the first equation of (6.12′)

$$\frac{p}{p_0} = \frac{i}{Hgk_x} D_t^0 v_x \tag{6.15}$$

in which, relations (1.17) and (1.16′) are used. Secondly, substitute the above formula into the fourth formula of (6.12′) to get

$$\mathrm{i}\left[\gamma k_x + \frac{1}{Hgk_x}(D_t^0)^2\right] v_x = \left(\frac{1}{H} - \gamma\partial_z\right) v_z \tag{6.16}$$

Third, substitute (6.15) into the second formula of (6.12′) to get

$$\begin{aligned} D_t^0 v_z &= -g\frac{\rho}{\rho_0} + (1 - H\partial_z)\frac{\mathrm{i}}{Hk_x} D_t^0 v_x \\ &= -g\frac{\rho}{\rho_0} + \frac{\mathrm{i}}{Hk_x} D_t^0 v_x - \frac{\mathrm{i}D_t^0}{k_x}\left(\partial_z v_x + \frac{H'}{H} v_x\right) \end{aligned} \tag{6.17}$$

Fourth, eliminate ρ/ρ_0 between (6.17) and the third formula of (6.12′) to get

$$\mathrm{i}\left[\frac{(D_t^0)^2}{Hgk_x} - \frac{H'}{Hgk_x}(D_t^0)^2 + k_x - \frac{(D_t^0)^2}{gk_x}\partial_z\right] v_x = \left[\frac{(D_t^0)^2}{g} + \frac{1}{H} + \frac{T'}{T} - \partial_z\right] v_z \tag{6.18}$$

Compose Eqs. (6.16) and (6.18) as a one-order coupled differential equation group. To obtain the solution, we substitute (6.16) and its differential quotient into (6.18), where a complete differential equation of $V(z)$ will be given as follows

$$\begin{aligned} V'' &- \frac{1}{H}\left[1 + \frac{k_x^2 c^2}{k_x^2 c^2 - \omega^{*2}} H'\right] V' + \left[\frac{k_x^2}{\omega^{*2}}\left\{\frac{(\gamma-1)g^2}{c^2} + \frac{g}{H}H'\right\}\right. \\ &\left. + \frac{\omega^{*2}}{c^2} - k_x^2 + \frac{k_x^2 c^2}{k_x^2 c^2 - \omega^{*2}} \frac{H'}{\gamma H^2}\right] V = 0 \end{aligned} \tag{6.19}$$

Using a similar method, we can get slightly complicated two-order differential equations about v_x, p/p_0 and ρ/ρ_0. In the particular case of an isothermal atmosphere, all of these equations will degenerate into the same one. For example, about $V(z)$ we have

$$V'' - \frac{1}{H}V' + \left[\frac{k_x^2}{\omega^{*2}} W_g^2 + \frac{\omega^{*2}}{c^2} - k_x^2\right] V = 0 \tag{6.19a}$$

Now if we suppose

$$V(z) = A_0 Z\exp(\mathrm{i}k_z z)$$

then (6.19a) becomes the dispersion relation (6.6), but now ω is substituted by ω^*, which is consistent with our original intuition and expectation.

6.2.2 Transition to isothermal atmosphere, slowly-varying atmosphere

Rewrite (6.19) in a more common form

$$V'' + f(z)V' + r(z)V = 0 \tag{6.19′}$$

by using the standard transform

$$V = \bar{V}\exp\left[-\frac{1}{2}\int^{z} f(z)\mathrm{d}z\right]$$

This can be transformed into the common form of the wave equation

$$\bar{V}'' + q^2\bar{V} = 0$$

where

$$q^2 = r - \frac{1}{4}f^2 - \frac{1}{2}f' \tag{6.20}$$

When $q^2 > 0$, $\bar{V}$ has the solution of a vibrating mode, while $q^2 < 0$, has the solution of an exponential or hyperbolic mode. With some reservation[2] the parameter q can be understood as the vertical component of the wave number.

Making a comparison between (6.19) and (6.19′), we can see that in the latter formula the term $f(z)$ is brought about by the change of atmospheric density. Undisturbed atmospheric pressure can be provided in a simple form like (1.14), but for density, under the non-isothermal condition, we can suppose a slightly different form (to compare with (1.15) which is under the isothermal condition)

$$\rho_0(z) = (\rho_0 H)_{z=0}\frac{1}{H}\exp\left\{-\int_0^z \frac{\mathrm{d}z}{H}\right\} \tag{6.21}$$

In fact $f(z)$ can be expressed as the following form

$$f(z) = d_z\ln\left(\frac{\rho}{b^2}\right) \tag{6.22}$$

where $b^2 \equiv \omega^{*2}/c^2 - k_x^2$, and $r(z)$ can be written as

$$r(z) = b^2 - \frac{k_x^2 g}{\omega^{*2}}\left(f(z) + \frac{g}{c^2}\right) \tag{6.23}$$

If we suppose that the atmospheric characteristics change very slowly, so that both H'^2 and H'' can be neglected, the vertical wave number can be treated as a constant. In this case, the "isothermal" dispersion equation (6.6) of an internal wave can be used, and we only need to substitute the corresponding ω, ω_g and ω_α of an isothermal atmosphere with the relevant values of ω^*, ω_{B} and $\omega_{\alpha n}$ of a slowly varying atmosphere. Here ω_{B} is related to ω_g by (1.28a′), and $\omega_{\alpha n}$ is the cut-off frequency of the sound wave under the non-isothermal condition. We rewrite (6.20) as

$$q^2 = r_0 + \Delta - \frac{\omega_{\alpha n}}{c^2} \tag{6.20'}$$

2 The reservation arises when using this procedure, where the parameters v_x, v_z, p/p_0 and ρ/ρ_0 will give different vertical wave numbers.

where

$$r_0 = r - \Delta$$

$$\Delta = \frac{k_x^2 c^2}{k_x^2 c^2 - \omega^{*2}} \frac{H'}{\gamma H^2}$$

$\Delta \to 0$ when $k_x \to 0$, and the higher-order differential quotient of H can be omitted, so

$$\omega_{an}^2 = \frac{c^2}{4} f^2 + \frac{c^2}{2} f' = \frac{c^2}{4H^2} H' = \omega_\alpha^2 + \frac{\gamma g}{2T} T'$$

From the above substitution we can see that in a non-isothermal atmosphere, and different to an isothermal atmosphere, the V-B frequency may be higher than the cut-off frequency of a sound wave. The definition of ω_B and ω_{an} indicates the condition of occurrence is

$$T' > \frac{2-\gamma}{2\gamma} \frac{gM}{R}$$

If we substitute the value of $M = 29$ and $\gamma = 7/5$, we can see that when the temperature gradient T' exceeds 7.3 K/km, ω_B exceeds ω_{an}. This value of the temperature gradient provides a rough criterion for the so called "slowly-varying atmosphere". If $T' > 7.3$ K/km, the atmosphere is no longer slowly varying, and a more accurate solution to the differential equations must be sought.

6.2.3 Velocity divergence equation

When replacing the differential equations mentioned in the second section above to describe the vertical velocity component v_x, it is often more advantageous to get an equation that describes the velocity divergence, which is exceedingly obvious in the applications of atmospheric acoustics. According to its definition, the velocity divergence is

$$\chi = \nabla \cdot \boldsymbol{v} = \partial_x v_x + \partial_z v_z \tag{6.24}$$

We then introduce the y-component of the velocity vorticity $\nabla \times \boldsymbol{v}$

$$\zeta_y = \partial_z v_x - \partial_x v_z \tag{6.25}$$

so after operation with a two-dimension Laplace operator, $\nabla^2 v_z$ can be written as

$$\nabla^2 v_z = \partial_x^2 v_z + \partial_z^2 v_z = \partial_x \chi - \partial_x \zeta_y$$

Under the condition of an isothermal atmosphere, Eq. (6.12′) can be written as matrix form

$$D_t^0 \boldsymbol{M}_A = \boldsymbol{M}_B \boldsymbol{M}_A \tag{6.26}$$

where

$$
\boldsymbol{M}_A \equiv \begin{bmatrix} \chi \\ \zeta_y \\ \dfrac{\rho}{\rho_0} - \dfrac{p_1}{p_0} \\ \dfrac{RT}{M}\nabla^2\left(\dfrac{p}{p_0}\right) \end{bmatrix}
$$

$$
\boldsymbol{M}_B \equiv \begin{bmatrix} 0 & 0 & -g\partial_z & -1 \\ 0 & 0 & g\partial_x & 0 \\ r-1 & 0 & 0 & 0 \\ g\partial_z - \dfrac{rRT}{M}\nabla^2 & -g\partial_x & 0 & 0 \end{bmatrix} \tag{6.27}
$$

From these matrices we can get the dispersion relation. In the non-isothermal case, the differential equation about χ can also be obtained by using the same method as in the upper section.

In [149], by considering the supposition that χ changes periodically and windless, we get the following equation

$$
\partial_z^2\chi - \frac{1-H'}{H}\partial_x\chi + \left[\frac{\omega^2}{c^2} - k_x^2 + \frac{k_x^2}{\omega^2}\left(\frac{gH'}{H} + \frac{(\varkappa-1)g^2}{c^2}\right)\right]\chi = 0 \tag{6.28}
$$

Handling the relevant equations in quite a similar way as mentioned in the previous two sections in [150], Eq. (6.28) is extended to consider winds and temperatures with arbitrary vertical changes. The authors proved that v_x, v_z and p/p_0 are related to the velocity divergence according to the following equation

$$
\left.\begin{aligned}
\frac{iG}{k_x c^2}v_x &= \left(\omega^{*2} - \frac{g}{H} + \frac{gH'}{H}\right)\chi + g\partial_z\chi \\
\frac{G}{c^2}v_z &= \left(k_x^2 g - \frac{\omega^{*2}}{H} + k_x\omega^*\partial_z v_x\right)\omega^{*2}\partial_z\chi \\
\frac{iG}{\gamma\omega^*}\frac{p}{p_0} &= \left(\omega^{*2} - \frac{g}{H} + \frac{gH'}{H}\right)\chi + g\partial_z\chi
\end{aligned}\right\} \tag{6.29}
$$

where

$$
G \equiv k_x^2 g^2 + (gk_x\partial_z v_x - \omega^{*2})\omega^* - \frac{\omega^{*2}gH'}{H} \tag{6.30}
$$

6.2.4 Energy density and Lagrange density

In section 1.3 Chapter 2, we discussed the Lagrangian density of sound waves. Now we will correspondingly discuss the event of gravity waves. Many previous results can be obtained more simply by using this method.

First we introduce three components ξ, η, ζ related to the displacement of particles of waves that deviate from its equilibrium position in rectangular coordinates, thus

$$v_x = \partial_t \xi, \quad v_y = \partial_t \eta, \quad v_z = \partial_t \zeta$$

so the volume increment, or strain, of the fluid is

$$\varepsilon = \partial_x \xi + \partial_y \eta + \partial_z \zeta$$

Then the Lagrangian density of gravity waves is [151] (compare with Eq. (2.27))

$$L = \frac{1}{2}\rho[(\partial_t \xi)^2 + (\partial_t \eta)^2 + (\partial_t \zeta)^2] - \frac{1}{2}\lambda \varepsilon^2 + \frac{1}{2} g\rho' \zeta^2 + \rho g \varepsilon \zeta \tag{6.31}$$

where $\lambda = \rho c^2$ is the volume coefficient of elasticity, and it is the reciprocal to the compressibility coefficient k in Eq. (2.27).

The first term of (6.31) is the kinetic energy density, while the remaining terms respectively represent the scalar potential function. Thereby we can deduce the conservative force system acting on the fluid. The second term represents the elasticity potential energy. The third term represents the gravitational potential energy in the fluid which cannot be compressed by degrees, and the last term represents the potential energy related to the displacement of a compressible fluid.

Therefore, by applying the Lagrangian motion equation (2.26) to the Lagrangian density (6.31), we get

$$\left.\begin{aligned} \rho\partial_t^2 \xi - \partial_x(\lambda\varepsilon) + \rho g \partial_x \zeta = 0 \\ \rho\partial_t^2 \eta - \partial_y(\lambda\varepsilon) + \rho g \partial_y \zeta = 0 \\ \rho\partial_t^2 \zeta - \partial_z(\lambda\varepsilon) - \rho g(\partial_x \xi + \partial_y \eta) = 0 \end{aligned}\right\} \tag{6.32}$$

Suppose that ρ and λ are only functionss of z, thus each above-mentioned equation is separable. For a horizontal plane wave, we can choose axis x as the direction of motion, then Eqs. (6.32) can be simplified. The first equation remains unchanged, the second equation vanishes because component y does not exist, while in the third equation $\partial_y \eta = 0$. In the appendix of [151], it has been proved that the two equations obtained in this way are equivalent to the first-order perturbation form of Euler's equation. The principal advantages of the equation expressed in Lagrangian form lie in unnecessary to introduce the thermodynamic concepts, and the realistic motion of particles in every typical wave motion can be determined directly. Suppose ξ and ζ are both exponential functions $\exp[\mathrm{i}(k_x x - \omega t)]$, then the simplified equations (6.32) become

$$\left.\begin{aligned} \rho\omega^2 \xi - \lambda k_x^2 \xi + \mathrm{i}k_x(\rho g \zeta - \lambda \zeta') = 0 \\ \rho\omega^2 \zeta + \lambda'(\zeta' - \mathrm{i}k_x \xi) + \lambda(\zeta'' - \mathrm{i}k_x \xi') - \mathrm{i}k_x \rho g \xi = 0 \end{aligned}\right\} \tag{6.33}$$

By substituting $\lambda = \rho c^2$ into the first equation of (6.33), we get

$$\xi = \mathrm{i}k_x \frac{c^2 \zeta' - g\zeta}{\omega^2 - k_x^2 c^2}$$

The equation connecting ξ and ζ is well matched with Eq. (6.16), which links v_x and v_z. When substituting it in the second equation of (6.33), we get the following two-order differential equation about ζ

$$\zeta'' + \left\{d_z\ln\left(\frac{\rho}{b^2}\right)\right\}\xi' + \left\{b^2 - \frac{k_x^2}{\omega^2}g d_z\ln\left(\frac{\rho}{b^2}\right) - \frac{k_x^2}{\omega^2}\frac{g^2}{c^2}\right\}\zeta = 0 \tag{6.34}$$

where b^2 is the same as in (6.22), where ω replaces ω^*. Eq. (6.34) is matched with (6.19). In an isothermal atmosphere, c = constant, thus $d_z(-\ln b^2) = 0$, and Eq. (6.34) becomes

$$\zeta'' + \{d_z(\ln\rho)\}\zeta' + \left(\frac{\omega^2}{c^2} - k_x^2 + \frac{k_x^2\omega_B^2}{\omega^2}\right)\zeta = 0 \tag{6.34a}$$

In which the expression (1.28) of ω_B is used. Let $\zeta = h\rho^{-1/2}$, then the above equation can be transformed into the following wave equation

$$h'' + k_z^2 h = 0$$

where

$$k_z^2 = \frac{\omega^2}{c^2} - k_x^2 + \frac{k_x^2}{\omega^2}\omega_B^2 - \frac{1}{4}(d_z\ln\rho)^2 - \frac{1}{2}d_z^2\ln\rho \tag{6.35}$$

If ρ is the exponential function (Eq. (1.15)) of z, then ω_B is constant and $d_z^2\ln\rho = 0$, therefore

$$k_z^2 = \frac{\omega^2}{c^2} - k_x^2 + \frac{k_x^2}{\omega^2}\omega_B^2 - \frac{1}{4H^2} \tag{6.35a}$$

We can see it is the same as (6.6a′).

The Lagrangian density can also be used as a starting point to study the fluid mechanics energy-momentum tensor [152]. The presentation of the tensor can give a single popularized expression equation for energy conservation, momentum conservation and angular momentum conservation, and a way to compute the energy exchange and momentum exchange between the wave and the background medium.

6.3 Polarization relations

6.3.1 Phase relations between perturbed quantities

At the very beginning of this chapter, we supposed that each perturbed quantity changed sinuously. Now we will further study the phase relations between them. For this purpose the perturbed quantities can be written as

$$\frac{v_x}{X} = \frac{v_z}{Z} = \frac{1}{P}\left(\frac{p}{p_0}\right) = \frac{1}{R}\left(\frac{\rho}{\rho_0}\right) = A_0\exp[i(k_x x + k_y y + k_z z - \omega^* t)] \tag{6.36}$$

where X, Z, P, R, and so forth, are called polarization coefficients, which can be conceived as complex amplitudes that determine the phase relations between the perturbed quantities. In order to determine these coefficients under the non-isothermal atmosphere condition, we must use the algebraic operation mentioned in section (2.1). For example, the phase difference between p/p_0 and v_x given in Eq. (6.15) can be written with polarization coefficients as

$$\frac{P}{X} = \frac{\omega^*}{Hgk_x} \tag{6.15a}$$

Computing the remaining ratios in this way is quite computationally heavy and complicated. But if we linearize the relevant equation and express it in matrix form, then it may be accepted as a correct and simple way to determine the polarization relations. That is Eq. (6.5), which means we can do this way for isothermal atmosphere. We now promote this procedure forward and consider the so-called "pseudo-thermal" atmosphere [153]. The characteristics of this kind of atmosphere are: except for one of the relevant equations (such as the third equation of (6.12′)), for which the temperature gradient T' obviously appears, we can regard as the atmosphere as being isothermal (both H and T are constant). The results in this way, to a certain extent, contribute to our understanding of how the temperature gradient influences the isothermal polarization. However, they should not be regarded as a strict solution.

Then, similar to (6.5), we can get[3]

$$\begin{bmatrix} -\mathrm{i}\omega^* & 0 & 0 & \mathrm{i}k_x gH \\ 0 & -\mathrm{i}\omega^* & g & g + \mathrm{i}k_z gH \\ \mathrm{i}k_x & -\frac{1}{H} - \frac{T'}{T} + \mathrm{i}k_z & \mathrm{i}\omega^* & 0 \\ \mathrm{i}\gamma k_x & -\frac{1}{H} + \mathrm{i}\gamma k_x & 0 & \mathrm{i}\omega^* \end{bmatrix} \begin{bmatrix} X \\ Z \\ R \\ P \end{bmatrix} = 0 \tag{6.37}$$

Owing to the fact that the equation group represented with (6.37) is linear and homogeneous, there will be an unlimited number of linear correlation solutions. This means that a wave with an arbitrary amplitude will satisfy the equation group, provided that the perturbations similarly (first-order perturbation values are very small) remain correct. The amplitude of an atmospheric wave is determined by the source term, after the source term is determined, the corresponding equation group will be inhomogeneous, thus the derived equation in this chapter is effective only in regions far from an atmospheric wave source.

In order to determine the polarization relations (that is, to find the ratios of the four unknown variables X, Z, R and P), we must regard one of these variables as

3 Attention should be paid: under windless isothermal circumstances ($\omega^* \to \omega$, $T' = 0$) the matrix elements in the last row of the 4×4 matrix in (6.37) are not the same as in (6.5), which is due to the different form of the continuity equation adopted when determining (6.37). This indicates that when we multiply a row (or a column) by a constant (now it is γ) and when we add each resulted element to any other row (or column), the value of the determinant is unchanged.

known and use the remaining three equations. There are four different ways to do this, and no matter which is used, the the final results are the same. The simplest combination is to regard Z as known, and seek the solutions of X/Z, R/Z and P/Z

$$\begin{bmatrix} -i\omega^* & 0 & ik_x gH \\ ik_x & -i\omega^* & 0 \\ i\gamma k_x & 0 & -i\omega^* \end{bmatrix} \begin{bmatrix} X \\ R \\ P \end{bmatrix} = Z \begin{bmatrix} 0 \\ \frac{1}{H} + \frac{T'}{T} - ik_z \\ \frac{1}{H} - i\gamma k_x \end{bmatrix} \tag{6.38}$$

According to Cramer's principle, the denominator of these three solutions is always the determinant of the coefficient matrices, so it is directly proportional to Z

$$Z \propto \omega^{*3} - c^2 k_x^2 \omega^* \tag{6.39}$$

Then we have

$$X \propto \omega^* c^2 k_x k_z - i\omega^* g k_x \tag{6.40}$$

$$R \propto \omega^{*2} k_z - i\omega^{*2}\left(\frac{\gamma g}{c^2} + \frac{T'}{T}\right) + ik_x^2\left[g(\gamma - 1) + \frac{c^2 T'}{T}\right] \tag{6.41}$$

$$P \propto -i\omega^{*2}\gamma g/c^2 + \omega^{*2}\gamma k_z \tag{6.42}$$

Now let us investigate how the "pseudo-thermal" atmosphere is similar to the realistic non-isothermal atmosphere. After a simple operation, we can again obtain the strict form (6.36) of P/X under the non-isothermal condition by using (6.40) and (6.42). As a result, P and X are accurate. By (6.16) the strict form about X/Z can be obtained

$$\frac{X}{Z} = \frac{k_x g + ic^2 k_x k_z}{i(\omega^{*2} - c^2 k_x)} \tag{6.43}$$

This indicates that Z is also accurate. And yet, the equation (6.41) of R lacks some terms that contain the temperature gradient. Even still, Eq. (6.41) still offers an approximation, which can be simply derived and is applicable.

After successfully applying the pseudo-thermal model to process the polarization relations, the question is then of whether we can try using it to derive the dispersion equation? Unfortunately, the answer is a no. In this case, we obtain a second-order differential equation (compare with (6.19a))

$$V'' - \frac{1}{H}V' + \left[k_x^2 \frac{\omega_B^2}{\omega^{*2}} + \frac{\omega^{*2}}{c^2} - k_x^2 - \frac{g}{c^2}\frac{T'}{T}\right] V = 0 \tag{6.44}$$

where ω_B is always larger than ω_{an} and their difference $\omega_B^2 - \omega_{an}^2$ is a constant. However, either the isothermal processing with ω_B to substitute ω_g, or pseudo-thermal processing will include considerably more complexity of the dispersion relations in a realistic atmosphere.

6.3.2 Air-parcel orbits

The real motion of an air mass is given by Eq. (6.43). For an acoustic wave with a high frequency $\omega/k_x \to c$, so $Z \to 0$ (Eq. (6.39)), thus the wave is a longitudinal wave. For gravity waves, the air-parcel orbits can be obtained via the expression of Z by stressing the terms containing c^2. This is a very good approximation when the compressibility effect is very small (as in the case of gravity waves).

After omitting the exponential spatial variation

$$v_z \propto -A_0 c^2 k_x^2 \omega^* \exp(-i\omega^* t)$$

and considering (6.40), we get

$$v_x \propto A_0(\omega^* c^2 k_x k_z - i\omega^* g k_x)\exp(-i\omega^* t)$$

We use v_z as a reference of the measuring phase, which can be written as $v_z \propto \cos\omega^* t$, thus we get

$$v_x \propto \frac{1}{k_x}\left(\mathrm{Im}k_z - \frac{g}{c^2}\right)\sin\omega^* t - \frac{k_z}{k_x}\cos\omega^* t$$

after eliminating time, then we obtain

$$\left(v_x + \frac{k_z}{k_x}v_z\right)^2 = \frac{1}{k_x^2}\left(\mathrm{Im}k_z - \frac{g}{c^2}\right)^2(1 - v_z^2) \tag{6.45}$$

in an isothermal atmosphere, from (6.7b) and (6.8), we have $\omega_\alpha/c = \mathrm{Im}k_z$. Therefore, if we use the dispersion equation, such as (6.6a) (by replacing ω with ω^* and replacing ω_g with ω_B), and once again paying attention to $\omega^{*2} \ll c^2 k_x^2$, then the particle velocity equation (6.45) can be written as

$$v_x^2 + v_z^2\left(\frac{\omega_B^2}{\omega^{*2}} - 1\right) + \frac{2k_z}{k_x}v_x v_z = \frac{1}{k_x^2}\left(\frac{\omega_\alpha}{c} - \frac{g}{c^2}\right)^2 \tag{6.46}$$

Because particle displacement is given by the time integration of particle velocity, so the equation form of particle track is the same as (6.46).

When $\omega^* < \omega_B$, (6.46) is the equation of an ellipse. At the long-wave limit, i.e. $k_x \to 0$ (from (6.40) we can see that $X \to 0$), the movement of particles becomes purely vertical. At short wave limit, when $\omega^* \to \omega_B$ and $k_x \to \infty$, the movement becomes vertical as well.

For purely horizontal propagation ($k_z = 0$), when $\omega^* = \omega_B/\sqrt{2}$, (6.46) is the equation of a circle, thus the particles' orbits are also circular. A gravity wave with this frequency is thus circularly polarized. When $\omega_B > \omega^* > \omega_B/\sqrt{2}$, the movement is elliptical, and the vertical axis is longer than the horizontal axis. However, at low frequencies when $\omega^* < \omega_B/\sqrt{2}$, the horizontal axis is longer than the vertical axis.

The typical air-parcel orbit of a gravity wave is shown graphically in Fig. 40 [29], which is a supplement to the propagating curved surface in Fig. 37(c).

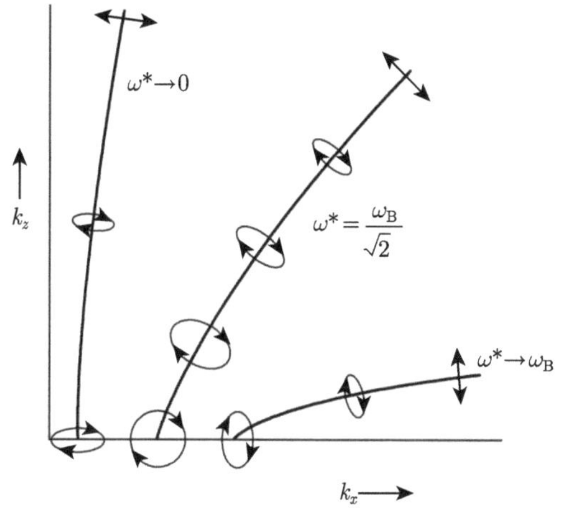

Fig. 40: Movement orbit of air-parcel in gravity waves.

To be simple and clear, only the first quadrant is presented. In [154], the air-parcel moving orbit graph is similar to Fig. 40 for evanescent waves and other internal waves.

When considering the earth's rotation, because it is necessary to introduce the dependence of different x and y of the Coriolis force, the problem becomes a three-dimensional one. When we do neglecting the earth's curvature, Eckart proved [155] that when $\omega_B > 2\Omega_E$, there will not be any propagation of internal waves with frequency $\omega^* < 2\Omega_E$. But in the next section we will see that, due to the earth's curvature the variation of the Coriolis force with latitude can generate long-period (> 24 h) Rossby waves. In the case of $\omega^* < 2\Omega_E$, when we do not considering earth's curvature, an internal wave with $2\Omega_E > \omega^* > \omega_B$ may exist, but a wave with a very long horizontal wavelength with $\omega^* < \omega_{an}$ cannot exist. The elliptical orbit of an air-parcel is no longer in the vertical plane. Tolstoy called this kind of wave a "gyroscopic wave", and found it has only one tolerable propagating direction [151].

6.3.3 Complex polarization terms

It is easy to see that most polarized terms are complex. As such, the phase relationship between any two parameters of waves can be determined. Because any complex $A + iB$ can be written as exponential form

$$(A^2 + B^2)^{1/2} \exp[i \arctan(B/A)]$$

then, for example, for X/Z determined via formula (6.43), if k_x is a pure real number and $k_z = \mathrm{Re}k_z + i/2H$, then the phase difference between X and Z (which corresponds to the phase difference between v_x and v_z) can be given as $\arctan[(r-2)/2\gamma(\mathrm{Re}k_z)H]$. For an evanescent wave, when $\mathrm{Re}k_z = 0$, the phase difference of movement between the two directions is 90°, therefore the movement of particles is circularly polarized.

6.4 Rossby waves

6.4.1 Geostrophic wind

As mentioned at the beginning of this chapter, when considering the effect of earth's rotation, the right side of Euler's equation (1.30a) must be added with a term that corresponds to the Coriolis force

$$D_t\boldsymbol{v} = -\frac{1}{\rho}\nabla p + \boldsymbol{g} + 2\boldsymbol{\Omega}_{\mathrm{E}} \times \boldsymbol{v} \tag{6.47}$$

In fact here we have made three suppositions: Firstly, the centrifugal acceleration (the third term in (1.43)) generated by geostropy is omitted, which is rational when air moves in large scales and without strong crimping movement (such as hurricane). Secondly, the horizontal component of the Coriolis vector $\boldsymbol{\Omega}_{\mathrm{E}}$ can be neglected[4] (i.e. the vertical component of $2\boldsymbol{\Omega}_{\mathrm{E}} \times \boldsymbol{v}$ can be neglected). This neglection is known as the "traditional approximation". Thirdly, suppose that the fluid is in static equilibrium in the vertical direction, i.e. formula (1.13) is still tenable. Under these suppositions the motion equation gives $\mathrm{d}_t v_z = 0$, so the vertical component of the wind is constant and can be determined by the boundary conditions. Because there is no possibility of a large-scale vertical wind at the ground, so in the atmosphere as a whole we have $v_z = 0$. But there exists small-scale deviations from static equilibrium, so vertical movement is formed trying to restore equilibrium.

Now we can use the equation of motion (6.47) to get the wind $\boldsymbol{v}_g$ (i.e. the so-called geostrophic wind) when the pressure gradient force is balanced by the Coriolis force. The component at mid-level latitude is

$$v_{ge} = -\frac{1}{f\rho}\partial_n p, \quad v_{gn} = \frac{1}{f\rho}\partial_e p, \quad v_{gz} = 0 \tag{6.48}$$

Here the introduced rectangular system is: e is pointed eastward, n is pointed northward, and z is pointed vertically upward. The Coriolis parameter is defined by

$$f = 2\Omega_{\mathrm{E}} \sin\theta \tag{6.49}$$

4 If we do not do this, the solved motion equation will indicate that an airflow is horizontal at first but after six hours becomes completely vertical. This is by no means what actually happens of course, because the airflow will move along the spherical surface of the curved earth's ground.

which describes the functional relationship between the Coriolis vector Ω_E and the latitude θ. In the northern hemisphere $f > 0$, while in the southern hemisphere $f < 0$. At the equator $f = 0$, thus there does not exist any geostrophic wind, which in turns implies that the airflow must travel directly across the isobaric line.

The geostrophic wind equation (6.48) describes the air-parcel movement at the start, which travels directly from a high-pressure area to a low-pressure one. But the movement is rather slow, because the Coriolis force will make the air-parcel tend to move parallel to the isobaric line. Just as the case where a temporary deviation from static equilibrium causes a vibration around the equilibrium condition with a period of several minutes, the deviation from geostrophic equilibrium will cause a vibration with the period of several hours or even longer – this vibration is called a planetary wave.

6.4.2 Formation of Rossby wave

Geostrophic movement can impart wave motions in an atmosphere. In the previous section we saw that for a geostrophic wind $D_t \boldsymbol{v}_g = 0$, thus the wind component in the reference system with velocity $\boldsymbol{v}_0$ that moves together with the fluid remains constant. At the starting point we take $\boldsymbol{v}_g = (v_{ge}, 0, 0)$, and if a certain horizontal pressure distribution exists, then together with a pressure gradient that extends from a low-pressure area to a high-pressure area, a longitudinal wind v_{gn} will form. Since the movement is pointing eastward, v_{ge} and v_{gn} both remain constant.

However, the earth's surface is a closed curve, thus low-pressure and high-pressure areas will certainly exist alternately. After connecting the circuit from east to west, there will be an area in which v_{gn} will blow in the opposite direction. For an observer at the ground surface, this is expressed as the drift of both the pressure and the longitudinal wind through a series of periods. But in reality the atmosphere is not so simple. In fact the geostrophic wind does not form this way. If the distance between a high-pressure area and a low-pressure area is very small, then rotation effects becomes more important. The centrifugal acceleration cannot be neglected and must be included when determining the gradient wind. In addition, the shear induced by a longitudinal wind may cause unsteady cyclone waves. Conversely, if the distance between the high-pressure area and low-pressure area is very large, the rotational longitudinal wind can be extremely weak, and the longitudinal variation of the Coriolis parameter f is exceedingly important. Thus a Rossby wave is formed.

A geostropic wind $\boldsymbol{v}_g$ is without dispersion, i.e. $\Delta \boldsymbol{v}_g = 0$. This is correct for any wind in incompressible homogeneous atmosphere. Similarly, the simplest atmospheric model that can generate a Rossby wave is a non-viscous, isothermal, non-dispersion incompressible atmosphere. This is the starting point that Rossby initially studied this kind of long wave [156], which represents a prototypical atmosphere.

If the wave motion is non-dispersive, the particle velocity $\boldsymbol{v}_1$ will have[5]

$$\nabla \cdot \boldsymbol{v}_1 = 0$$

So if there exists a plane wave, the wave vector $\boldsymbol{k}$ should satisfy

$$\boldsymbol{k} \cdot \boldsymbol{v}_1 = 0$$

Therefore the wave is a transverse wave, while for any arbitrary steady movement whose $\boldsymbol{k}$ is a pure real number, the waves are all linearly polarized.

If we want to investigate a Rossby wave in a plane, the variation of the Coriolis parameter f with latitude must be taken into account. In order to do that, we can regard f as a linear function: $f = f_0 + \beta_n, \beta \equiv \partial_n f$. This plane is called the β plane, at which we take the following rectangular coordinates: the x axis points in the phase-propagation direction of the wave; while the y axis points to the north. The direction of x and y does not need to correspond to the latitudinal and longitudinal directions. The reason is that whether for a Rossby wave or a planetary wave with the form of extending on the spherical surface, the phase-propagation direction can either be vertical to the axis of rotation, or forms an oblique angle with it. After choosing the horizontal and vertical axes of x and y, the particle velocity components of the wave are

$$\begin{aligned} v_{1x} &= 0 \\ v_{1y} &= Y\exp[\mathrm{i}(kx - \omega t)] \end{aligned} \tag{6.50}$$

In many situations, expressing the horizontal wind field with rotational characteristics is advantageous. For any fluid, if

$$\boldsymbol{v} = \boldsymbol{\Omega} \times \boldsymbol{r}$$

then

$$\boldsymbol{\Omega} = \frac{1}{2}\nabla \times \boldsymbol{v} \equiv \frac{1}{2}\boldsymbol{\zeta} \tag{6.51}$$

The revolving degree of wind velocity is a measure of wind revolution (also named as vorticity $\boldsymbol{\zeta} = \nabla \times \boldsymbol{v}$). Because we only concerned with the vertical component of f, we are in turn mainly interested in the vertical component ζ_z of the vorticity. f is the vorticity possessed by a static fluid relative to the rotating earth, and the absolute vertical vorticity of a fluid element is $\zeta_\alpha = \zeta_z + f$. For the wave motion to be processed, we take $\zeta_z = (\nabla \times \boldsymbol{v}_1)_z = \partial_x \boldsymbol{v}_{1y}$.

5 Here we include the subscript 1 to indicate the disturbance velocity, in order to distinguish the fluid velocity $\boldsymbol{v}$ (or $\boldsymbol{v}_g$) without any subscript.

Now we need to get the equation that describes the variation of the vorticity. Starting from the motion equation (6.47), omitting gravity, and according to the e-n coordinate system, we can write the component equation as follows

$$D_t v_e = -\partial_e p + f v_n, \quad D_t v_n = -\partial_n p - f v_e$$

After taking the vorticity of the motion equation

$$D_t(\partial_e v_n - \partial_n v_e) = -f(\partial_e v_e + \partial_n v_n) - v_n \partial_n f - v_e \partial_e f$$

Because $\partial_t f = 0$, we can add $\partial_t f$ to the left-hand side of the above formula, which then becomes

$$D_t(\zeta_z + f) = -f\nabla \cdot \mathbf{v}$$

For an incompressible atmosphere, $\nabla \cdot \mathbf{v} = 0$, so we have

$$D_t(\zeta_z + f) = D_t \zeta_\alpha = 0 \tag{6.52}$$

Which indicates that the absolute vorticity is conservative. We then expand the above formula into

$$\partial_t \zeta_z + v_{0x}\partial_x \zeta_z + v_{0y}\partial_y \zeta_z + v_{1y}\partial_y f = \partial_t(\partial_x v_{1y}) + v_{0x}\partial_x^2 v_{1y} + \beta_y v_{1y} = 0 \tag{6.53}$$

among this, we let $\beta_y \equiv \partial_y f$. Substituting (6.50) into (6.35), we can see that the frequency of a plane wave is

$$\omega = v_{0x} k - \frac{\beta_y}{k} = \left(v_{0e} k - \frac{\beta}{k}\right) \cos\alpha \tag{6.54}$$

where α is the angle between e and x. Generally arranged, $\beta_y > 0$ and $|\alpha| < \pi/2$.

6.4.3 Properties of Rossby wave

The frequency–dispersion relation (6.54) consists of two parts: the first term represents the convection of latitudinal winds, while the second term implies that the phase propagation of a Rossby wave always has a westward component. In fact, this latitudinal component of phase velocity has nothing to do with angle α, and is always equal to

$$V_{\mathrm{Pe}} = v_{0e} - \frac{\beta}{k^2}$$

While the longitudinal component of the phase velocity $V_{\mathrm{Px}} = V_{\mathrm{Pe}} \cos\alpha$ can properly be regarded as the result of the westward drift.

In [157], an interesting explanation is given for the westward drift: If we regard the variable Coriolis parameter as a field of force lines from east to west, then we can say that when cutting these force lines, the transverse velocity of the wave will generate an array that is alternately composed of induced vorticity and corresponding inducing velocities. The resulting phase velocity always has a westward component – this anisotropy originates from the fact that the gradient of earth's vorticity (Coriolis parameter) points to the north. When the transverse velocity cuts the force lines at the most rapid speed, the strongest inductive effect can be obtained. This condition occurs when the wave peak is orientated from south to north. Conversely, when the wave peak is orientated from east to west, the induction will disappear and the wave will no longer move.

From (6.54), we can obtain the longitudinal component of the group velocity

$$V_{Gx} = \partial_k \omega = \left(v_{0_e} + \frac{\beta}{k^2} \right) \cos \alpha \tag{6.55a}$$

while the transverse component is

$$V_{Gy} = \left(\frac{1}{k} \right) \partial_\alpha \omega = \left(\frac{\beta}{k^2} - v_{0x} \right) \sin \alpha \tag{6.55b}$$

Thereupon the group velocity also consists of two parts: the convection of the latitudinal wind v_{0e} and the wave group velocity with magnitude β/k^2 and angle 2α (see Fig. 41).

Rossby waves occur in non-dispersive atmospheres, which is sure to restrict that they only can happen at altitudes above the Ekman boundary layer (which consists of the lowest atmospheric layer at 1 km).

Rossby waves can be divided into two types: free Rossby waves and forced Rossby waves. The former are formed by random deviations related to geostrophic equilibrium. For example, they can be induced via oblique pressure unsteadinesses, positive-pressure unsteadinesses, or thermal disturbances. The latter are formed from deviations related to geostrophic equilibrium induced by mountains and mountain ranges. The former can be created by either eastward or westward air currents, while the latter can only be created by eastward ones.[6] If the wind v_{0x} blows westward, then it certainly will weaken the formation of a wave. This is very easy to imagine by noticing that in the northern hemisphere, the Coriolis force tries to make the wind deviate to the right side of its moving direction, while in the south hemisphere, to the left. Therefore, if we consider the disturbances of eastward currents at the equator, there will always be a restoring Coriolis force that points to the equator, so that a wave is generated. Conversely, there will be always a restoring force that departs from the equator in the westward current, so that a wave cannot be formed. A similar mechanism is also applicable at mid-latitudes.

A hot wind in the upper layers of the convection layer blows eastwards, and a Rossby wave above the Ekman boundary layer is certainly dominant. On the other

6 See the discussion about the mountain-back wave in Chapter 10.

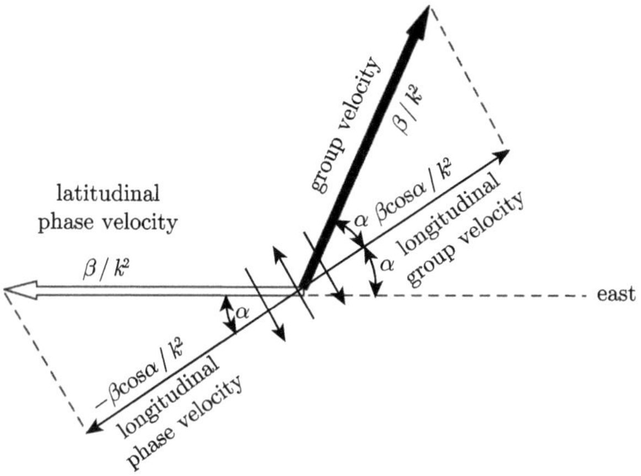

Fig. 41: Velocity relationships in Rossby waves.

hand, a temperature gradient may generate by a strong enough eastward air current so that Rossby wave motion can be neglected. This case occurs in the torrent of the convection layer with a fast eastwards airflow. In this respect, perhaps we should emphasize again the fact that Rossby waves are the inevitable result of the variation of Coriolis parameters with longitude. Without the latitudinal variation of the Coriolis force, Rossby waves should be degenerated into the geostrophic wind: in which Coriolis force balances with the pressure gradient.

6.5 External waves

6.5.1 Characteristic surface waves

According to the mode of energy storage, external waves can be divided into two types. The existence of any wave relies on two or more energy storage types, where energy exchange occurs repeatedly between them. The label "surface waves" simply indicates any waves whose mode of energy storage only exists at the medium surface or at a discontinuous surface. In the boundary layer between two different media, a discontinuity of density exists, where the formed wave is called a Helmholtz wave;[7] if the discontinuity depends on other parameters except density (such as temperature, wind speed, etc.), then the formed surface wave is called a boundary wave.

An evanescent wave not only exists on the surface of the fluid but also inside of it, where the most noted example is above-mentioned (see section 1.2) Lamb wave. It is

7 For example, a wave formed by the restoring force via surface tension generated by the fluid surface belongs to this type, which is called a surface tension wave or capillary wave.

a non-frequency-dispersive wave that satisfies the sound wave relation $\omega = ck_x$ in an isothermal atmosphere for all frequencies. A Lamb wave is deduced directly under the lower boundary condition of rigid ground surface. Similar to other evanescent waves, most of the energy of a Lamb wave is concentrated near the edge of non-homogeneities formed at the boundary. As such, the term "edge wave" was fabricated to describe a wave that was matched with a Lamb wave in a non-isothermal atmosphere that includes wind [158].

A stationary wave formed via reflection from two fixed boundary layers is called an "annular wave", or a "cellular wave". If we take the boundary layers at two altitudes $z = 0$ and $z = h$, then the wave function at these two altitudes must be zero everywhere, so that the wave function of the formed annular wave is

$$A_0 \sin\left(\frac{2\pi}{h}z\right) \exp[k_x x - \omega t + \mathrm{iIm}k_z z]$$

Internal waves whose phase without vertical changes can exist as well, in isothermal atmosphere, the characteristics of internal wave is Im $k_z = 1/2H$, so Re $k_z = 0$. This kind of internal wave determines the boundary of an internal wave area in the frequency–dispersion diagram.

The counting boundary condition (or source term) means that in the boundary-dispersion diagram there will not be an area that can represent an evanescent wave; while the criterion $k_z = 0$ can be represented by a straight line on the diagram. The equation of a straight line, which is often present in the isothermal atmosphere is

$$\omega^{*2} = gk_x\sqrt{\gamma - 1} \tag{6.56}$$

More commonly, in isothermal atmosphere, it is

$$\omega^{*2} = c\omega_B k_x \tag{6.56$'$}$$

This equation is similar to the gravity wave equation of a deep water surface $\omega^2 = gk_x$, thus the represented wave is called a characteristic surface wave.

A characteristic surface wave can always be separated from the sound wave area to the gravity wave area. The reason for third is that formula (6.56′) represents the atmospheric wave equation when the impedance is infinite, thus it is impossible that the gravity wave in an atmosphere of a certain area can be coupled with a sound wave of the atmosphere with different properties in another area. Similarly, an evanescent wave area can be divided into a sound wave and a gravity wave without any interaction. Therefore, an evanescent wave can interact with an internal wave of the same type.

A characteristic surface wave is elliptically polarized. By means of the results from section (6.3) we can see that for all evanescent waves, Z is a pure real number, while X is a pure complex, thus evanescent wave is elliptically polarized – but there is an exception. The exception is the Lamb wave: a Lamb wave has no variation in the vertical phase, so $Z = 0$ at all altitudes, hence

$$\omega / k_x = c \tag{6.57}$$

Therefore a Lamb wave can be judged as a body acoustic wave propagating in the medium, thus when $Z = 0$, it is linearly polarized. Substituting it into the frequency–dispersion relation (6.6), we get for a Lamb wave,

$$\mathrm{Im} k_z = \frac{g}{2c^2}[r \pm (r - 2)] \tag{6.58}$$

There exist two types of Lamb wave, where the polarization relations of these two are given by $P = \gamma$, $R = 1$, $X = c$ and $Z = 0$. The second type of Lamb wave which satisfies the relation $\mathrm{Im} k_z = g/c^2$ is the wave that exists in the lower side of the energy source; if the energy completely travels upward, it will not exist. Other important characteristics of a Lamb wave is that the total energy density ε of a wave in an infinitely large air column is limited, while in an isothermal atmosphere (potential energy omitted) we get

$$\varepsilon \propto \frac{1}{2}\rho_0 v_{1x}^2 \propto \int_0^\infty \exp[(\gamma - 2)g/c^2]\mathrm{d}z \propto c^2/(\gamma - 2)g \tag{6.59}$$

The curves of (6.56) and (6.57) will intersect in the $\omega - k_x$ plane where ω is equal to the V-B frequency ω_g, and

$$k_x = \sqrt{\gamma - 1} g/c^2 \tag{6.60}$$

(see Fig. 42). Thus, between characteristic surface waves and Lamb waves, there is the possibility that energy interactions may occur. However, any direct interaction between a linearly polarized Lamb wave and an elliptically polarized characteristic surface wave can be excluded immediately.

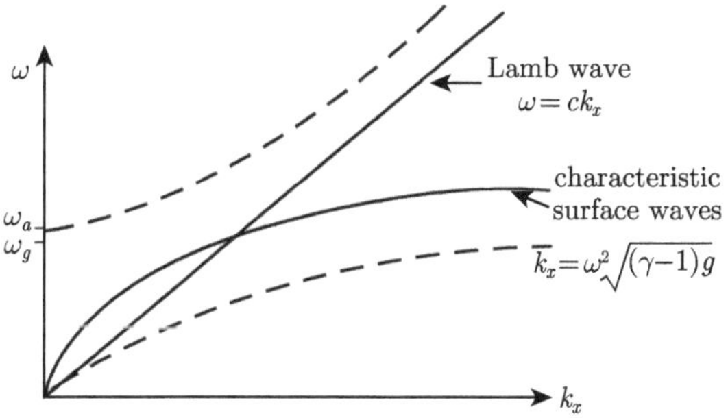

Fig. 42: Frequency-dispersion diagram of Lamb waves and characteristic surface waves.

6.5.2 Comparison with internal waves

The scheme of propagate figures shown in [159] is particularly useful for comparing the properties of internal waves and outer ones. The index of refraction is defined as

$$n_x = ck_x/\omega^*, \quad n_z = ck_z/\omega^* \tag{6.61}$$

and we can mark and indicate the curve at which ω^* is constant on the (n_x^2, n_z^2) plane. In an isothermal atmosphere, these curves become straight lines (see Fig. 43). The area under the n_z^2 axis corresponds to the imaginary part of n_z, which represent evanescent waves. The line connecting (1,0) and (0,1) represents $\omega^* = \infty$ and corresponds to sound waves under the high-frequency limit. Just as in the shadowed part in Fig. 37(a), where waves cannot exist, in the shadowed part in Fig. 43, waves also cannot exist – neither internal waves, external waves nor body waves. To mark and indicate index of the refraction is uncommonly useful for processing the isothermal atmosphere, but some properties of a non-isothermal atmosphere may be covered up [160]; under this case the better method is to mark and indicate the relationship between k_x and k_z.

For an isothermal atmosphere, we can easily see that when $\omega^* \to 0$, all of the waves have the same horizontal phase velocity $V_{Px} = 2c\sqrt{\gamma - 1}/\gamma \approx 0.9c$ (as $\gamma = 1.4$). When ω^* increases, the horizontal phase velocity of internal waves decreases, while the horizontal phase velocity of evanescent waves varies in a more complicated way. As shown in Fig. 44, the phase velocity and group velocity of isothermal atmospheric waves can be written as follows

$$V_{Px} = c/n_x \tag{6.62}$$

$$V_{Gx} = V_{Px}(\omega^{*2} - \omega_B^2) \Big/ \left(\frac{\omega^{*2}}{n_x^2} - \omega_B^2\right) \tag{6.63}$$

Then we can see that for certain waves (finally proved to be evanescent waves), $V_{Gx} > V_{Px}$, while V_{Gx} may become infinity; this happens when

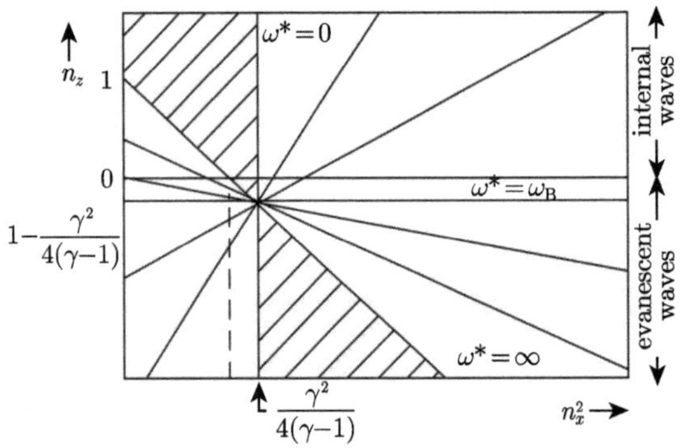

Fig. 43: The constant period curve in a refraction square domain.

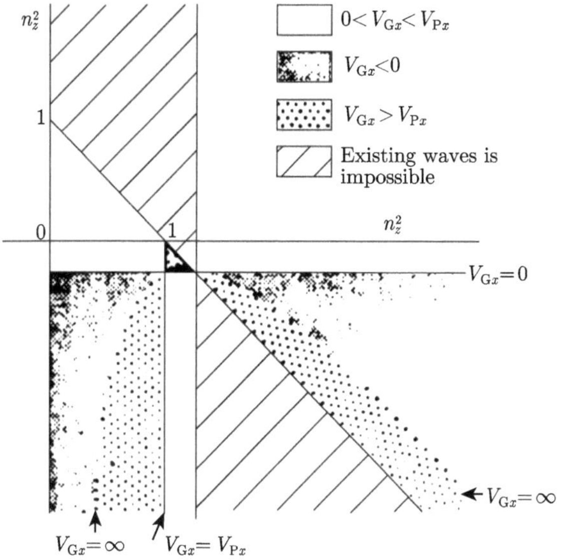

Fig. 44: The relationship between the horizontal phase velocity and the group velocity in a refraction square domain.

$$\omega^{*2} = ck_x\omega_B \tag{6.64}$$

These are just characteristic surface waves.

We can draw an analogy as follows: when a barrier moves in motionless water, a wave will be formed in front of the barrier. This "bow-shaped wave" may remain fixed relative to the barrier. Therefore the phase velocity of the bow-shaped wave is equal to the moving velocity of the barrier; but the group velocity of the bow-shaped wave must exceed the velocity of the barrier in order to transmit energy to points in front of the barrier. This indicates that evanescent waves with a high-group velocity are related to the movement of the atmospheric "barrier" with a high impedance (such as meteorite, aircraft etc.). When the group velocity becomes infinite, there will be the possibility of unsteadiness [161].

In the troposphere, it has been proved that evanescent waves can be generated by wind shear, which is original at horizontal wind speed varying gradually in the vertical direction [162]. At ionospheric altitudes, the existence of evanescent waves has not yet been firmly proved. It has been thought that noctilucent cloud may be surface waves generated by a discontinuity of temperature at the top of the mid-layer [163], and their existence in the E layer and the F layer has also been somewhat confirmed [164, 165].

On the other hand, Lamb waves and edge waves can only exist as forced vibrations above the stratosphere [166]. More than 99% of the energy of an evanescent wave is at altitudes below 110 km [158]. But it seems that most observations of micro-pressure of different pressure pulses are included by edge waves [158, 167]. These kinds of pulses are noted in history as a series of nature "explosions" such as the noted Krakatoa volcano eruption in 1883, the fall of huge meteorite in Siberia in 1908, the strong earthquake in Alaska in 1964, and artificial explosions such as nuclear tests in the atmosphere in previous years, and the tragic explosion of the U.S. space shuttle "Challenger" in 1986 (see Chapter X, section 10.2 and 10.3).

6.5.3 Boundary waves

Now let us investigate waves at the boundary between two isothermal layers (with different densities and temperatures, respectively) in the atmosphere. We set the origin in the boundary layer, where we indicate the lower layer with the subscript $l(z < 0)$ and the upper layer with $u(z > 0)$.

At the boundary, continuity of pressure and the vertical component of velocity are required, but due to the boundary layer moving arbitrarily, we cannot simply demand that the pressure and velocity are equal at both sides of $z = 0$. Even if the movement is arbitrarily small, there will be a zeroth-order gradient of pressure ($\mathrm{d}_z p_0$) in the vertical direction. For this we must apply the boundary condition

$$z = 0 : \begin{cases} D_t p|_l = D_t p|_u & (6.65) \\ D_t v_z|_l = D_t v_z|_u & (6.66) \end{cases}$$

where

$$D_t p = \partial_t p_1 + v_{1z}\partial_z p_0$$

and

$$D_t v_z = \partial_t v_{1z}$$

Applying formula (6.36) to (6.65), we get

$$\frac{\rho_l c_l^2(\mathrm{i}\omega^2 k_{zl} + k_x^2 g)}{\rho_u c_u^2(\mathrm{i}\omega^2 k_{zu} + k_x^2 g)} = \frac{A_u}{A_l} \tag{6.67}$$

and applying (6.36) to (6.66), then we get

$$\frac{\omega^2 - c_l^2 k_x^2}{\omega^2 - c_u^2 k_x^2} = \frac{A_u}{A_l} \tag{6.68}$$

where the horizontal wave number k_{xu} and k_{xl} in the upper and lower layers are both replaced with k_x, so that the wave-fronts are matched.

We can apply the frequency–dispersion equation (6.6) to each layer to seek k_{zl} and k_{zu}. If we suppose that the surface wave is an external wave and $k_z = \mathrm{i}\Gamma$, then in each layer

$$\Gamma = \frac{1}{2H} \pm \left[\left(\frac{1}{2H}\right)^2 - \frac{\omega^2}{c^2} + k_x^2\left(1 - \frac{\omega_g^2}{\omega^2}\right)\right]^{1/2} \tag{6.69}$$

By making (6.67) and (6.68) two formulae equal, we obtain

$$\frac{\rho_l c_l^2 (k_x^2 g - \omega^2 \Gamma_l)}{\rho_u c_u^2 (k_x^2 g - \omega^2 \Gamma_u)} = \frac{\omega^2 - c_l^2 k_x^2}{\omega^2 - c_u^2 k_x^2} \tag{6.70}$$

The above formula is the case without a background wind. If we count the horizontal background wind, then only using ω_l^* and ω_u^* to replace ω respectively will do.

If in the boundary layer $c_l = c_u = c$, then (6.70) becomes

$$\rho_l (k_x^2 g - \omega^2 \Gamma_l) = \rho_u (k_x^2 g - \omega^2 \Gamma_u)$$

Therefore, for a surface gravity wave above an incompressible homogeneous half-space, its characteristic equation can be given by the following steps: take $c = \infty$, $H_l = \infty$, so $\Gamma_l = k_x$, then set $k_x \gg 1/2H_u$, hence $\Gamma_u = -k_x$ (the sign is selected in order to make the energy gradually weaken when leaving from the boundary), and the phase velocity of wave can be given by the following formula

$$V_{\mathrm{P}x}^2 = \left(\frac{\omega}{k_x}\right)^2 = \frac{g(\rho_l - \rho_u)}{k_x(\rho_l + \rho_u)} \tag{6.71}$$

The above formula is a well-known equation used for a surface wave when it is in the boundary layer between two different incompressible fluids (such as oil and water). When $\rho_l \gg \rho_u$, as on the boundary layer of air and water, this equation can be further simplified as follows

$$V_{\mathrm{P}x}^2 = g/k_x \tag{6.71a}$$

Which gives the frequency–dispersion equation of a surface gravity wave in deep water

$$\omega^2 = g k_x \tag{6.71a'}$$

This equation is not under the influences of compressibility or density stratification in the lower half-space, but it is related to the sudden non-continuity of density between the two layers. However, in the compressible lower layer, there is a second solution that corresponds to a Lamb wave [151]. Under this case, it is unnecessary to cite the lower-boundary condition, because a Lamb wave is a sound wave piloted by the boundary layer. There is no interaction between the surface wave mode and the sound (Lamb) wave mode, because the surface wave is circularly polarized and vortex-free, conversely, the Lamb wave is linearly polarized and is not vortex-free.

Nevertheless, except for the lowermost boundary layer, it is not possible for a density non-continuity to exist in the atmosphere. So we have every reason to suppose that the surface gravity wave which follows from (6.71a) can extend to the atmosphere above the sea, but its amplitude is extremely small and not of any meteorological importance.

For more realistic atmosphere, we take boundary condition: when $z = 0$, $\rho_l = \rho_u$, we can prove that in an isothermal atmosphere $\Gamma = 1/2H$, the energy flow at all altitudes is constant. In order to make the energy carried by boundary wave be limited, above the boundary Γ must be less than $1/2H$, while under the boundary Γ must be more than $1/2H$. That is, the energy flow must be concentrated at the boundary, thus in the expression of Γ, the negative sign is used to denote an upper area, while a positive sign used for a lower area.

Under the low-frequency limit, we set $(1/2H) > k_x$, then we have

$$\Gamma_l = \frac{1}{H_l} - \frac{(\gamma-1)gk_x^2}{\gamma\omega^2}$$

while

$$\Gamma_u = \frac{(\gamma-1)gk_x^2}{\gamma\omega^2}$$

so

$$c_l^2\left[(2\gamma-1)gk_x^2 - \frac{\omega^2\gamma}{H_l}\right](\omega^2 - c_u^2k_x^2) = c_u^2gk_x^2(\omega^2 - c_l^2k_x^2) \tag{6.72}$$

or expressed by V_{Px}

$$c_l^2\left[(2\gamma-1)g - \frac{\gamma V_{Px}^2}{H_l}\right][V_{Px}^2 - c_u^2] = c_u^2g[V_{Px}^2 - c_l^2] \tag{6.72'}$$

Which gives a second-order equation of V_{Px}^2

$$V_{Px}^4\gamma^2g - V_{Px}^2[c_l^2(2\gamma-1)g + c_u^2(\gamma^2-1)g] + c_u^2c_l^2(2\gamma-2)g = 0 \tag{6.73}$$

By assuming $c_u > c_l$, the two solutions of V_{Px}^2 are approximately

$$V_{Px}^2 = \frac{2c_l^2}{\gamma+1}\left[1 - \frac{(2\gamma-1)c_l^2}{(\gamma^2-1)c_u^2}\right] \tag{6.74a}$$

and

$$V_{Px}^2 = \frac{\gamma^2-1}{\gamma^2}c_u^2\left[1 + \frac{c_l^2}{(\gamma+1)^2c_u^2}\right] \tag{6.74b}$$

Basically there are two possible types of boundary waves: one is movement associated with the sound speed near lower layer, the other is movement associated with the sound speed near the upper layer. In the common n-layer model of the atmosphere, the equations that describe phase velocity are of $2(n-1)$ order, so the possible number of boundary waves is also $2(n-1)$.

6.6 Atmospheric tides

6.6.1 Outlines

Under the action of periodic external forces (an attraction force and a "heat force"; where among them the heat effect of the sun is most important) applied by the sun and the moon, the earth's atmosphere can provoke a response analogous to forced vibration. This kind of response, whose period is the divisor of solar day, is called an atmospheric tide. In the theory of atmospheric wave motion, a half-day period holds the central position, as its action is just like a cut-off frequency. For a wave (inertial wave) in a massive rotating homogeneous incompressible fluid, at the two poles of the earth, the action of the half-day frequency $2\Omega_E$ is like a high cut-off frequency. The reason is that when the Coriolis parameter remains constant, a small variation of vorticity can be given from (6.51) as follows

$$\partial_t(\nabla \times \boldsymbol{v}) = 2\Omega_E \partial_z \boldsymbol{v} \tag{6.75}$$

Let us once more seek the vorticity and take the incompressible homogeneous fluid $\nabla \cdot \boldsymbol{v} = 0$ into account. Then, we have

$$-\partial_t^2(\nabla^2 \boldsymbol{v}) = 4\Omega_E^2 \partial_z^2 \boldsymbol{v} \tag{6.76}$$

Equation (6.76) gives the solution of plane wave only when

$$\omega^2(k_x^2 + k_y^2 + k_z^2) = 2\Omega_E^2 k_z^2 \tag{6.77}$$

Thus the vibration period must be longer than the half period of the earth's rotation. Just as we have seen in the previous section, planetary waves are inertial waves when the Coriolis parameter has longitudinal variation, and their period must be longer than the half-day period.

On the other hand, gravity waves have a low cut-off frequency at the half-day frequency. For a homogeneous fluid with a free surface at $z = h$ and with $z = 0$ at the bottom, Eq. (6.77) can still determine the waves when $\omega < 2\Omega_E$ and k_x, k_y, k_z are all real numbers. However, a free surface also can make possible to obtain a solution for k_z as pure imaginary. Thus if $\omega > 2\Omega_E$, we get

$$\Gamma^2(\omega^2 - 4\Omega_E^2) = \omega^2(k_x^2 + k_y^2) \tag{6.78}$$

These solutions have a frequency–dispersion relationship

$$\omega^2 = g\Gamma \tanh(\Gamma h) \tag{6.79}$$

Formulae (6.78) and (6.79) indicate that, for $\omega > 2\Omega_E$, when inertial waves cannot propagate inside, surface waves may be generated. Then the frequency–dispersion relation will be rectified by earth's rotation. These kinds of waves have a low cut-off frequency at the half-day frequency, otherwise Rossby waves with very large values l can be regarded as surface waves with $\omega < 2\Omega_E$.

The disturbance value of a vertical vorticity is generated by the variation of depth induced by surface waves, which is equal to f multiplied by the ratio between the true depth and the undisturbed depth. The existence of an disturbed vorticity makes another component (such as k_y) of wave vector, as a pure imaginary solution, possible. In this case, the boundary condition can be satisfied by the vorticity exponentially decreasing with the distance y away from the boundary, so that the vorticity travels as a one-dimensional propagation along the boundary with the rise and fall of depth. This is a Kelvin wave.

To analyze an internal gravity wave in the rotating fluid, we must begin with formula (6.75). We then have to add one term, which expresses the productivity of the horizontal vorticity caused by the potential temperature gradient. The component z in (6.76) becomes

$$-\partial_t^2(\nabla^2 v_z) = 4\Omega_{\mathrm{E}}^2 \partial_z^2 v_z + \omega_{\mathrm{B}}^2(\partial_x^2 + \partial_y^2)v_z \tag{6.80}$$

The plane-wave solution of above formula should satisfy

$$\omega^2(k_x^2 + k_y^2 + k_z^2) = 4\Omega_{\mathrm{E}}^2 k_z^2 + \omega_{\mathrm{B}}^2(k_x^2 + k_y^2) \tag{6.81}$$

Because these three wave number components all include real numbers, the plane-wave solution only exists when ω is located between $2\Omega_{\mathrm{E}}$ and ω_{B}. When $\omega_{\mathrm{B}} < 2\Omega_{\mathrm{E}}$, the returning wave of $\omega_{\mathrm{B}} < \omega < 2\Omega_{\mathrm{E}}$ is the mere survivor, therefore we can see that the internal gravity wave whose period is longer than half-day period cannot exist, which is suitable for all latitudes.

This indicates that except for the full-day component, all of the atmospheric tides are in gravity mode. In the region between latitudes ±30°, a full-day atmospheric tide may be a gravity wave as well, but if it exists at higher latitudes, it must be a Rossby mode. In a first-degree approximation, the atmospheric tide mode at low latitudes is evanescent, the reason is that the wavelength of long-period evanescent waves satisfies the following condition (see section 5.2)

$$\lambda > 0.9(2\pi)c/\omega \tag{6.82}$$

All waves with lengths that do not satisfy this condition belong to the class of internal gravity waves. Atmospheric tide mode wavelengths are given by $2\pi R_{\mathrm{E}}/m$ and frequencies are $\omega = 2\pi m/24(\mathrm{h})^{-1}$, so the premise to satisfy condition (6.82) is that the atmospheric temperature affecting sound velocity should be less than 750K. But this is not correct in the upper atmosphere; at heights above 150 km, finding a vertical phase structure in an atmospheric tide mode should be expected. So intuitively we can expect that there is no vertical phase variations from the ground surface to the altitude equal to the equivalent depth; similarly we can also expect to find vertical phase variations in the upper atmosphere.

6.6.2 Theory

In spherical coordinates, the motion equation (6.47) of the atmosphere of the rotating earth can be written as

$$\left.\begin{aligned}
D_t v_\phi &= \alpha_\phi - \left(2\Omega_{\mathrm{E}} + \frac{v_\phi}{r\sin\theta}\right)(v_\theta + v_r\sin\theta)\\
D_t v_\theta &= \alpha_\theta + \left(2\Omega_{\mathrm{E}} + \frac{v_\phi}{r\sin\theta}\right)v_\phi\cos\theta - \frac{v_r v_\theta}{r}\\
D_t v_r &= \alpha_r - g + \left(2\Omega_{\mathrm{E}} + \frac{v_\phi}{r\sin\theta}\right)v_\phi\sin\theta + \frac{v_\theta^2}{r}
\end{aligned}\right\} \tag{6.83}$$

Among them, the centrifugal acceleration has been merged into g, α is the acceleration caused by external pressure and friction forces, and the velocity components are defined as follows

$$v_r = D_t r, \quad v_\theta = rD_t\theta, \quad v_\phi = r\sin\theta D_t\phi$$

For the atmospheric tidal movement, the whole atmospheric layer can be regarded as very "shallow", so we can take $r = R_\mathrm{E}$ (radius of the earth), and use ∂_z instead of ∂_r. When $r = R_\mathrm{E}$, the coordinate is 0. Then we can suppose that $v_{0r} = 0$ [175], thus the motion equation of disturbance can be written as[8]

$$\left.\begin{aligned} \partial_t v_{1\theta} - 2\Omega_\mathrm{E} v_{1\phi}\cos\theta &= -\frac{1}{R_\mathrm{E}}\partial_\theta\left(\frac{p_1}{\rho_0} + \Phi\right) \\ \partial_t v_{1\phi} + 2\Omega_\mathrm{E} v_{1\theta}\cos\theta &= -\frac{1}{R_\mathrm{E}\sin\theta}\partial_\phi\left(\frac{p_1}{\rho_0} + \Phi\right) \end{aligned}\right\} \tag{6.84}$$

while

$$\partial_z p_1 = -g\rho_1 - \rho_0\partial_z\Phi \tag{6.85}$$

Because

$$D_t = \partial_t + v_r\partial_z + \frac{v_\theta}{R_\mathrm{E}}\partial_\theta + \frac{v_\phi}{R_\mathrm{E}\sin\theta}\partial_\phi$$

In order to simplify the problem, we have omitted the background prevailing wind v_0, Φ is a scalar potential to describe the gravitational tide-generating force, and the loss effect of viscosity and thermal conductivity has been omitted [149].

Now we can take the form of the continuity equation as

$$D_t\rho = \partial_t\rho_1 + v_{1z}\partial_z\rho_0 = -\rho_0\chi \tag{6.86}$$

and the adiabatic equation is

$$\frac{R}{M(\gamma-1)}D_t T = \frac{R}{M(\gamma-1)}(\partial_t T_1 + v_{1z}\partial_z T_0) = \frac{gH}{\rho_0}D_t\rho + J \tag{6.87}$$

where the velocity divergence equation is

$$\chi = \nabla\cdot\boldsymbol{v} = \frac{1}{R_\mathrm{E}\sin\theta}\partial_\theta(v_{1\theta}\sin\theta) + \frac{1}{R_\mathrm{E}\sin\theta}\partial_\phi v_{1\phi} + \partial_z v_{1z} \tag{6.88}$$

J is the "thermo-tidal heating" of a unit mass in unit time, indicating the periodic driving force acting on the free atmosphere vibration.

Now it is most convenient for us to seek the periodic solution of the quantity $-(\gamma p_0)^{-1}D_t p$ in the above equations. For a free vibration without driving force J, this

8 Here we resume using subscript 0 and 1, whose implied meaning is as before.

quantity corresponds to the velocity divergence χ. However, when counting J, it is no longer equal to χ, and we can write it down as

$$-\frac{1}{\gamma p_0} D_t p = G(\theta, z)\exp[\mathrm{i}(\omega t + m\phi)]$$

where $2\pi/\omega$ represents either a solar day or a lunar day, and it can also represent an appropriate portion of them, while $m = 0, \pm 1, \pm 2, \ldots$.

All of the variables of the wave field have the same exponential dependency, so all the terms including ∂_t and ∂_ϕ can be omitted. Thus we can directly derive an expression for the velocity functions $v_{1\theta}(\theta, z)$ and $v_{1\phi}(\theta, z)$, as indicated by pressure disturbance $p_1(\theta, z)$.[9] On the premise that $\partial_z^2 \Phi$ can be neglected, we can get the equation of $G(\theta, z)$

$$H\partial_z^2 G + (H' - 1)\partial_z G = \frac{g}{4R_\mathrm{E}^2 \Omega_\mathrm{E}^2} \hat{F}\left[\left(H' + \frac{\gamma - 1}{\gamma}\right) G - \frac{(\gamma - 1)J(\theta, z)}{\gamma^2 g H}\right] \tag{6.89}$$

where $\hat{F}$ is an operator

$$\hat{F} \equiv \frac{1}{\sin\theta}\partial_\theta\left(\frac{\sin\theta}{f^2 - \cos^2\theta}\partial_\theta\right) - \frac{1}{f^2 - \cos^2\theta}\left(\frac{m}{f}\frac{f^2 + \cos^2\theta}{f^2 - \cos^2\theta} + \frac{m^2}{\sin^2\theta}\right) \tag{6.90}$$

while

$$f = \frac{\omega}{2\Omega_\mathrm{E}}, \quad J = J(\theta, z)\exp[\mathrm{i}(-\omega t + m\phi)]$$

Eq. (6.89) can be solved by using the separation-of-variables method. If we let

$$G(\theta, z) = \sum_n L_n(z)\Theta_n(\theta) \tag{6.91}$$

Then we can expand J, where Θ_n is commonly called the Hough function

$$J(\theta, z) = \sum_n J_n(z)\Theta(\theta) \tag{6.92}$$

In this way, (6.89) is split up into two equations: a Laplacian tidal equation and a vertical structure equation.

(1) Laplacian tidal equation

The separation-of-variables method splits the partial differential equation into two or more parts, where each part is a function of only one variable. In the present case, the two parts are equal, so each part of (6.89) must be equal to a constant. Let us write the constant as h_n, then we have

$$\hat{F}(\Theta_n) = -\frac{2R_\mathrm{E}^2\Omega_\mathrm{E}^2}{g h_n}\Theta_n \tag{6.93}$$

9 The details of deriving can be seen in [170].

A group of h_n values is a group of eigen-values of the Laplacian tidal equation. Analogous to ocean tide theory, h_n is called the equivalent depth of the atmosphere, which is obtained by substituting (6.90) into (6.93)

$$d_\mu\left(\frac{1-\mu^2}{f^2-\mu^2}d_\mu\Theta\right)-\frac{1}{f^2-\mu^2}\left[\frac{m(f^2+\mu^2)}{f(f^2-\mu^2)}+\frac{m^2}{1-\mu^2}\right]\Theta_n+\frac{4R_E^2\Omega_E^2}{gh_n}\Theta_n=0 \qquad (6.94)$$

where $\mu = \cos\theta$. This equation has a simple solution only under the following three specified conditions: irrotational earth, infinite equivalent depth, and a half-day tide when $m = 0$.

In the last condition, $m = 0$ and $f = 1$, then we have

$$d_\mu^2\Theta_n+\frac{4R_E^2\Omega_E^2}{gh_n}\Theta_n=0 \qquad (6.95)$$

These equivalent depths can be obtained either by using mathematical conditions – Θ is a monotonic function with a continuously different quotient and with only a single determined value; or can be determined by using physical conditions – $v_{1\theta}$ at both the south and north poles it should be zero. These conditions can be satisfied only when $2R_E\Omega_E/\sqrt{gh_n} = n\pi/2$, at that time the equivalent depth is $h_n = \frac{16R_E^2\Omega_E^2}{n^2\pi^2 g}\Theta_n = \infty$, 35.04 km, 8.76 km, 3.89 km ... (corresponding, to $n = 0, 1, 2, 3, \ldots$ respectively).

However, generally speaking, it is possible to obtain a third-order recursive relationship. To do this we must put the Hough function expressed in the sum of related Lagrange polynomial expressions

$$\Theta_n^{\omega,m}=\sum_{l=m}^{\infty}C_{n,l}^{\omega,m}P_l^m(\cos\theta)$$

Where S_s^m represents the solar tidal mode, while L_s^m represents the lunar tidal mode, where s satisfies $\omega = 2\pi s/24$, the units of ω are the reciprocal of the sidereal hour (59′4.09″) and the lunar hour (62′5.00″), respectively. The most important three atmospheric tides are: half-day solar tide S_2^2 with $m = 2$ and $f = 1$,[10] whole-day solar tide S_1^1 with $m = 1$ and $f = 0.5$, and half-day lunar tide L_2^2 with $m = 2$ and $f = 0.965$.

(2) Vertical structure equation

Using the separation-of-variables method from (6.89) we can get a second equation

$$HL_n''+(H'-1)L_n'+\frac{1}{h_n}\left(H'+\frac{\gamma-1}{\gamma}\right)L_n=\frac{\gamma-1}{\gamma^2 gHh_n}J_n \qquad (6.96)$$

in order to transform the equation into standard form, we change the altitude to scalar-height units

$$x=\int_0^z\frac{dz}{H}$$

10 In fact $f = 0.99727$, because the period of S_2^2 mode is 12 sidereal hours, while the period of Ω_E is 12 solar hours (60′00″). Analogical indication is applicable to all solar tides.

Then we have

$$d_x^2 L_n - d_x L_n + \frac{1}{h_n}\left[d_x H + \frac{(\gamma-1)H}{\gamma}\right] L_n = \frac{\gamma-1}{\gamma^2 g h_n} J_n$$

By means of the standard transform, such as (6.20)

$$L_n = y_n \exp\left(+\frac{x}{2}\right)$$

we can transform the above formula into

$$d_x^2 y_n - \frac{1}{4}\left[1 - \frac{4}{h_n}\left(\frac{(\gamma-1)H}{\gamma} + d_x H\right)\right] y_n = \frac{(\gamma-1)J_n}{\gamma_2 g h_n}\exp\left(-\frac{x}{2}\right) \tag{6.97}$$

For the remaining atmospheric parameters, we can use the same way as mentioned in section 2.6 to derive the expression. When using the Hough function to expand the tidal potential, we have

$$\Phi = \sum_n \Phi_n(x)\Theta_n$$

then we can get (the exponential factors $\exp[i(-\omega t + m\phi)]$ are all neglected)

$$p_1 = \sum_n \frac{p_0(0)}{H(x)}\left[-\frac{\Phi_n}{g}\exp(-x) + \frac{\gamma h_n}{i\omega}\exp\left(-\frac{x}{2}\right)\left(d_x y_n - \frac{y_n}{2}\right)\right]\Theta_n$$

$$v_{1z} = \sum_n \left[\gamma h_n \exp\left(\frac{x}{2}\right)\left\{d_x y_n + \left(\frac{H}{h_n} - \frac{1}{2}\right)y_n\right\} - \frac{i\omega}{g}\Phi_n\right]\Theta_n$$

$$v_{1\theta} = \sum_n \frac{\gamma g h_n}{4R_E\Omega_E^2}\exp\left(\frac{x}{2}\right)\left(d_x y_n - \frac{y_n}{2}\right)\frac{1}{f^2 - \cos^2\theta}\left(d_\theta + \frac{m\cos\theta}{f}\right)\Theta_n$$

$$v_{1\phi} = \sum_n \frac{i\gamma g h_n}{4R_E\Omega_E^2}\exp\left(\frac{x}{2}\right)\left(d_x y_n - \frac{y_n}{2}\right)\frac{1}{f^2 - \cos^2\theta}\left(\frac{\cos\theta}{f}d_\theta + \frac{m}{\sin\theta}\right)\Theta_n$$

$$\rho_1 = \sum_n \frac{p_0(0)}{(gH)^2}\left\{-\Phi_n\exp(-x)\left(1 + \frac{1}{H}d_x H\right) + \frac{\gamma g h_n}{i\omega}\exp\left(-\frac{x}{2}\right)\left[\left(1 + \frac{1}{H}d_x H\right)\right.\right.$$
$$\left.\left.\times\left(d_x y_n - \frac{y_n}{2}\right) + \frac{H}{h_n}\left(\frac{\gamma-1}{\gamma} + \frac{1}{H}d_x H\right)y_n\right] - \frac{\gamma-1}{i\omega\gamma}J_n\right\}\Theta_n$$

$$T_1 = \sum_n \frac{M}{R}\left\{\frac{\Phi_n}{H}d_x H\exp(-x) - \frac{\gamma g h_n}{i\omega}\exp\left(-\frac{x}{2}\right)\left[\frac{\gamma-1}{\gamma h_n}H + \frac{1}{H}d_x H\right.\right.$$
$$\left.\left.\times\left(d_x + \frac{H}{h_n} - \frac{1}{2}\right)\right]y_n + \frac{\gamma-1}{i\omega\gamma}J_n\right\}\Theta_n \tag{6.98}$$

The boundary condition $v_{1x} = 0$ at $z = 0$ means

$$d_x y_n + \left(\frac{H_g}{h_n} - \frac{1}{2}\right)y_n = \frac{i\omega}{\gamma g h_n}\Phi_n \tag{6.99}$$

while the "upper boundary condition" can be obtained from the "vertical structure equation" (6.97). The general solution, in an isothermal atmosphere, of this equation is

$$y = A\exp(iqz) + B\exp(-iqz)$$

where

$$q = \frac{1}{H}\sqrt{\frac{(\gamma - 1)H}{\gamma h} - \frac{1}{4}}$$

The standard boundary condition requires radiation conditions and to set $B = 0$. In this case, the action of $q = \sqrt{\frac{1}{Hh}\left[\frac{\gamma - 1}{\gamma} + \mathrm{d}_z H\right] - \frac{1}{4H^2}}$ is just like a vertical wave number. If q at any given altitude are real numbers, that is the internal wave mode and the vertical propagation of energy is possible, but for the waves with imaginary q, they are evanescent waves and the vertical propagation of energy is forbidden.

Chapter 7
Computational Atmospheric Acoustics[1]

What is discussed in the foregoing chapters are basic problems in atmospheric acoustics, while in this chapter, some different approaches will be introduced to discuss this fundamental subject as deduced from these basic problems: after an acoustic wave is emitted from a given source it experiences a series of physical processes in the atmosphere, such as geometrical spreading, reflection from a boundary with complex impedance, refraction and diffraction in the atmosphere with a given profile, scattering from various non-homogeneities, absorption by the atmosphere, etc., and how to estimate accurately their transmission loss (TL) at the receiving point. In these general analyses, for the sake of convenience in handling, influences from various physical processes are considered separately, while in practice, of course, these influences will produce effects simultaneously and synthetically, thus making the original, very complicated, problem even more complicated.

As pointed out in chapter II, most of the propagation problems are reduced to solving the wave equation or Helmholtz equation under a given profile and solution-defining conditions. In recent years, increasing attention have been paid to problems concerning synthetic effects from refraction in the atmosphere with a finite ground impedance. In this situation the full-wave analytic algorithm [172, 173] or the normal-mode algorithm may after all be accepted as the most fundamental method. Namely, when solving the wave equation by the separation of variables, after using Hankel integral transforms, the field can be expressed in terms of the sum of normal modes via the residue theorem. At high frequencies, one must calculate the normal modes of higher orders, and thus larger amounts of computer hours are required. On the other hand, these methods are applicable only to certain types of atmospheric profiles above impedance surfaces, and it is quite difficult to obtain solutions for more general profiles.

---

1 In the first edition of this book, the author wrote a footnote here: "this term (i.e. 'computational atmospheric acoustics') has not appeared in any other literature until now, although in some allied fields the similar given names has been used one after another, such as computational fluid mechanics, computational acoustics and computational hydroacoustics. At any rate, in recent years in this category, the number o practical works have increased with each passing day, so it seems that the formation of a new discipline will be realized very soon." After this book was published for four years, in 2001 a monograph of atmospheric acoustics came off the press [171]. This indicates that the author's prediction was entirely confirmed. However, that monograph only discussed emphatically the FFP and PE methods, while the explicated methods such as the previous two sections in this chapter were not involved. Therefore, according to the above-mentioned monograph and other recent literature, the author particularly gave the former two sections a comprehensive correction and supplement.

Ray tracing [174] is another algorithm that is quite commonly used. This method approximately describes the paths of refraction in an atmosphere, but does not easily manifest low-frequency wave phenomena. The conditions for which the ray-tracing method is applicable can be expressed as follows [175]:

(1) The index of refraction n, its rate of change n' and the sound wavelength λ must satisfy the relationship $n^2 \gg \lambda n'$;
(2) The rate of spatial change in the logarithm of acoustic amplitude A must satisfy $(\ln A)' < 1/\lambda$;
(3) The propagation range R and the grazing angle θ must satisfy $(\lambda/2\pi R)^{1/3} < \theta$.

It can be seen that each condition becomes increasingly difficult to meet as λ increases (and frequency decreases). The most common sources of low-frequency sound are explosions. Pressure waves produced by both natural and artificial high energetic explosions (see section 10.3 in Chap X) are refracted gradually to the upper atmosphere, so that n' is always very small. In these events, ray-tracing routines accurately predict "quiet zones" relatively near both sources, and "audible regions"much further away due to refractive focusing. Sound sources with weaker strengths cannot produce sufficient energies that result in detectable sound levels after reflection by the upper atmosphere. In these cases, n' can be quite large, thus ray tracing may lead to poor predictions of large range propagation and a numerical solution to the wave equation becomes necessary.

There are various forms for such numerical algorithms, each appropriate to different situations and different scopes, and each has its own advantages and disadvantages. For example, the most accurate one is the finite-element solution, but the computation is too verbose. For propagation distances greater than a few meters, such algorithms take many hours of supercomputer time. Therefore, in practice, they are usually used as a supplementary means and play a supporting role to other algorithms. The remaining main algorithms include the Fast Field Program (FFP), the Parabolic Equation (PE) approach and modified ray-tracing routines. The fundamental concepts of these algorithms will be briefly interpreted in the main points in the successive sections.

7.1 Fast field program (FFP)

The Fast Field Program (FFP) stands as a kind of numerical method for computing the sound field from a monopole source in the stratified atmosphere above the earth surface. The ground is characterized by the impedance of the earth's surface, and the atmosphere is described by its temperature gradient and wind velocity gradient. In FFP processing, the continuous variation of sound speed with height (profile) is replaced by a stratified variation, and each stratum is assumed to have a constant sound speed. The acoustic pressure at the receiver is calculated as a function of the horizontal wave number k, and a fast Fourier transform (FFT) is then used as an approximation to the Hankel transform to give the acoustic pressure as a function of distance r from the source. The FFP is potentially capable of producing field predictions under a wide range of meteorological conditions while taking proper account of ground reflection. In particular, this algorithm can give reliable predictions when shadow and enhanced conditions occur in the same projection plane. Since the sound field in the spatial domain is calculated by the inverse Fourier transform from that in the horizontal wave number domain, the FFP is sometimes called the "wave number integration method" [176].

Like all linear methods, its solution must produce errors in zones where ray crossings occur. However, the nature of the solution is such that these errors are likely to be much smaller than those generated by the "ray-invariant model".[2]

The FFP method was originally developed for hydroacoustics,[3] and it was adopted for atmospheric propagation about 30 years ago [178–180]. Raspet [181] and Lee et al. [182] developed one kind of FFP method that is available for atmospheric acoustics. This is a two-dimensional algorithm suitable for an axisymmetric atmosphere, in which the effects from wind is accounted for by introducing an "effective sound velocity". Nijs-Wapenaar [183] and Wilson [184] developed a three-dimensional algorithm for a moving atmosphere, which can be regarded as a generalized FFP method. The matrix formulation adopted in [178] generates large numerical errors. An electrical ladder network was used in [179] to replace the atmospheric strata, which permitted the use of a more stable and less error-prone numerical procedure. This method was further elaborated afterwards. What is especially worth mentioning is Wilson's unpublished paper [185], which included a new modification to the method developed in [179], and permits the sound speed to be a linear function of height instead of a constant in each stratum.

7.1.1 Helmholtz equation, axial symmetric approximation

We first recall the continuous equation (1.31′) – the first approximation after being linearized:

$$\partial_t \rho + v_{0x}\partial_x \rho + v_{0y}\partial_y \rho + v_{0z}\partial_z \rho_0 + \rho_0(\partial_x v_x + \partial_y v_y + \partial_z v_z) = 0 \tag{7.1}$$

2 This means a propagation model fully controlled by ray theory, while the sound level prediction is given by a ray. The corresponding premises lie in: (i) reflection from the ground can be omitted; (ii) ray curvature varies only for distances much greater than the wavelengths; (iii) propagation behavior is independent of frequency. It turns out that, such a model can give reliable predictions for long-range propagation in the "strengthened" regions where the effects from meteorological effects dominate [177].

3 See, e.g., H. W. Kutschale, Report No. CU-1-70, Columbia Univ., New York: Palisades. 1970; F. R. Di Napoli, NUSC Tech. Report 4103, Washington D. C.: Naval Underwater Systems Center, 1971.

Secondly, we rewrite the equation of motion of sound waves (1.30a) in three component equations

$$\begin{aligned}\partial_t v_x + v_{0x}\partial_x v_x + v_{0y}\partial_y v_x + v_z\partial_z v_{0x} + (1/\rho_0)\partial_x p &= 0\\ \partial_t v_y + v_{0x}\partial_x v_y + v_{0y}\partial_y v_y + v_z\partial_z v_{0y} + (1/\rho_0)\partial_y p &= 0\\ \partial_t v_z + v_{0x}\partial_x v_z + v_{0y}\partial_y v_z + (1/\rho_0)\partial_z p &= 0\end{aligned} \tag{7.2}$$

Finally, we rewrite the equation of state (1.32) at a first approximation

$$\partial_t p + v_{0x}\partial_x p + v_{0y}\partial_y p + v_z\partial_z p_0 + (1/\rho_0)\partial_x p = c^2(\partial_t\rho + v_{0x}\partial_x\rho + v_{0y}\partial_y\rho + v_z\partial_z\rho_0) \tag{7.3}$$

We then apply a Fourier transformation to various related variables

$$p \to P, \quad \rho \to \Omega, \quad v_x \to U, \quad v_y \to V, \quad v_z \to W$$

After making the Fourier transform to Eqs. (7.1), (7.2) and (7.3), we obtain

$$\begin{aligned}&\eta\Omega + \rho_0' W + \rho_0(\mathrm{i}k_x U + \mathrm{i}k_y V + W') = 0\\ &\eta U + v_{0x}' W + \rho_0^{-1}\mathrm{i}k_x P = 0\\ &\eta V + v_{0y}' W + \rho_0^{-1}\mathrm{i}k_y P = 0\\ &\eta W + \rho_0^{-1}P' = 0\\ &\eta P = c^2(\eta\Omega + \rho_0' W)\end{aligned} \tag{7.4}$$

where $\eta \equiv -\mathrm{i}\omega + \mathrm{i}k_x v_{0x} + \mathrm{i}k_y v_{0y}$, and the primes denote derivatives with respect to z. Using the second to fifth equations listed above, we can express U, V, W and Ω in terms of P. Substituting these results into the first equation, and neglecting several small quantities yield

$$P'' - [(2\eta'/\eta) + (\rho_0'/\rho_0)]P' - [(\eta^2/c^2) + k_x^2 + k_y^2]P = 0 \tag{7.5}$$

Introducing the wave number

$$k_m = k - k_x m_x - k_y m_y$$

where $m_x = v_{0x}/c$ and $m_y = v_{0y}/c$, thus

$$k_{mz}^2 = k_m^2 - k_x^2 - k_y^2$$

and Eq. (7.5) can be rewritten as

$$P'' - 2(k_m'/k_m)P' + k_{mz}^2 P = 0 \tag{7.6}$$

It can be also written as

$$k_m^2\partial_z(k_m^{-2}\partial_z P) + k_{mz}^2 P = 0 \tag{7.6'}$$

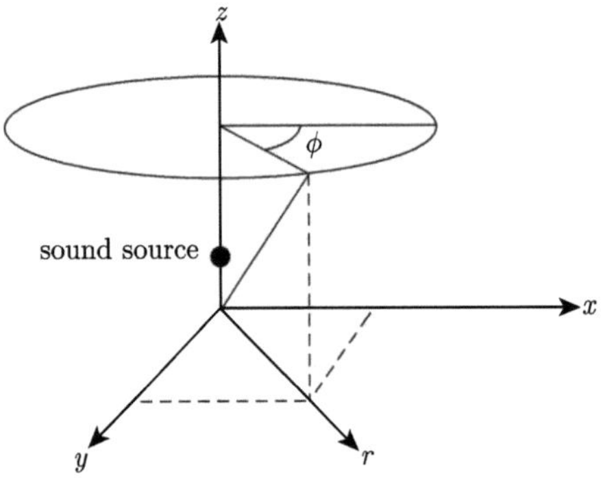

Fig. 45: Rectangular coordinates vs. cylindrical coordinates.

This equation can be regarded as the Helmholtz equation in the horizontal wave number domain, which stands as the fundamental equation in FFP method. If there is a monopole sound source with unit amplitude at the location $\boldsymbol{r}_S = (0, 0, z_S)$, the corresponding inhomogeneous Helmholtz equation is then

$$k_m^2 \partial_z (k_m^{-2} \partial_z P) + k_{mz}^2 P = -4\pi\delta(z - z_S) \tag{7.7}$$

We also can transform Eq. (7.6′) into the Helmholtz equation in the spatial domain. In this case we have $m_x = 0$, $m_y = 0$, $k_m = k$, and then Eq. (7.6′) becomes

$$k^2 \partial_z (k^{-2} \partial_z P) + (k^2 - k_x^2 - k_y^2) P = 0 \tag{7.8}$$

The inverse Fourier transformation yields the Helmholtz equation for a complex sound pressure amplitude

$$k^2 \nabla \cdot (k^{-2} \nabla p_c) + k^2 p_c = 0 \tag{7.9}$$

The axisymmetric approximation simply neglects variations in the sound field with an azimuth angle ϕ, which is always a nice approximation under quite a real practical circumstances, since variations with azimuth angle ϕ for both temperature and wind are much smaller than their variations with altitude z. This approximation is the foundation for elucidating two-dimensional problems in an atmosphere, and has been applied extensively in various two-dimensional numerical algorithms. To this end, we will describe the axisymmetric approximation for a sound field in a stratified atmosphere (the effects from wind is accounted via the "effective sound velocity" or the corresponding "effective wave number" k_{eff}).

Assuming that a sound field is generated by a point source, we adopt cylindrical coordinates with the z axis along a vertical line through the source (Fig. 45), then Eq. (7.9) becomes

$$(1/r)\partial_r (r \partial_r p_c) + k_{\text{eff}}^2 \partial_z (k_{\text{eff}}^{-2} \partial_z p_c) + (1/r^2) \partial_\varphi^2 p_c + k_{\text{eff}}^2 p_c = 0 \tag{7.10}$$

Under the axisymmetric approximation, the variation in sound field with azimuth angle is neglected, thus the third term on the left-hand side of Eq. (7.10) vanishes. Introducing a new quantity, defined by the following equation to replace p_c,

$$q_c = p_c \sqrt{r} \tag{7.11}$$

Eq. (7.10) can be written as

$$\partial_r^2 q_c + (1/4r^2)q_c + k_{\text{eff}}^2 \partial_z(k_{\text{eff}}^{-2}\partial_z q_c) + k_{\text{eff}}^2 q_c = 0 \tag{7.10'}$$

Applying the far-field approximation $r \gg k_{\text{eff}}^{-1}$, such that the second term on the left-hand side can be neglected, the two-dimensional Helmholtz equation is then obtained

$$\partial_r^2 q_c + k_{\text{eff}}^2 \partial_z(k_{\text{eff}}^{-2}\partial_z q_c) + k_{\text{eff}}^2 q_c = 0 \tag{7.10a}$$

Now we will derive the corresponding inhomogeneous Helmholtz equation in the horizontal wave number domain. From Eq. (7.7) we have

$$k_{\text{eff}}^2 \partial_z(k_{\text{eff}}^{-2}\partial_z P) + (k_{\text{eff}}^2 - k_x^2 - k_y^2)P = -4\pi\delta(z - z_S) \tag{7.12}$$

where the Fourier transform P is given by

$$P(k_x, k_y, z) = \int_{-\infty}^{\infty}\int_{-\infty}^{\infty} \exp(-\mathrm{i}k_x x - \mathrm{i}k_y y)p_c(\mathbf{r})\mathrm{d}x\mathrm{d}y$$

Under the axisymmetric approximation, p_c is a function of r and z only, thus the above equation becomes

$$P(k_x, k_y, z) = \int_0^{2\pi}\int_0^{\infty} \exp(-\mathrm{i}k_x r\cos\varphi - \mathrm{i}k_y r\sin\varphi)p_c(\mathbf{r})r\mathrm{d}r\mathrm{d}\varphi \tag{7.13}$$

Substituting it into Eq. (7.12) and considering Eq. (7.11), we get

$$\begin{aligned}&\int_0^{2\pi}\int_0^{\infty} \exp(-\mathrm{i}k_x r\cos\varphi - \mathrm{i}k_y r\sin\varphi)[k_{\text{eff}}^2 \partial_z(k_{\text{eff}}^{-2}\partial_z q_c)\\ &+ (k_{\text{eff}}^2 - k_x^2 - k_y^2)q_c]\sqrt{r}\mathrm{d}r\mathrm{d}\varphi = -4\pi\delta(z - z_S)\end{aligned} \tag{7.14}$$

As $k_x = k_r\cos\varphi$, $k_y = k_r\sin\varphi$, we have

$$\begin{aligned}&\int_0^{\infty} [k_{\text{eff}}^2 \partial_z(k_{\text{eff}}^{-2}\partial_z q_c) + (k_{\text{eff}}^2 - k_r^2)q_c]\sqrt{r}\int_0^{2\pi} \exp[-\mathrm{i}k_r r\cos(\varphi - \psi)]\mathrm{d}\varphi\mathrm{d}r\\ &= -4\pi\delta(z - z_S)\end{aligned} \tag{7.14'}$$

The integral over φ is just the Bessel function, of order zero, $J_0(k_r r)$, and the following asymptotic expression is available for large $k_r r$

$$J_0(k_r r) = (2/\pi k_r r)^{1/2}\cos(k_r r - \pi/4)$$

In this situation, Eq. (7.14′) becomes

$$\int_0^{\infty} [k_{\text{eff}}^2 \partial_z(k_{\text{eff}}^{-2}\partial_z q_c) + (k_{\text{eff}}^2 - k_r^2)q_c]^2\sqrt{(2\pi/k_r)}\cos(k_r r - \pi/4)\mathrm{d}r = -4\pi\delta(z - z_S) \tag{7.15}$$

Defining

$$Q(k_r, z) = \int_0^\infty q_c \cos(k_r r - \pi/4) \mathrm{d}r = \int_0^\infty [(q_c/\sqrt{2})\cos(k_r r) + (q_c/\sqrt{2})\sin(k_r r)] \mathrm{d}r \quad (7.16)$$

We find

$$k_{\text{eff}}^2 \partial_z (k_{\text{eff}}^{-2} \partial_z Q) + (k_{\text{eff}}^2 - k_r^2) Q = -\sqrt{(2\pi k_r)} \delta(z - z_S) \quad (7.17)$$

Using the FFP method to solve the above equation numerically we can compute the function $Q(k_r, z)$, and then the sound field $q_c(r, z)$ can be obtained from the inverse Fourier transformation that corresponds to Eq. (7.16)

$$q_c = (\sqrt{2}/\pi) \int_{-\infty}^{\infty} Q(k_r, r) \cos(k_r r) \mathrm{d}k_r = (\sqrt{2}/\pi) \int_{-\infty}^{\infty} Q(k_r, r) \sin(k_r r) \mathrm{d}k_r \quad (7.18)$$

By estimating the integral over azimuth angle φ in Eq. (7.14′), we can reduce the three-dimensional problem to a two-dimensional one. In the derivation, we have assumed that p_c is independent of the azimuth angle φ; this assumption is valid for a non-moving stratified atmosphere but is invalid for a moving atmosphere since sound waves travel faster in the downwind direction than in the upwind direction. As an approximation, however, a moving atmosphere can be replaced by a non-moving atmosphere with an effective sound velocity, so that the assumption of axial symmetry is still applicable.

7.1.2 Solutions of the Helmholtz equation

First divide the atmosphere characterized by wave number $k(z)$ into a series of horizontal homogeneous layers (Fig. 46). The heights of interfaces between adjacent layers are denoted as $z_j (j = 1, 2, \cdots, N)$. The ground corresponds to $z_1 = 0$, and the height of the sound source coincides with the interface m, such that $z_S = z_m$. The inhomogeneous Helmholtz equation then can be written as

$$(\partial_z^2 + k_z^2) P = -S_\delta \delta(z - z_S) \quad (7.19)$$

where $S_\delta = 4\pi$, $k_z^2 = k^2 - k_x^2 - k_y^2$ in the three-dimensional case; and $S_\delta = \sqrt{(2\pi k_r)}$, $k_z^2 = k^2 - k_r^2$ in the two-dimensional case. The solution to Eq. (7.19) in layer j can be written as

$$P_j = A_j \exp(\mathrm{i} k_{zj} z) + B_j \exp(-\mathrm{i} k_{zj} z), \quad z_j \leqslant z \leqslant z_{j+1} \quad (7.20)$$

where k_{zj} is the value of k_z in layer j, and A_j and B_j are constants, which are determined by the boundary conditions at the interfaces, respectively. At the top of the highest layer between z_{N-1} and z_N we set $B_{N-1} = 0$ such that there is an upwards traveling wave only. This implies that the highest layer should be chosen above the region

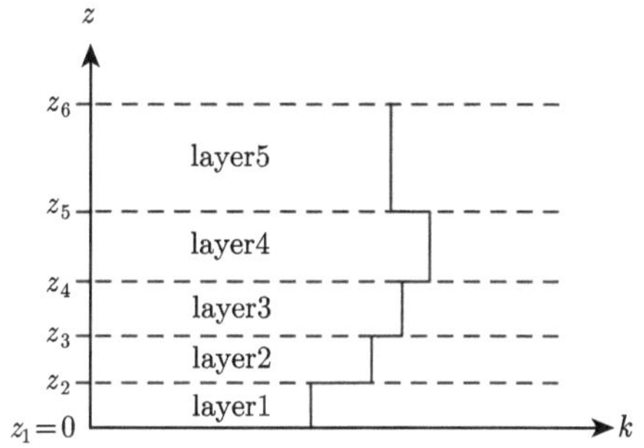

Fig. 46: Divide the atmosphere into a series of horizontal homogeneous layers. The wave number k is a constant in each of them in order to approximate the profile $k(z)$.

where sound is refracted downward to the receiver. The height of this region usually increases with increasing distance between the source and the receiver. The solution to Eq. (7.19) in the region below the ground surface ($z \leqslant 0$) can be written as

$$P_O = B_O \exp(-ik_O z), \quad z \leqslant 0 \tag{7.21}$$

where B_O is a constant and k_O is the complex wave number.

The sound pressure, as well as the normal velocity, is continuous at all interfaces except at the source height z_s, where only the pressure is continuous but the normal velocity is not. The boundary conditions at each interface can thus be obtained, where the third equation of Eq. (7.4) yields

$$\begin{aligned}
&\text{for} \quad j = 1, 2, \cdots, N \qquad P_j(z_j) = P_{j-1}(z_j) \\
&\text{for} \quad j = 2, 3, \cdots, N(j \neq m) \qquad \eta_j^{-1}\rho_j^{-1}\partial_z P_j(z_j) = \eta_{j-1}^{-1}\rho_{j-1}^{-1}\partial_z P_{j-1}(z_j) \\
&\text{for} \quad j = m \qquad \partial_z P_j(z_j) = \partial_z P_{j-1}(z_j) - S_\delta \\
&\text{for} \quad j = 1 \rho_j^{-1}\partial_z P_j(z_j) = \rho_{j-1}^{-1}\partial_z P_{j-1}(z_j)
\end{aligned} \tag{7.22}$$

where η_j and ρ_j are values of η and ρ_O in layer j, respectively. The discontinuity for profile $k(z)$ at $z = z_m$ can be ignored in the derivation under a type of "staircase approximation", since the real profile is continuous there.

The effects of the factors in the second and fourth equations of Eq. (7.22) are often very small and can be omitted. For the two-dimensional case of a non-moving atmosphere with an effective sound velocity we have $\eta = i\omega$ and $\eta_j^{-1} = \eta_{j-1}^{-1}$, and too for the three-dimensional case we have $\eta = -ick_m$ and $\eta_j^{-1} \approx \eta_{j-1}^{-1}$. As $\eta_1 = \eta_O$ since $v_{Ox} = v_{Oy} = 0$ at the ground surface $z_1 = 0$, the second equation is consistent with the fourth one.

The boundary conditions at $z_1 = 0$ give the relation

$$A_j = R(k_{z1})B_1 \tag{7.23}$$

where

$$R(k_{z1}) = [k_{z1} - k(z_1)/Z_S]/[k_{z1} + k(z_1)/Z_S] \tag{7.24}$$

is the reflection coefficient for a plane wave, and Z_S is the normalized impedance of the ground surface. For a locally reacting ground surface, Z_S is equal to the normalized characteristic impedance of the ground material.

To derive a set of equations that determine constants A_j and B_j, we can apply the rest-boundary conditions, however there is an easier approach. From Eq. (7.20) we find the relations

$$\begin{aligned} P_j(z+\Delta z) &= \cos(k_{zj}\Delta z)P_j(z) + k_{zj}^{-1}\sin(k_{zj}\Delta z)P_j'(z) \\ P_j'(z+\Delta z) &= -k_{zj}\sin(k_{zj}\Delta z)P_j(z) + \cos(k_{zj}\Delta z)P_j'(z) \end{aligned} \tag{7.25}$$

for z and $z+\Delta z$ in layer j, where $\cos\omega$ and $\sin\omega$ are defined for complex ω as $(1/2)(e^{i\omega}+e^{-i\omega})$ and $(1/2i)(e^{i\omega}-e^{-i\omega})$ respectively, and these relations will be used for determining the quantities $P_j\,(z_j)$.

Now we extrapolate to the sound source from the ground to the top, respectively. We first start from the ground surface at $z_1 = 0$. Arbitrarily setting $B_1 = 1$ for the moment, from Eqs. (7.20) and (7.23) we obtain

$$\begin{aligned} P_1(z_1) &= R(k_{z1}) + 1 \\ P_1'(z_1) &= ik_{z1}[R(k_{z1}) - 1] \end{aligned} \tag{7.26}$$

By applying Eq. (7.26), (7.25) (in which letting $z = z_j$ and $\Delta z = z_{j+1} - z_j$) and the first and second equations of Eq. (7.22), the values of $P_{j-1}(z_j)$ and $P_{j-1}'(z_j)$ can be determined for $j = 2, 3, \cdots, m$ successively. Denote the final values of $P_{j-1}(z_m)$ and $P_{j-1}'(z_m)$ at the source height z_m as $P_{m\mathrm{L}}$ and $P_{m\mathrm{L}}'$, respectively, where the index "L" denotes "Lower region", i.e., the region below the source.

We next start from the top at height z_N. Arbitrarily setting $P_{N-1}(z_N) = 1$ for the moment, and letting $B_{N-1} = 1$ in Eq. (7.20) we find

$$\begin{aligned} P_{N-1}(z_N) &= 1 \\ P_{N-1}'(z_N) &= ik_{z_N} \end{aligned} \tag{7.26$'$}$$

By applying Eq. (7.26′), (7.25) (in which letting $z = z_{j+1}$ and $\Delta z = z_j - z_{j+1}$) and the first and second equations of Eq. (7.22), the values of $P_j\,(z_j)$ and $P_j'(z_j)$ can be determined for $j = N-1, N-2, \cdots, m$ successively. Denote the final values of $P_m(z_m)$ and $P_m'(z_m)$ at the source height z_m as $P_{m\mathrm{U}}$ and $P_{m\mathrm{U}}'$, respectively, where the index "U" denotes the "Upper region", i.e., the region above the source.

7.1.3 Field at the receiver

As the settings $B_1 = 1$ and $P_{N-1}(z_N) = 1$ were arbitrarily set, we cannot confirm that the numerical values of P_j and P_j' are correct yet, but only that the ratios P_j'/P_j are correct. At the source height z_m, from the third equation of Eq. (7.22) we have

$$(P_{m\mathrm{U}}'/P_{m\mathrm{U}})P_m - (P_{m\mathrm{L}}'/P_{m\mathrm{L}})P_m = S_\delta \tag{7.27}$$

where P_m is the correct value of P at z_m. This gives

$$P_m = -S_\delta/[(P'_{mU}/P_{mU}) - (P'_{mL}/P_{mL})] \tag{7.28}$$

Now the values of P_j determined by the above calculations can be scaled correctly by multiplying by the following factor

$$\begin{aligned} &P_m/P_{mU} \quad &&\text{for } z_j > z_m, \\ &P_m/P_{mL} \quad &&\text{for } z_j < z_m \end{aligned} \tag{7.29}$$

Finally, applying an inverse Fourier transformation to the quantities P_j yields the complex pressure amplitude in the spatial domain. In the three-dimensional case we have

$$P_c(x, y, z) = 1/(2\pi)^2 \int_{-\infty}^{\infty}\int_{-\infty}^{\infty} \exp(\mathrm{i}k_x x + \mathrm{i}k_y y)P(k_x, k_y, z)\mathrm{d}k_x k_y \tag{7.30}$$

While in the two-dimensional case we have directly from Eq. (7.18), by replacing P by Q,

$$Q_c(r, z) = 1/(\sqrt{2}\pi) \int_{-\infty}^{\infty} [\exp(\mathrm{i}k_r r) + \exp(-\mathrm{i}k_r r)]Q(k_r, z)\mathrm{d}k_r \tag{7.31}$$

First we consider the two-dimensional case. The integrand in Eq. (7.31) has poles on the integration path near $k_r = -k$ and $k_r = k$, as can be seen from the analytical solution for the ground without a homogeneous-atmosphere case

$$Q(k_r, z) = \mathrm{i}\sqrt{(\pi/2)}(\sqrt{k_r}/k_z)\exp(\mathrm{i}k_z|z - z_s|),\ k_z^2 = k^2 - k_r^2 \tag{7.32}$$

Of course, the solution for an inhomogeneous atmosphere is slightly different, but the integral in Eq. (7.31) is still dominated by the regions near the poles. To avoid the poles we do not take the integration along the real axis but along the path shown in Fig. 47 instead. For positive k_r we include a small imaginary term $-\mathrm{i}k_t$, and for negative k_r we include the opposite imaginary term $+\mathrm{i}k_t$, where k_t is a positive small number. Such a choice corresponds to a positive imaginary part of k_z for $k_r^2 > k^2$, so that we have $Q \to 0$ for $k_r \to \pm\infty$, as can be seen from Eq. (7.32). It should be noted that the factor exp $(\pm k_r r)$ can be taken out of the integral $\int \exp(\pm k_r r)Q\mathrm{d}k_r$, therefore the integrals can still be performed via standard Fourier techniques.

By using the relation $Q(-k_r, z) = \pm\mathrm{i}Q(k_r, z)$, the integral in Eq. (7.31) can be transformed into an integral over the positive wave number k_r only. As we will see in section 7.1.5, we must use a minus sign here. Therefore Eq. (7.31) can be written as

$$Q_c(r, z) = (1 - \mathrm{i})/(\sqrt{2}\pi) \int_{0}^{\infty} [\exp(\mathrm{i}k_r r) + \exp(-\mathrm{i}k_r r)]Q(k_r, z)\mathrm{d}k_r \tag{7.31a}$$

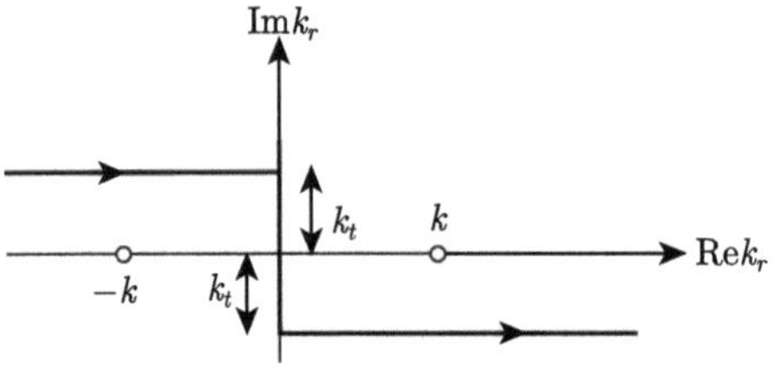

Fig. 47: Integration path that avoid the poles at $k_r = \pm k$ on the real axis.

In order to evaluate this integral numerically, we discretize the integration variable k_r as follows

$$K_{r,n} = k_{s,n} - \mathrm{i}k_t \qquad (n = 1, 2, \cdots, M)$$

where k_t, as mentioned above, is a small positive quantity, and $k_{s,n} = (1/2)\Delta k$, $(3/2)\Delta k, \cdots, k_{s,M}$ are the real parts. Δk is the wave number spacing. Owing to the discretization, the solution to q_c, given by Eq. (7.31a), becomes periodic to r with periodic distance $2\pi/\Delta k$. Δk should be chosen small enough (for example, to choose $2\pi/\Delta k \geqslant 3r$) to ensure that the value of q_c at the receiver is not affected by the periodicity. The choice of the maximum wave number $k_{s,M}$ depends on the frequency since the integral is dominated by wave numbers near the pole at $k_r = k = \omega/c_{\mathrm{eff}}$. One can use $k_{s,M} \approx 3\omega/c_0$, where c_0 is the sound velocity at $z = 0$. As for the positive small number k_t one can use the value of Δk.

The truncation of the integration interval in Eq. (7.31a) at the maximum wave number $k_{s,M}$ induces rapid small oscillations of q_c as a function of r. These can be easily eliminated by including a window function as a factor in the integrand of Eq. (7.31a); this function is equal to 1 except near the integration limits where the function smoothly approaches 0.

Next we consider Eq. (7.30) for the three-dimensional case. Just as in the two-dimensional case, the solution to a homogeneous atmosphere is directly proportional to k_z^{-1}. This implies that there are poles on the real $k_x k_y$ integration plane. The location of these poles depends on the receiver coordinates x and y. For $y = 0$ the poles are on the k_x axis, at $(k_x, k_y) = (-k, 0)$ and $(k_x, k_y) = (k, 0)$. In this case the k_x integration interval can be chosen as $[-3k_0, 3k_0]$ and the k_y integration interval as $[-k_0, k_0]$, $k_0 = \omega/c_0$. Also, the integration variables are discretized and the poles must be avoided by including a small imaginary term in k_x and k_y. For $y = 0$ the poles are on the k_x axis and the imaginary part of k_x must be negative for $k_x > 0$ and positive for $k_x < 0$ (see Fig. 47), thus a possible choice would be [184]

$$\begin{aligned} k_{x,m} &= k_{sx,m} - \mathrm{i}k_t\cos\theta_{mn} \\ k_{y,n} &= k_{sy,n} - \mathrm{i}k_t\sin\theta_{mn} \end{aligned} \tag{7.33}$$

where $k_{sx,m}$ and $k_{sy,n}$ are the discretized variables in the real $k_x k_y$ integration plane, $\theta_{mn} = \arctan(k_{sy,n}/k_{sx,m})$ is the polar angle of the vector $(k_{sx,m}, k_{sy,n})$ running from 0 to 2π, and k_t is a small positive number. As in the two-dimensional case, a window function must be used to eliminate spurious effects from the truncation of the integration intervals.

7.1.4 Improvements to the accuracy of numerical evaluations

The efficiency of the computational method described above is decided by the number of horizontal layers adopted, which depends on both the vertical profiles of the adiabatic sound speed and of the wind velocity. At heights where vertical gradients are large (such as near the ground surface), the layers must be sufficiently thin, while at heights where the vertical gradients are small, the layers can be thicker. In the case of a homogeneous atmosphere, only two layers may be sufficient.

In cases where the number of layers is large, the numerical errors that result from calculating P_j by repeated application of Eq. (7.25) can be quite considerable. These errors originate predominantly from a factor of $\exp(\pm \mathrm{i} k_{zj} \Delta z)$. When calculating the unscaled quantities P_j for $j = 2, 3, \cdots, m$, P_j may become large with increasing j owing to a factor of $\exp(-\mathrm{i} k_{zj} \Delta z)$ with imaginary k_{zj}. Similarly, when calculating P_j for $j = N-1, N-2, \cdots, m$, P_j may become large near the source owing to a factor of $\exp(+\mathrm{i} k_{zj} \Delta z)$ with imaginary k_{zj}. This problem can be solved by multiplying P_j and P_j' by a factor of

$$\begin{array}{ll} \exp(\mathrm{i} k_{z0} \Delta z) & \text{for } \Delta z > 0 \\ \exp(-\mathrm{i} k_{z0} \Delta z) & \text{for } \Delta z < 0 \end{array}$$

after each step $z \to z + \Delta z$. These fractions will not affect the ratios P_j'/P_j. In the final scaling to correct values of P_j, the reciprocal factors should be included; this gives an additional factor of $\exp(\mathrm{i} k_{z0} |z - z_0|)$ in the scale factors in (7.29). As for the wave number k_{z0}, the vertical wave number at the source height, for example, can be used.

However, an even better approach, which is equivalent to the "admittance extrapolation approach" described in [186], can be applied to layers above the highest location where the field is to be computed. In this case one can simply divide P_j and P_j' by P_j, and thus P_j and P_j' are replaced by 1 and P_j'/P_j, respectively.

7.1.5 FFP solutions in homogeneous atmosphere in two dimensions

Through this simple example one can have a better understanding of the quintessence of the FFP method. In this case we use only two layers: layer 1 between $z_1 = 0$ and $z_2 = z_S (m = 2)$; layer 2 between z_2 and $z_3 > z_2 (N = 3)$.

We recall once again that in two-dimensional problems the quantities $P \to Q$ and Eq. (7.26) should be replaced by

$$\begin{aligned} Q_1(z_1) &= R + 1 \\ Q_1'(z_1) &= \mathrm{i} k_z (R - 1) \end{aligned}$$

$$\begin{aligned} Q_{mL} &= \exp(ik_z z_s)R + \exp(-ik_z z_s) \\ Q'_{mL} &= ik_z[\exp(ik_z z_s)R - \exp(-ik_z z_s)] \end{aligned} \tag{7.34}$$

And Eq. (7.26′) become

$$\begin{aligned} Q_{N-1}(z_N) &= 1 \\ Q'_{N-1}(z_N) &= ik_z \end{aligned}$$

Applying Eq. (7.25), where $z = z_N$ and $\Delta z = z_s - z_N$ we get

$$\begin{aligned} Q_{mU} &= \exp[ik_z(z_s - z_N)] \\ Q'_{mU} &= ik_z\exp[ik_z(z_s - z_N)] \end{aligned} \tag{7.34′}$$

Eq. (7.28) yields

$$Q_m = i\sqrt{(\pi/2)}\sqrt{k_r}/k_z[1 + R\exp(2ik_z z_s)] \tag{7.35}$$

Considering the scale factors in Eq. (7.29), the correct values of $Q_j(z_j)$

$$Q_j(z) = i\sqrt{(\pi/2)}\sqrt{k_r}/k_z\{\exp[ik_z|z - z_s|] + R\exp[ik_z|z + z_s|]\} \tag{7.36}$$

are valid for arbitrary $z > 0$.

For the case without a ground surface, i.e., $R = 0$, then the second term in above Eq. vanishes. Eq. (7.31a) then gives

$$\begin{aligned} q_c = &\frac{i(1-i)}{2\sqrt{\pi}} \int_0^\infty \exp(ik_r r + ik_z|z - z_s|)\frac{\sqrt{k_r}}{k_z}dk_r \\ &+ \frac{i(1-i)}{2\sqrt{\pi}} \int_0^\infty \exp(-ik_r r + ik_z|z - z_s|)\frac{\sqrt{k_r}}{k_z}dk_r \end{aligned} \tag{7.37}$$

Rewriting the first integral on the right-hand side of the above equation in the form

$$I_1 = \int_0^\infty \exp[iF(k_r)]G(k_r)dk_r$$

For large $k_r r$, the above integral can be calculated via the stationary phase method [187]

$$I_1 = \sqrt{\frac{2\pi}{F''(k_r, 0)}}G(k_r, 0)\exp\left[iF(k_r, 0) + i\mu\frac{\pi}{4}\right] \tag{7.38}$$

with $\mu = \text{sign}[F''(k_{r,0})]$, where sign $[x] = 1$ for $x > 0$ and $= -1$ for $x < 0$. $k_{r,0}$ is the wave number at the stationary phase point, i.e., the solution to equation $F'(k_{r,0}) = 0$. We find $k_{r,0} = rk/R_1$, and $R_1 = \sqrt{[r^2 + (z - z_s)^2]}$, the integral becomes

$$I_1 = \sqrt{\pi}(1 - i)\sqrt{r}\frac{\exp(ikR_1)}{R_1} \tag{7.38a}$$

The second integral on the right-hand side of Eq. (7.37) can be treated similarly. In this case the stationary phase point is at $k_{r,0} = -rk/R_1$, which is outside of the integration interval, and this means that the second integral can be neglected as compared with the first one. Therefore Eqs. (7.37) and (7.38a) give

$$q_c = \sqrt{r}\frac{\exp(\mathrm{i}kR_1)}{R_1} \tag{7.39}$$

Correspondingly,

$$p_c = \frac{\exp(\mathrm{i}kR_1)}{R_1} \tag{7.40}$$

In summary, the FFP method is the only model capable of providing sound pressure level predictions that take proper account of ground reflections and detailed meteorological behavior throughout the atmosphere. This algorithm does, however, suffer from significant errors at large ranges due to the inadequacy of the generalized damping method and the stratum quantization error. The damping procedure used has differing effects on each of the main spikes in the k spectrum, which are dependent on their proximity to a k sample value. This k sampling error therefore enhances the effect of some of the strata at the expense of others. The stratum quantization error may be reduced by using much larger numbers of strata, especially where sound speed gradients are large. However, this causes increased numerical error and reduces long-range accuracy.

The FFP is also computationally very demanding, and runs very slowly on desktop machines. Clearly there are still many occasions when the simple models are operationally more viable and capable of giving predictions which are sufficiently reliable. These occasions are when the conditions are clearly definable either as shadow or enhancement. In interface regions between these zones FFP must be used. As the FFP is range-independent, it cannot give predictions when propagation occurs through an anon-stratified atmosphere or over a ground with a spatially varying impedance.

7.2 Parabolic equation (PE) method I: Crank-Nicholson parabolic equation (CNPE) method

The parabolic-equation (PE) method was proposed by Леонтович and Фок half century ago and applied to radio propagation in the troposphere.[4] Thereafter, it was widely applied to various kinds of fields, such as quantum mechanics, plasma physics,

4 Refer to J. Phys. USSR, 1946, 10(1): 13. Fock V. A. *Electromagnetic Diffraction and Propagation Problems*, New York: Pergamon. 1965.

earthquake wave propagation, optics and hydroacoustics.[5] As for in atmospheric acoustics, only about 30 years ago did several potential applications appear [83, 188].

The requirements of the PE equation for an atmospheric model is easy to be met by allowing the acoustic velocity to vary point by point; that is to say, we can specify the variation of the acoustic velocity at each point of the altitude horizontal distance lattice [189], and we can use the PE method to estimate the effect of atmospheric turbulence and irregular terrain on sound propagation. In the PE method, we first neglect the contributions of large elevation angles to the acoustic field in order to obtain the wave equation, then to compute the acoustic field via the achieved parabolic equation. The PE method can give accurate results in the range specified by the largest elevation angle. Generally, the largest elevation angle is from 10° to 70° or higher, depending on the adopted "minimum approximation" [176] used when deriving the parabolic equation. The largest elevation angle is usually sufficient for sound sources and receivers located near the ground.

A crucial step in the development of the PE method in atmospheric acoustics was completed by Gilbert & White [190] in 1989, this kind of methods is called Crank-Nicholson parabolic equation (CNPE) method. Additionally, there is the "Green function parabolic equation (GFPE) method". They are mainly applied when computing the acoustic field of a monopole source refracted in an atmosphere, and all two-dimensional methods are based on an axisymmetric approximation. In recent years, this pair of three-dimensional PE methods have been further developed [191, 192]. CNPE can give a finite difference solution to the wide-angle parabolic equation, where the solution is accurate to about 35° of the elevation angle. In the case of wide-angle propagation and a large acoustic velocity gradient, GFPE is not as accurate as CNPE, but in most cases, GFPE is still accurate enough. The merit of GFPE lies in its efficiency, which is faster than CNPE.

In the most common form, the solution to PE can also be used to solve the Helmholtz equation in a far field ($kr \gg 1$). Using PE is more advantageous than directly applying the Helmholtz equation. Because, in PE we neglect the backward dissipation field and PE is completely expandable [193], thus to greatly reduce links between neighboring grid points in the numerical computation, therefore the demands for operating time of computer and storage space can be moderated.

Nevertheless, before using the PE method to perform numerical computation, in fact the approximation we have done will restrict the accuracy of larger angle propagation mode. The anxiety for the restriction can be partly eliminated via a certain PE that can give accurate results [194] when propagating at 45° to the horizontal plane,

5 Refer to, for example, Tappert F. D., *The parabolic approximation method. Wave Propagation and Underwater Acoustics*, (ed. J. B. Keller and J. S. Paradkis), Lecture Notes in Physics, 70, Berlin: Springer Verlag. Bobertson, J. S. et al., Acoust. Soc. Amer., 1987, 82:559.

and in later papers [195] a higher order PE was reported, where it seems that the half-width of rays can be expanded to 60°.

7.2.1 Derivation of narrow-angle PE and wide-angle PE

Still, in views of the axisymmetric approximation, the second term $k_{\text{eff}}^2\partial_z(k_{\text{eff}}^{-2}\partial_z q_c)$ can be approximated as $\partial_z^2 q_c$. Numerical computation indicates that the effect brought about by the approximation can usually be neglected, so Eq. (7.10a) becomes

$$(\partial_r^2 + \partial_z^2 + k_{\text{eff}}^2)q_c = 0 \tag{7.41}$$

Later the subscript "eff" in k_{eff}^2 and the subscript "c" in q_c will be omitted.

In the CNPE method, the acoustic field $q(r, z)$ is computed among the lattice in the rz plane (see Fig. 48). Computation starts from $r = 0$ to represent the initial function $q(0, z)$. The extrapolated step from r to $r + \Delta r$ can be written as

$$q(r, z) \to q(r + \Delta r, z)$$

Therefore we can obtain the value q at the grid $r+\Delta r$ by computing the value q at r. But only when the horizontal space Δr and vertical space Δz of the grids both do not exceed 1/10 of the average wavelength, can accurate results be given out. The lattice has a confined height – at the top we use an artificial absorption layer to eliminate refraction from the top. Ground effect is counted by the boundary terms with parameters of complex ground impedance. The dependency of the wave number k on distance r can be considered section by section; in an extrapolation space, this dependency can be neglected, so k is only a function of z.

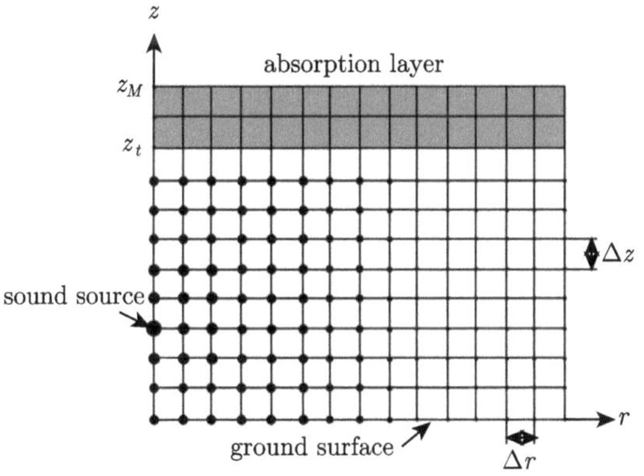

Fig. 48: Lattice on rz plane used in the PE method. The acoustic pressure amplitude at each grid graphically denotes the size of the circles in the corresponding grid. In fact, the number of grids along z axis is usually 1000 or more.

The solution to Eq. (7.41) can be written as

$$q(r, z) = \psi(r, z)\exp(\mathrm{i}k_\alpha r) \tag{7.42}$$

where k_α is a certain average altitude or just the wave number at the ground surface, the factor of $\exp(\mathrm{i}k_\alpha r)$ expresses the propagating wave in the positive r direction and oscillates rapidly as function of r; but function $\psi(r, z)$ generally varies slowly with r. Substituting (7.42) into Eq. (7.41) gives

$$\partial_r^2\psi + 2\mathrm{i}k_\alpha\partial_r\psi + \partial_z^2\psi + (k^2 - k_\alpha^2)^2\psi = 0 \tag{7.43}$$

As $\psi(r, z)$ varies slowly with r, so the first term on the left-hand side compared with the second term can be omitted, thus the obtained equation is called a narrow-angle parabolic equation.

The narrow-angle parabolic equation can also be derived via the other method. Let

$$k^2(z) - k_\alpha^2 \equiv \delta k^2(z) \tag{7.44}$$

and introduce the following operator

$$H_2(z) \equiv k^2(z) + \partial_z^2 = k_\alpha^2 + \delta k^2(z) + \partial_z^2 \equiv k_\alpha^2(1 + s) \tag{7.45}$$

Eq. (7.41) can be written as

$$\partial_r^2 q + H_2(z)q = [\partial_r - \mathrm{i}H_1(z)][\partial_r + \mathrm{i}H_1(z)]q = 0 \tag{7.46}$$

where

$$H_1(z) = k_\alpha\sqrt{1 + s} \tag{7.47}$$

is a square-root operator. The square root of differential operator $1+s$ is determined by expansion $\sqrt{1 + s} = 1+\frac{1}{2}S-\frac{1}{8}S^2+\cdots$, similar to the expansion of ordinary square-root function [194].

When deriving Eq. (7.46), we have actually assumed that H_1 and ∂_r are reciprocal, i.e. we have $H_1\partial_r = \partial_r H_1$. Strictly speaking, it is only valid for a stratified atmosphere ($k = k(z)$), as an atmosphere ($k = k(r, z)$) whose wave number relies on distance is only an approximation.

The factor of $[\partial_r - \mathrm{i}H_1(z)]$ in Eq. (7.46) expresses the propagation wave in the positive r direction, while $[\partial_r + \mathrm{i}H_1(z)]$ expresses the wave propagating in the opposite direction. Here we assume that the sound source is located at $r = 0$ and the receiver is located at $r > 0$, and we only consider the propagating wave in the positive r direction without considering back dissipation. Eq. (7.46) then becomes a single-direction wave equation

$$\partial_r q - \mathrm{i}H_1(z)q = 0 \tag{7.48}$$

If we use the following equation to approximate the square-root operator (7.47)

$$H_1(z) = k_\alpha \left(1 + \frac{1}{2}s\right) \tag{7.49}$$

then we can obtain the narrow-angle parabolic equation

$$\partial_r q - \mathrm{i}k_\alpha q - \frac{\mathrm{i}}{2k_\alpha}(\partial_z^2 + \delta k^2(z))q = 0 \tag{7.50}$$

If we use ψ defined in (7.42) to substitute in for q, we then obtain

$$\partial_r \psi = \frac{\mathrm{i}}{2} k_\alpha s \psi \tag{7.51}$$

Which is consistent with Eq. (7.43) only to its first term.

The approximation (7.49) of the square-root operator is accurate only when elevation angle smaller than about 10°, then a more accurate expansion is [194]

$$H_1(z) = k_\alpha \frac{1 + (3/4)s}{1 + (1/4)s} \tag{7.52}$$

Then single-direction wave equation (7.48) becomes

$$\left(1 + \frac{1}{4}s\right) \partial_r q = \mathrm{i}K_\alpha \left(1 + \frac{3}{4}s\right) q \tag{7.53}$$

This is the wide-angle parabolic equation. The corresponding equation for ψ defined by (7.42) is

$$\left(1 + \frac{1}{4}s\right) \partial_r \psi = \frac{\mathrm{i}}{2} k_\alpha s \psi \tag{7.53$'$}$$

7.2.2 Finite-difference solutions of narrow-angle PE and wide-angle PE

The above-derived narrow-angle PE and wide-angle PE can both be used for the finite-difference method to substitute the differential quotient approximately in order to solve the value [196]. In order to solve narrow-angle PE, we rewrite Eq. (7.43), with the first term omitted as follows

$$\psi = \alpha \partial_z^2 \psi + \beta \psi \tag{7.54}$$

where $\alpha = \mathrm{i}/2k_\alpha$, and $\beta = \mathrm{i}(k^2 - k_\alpha^2)/2k_\alpha = \mathrm{i}\delta k^2/2k_\alpha$. Adopting the lattice shown in Fig. 48, the height where the grid is located

$$z_j = j\Delta z, \quad j = 1, 2, \cdots, M \tag{7.55}$$

We record the field ψ at a distance r as the vector $\psi_j = \psi(r, z_j)$ of cardinal elements $\boldsymbol{\psi}$. Using the center difference formula

$$(\partial_z^2 \psi)_{z_j} = \frac{\psi_{j+1} - 2\psi_j + \psi_{j-1}}{(\Delta z)^2} \tag{7.56}$$

(7.54) can be rewritten as

$$\partial_r \begin{bmatrix} \psi_1 \\ \psi_2 \\ \psi_3 \\ \vdots \\ \vdots \\ \psi_{M-1} \\ \psi_M \end{bmatrix} = \left[\gamma \begin{bmatrix} -2 & 1 & & & & & \\ 1 & -2 & 1 & & & & \\ & 0 & -2 & 1 & & & \\ & & \ddots & \ddots & \ddots & & \\ & & & \ddots & \ddots & \ddots & \\ & & & & 1 & -2 & 1 \\ & & & & & 1 & -2 \end{bmatrix} + \begin{bmatrix} \beta_1 & & & & & & \\ & \beta_2 & & & & & \\ & & \beta_3 & & & & \\ & & & \ddots & & & \\ & & & & \ddots & & \\ & & & & & \beta_{M-1} & \\ & & & & & & \beta_M \end{bmatrix} \right] \begin{bmatrix} \psi_1 \\ \psi_2 \\ \psi_3 \\ \vdots \\ \vdots \\ \psi_{M-1} \\ \psi_M \end{bmatrix} + \gamma \begin{bmatrix} \psi_0 \\ 0 \\ 0 \\ \vdots \\ \vdots \\ 0 \\ \psi_{M+1} \end{bmatrix} \tag{7.57}$$

where $\gamma = \alpha/(\Delta z)^2$ and $\beta_j = \beta(z_j)$. On the right-hand side of each matrix, only non-zero matrix elements are listed. Vector equation (7.57) represents an equation group that includes M times equations, where each equation links a matrix element $\partial_r \psi_j$ with matrix elements ψ_{j-1}, ψ_j and ψ_{j+1} in the same matrix. The last term on the right-hand side of Eq. (7.57) includes the field ψ_0 at $z_0 = 0$ at the ground surface and the field ψ_{M+1} at the height $z_{M+1} = (M+1)\Delta z$.

For the field ψ_0 at the ground surface $z_0 = 0$, we can use the relation

$$\psi_0 = \sigma_1 \psi_1 + \sigma_2 \psi_2 \tag{7.58}$$

where the coefficients σ_1 and σ_2 rely on the ground impedance. For the field ψ_{M+1} we can use the relation

$$\psi_{M+1} = \tau_1 \psi_M + \tau_2 \psi_{M-1} \tag{7.59}$$

then (7.57) can be rewritten as

$$\partial_r \boldsymbol{\psi} = (\gamma T + D) \boldsymbol{\psi} \tag{7.60}$$

where T is triple diagonal matrix, while D is diagonal matrix; they are given respectively as

$$T = \begin{bmatrix} -2+\sigma_1 & 1+\sigma_2 & & & & \\ 1 & -2 & 1 & & & \\ & 1 & -2 & 1 & & \\ & & \ddots & \ddots & \ddots & \\ & & & 1 & -2 & 1 \\ & & & & 1+\tau_2 & -2+\tau_1 \end{bmatrix} \tag{7.61}$$

$$D = \begin{bmatrix} \beta_1 & & & & & \\ & \beta_2 & & & & \\ & & \beta_3 & & & \\ & & & \ddots & & \\ & & & & \beta_{M-1} & \\ & & & & & \beta_M \end{bmatrix} \tag{7.62}$$

Integrating (7.6) over r to $r + \Delta r$ gives

$$\boldsymbol{\psi}(r+\Delta r) - \boldsymbol{\psi}(r) = (\gamma T + D) \int_r^{r+\Delta r} \boldsymbol{\psi} \mathrm{d}r \tag{7.63}$$

We approximately substitute the integral on the right-hand side by the product $\frac{1}{2}[\boldsymbol{\psi}(r+\Delta r)+\boldsymbol{\psi}(r)]\Delta r$ of the average value of the integrand and the integration interval Δr. This approximation is called the Crank-Nicholson approximation. We obtain

$$M_2 \boldsymbol{\psi}(r+\Delta r) = M_1 \boldsymbol{\psi}(r) \tag{7.64}$$

where M_1 and M_2 are both triple diagonal matrices, and are given as the following two equations

$$\begin{aligned} M_1 &= 1 + \frac{1}{2}\Delta r(\gamma T + D) \\ M_2 &= 1 - \frac{1}{2}\Delta r(\gamma T + D) \end{aligned} \tag{7.65}$$

The step $\boldsymbol{\psi}(r) \to \boldsymbol{\psi}(r+\Delta r)$ of the PE then becomes the solution to Eq. (7.64) (an equation group consists of M equations with M unknowns $\boldsymbol{\psi}_j(r+\Delta r)$). Because M_1 and M_2 are both triple diagonal matrices, so the solution can be obtained effectively by using the Gauss elimination method [197].

The difference between the wide-angle PE (7.53′) and the narrow-angle PE (7.51) is a factor of $(1 + s/4)$ on the left-hand side. Comparing (7.51) with (7.60) indicates that the finite-difference matrix form of the operator $\mathrm{i}k_\alpha s/2$ is $\gamma T + D$. Therefore, the factor of $(1+s/4)$ in wide-angle PE will make the left-hand side of Eq. (7.63) appear factor

$[1 + (\gamma T + D)/(2ik_\alpha)]$. This causes the two matrices M_1 and M_2 in matrix Eq. (7.64) to be rectified as follows

$$\begin{aligned} M_1 &= 1 + \frac{1}{2}\Delta r(\gamma T + D) + \frac{\gamma T + D}{2ik_\alpha} \\ M_2 &= 1 - \frac{1}{2}\Delta r(\gamma T + D) + \frac{\gamma T + D}{2ik_\alpha} \end{aligned} \tag{7.66}$$

7.2.3 Effects of density profile

As mentioned above, the term $\partial_z^2 q$ in Eq. (7.41) is approximated from $k_{\text{eff}}^2 \partial_z(k_{\text{eff}}^{-2}\partial_z q)$. If we do not make this approximation, then for example, for a motionless equi-pressure atmosphere this term is $\rho\partial_z(\rho^{-1}\partial_z q)$, here $\rho = \rho(z)$ is the density of atmosphere. Thus Eq. (7.41) becomes

$$\partial_r^2 q + \rho\partial_z(\rho^{-1}\partial_z q) + k^2 q = 0 \tag{7.67}$$

The finite difference form of the differential operator $\rho\partial_z(\rho^{-1}\partial_z)$ is given by the following formula

$$\begin{aligned} \rho\partial_z(\rho^{-1}\partial_z\psi)_{z_j} = \frac{1}{(\Delta z)^2} \times \Bigg[& \frac{1}{2}\left(1 + \frac{\rho_j}{\rho_{j+1}}\right)\psi_{j+1} - \frac{1}{2}\left(\frac{\rho_j}{\rho_{j+1}} + 2 + \frac{\rho_j}{\rho_{j-1}}\right)\psi_j \\ & + \frac{1}{2}\left(1 + \frac{\rho_j}{\rho_{j-1}}\right)\psi_{j-1}\Bigg] \end{aligned} \tag{7.68}$$

The unique effect of using $\rho\partial_z(\rho^{-1}\partial_z)$ as a substitute for ∂_z^2 is that the matrix T in (7.65) and (7.66) should be modified as follows

$$\begin{aligned} T = & \begin{bmatrix} -2\gamma_{0,1} & \gamma_{1,1} & & & & \\ \gamma_{-1,2} & -2\gamma_{0,2} & \lambda_{1,2} & & & \\ & \gamma_{-1,3} & -2\gamma_{0,3} & \gamma_{1,3} & & \\ & & \ddots & \ddots & \ddots & \\ & & & \gamma_{-1,M-1} & \gamma_{0,M-1} & \gamma_{1,M-1} \\ & & & & \gamma_{-1,M} & -2\gamma_{0,M} \end{bmatrix} \\ & + \begin{bmatrix} \gamma_{-1,1}\sigma_1 & \gamma_{-1,1}\sigma_2 & & & & \\ 0 & 0 & 0 & & & \\ & 0 & 0 & 0 & & \\ & & \ddots & \ddots & \ddots & \\ & & & 0 & 0 & 0 \\ & & & & \gamma_{1,M}\tau_2 & \gamma_{1,M}\tau_1 \end{bmatrix} \end{aligned} \tag{7.69}$$

where

$$\begin{aligned}
\gamma_{1,j} &= \frac{1}{2}\left(1 + \frac{\rho_j}{\rho_{j+1}}\right) \\
\gamma_{0,j} &= \frac{1}{4}\left(\frac{\rho_j}{\rho_{j+1}} + 2 + \frac{\rho_j}{\rho_{j-1}}\right) \\
\gamma_{-1,j} &= \frac{1}{2}\left(1 + \frac{\rho_j}{\rho_{j-1}}\right)
\end{aligned} \tag{7.70}$$

Numerical computation indicates that in Eq. (7.41) the effect caused by approximating the term $k_{\text{eff}}^2 \partial_z (k_{\text{eff}}^{-2} \partial_z q)$ to $\partial_z^2 q$ is very small. The effect of an atmospheric gradient on sound propagation is almost completely determined by the third term on the left-hand side of Eq. (7.67).

7.2.4 Finite-element solutions

Gilbert & White proposed a more accurate finite-element approximation solution [190], in which we can use an unevenly distributed space of vertical grids, that is to say, $z_j - z_{j-1}$ is allowed to vary with j. The foundation of the finite-element solution is Eq. (7.67), thus the solution will also include a density gradient.

The resulting form of the finite-different method is a triple diagonal matrix as in (7.64). The finite-element method finally leads to a triple diagonal matrix in the same form

$$M^{-}\boldsymbol{\psi}(r + \Delta R) = M^{+}\boldsymbol{\psi}(r) \tag{7.71}$$

Here the triple diagonal matrix $M^{\pm}$ is given by the following formula [190]

$$M^{\pm} = \begin{bmatrix}
M_{0,1}^{\pm} & M_{1,1}^{\pm} & & & & \\
M_{-1,2}^{\pm} & M_{0,2}^{\pm} & M_{1,2}^{\pm} & & & \\
& M_{-1,3}^{\pm} & M_{0,3}^{\pm} & M_{1,3}^{\pm} & & \\
& & \ddots & \ddots & \ddots & \\
& & & M_{-1,M-1}^{\pm} & M_{0,M-1}^{\pm} & M_{1,M-1}^{\pm} \\
& & & & M_{-1,M}^{\pm} & M_{0,M}^{\pm}
\end{bmatrix}
+ \begin{bmatrix}
M_{-1,1}^{\pm}\sigma_1 & M_{-1,1}^{\pm}\sigma_2 & & & & \\
0 & 0 & 0 & & & \\
& 0 & 0 & 0 & & \\
& & \ddots & \ddots & \ddots & \\
& & & 0 & 0 & 0 \\
& & & & M_{1,M}^{\pm}\tau_2 & M_{1,M}^{\pm}\tau_1
\end{bmatrix} \tag{7.72}$$

where $M^{\pm}_{m,j}$ is given by the following formula

$$M^{\pm}_{m,j} = \sum_{n=1}^{3} c^{\pm}_n H^{(n)}_{j,j+m} \tag{7.73}$$

where $m = -1, 0, 1; j = 1, 2, \cdots, M$. Each coefficient $c^{\pm}_n$ is given by the formulae

$$\begin{aligned} c^{\pm}_1 &= \frac{3}{4} \mp \frac{1}{4} \mathrm{i} k_\alpha \Delta r \\ c^{\pm}_2 &= \frac{1}{4} \pm \frac{1}{4} \mathrm{i} k_\alpha \Delta r \\ c^{\pm}_3 &= c^{\pm}_2 / k^2_\alpha \end{aligned} \tag{7.74}$$

The value $H^{(n)}_{j,j+m}$ when $n = 1, 2$ is as follows

$$\begin{aligned} H^{(n)}_{j,j} &= \frac{1}{12}[(z_j - z_{j-1}) f_n(z_{j-1}) + 3(z_{j+1} - z_{j-1}) f_n(z_j) + (z_{j+1} - z_j) f_n(z_{j+1})] \\ H^{(n)}_{j,j\pm'} &= \pm \frac{1}{12}[f_n(z_j) + f_n(z_{j\pm 1})](z_{j\pm 1} - z_j) \end{aligned} \tag{7.75a}$$

When $n = 3$, we have

$$\begin{aligned} H^{(3)}_{j,j\pm 1} &= \pm \frac{1}{2}[f_1(z_j) + f_1(z_{j\pm 1})]/(z_{j\pm 1} - z_j) \\ H^{(3)}_{j,j} &= -[H^{(3)}_{j,j+1} + H^{(3)}_{j,j-1}] \end{aligned} \tag{7.75b}$$

and $f_1(z) = \rho^{-1}(z); f_2(z) = \rho^{-1}(z) k^2(z)/k^2_\alpha$.

7.3 Parabolic equation (PE) method II: Green function parabolic equation (GFPE) method

7.3.1 Unbounded non-refracting atmosphere

The GFPR method recounted in this section is still based on the axisymmetric approximation, and we will now proceed from Eq. (7.41) (where we still neglect the subscript "eff" of k).

To make it simple and clear, we firstly consider an unbounded non-refracting atmosphere, thus the wave number k is a constant. We then perform a Fourier transform $z \to k_z$ over (7.41), that is to multiply $\exp(-\mathrm{i}k_z z)$ and integrate over z from $-\infty$ to $+\infty$, from this we obtain

$$[\partial^2_r + (k^2 - k^2_z)]Q = 0 \tag{7.76}$$

where

$$Q(r, k_z) = \int_{-\infty}^{+\infty} \exp(-\mathrm{i}k_z z) q(r, z) \mathrm{d}z \tag{7.77}$$

It is the Fourier transform of q. Now we rewrite Eq. (7.37) as following

$$(\partial_r - \mathrm{i}\sqrt{k^2 - k_z^2})(\partial_r + \mathrm{i}\sqrt{k^2 - k_z^2})Q = 0 \tag{7.78}$$

From the first factor we get the solution $Q(r, k_z) = Q(0, k_z)\exp(\mathrm{i}r\sqrt{k^2 - k_z^2})$, which indicates that the wave is traveling along the direction $+r$ (of course, we should add the once-neglected time factor $\exp(\mathrm{i}\omega t)$, while from the second factor we get the solution $Q(r, k_z) = Q(0, k_z)\exp(-\mathrm{i}r\sqrt{k^2 - k_z^2})$, which indicates that the wave is traveling along the direction $-r$. Now we confine the propagating wave so that it only travels along the direction $+r$, so the first solution is adopted and it can be written as

$$Q(r + \Delta r, k_z) = Q(r, k_z)\exp(\mathrm{i}\Delta r\sqrt{k^2 - k_z^2}) \tag{7.79}$$

Proceeding the Fourier inversion we get the function $q(r + \Delta r, k_z)$

$$q(r + \Delta r, z) = \frac{1}{2\pi}\int_{-\infty}^{\infty} \exp(\mathrm{i}k_z z)\exp(\mathrm{i}\Delta r\sqrt{k^2 - k_z^2})Q(r, k_z)\mathrm{d}k_z \tag{7.80}$$

Therefore, using both a forward and inverse Fourier transform we can extrapolate the acoustic field from r to $r + \Delta r$.

The Eq. (7.80), which is applicable to an unbounded non-refracting atmosphere, can be extended to a refracting atmosphere above a finite impedance ground [198]. Its foundation is the two-dimensional variant of the Kirchhoff-Helmholtz equation.

In a passive three-dimensional volume V (enveloped by a surface S) at point $\boldsymbol{r}_1 = (x_1, y_1, z_1)$, the complex acoustic pressure amplitude can be given via the Kirchhoff-Helmholtz integral equation [88].

$$p(\boldsymbol{r}_1) = \frac{1}{4\pi}\iint_S [g_3(\boldsymbol{r}, \boldsymbol{r}_1)\nabla p(\boldsymbol{r}) - p(\boldsymbol{r})\nabla g_3(\boldsymbol{r}, \boldsymbol{r}_1)] \cdot \boldsymbol{n}\mathrm{d}S \tag{7.81}$$

where $\boldsymbol{r} = (x, y, z)$, $\nabla \equiv \partial_r = (\partial_x, \partial_y, \partial_z)$, $\boldsymbol{n}$ is the outward unit normal vector on surface S, while the three-dimensional Green Function $g_3(\boldsymbol{r}, \boldsymbol{r}_1)$ is the solution to the following inhomogeneous Helmholtz equation

$$\nabla^2 g_3(\boldsymbol{r}, \boldsymbol{r}_1) + k^2(\boldsymbol{r})g_3(\boldsymbol{r}, \boldsymbol{r}_1) = -4\pi\delta(\boldsymbol{r} - \boldsymbol{r}_1) \tag{7.82}$$

while $p(\boldsymbol{r})$ satisfies the homogeneous Helmholtz equation

$$\nabla^2 p(\boldsymbol{r}) + k^2(\boldsymbol{r})p(\boldsymbol{r}) = 0 \tag{7.83}$$

In the two-dimensional case, (7.81) can be derived as

$$p(\boldsymbol{r}_1) = \frac{1}{4\pi}\int_C [g_2(\boldsymbol{r}, \boldsymbol{r}_1)\nabla p(\boldsymbol{r}) - p(\boldsymbol{r})\nabla g_2(\boldsymbol{r}, \boldsymbol{r}_1)] \cdot \boldsymbol{n}\mathrm{d}S \tag{7.84}$$

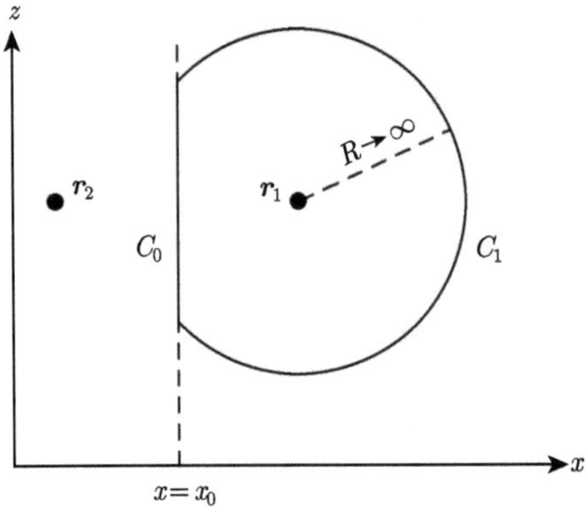

Fig. 49: Geometry of the two-dimensional Kirchhoff-Helmholtz integral equation.

where $\boldsymbol{r} = (x, z)$, $\boldsymbol{r}_1 = (x_1, z_1)$, $\nabla \equiv \partial_r = (\partial_x, \partial_z)$, the integration is over the circuit C that envelops S, while the two-dimensional Green function is

$$g_2(\boldsymbol{r}, \boldsymbol{r}_1) = \int_{-\infty}^{\infty} g_3(\boldsymbol{r}, \boldsymbol{r}_1) \mathrm{d}y \tag{7.85}$$

Which satisfies the inhomogeneous Helmholtz equation

$$\nabla^2 g_2(\boldsymbol{r}, \boldsymbol{r}_1) + k^2(\boldsymbol{r}) g_2(\boldsymbol{r}, \boldsymbol{r}_1) = -4\pi\delta(\boldsymbol{r} - \boldsymbol{r}_1) \tag{7.86}$$

Now, we choose an integral circuit C for Eq. (7.84) as shown as in Fig. 49. C consists of the line C_0 at $x = r_0$ and the circumference C_1 whose center is at $\boldsymbol{r}_1$ and radius $R \to \infty$. When $R \to \infty$ we have $p \to 0$, and so $g_2 \to 0$, therefore the contribution of C_1 to the circuit integral approaches zero. When using the Green function g_2 we have considerable freedom; the only condition is that g_2 must satisfy is Eq. (7.86), so g_2 includes a contribution from the monopole acoustic source at $\boldsymbol{r}_1$.

We can also count the contribution of the source outside of the range enveloped by circuit C; here we choose g_2 to make $g_2 = 0$ on C_0. For this we can adopt a field of two opposite monopole sources at $\boldsymbol{r}_1$ and $\boldsymbol{r}_2$, respectively,

$$g_2(\boldsymbol{r}, \boldsymbol{r}_1) = g(\boldsymbol{r}, \boldsymbol{r}_1) - g(\boldsymbol{r}, \boldsymbol{r}_2) \tag{7.87}$$

where $g(\boldsymbol{r}, \boldsymbol{r}_1)$ and $-g(\boldsymbol{r}, \boldsymbol{r}_2)$ are the fields of the two opposite monopole sources at $\boldsymbol{r}_1$ and $\boldsymbol{r}_2$, which are the solutions to Eq. (7.86) after substituting $\boldsymbol{r}_1$ with $\boldsymbol{r}_2$.

Because $g_2 = 0$ on C_0, so the first term of the integrand in Eq. (7.84) was omitted. For the second term we have $\nabla g_2 \cdot \boldsymbol{n} = -\partial_z g_z = -2\partial_z g$ on C_0, there we used (7.87). Then (7.84) becomes

$$P(\boldsymbol{r}_1) = \frac{1}{2\pi} \int_{-\infty}^{\infty} (p(r)\partial_x g(r, r_1))_{x = x_0} \mathrm{d}z \tag{7.88}$$

the above integral is noted for the Rayleigh integral (see [36] Sec. 278, and [56, 199]).

Now we will return to variable q and setting the ground at $z = 0$, and we will consider the effects caused by the ground. The integral lower limit in (7.88) can be rectified as $z = 0$, while the reflecting effect of the ground can be expressed in terms of the Green function, so (7.88) becomes

$$q(r_1, z_1) = \frac{1}{2\pi} \int_0^\infty [q(r,z)\partial_r g(r,z;r_1,z_1)]_{r=r_0} \mathrm{d}z \tag{7.89}$$

The Green function $g(r, z; r_1, z_1)$ is just the acoustic field at (r, z) caused by the monopole source at (r_1, z_1), which satisfies the following equation

$$[\partial_r^2 + \partial_z^2 + k^2(z)]g(r,z;r_1,z_1) = -4\pi\delta(r - r_1)\delta(z - z_1) \tag{7.90}$$

The wave number k is only a function of z, and its dependency on r can be used on k to count with the variation between "step increments", thus the Green function depends on r and r_1 through the difference $\Delta r = r_1 - r$ only, so we can express it as $g(\Delta r; z, z_1)$. If we introduce a Fourier transform, then

$$G(k, z, z_1) = \int_{-\infty}^{\infty} g(\Delta r, z, z_1)\exp(-\mathrm{i}k_h\Delta r)\mathrm{d}(\Delta r)$$

where k_h is a horizontal wave number. Substituting the above inverse Fourier transform into (7.89) can find (with $\partial_r = -\partial_{\Delta r}$, and z_1 noted as z, and z noted as z')

$$q(r + \Delta r, z) = \frac{1}{4\pi^2\mathrm{i}} \int_{-\infty}^{\infty} \exp(\mathrm{i}k_h\Delta r)k_h\mathrm{d}k_h \int_0^\infty G(k_h, z', z)q(r, z')\mathrm{d}z' \tag{7.91}$$

The Green function $G(k_h, z', z)$ is the solution to Eq. (7.51) after applying a Fourier transformed. We then multiply (7.90) by $\exp(-\mathrm{i}k_h\Delta r)$ and integrate over Δr to obtain

$$[\partial_r^2 + k^2(z) - k_h^2]G(k_h, z', z) = -4\pi\delta(z - z') \tag{7.92}$$

from Eq. (7.91) and (7.92) we can deduce the fundamental equation of the GFPE method.

For a non-refracting atmosphere, wave number $k(z) = k_0 =$ const. In this case, if at $z = 0$ there is finite impedance ground, then the solution to Eq. (7.92) is [177, Sect. D. 4]

$$G(k_h, z', z) = \frac{2\pi\mathrm{i}}{k_v}\{\exp(\mathrm{i}k_v|z - z'|) + R(k_v)\exp(\mathrm{i}k_v[z + z'])\} \tag{7.93}$$

where k_v is a vertical wave number

$$k_v^2 = k^2(z) - k_h^2$$

While

$$R(k_v) = \frac{k_vZ - k_0}{k_vZ + k_0} \tag{7.94}$$

is the reflecting coefficient of a plane wave, and Z is the normalized impedance of the locally responding ground. Substituting (7.93) into (7.91) gives

$$q(r+\Delta r, z) = \frac{1}{\pi \mathrm{i}} \int_{-\infty}^{\infty} \exp(\mathrm{i}k_h \Delta r) k_h \mathrm{d}k_h \times \int_{0}^{\infty} \frac{\mathrm{i}}{2k_v} \{\exp(\mathrm{i}k_v |z-z'|) + R(k_v)\exp(\mathrm{i}k_v[z+z'])\} q(r, z')\mathrm{d}z' \tag{7.95}$$

By properly choosing the integral path and applying the residue theorem, we can finally obtain

$$\begin{aligned} q(r+\Delta r, z) &= \frac{1}{2\pi} \int_{-\infty}^{\infty} \exp\left(\mathrm{i}\Delta r\sqrt{k_0^2 - k_z^2}\right) \exp(\mathrm{i}k_z z)\mathrm{d}k_z \int_{0}^{\infty} \exp(-\mathrm{i}k_z z') q(r, z')\mathrm{d}z' \\ &+ \frac{1}{2\pi} \int_{-\infty}^{\infty} R(k_z) \exp\left(\mathrm{i}\Delta r\sqrt{k_0^2 - k_z^2}\right) \exp(\mathrm{i}k_z z)\mathrm{d}k_z \int_{0}^{\infty} \exp(\mathrm{i}k_z z') q(r, z')\mathrm{d}z' \\ &+ 2\mathrm{i}k_z \exp(-\mathrm{i}k_z z) \exp\left(\mathrm{i}\Delta r\sqrt{k_0^2 - k_z^2}\right) \int_{0}^{\infty} \exp(-k_z z') q(r, z')\mathrm{d}z' \end{aligned} \tag{7.96}$$

The three terms on the right-hand side of the above formula represent three different sound waves respectively. The first term represents the direct wave, the second term represents the reflected wave by ground, the third term represents the the so-called surface wave. If without a ground surface, then the second and third terms will disappear, and (7.96) (at this time the integral lower limit $z' = 0$ should be changed into $z' = -\infty$), which recovers to (7.80).

7.3.2 Refracting atmosphere

For a refracting atmosphere, the wave number k in Eq. (7.41) varies with altitude z, and the equation can be written as

$$\partial_r^2 q(r, z) = -H_2(z) q(r, z) \tag{7.97}$$

where the operator $H_2(z)$ is defined by in Eq. (7.45).

For a wave propagating in the positive r direction, the corresponding single wave equation is (7.48). The square-root operator $H_1 = \sqrt{H_2}$. $k^2(z)$ can be written as in Eq. (7.44), where k_α is a constant wave number, and we can take the wave number value at a certain average altitude or take the wave number value $k_0 \equiv k(0)$ on the ground. Then (7.45) becomes

$$H_2(z) = H_{2\alpha} + \delta k^2(z) \tag{7.98}$$

where $H_{2\alpha} = k_\alpha^2 + \partial_z^2$, and the square-root operator can be approximated to

$$H_1(z) = \sqrt{H_{2\alpha} + \delta k^2(z)} \approx H_{1\alpha} + \frac{\delta k^2(z)}{2k_\alpha} \tag{7.99}$$

where $H_{1\alpha}^2 = H_{2\alpha} \approx k_\alpha^2$. Substituting this into (7.48) gives

$$\partial_r q(r,z) = \mathrm{i}H_{1\alpha}q(r,z) + \mathrm{i}\frac{\delta k^2(z)}{2k_\alpha}q(r,z) \tag{7.100}$$

The first term on the right-hand side of the above equation represents the propagation in a non-refracting atmosphere with wave number k_α, which is a constant. The second term indicates the atmospheric refracting effect.

Integrating Eq. (7.100) from r to $r + \Delta r$, we obtain

$$q(r + \Delta r, z) = \exp\left[\mathrm{i}\Delta r\frac{\delta k^2(z)}{2k_\alpha}\right]\exp(\mathrm{i}H_{1\alpha}\Delta r)q(r,z) \tag{7.101}$$

The factor of $\exp(\mathrm{i}H_{1\alpha}\Delta r)q(r,z)$ is a formal expression for the non-refracting atmosphere solution to Eq. (7.100). In the above section we have seen that the solution of a non-refracting atmosphere is given by (7.96), and from (7.101) we can see that the atmospheric refraction can be counted by multiplying a phase facer to the solution after a series of extrapolation.

In order to increase the accuracy of the numerical computation, we use $\psi = \exp(\mathrm{i}k_\alpha r)\, q(r,z)$ as a substitute for $q(r,z)$. Counting the refraction factor included in (7.101), Eq. (7.96) then becomes

$$\begin{aligned}&\psi(r+\Delta r, z)\\ &= \exp\left[\mathrm{i}\Delta r\frac{\delta k^2(z)}{2k_\alpha}\right]\left\{\frac{1}{2\pi}\int_{-\infty}^{\infty}[\Psi(r,k_z) + R(k_z)\Psi(r_2 - k_z)]\exp(\mathrm{i}\Delta r[\sqrt{k_\alpha^2 - k_z^2} - k_\alpha])\right.\\ &\quad\left.\times\exp(\mathrm{i}k_z z)\mathrm{d}k_z + 2\mathrm{i}k_z\Psi(r,k_z)\exp(\mathrm{i}\Delta r[\sqrt{k_\alpha^2 - k_z^2} - k_\alpha])\exp(\mathrm{i}k_z z)\right\}\end{aligned} \tag{7.102}$$

where

$$\Psi(r,k_z) = \int_0^{\infty}\exp(-\mathrm{i}k_z z')\psi(r,z')\mathrm{d}z' \tag{7.103}$$

This is the spatial Fourier transform of $\psi(r,z)$. Thus Eqs. (7.102) and (7.103) are the fundamental equations of the GFPE method, besides the above deduction with the Kirchhoff-Helmholtz integral equation, which can also be derived via the spectrum theorem of functional analysis [200].

7.3.3 Three-dimensional GFPE method

Let us compute the acoustic field of a point sound source in the cake-slice shaped area shown in Fig. 50, the side of the slice sections $\Phi = 0$ and $\Phi = \delta$ added periodic

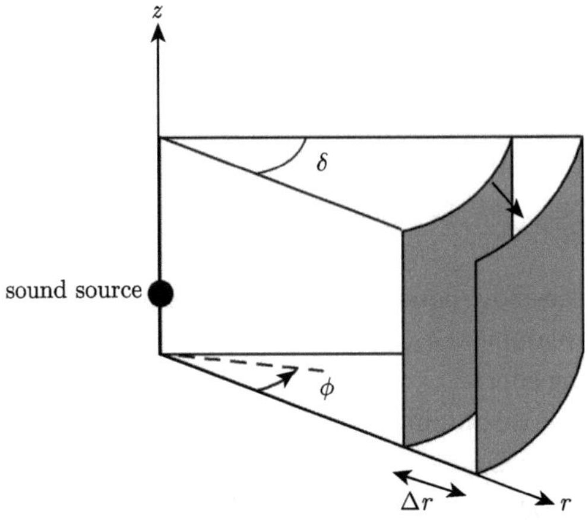

Fig. 50: Cake-slice shaped area used in the three-dimensional GFPE method.

boundary condition. To maintain an effective computation, we choose a very small value of δ. As in the two-dimensional GFPE method, we adopt impedance conditions on the ground as $z = 0$, and use an absorptive layer on the slice top.

The computational procedure is basically to repeat the extrapolating step upward in the positive r direction. In a given extrapolating step, the field at $r + \mathrm{d}r$ of the curved surface is computed from the field at r of the curved face. It can be proved that an extrapolating step in the three-dimensional GFPE method [191] demands an estimate of the forward two-dimensional fast Fourier transform (FFT) and an inverse two-dimensional fast Fourier transform; while an extrapolating step in the two-dimensional GFPE method demands a forward one-dimensional FFT and an inverse one-dimensional FFT.

Let us start from the following far-field Helmholtz equation in the cylindrical coordinate system $r\phi z$

$$(\partial_r^2 + r^{-2}\partial_\phi^2 + \partial_z^2 + k^2)q = 0 \tag{7.104}$$

The difference between this equation and the starting equation (7.41) of the two-dimensional GFPE method is the existence of the second term.

In the far-field approximation, we neglect the curvature of the cake-slice curved face, thus we can use the two-dimensional analogy in rectangular coordinates of the one-dimensional Rayleigh integral (7.88)

$$p(\boldsymbol{r}_1) = \frac{1}{2\pi}\int_{-\infty}^{\infty}\!\!\int (p(\boldsymbol{r})\partial_x g(\boldsymbol{r}, \boldsymbol{r}_1))_{x = x_0}\mathrm{d}y\mathrm{d}z \tag{7.105}$$

where $\boldsymbol{r} = (x, y, z)$, and $\boldsymbol{r}_1 = (x_1, y_1, z_1)$; here the Green function g satisfies Helmholtz equation.

$$(\partial_x^2 + \partial_y^2 + \partial_z^2 + k^2)g(\boldsymbol{r}, \boldsymbol{r}_1) = -4\pi\delta(x - x_1)\delta(y - y_1)\delta(z - z_1) \tag{7.106}$$

Let us extend the above-mentioned step in (7.91) deduced from the two-dimensional GFPE method to the three-dimensional case, to obtain

$$q(r+\Delta r, z) = \frac{1}{4\pi^2 \mathrm{i}} \int_{-\infty}^{\infty} \exp(\mathrm{i}k_h \Delta r) k_h \mathrm{d}k_h \int_0^{\infty} r\mathrm{d}\phi' \int_0^{\infty} \mathrm{d}z' G(k_h, \phi', \phi, z', z) q(r, \phi', z') \tag{7.107}$$

where the Green function G satisfies the following equation

$$[r^{-2}\partial_\phi^2 + \partial_z^2 + k^2(z) - k_h^2] G(k_h, \phi', \phi, z', z) = -4\pi\delta(r_\phi - r_{\phi'})\delta(z - z') \tag{7.108}$$

Eqs. (7.107) and (7.108) are in correspondence with Eq. (7.91) and (7.92) in the two-dimensional GFPE method, and they can also be deduced by the spectrum theorem of functional analysis.

Similar to the above discussion regarding the two-dimensional GFPE method, we also first consider the non-refracting case, there the wave number $\boldsymbol{k}$ in (7.108) is constant, so the Green function $G(k_h, \phi', \phi, z', z)$ only by the difference $\phi - \phi'$ relies on ϕ and ϕ'. Applying a Fourier transform $r\phi - r\phi' \to k_{r\phi}$, we obtain

$$G(k_h, \phi', \phi, z', z) = \frac{1}{2\pi} \int_{-\infty}^{\infty} \exp(\mathrm{i}k_{r\phi}[r\phi - r\phi']) G_\phi(k_h, k_{r\phi}, z', z) \mathrm{d}k_{r\phi} \tag{7.109}$$

where G_ϕ is the Fourier transform of G. Substituting the following relation

$$\delta(r\phi - r\phi') = \frac{1}{2\pi} \int_{-\infty}^{\infty} \exp(\mathrm{i}k_{r\phi}[r\phi - r\phi']) \mathrm{d}k_{r\phi} \tag{7.110}$$

and (7.109) into (7.108), we get

$$\partial_z^2 G_\phi + (k^2 - k_h^2 - k_{r\phi}^2) G_\phi = -4\pi\delta(z - z') \tag{7.111}$$

The solution of the above equation can be given by (7.93), where the vertical wave number k_v is given by the following formula

$$k_v^2 = k^2 - k_h^2 - k_{r\phi}^2 \tag{7.112}$$

Adopting similar steps as used in the two-dimensional case, results equivalent to (7.95) can be obtained

$$q(r+\Delta r, \phi, z) = \frac{1}{2\pi} \int_{-\infty}^{\infty} \exp(\mathrm{i}k_\phi \phi) \mathrm{d}k_\phi \times \left\{ \frac{1}{2\pi} \int_{-\infty}^{\infty} \exp(\mathrm{i}k_z Z) \mathrm{d}k_z \right.$$
$$\times \exp\left(\mathrm{i}\Delta r \sqrt{k_0^2 - k_z^2 - k_\phi^2 / r^2}\right) \int_0^{\delta} \exp(-\mathrm{i}k_\phi \phi') \mathrm{d}\phi' \int_0^{\infty} \exp(-\mathrm{i}k_z z')$$

$$\times q(r,\phi',z')\mathrm{d}z' + \frac{1}{2\pi}\int_{-\infty}^{\infty} R(k_z)\exp(\mathrm{i}k_z z)\mathrm{d}k_z$$

$$\times \exp(\mathrm{i}\Delta r\sqrt{k_0^2 - k_z^2 - k_\phi^2/r^2})\int_0^{\delta}\exp(-\mathrm{i}k_\phi\phi')\mathrm{d}\phi'\int_0^{\infty}\exp(\mathrm{i}k_z z')$$

$$\times q(r,\phi',z')\mathrm{d}z' + 2\mathrm{i}k_z\exp(-\mathrm{i}k_z z)\exp(\mathrm{i}\Delta r\sqrt{k_0^2 - k_z^2 - k_\phi^2/r^2})$$

$$\times \int_0^{\delta}\exp(-\mathrm{i}k_\phi\phi')\mathrm{d}\phi'\int_0^{\infty}\exp(-\mathrm{i}k_z z')q(r,\phi',z')\mathrm{d}z'\Big\} \tag{7.113}$$

where $k_\phi \equiv k_{r\phi}r$.

For a refracting atmosphere, and by adopting similar steps, the equivalent results as that derived in (7.102) can be obtained

$$\psi(r+\Delta r,\phi,z) = \exp(\mathrm{i}\Delta r\delta k)\frac{1}{2\pi}\int_{-\infty}^{\infty}\exp(\mathrm{i}k_\phi\phi)\mathrm{d}k_\phi\Big\{\frac{1}{2\pi}\int_{-\infty}^{\infty} rxp(\mathrm{i}k_z z)\mathrm{d}k_z$$

$$\times [\Psi(r,k_\phi,k_z) + R(k_z)\Psi(r,k_\phi,-k_z)]$$

$$\times \exp(\mathrm{i}\Delta r[\sqrt{k_\alpha^2 - k_z^2 - k_\phi^2/r^2} - k_\alpha]) + 2\mathrm{i}k_z\exp(-\mathrm{i}k_z z)$$

$$\times \Psi(r,k_\phi,k_z)\exp(\mathrm{i}\Delta r[\sqrt{k_\alpha^2 - k_z^2 - k_\phi^2/r^2} - k_\alpha])\Big\} \tag{7.114}$$

where

$$\Psi(r,k_\phi,k_z) = \int_0^{\delta}\exp(-\mathrm{i}k_\phi\phi')\mathrm{d}\phi'\int_0^{\infty}\exp(-\mathrm{i}k_z z')\psi(r,\phi',z')\mathrm{d}z' \tag{7.115}$$

This is the two-dimensional Fourier transform of $\psi(r,\phi,z)$. Eqs. (7.114) and (7.115) are just the fundamental equations of the three-dimensional GFPE method. In the axisymmetric case, k and ψ are independent to ϕ, while Eqs. (7.114) and (7.115) are consistent, with Eq. (7.92) and (7.93) respectively, in the two-dimensional GFPE method. The refracting factor $\exp\left[\mathrm{i}\Delta r\frac{\delta k^2(z)}{2k_\alpha}\right]$ is replaced by $\exp(\mathrm{i}\Delta r\delta k)$, while in Eq. (7.115), the relation $2\pi\Delta(k_\phi) = \int_0^{\delta}\exp(-\mathrm{i}k_\phi\phi')\mathrm{d}\phi'$ is to be used.

To sum up, the PE method is a powerful tool for studying acoustic propagation in an atmosphere. It retains the full-wave effect of the Helmholtz equation, and is therefore capable of predicting an acoustic field in the presence of refraction and diffraction. Moreover, it is applicable to arbitrary acoustic profiles and locally responding ground surfaces. It seems that for downward refraction, using the present PE method can give the most accurate results, while for slightly upward refraction only short ranges can give accurate results. For upward refraction at long ranges, the accuracy of the results are poor. The advantages of the PE method are also shown in its potential to handle various complicated effects such as turbulence and irregular terrain.

7.4 Ray tracing

As has been pointed out at the beginning of this chapter, the ray-tracing method is an important fundamental algorithm, and it possesses a special importance in atmospheric acoustics. It is the earliest numerical method used to this field.

The two approaches described above are based on the wave equation (or the Helmholtz equation), while ray-tracing takes the viewpoint from geometrical acoustics as its starting point. In this section, basing on work concerned with the treatment of moving media [201], the basic concepts of ray tracing will be illustrated in some detail, and by doing so discussions regarding the geometrical acoustics of a moving medium in section 6.2 of chapter 2 and in chapter 3 will be supplemented and extended.

7.4.1 Ray equations

Disregarding turbulence and viscosity but taking the effects of wind and gravity into account, the fundamental equations for describing acoustic perturbations in the atmosphere are described in Eq. (3.35), in which the first equation (Euler's equation) will be modified by adding gravitational acceleration $\boldsymbol{g}$ to the right-hand side. In rectangular coordinates, Eq. (3.35) (the third equation being excluded) can be written in their multicomponent forms

$$\begin{aligned} &D_t^0 v_{1x} + (1/\rho_0)\partial_x p_1 = 0 \\ &D_t^0 v_{1y} + (1/\rho_0)\partial_y p_1 = 0 \\ &D_t^0 v_{1z} + (1/\rho_0)\partial_z p_1 + (g/\rho_0)\rho_1 = 0 \\ &D_t^0 \rho_1 + \rho_0' v_{1z} + \rho_0 \nabla \cdot \boldsymbol{v}_1 = 0 \\ &D_t^0 p_1 + p_0' v_{1z} = c^2 (D_t^2 \rho_1 + \rho_0' v_{1z}) \end{aligned} \tag{7.116}$$

where, as usual, the subscript "0" refers to unperturbed quantities, while subscript "1" refers to perturbed quantities. The operator $D_t^0 \equiv \partial_t + \boldsymbol{v}_0 \nabla$ (cf. Eq. (1.29)), and primes represent derivatives with respect to the argument z.

We will assume that ρ_0 and p_0 are functions of z only. In these functions, small second order quantities are neglected, and g, c and v_0 ($\ll c$) are considered as constants "locally"within a certain region in the atmosphere.

Eliminating ρ_1 from the last two equations in (7.116) and utilizing the hydrostatic equation (1.13), we obtain

$$D_t^0 p_1 - \rho_0 g v_{1z} + \rho_0 c^2 \nabla \cdot \boldsymbol{v}_1 = 0 \tag{7.117}$$

Eliminating ρ_1 again from the 3rd and 5th equations of (7.116) yields

$$(D_t^0 + \omega_B^2) v_{1z} + (1/\rho_0) D_t^0 (\partial_z + g/c^2) p_1 = 0 \tag{7.118}$$

where ω_B is the V-B frequency defined in Eq. (1.28).

To eliminate the dependence on $\rho_0(z)$, we follow Eckart by introducing "field variables" [202][6]

$$\boldsymbol{V} \equiv (\rho_0/\rho_R)^{1/2} \boldsymbol{v}_1, P \equiv (\rho_0/\rho_R)^{1/2} p_1 \tag{7.119}$$

Where ρ_R is the density at a certain reference level (usually simply taken as the ground level). Then the first two equations in (7.116) and Eqs. (7.117) and (7.118) are transformed to

$$\begin{aligned} D_t^0 v_x + (1/\rho_R)\partial_x P &= 0 \\ D_t^0 v_y + (1/\rho_R)\partial_y P &= 0 \\ (1/c^2\rho_R) D_t^0 P + \partial_x V_x + \partial_y V_y + (\partial_z - \Gamma) V_z &= 0 \\ (D_t^{02} + \omega_B^2) V_z + (1/\rho_R) D_t^0 (\partial_z + \Gamma) P &= 0 \end{aligned} \tag{7.120}$$

where Γ is the Eckart's coefficient [155]

$$\Gamma \equiv (\rho_0'/2\rho_0) + (g/c^2) = (1/c)[(g/2c)(2-\gamma) - c'] \tag{7.121}$$

When transforming the first form into the second one, the state equation for a perfect gas (1.7) and the relation for adiabatic sound speed (1.17) were used. According to our assumption, and because $\rho_0(z)$ is an exponential function of z (Eq. (1.15)), now the coefficients in Eq. (7.120) are constants locally, and thus we can set their plane wave solutions to the form

$$P, V \propto \exp\{i\tau(r,t)\} \equiv \exp\{i(\boldsymbol{k} \cdot \boldsymbol{r} - \omega t)\} \tag{7.122}$$

where $\boldsymbol{k}$ is a wave vector, $\boldsymbol{r}$ is the vector radius, τ is the eikonal as defined in Eq. (2.82) and

$$\boldsymbol{k} = \nabla\tau, \omega = -\partial_t \tau \tag{7.123}$$

Substituting Eq. (7.122) into (7.120) yields

$$\rho_R \omega^* V_x - k_x P = 0$$

6 Our approach is slightly different to Eckart's: here we use $(\rho_0/\rho_R)^{1/2}$ instead of $(\rho_0 c_s)^{1/2}$, the aim of doing so is to keep the integration of dimension and be consistent with the approximated dimensional value of the basic field variables.

$$\rho_R\omega^* V_y - k_y P = 0$$

$$k_x V_x + k_y V_y + (k_z + i\Gamma)V_z - (\omega^*/c^2\rho_R)P = 0 \tag{7.124}$$

$$(\omega_B^2 - \omega^{*2})V_z + (\omega * /\rho_R)(k_z - i\Gamma)P = 0$$

where

$$\omega^* \equiv \omega - \boldsymbol{k} \cdot \boldsymbol{v}_0 \tag{7.125}$$

is the intrinsic frequency, i.e., the wave frequency observed in the coordinate system that is attached to the moving fluid, which we discussed in section 5.2 of chapter III.

Equalizing the characteristic determinant of the system (7.124) to zero, just as what was done in section 1.1 of chapter VI, we can obtain the "local" dispersion relation

$$(\boldsymbol{r}, t; \boldsymbol{k}, \omega) \equiv K^2(\omega_B^2 - \omega^{*2}) + \omega^{*2}(k^2 + \Gamma^2) = 0 \tag{7.126}$$

where

$$K^2 \equiv (\omega^*/c)^2 - k_x^2 - k_y^2 \tag{7.127}$$

As mentioned above, strictly speaking, Eq. (7.126) is valid only in the case where the coefficients of the differential equations (7.124) are constants. It is also approximately in some cases of practical interest, however, if the spatial and temporal derivatives of the wind velocity $\boldsymbol{v}_0$ are small compared with $\omega*$ and g, respectively. In fact, this requirement is equivalent to the requirement that variations of a medium's properties within the range of a wavelength are small and thus the WKB approximation is applicable (see section 4 of chapter II). Under this approximation, we can assume that the field variables in Eq. (7.119) have the form as expressed in Eq. (7.122), and $\boldsymbol{k}$ and ω in (7.123) are slowly varying functions of both time and space such that their derivatives can be neglected. Under this approximation, by analogy to the Hamilton-Jacob equation in analytical mechanics, the eikonal equation for three-dimensional rays can be derived as follows.

We introduce a new variable ν via the definition

$$d_t = d_\nu d_t \nu \tag{7.128}$$

The system of equations

$$d_\nu \boldsymbol{r} = \partial_{\boldsymbol{k}} H, \; d_\nu t = -\partial_\omega H$$

$$d_\nu \boldsymbol{k} = -\partial_{\boldsymbol{k}} H, \; d_\nu \omega = \partial_t H \tag{7.129}$$

describe the motion of a point in the eight-dimensional space $(\boldsymbol{r}, t; \boldsymbol{k}, \omega)$, such that if $H = 0$ for a given initial value of ν, then H equals to zero for all values of ν. As $\boldsymbol{k}$ and ω vary, these equations describe the motion of a point on the ray or a wave-packet in physical space.

Application of the relation

$$d_t \boldsymbol{r} = \partial_{\boldsymbol{k}} H/(-\partial_\omega H)$$

which is obtained from Eq. (7.123), (7.128) and (7.129), to (7.126), and by taking Eq. (7.125) into consideration and at the same time letting $v_{0z} = 0$, i.e., considering the case of horizontal winds only, we then obtain the parametric equations for the rays

$$\begin{aligned} d_t x &= [c^2(\omega_B^2 - \omega^{*2})k_x/\omega^*\Phi^2] + v_{0x} \\ d_t y &= [c^2(\omega_B^2 - \omega^{*2})k_y/\omega^*\Phi^2] + v_{0y} \\ d_t z &= -c^2\omega^* k_z/\Phi^2 \end{aligned} \tag{7.130}$$

where

$$\Phi^2 \equiv \omega_B^2 + c^2\Gamma^2 + c^2k^2 - 2\omega^{*2} = [(\gamma g/2c) + c']^2 + c^2k^2 - 2\omega^{*2} \tag{7.131}$$

$$k^2 \equiv k_x^2 + k_y^2 + k_z^2 \tag{7.132}$$

After eliminating the parameter t in Eq. (7.130), we obtain differential ray equations in the rectangular coordinate system[7]

$$\begin{aligned} d_z x &= [c^2(\omega^{*2} - \omega_B^2)k_x - \omega^*\Phi^2 v_{0x}]/c^2\omega^{*2}k_z \\ d_z y &= [c^2(\omega^{*2} - \omega_B^2)k_y - \omega^*\Phi^2 v_{0y}]/c^2\omega^{*2}k_z \end{aligned} \tag{7.133}$$

The rate of change in time of the wave vector $\boldsymbol{k}$ can be found via analogous procedures

$$d_t\boldsymbol{k} = \partial_{\boldsymbol{r}}H/\partial_\omega H$$

Under the assumption of a stratified atmosphere (v_{0x}, v_{0y} and c are functions of altitude z only), after some troublesome calculations and using the 3rd equation in (7.130) to eliminate dt, we obtain the rate of change for $\boldsymbol{k}$ with z

$$d_z k_x = 0, \qquad d_z k_y = 0$$

$$\begin{aligned} d_z k_z = {} & (1/c^2\omega^{*2}k_z)\{\omega^*\Phi^2[k_x v'_{0x} + k_y v'_{0y}] + (c'/c)[\omega^{*2}(\omega_B^2 - \omega^{*2}) \\ & + (\gamma - 1)g^2k^2 + (2-\gamma)g\omega^{*2}\Gamma] + (c'^2 - cc'')(gk^2 - \omega^{*2}\Gamma)\} \end{aligned} \tag{7.134}$$

The first two equations show that the x- and y- components of $\boldsymbol{k}$ do not vary with altitude. This is a logical and necessary result to the assumption of a stratified atmosphere.

Applying the dependence of c and $\boldsymbol{v}$ on z to a realistic atmosphere, i.e., atmospheric profiles, to Eq. (7.133) and (7.134) for numerical integration, the concrete ray paths can be obtained.

---

7 In [203], the ray equations in spherical coordinates were obtained via a similar method. Later the author derived an extremely complicated corresponding ray equations that simultaneously counted the most common condition of gravity, motion and the absorption by the medium. See section 7.4A for details.

7.4.2 Concrete example for numerical integration – ray tracing for the infrasonic waves generated by typhoon

We will take the infrasonic waves generated by a typhoon as an example for the application of the above theoretical results. The sea wave created by a typhoon (or storms) is an important natural source of infrasound (see chapter X), and the periods of such infrasound lie between 3 to 8 seconds (the corresponding frequency will be taken as 0.2 Hz in the following calculations). The atmospheric profiles adopted are sketched in Fig. 51(a), and their fundamental shapes resemble the U. S. "standard atmosphere" [204], except that the data below an altitude of 1.7 km have been modified according to the average spot meteorological material from the West Pacific to the seaboard of Fukien Province, China when observing typhoon No. 6 ("Wendy" of July 1978).

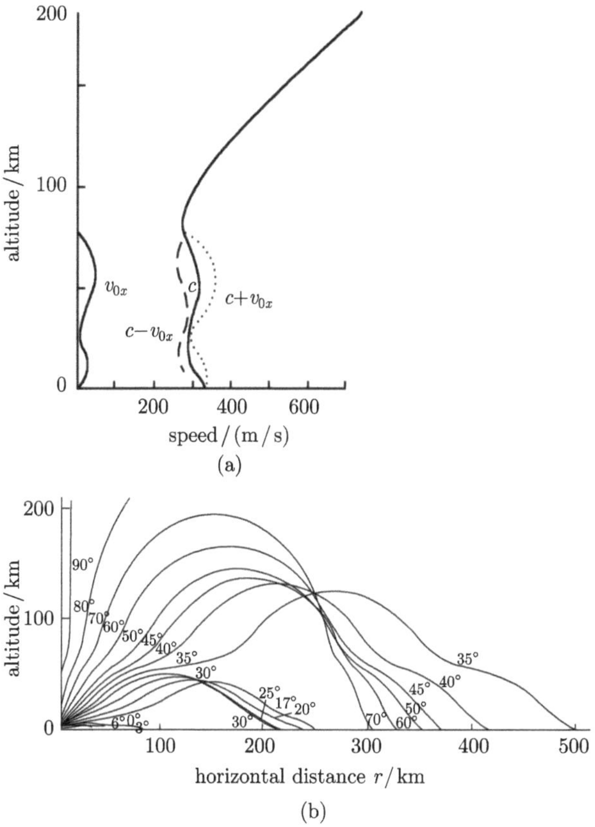

Fig. 51: Infrasonic ray tracing by a typhoon. (a) atmospheric profiles used for sound ray computation (c: sound speed; v_{0x}: x direction (west) wind speed); (b) sound ray pattern computed by the atmospheric profile in Fig. (a) (the marked angular number on the curve is the corresponding initial angle χ_0).

We chose the positive x-axis as pointing to the west and set $v_{0y} = 0$, since an easterly wind was predominant. The continuous function $c(z)$ and $v_{0x}(z)$ are divided into a series of uniform layers, in which c and v_{0x} are constants, with step of size 2.5 km. We then applied the fourth order Runge-Kutta method to integrate Eqs. (7.134) and (7.133) numerically on a microcomputer TRS-80, with an integration step of 100 m. Since $v_{0y} = 0$ and $\mathrm{d}_z k_x = d_z k_y = 0$, the acoustic rays are planar curves in the x-z plane, therefore the ray direction can be described by the angle χ between the wave vector $\boldsymbol{k}$ and the positive x-axis. $k_x(= (\omega/c_0)\cos\chi_0)$ and $k_y(= 0)$ are constants, and the initial value of k_z is $k_{z0} = (\omega/c_0)\sin\chi_0$. The infrasonic frequency was computed as 0.2 Hz. After finding out the z-dependence of k_z by integrating Eq. (7.134), we can substitute it into Eq. (7.133) to calculate various rays that correspond to different values of χ_0. It must be considered in the computer program that after the rays pass turning points where $\mathrm{d}_z x = 0$, the values of z will decrease ($\mathrm{d}z < 0$) as k_z becomes negative to correspond to the fact that the rays are bending downwards. The rays bend downwards continuously until the earth's surface (sea surface or ground) is touched, at which reflections occur (with reflection angles equal to the projecting angles χ_0's) and the rays propagate upwards again. This of course will repeat, and cycle endlessly.

The resulting ray pattern is shown in Fig. 51(b), in which the reflected rays are omitted for simplicity. It can be seen that all of the rays have an obvious symmetry, i.e., the shapes of the upwards-traveling parts and downwards-traveling parts are about the same (subtle differences are mainly due to the fact that the steps adopted when calculating the turning points of the rays are not sufficient small), this is certainly to be expected. If one superposes the profile of v_{0x} on that of c (see the dashed curve in Fig. 51(a), which is actually the profile of the "equivalent sound velocity"), we can find that there are three sound channels in the downwind case: the channel axes lie on the zero level (surface wave guide), the altitudes of about 25 km (first channel) and about 85 km (second channel), respectively. The acoustic rays can be divided accordingly into three groups: the first group (which corresponds to $\chi_0 \leqslant 16°$) is entirely controlled by the surface channel; the second one (which corresponds to $16° < \chi_0 \leqslant 33°$) and the third one (which corresponds to $\chi_0 > 33°$) are mainly affected by the first and second channels, respectively. For the case of contrary winds ($90° < \chi_0 <= 180°$), we performed the computation as well, but the results were only meaningful for observation stations east of the typhoon center, so the corresponding ray pattern is not given here. But it should be pointed out that the equivalent velocity profile (dotted line in Fig. 51(a)) here is quite different to the case of downwind (dashed line). Firstly, only two sound channels exist, the surface wave guide disappears and the position of the first channel moves downward. Secondly, the variation of the equivalent sound speed with z is somewhat "flatten-out" (due to the sound velocity "offset" partly by the wind's velocity), the difference between the maximum and the minimum decreases, so that the action of the sound channels is also weakened. These factors lead to remarkable changes in the ray's pattern, where the manifestation lies in a decrease of the "critical" values of χ_0's and an increase of the corresponding "span" of the ray (i.e., the horizontal distance traversed by the same ray between the origin and the reflection point, or between two successive reflections) for a definite χ_0.

Here the author would like to replenish this chapter with the following section as a supplement to further consider the ray theory for an absorbing atmosphere.

7.4A Ray theory for an absorbing atmosphere [A1]

Now we will extend the above theory to lossy media. The problem of ray theory for lossy media has been investigated quite thoroughly in the fields of optics [A2] and radio waves [A3], especially those in the ionosphere [A4], while in the field of acoustics, it seems to have been ignored.

Damping effects on acoustic waves propagating in an atmosphere, however, even acoustic-gravity waves with extra-low frequencies, cannot always be neglected. In particular, wave damping generally becomes an important factor at some later stage in propagation. This is due firstly to the fact that the small effect of damping will be cumulative and will eventually become significant. This situation occurs in the case of guided modes, e.g., near-ground modes with periods of a few minutes excited from nuclear explosions or intense seisms, which can run around the globe for several cycles. In the second place, waves may propagate into a region with much stronger dissipation. This is the case when waves propagate upwards from the troposphere into the ionosphere, since molecular kinematic viscosity and thermal conductivity increase by orders of magnitude every few tens of kilometers in the vertical direction. Moreover, the wave may be deformed by the medium due to refraction, scattering or nonlinear interactions with other modes as it propagates, such that its spectral density would be shifted towards shorter wavelengths and slower phase velocities, and thereby the dissipative processes from viscosity and thermal conductivity would play a more effective role. Finally, temporal variations in the properties of the medium may also lead to additional dissipations [202].

We intend to treat the acoustic case with a "quasi-optical" technique, since the problems in both fields are similar. It is well-known that the ray paths are determined by Fermat's principle, which can be extended to the case of a lossy medium by formally introducing a complex index of refraction to substitute for the ordinary real one. Applying this generalized Fermat's principle, one can derive the corresponding generalized Snell's law of refraction for cases of sharp boundaries as well as the Haselgrove equations for cases of a continuously stratified medium. The resulting path would have complex coordinates and can contribute automatically the phase integral in the complex plane.

Alternately, an entirely equivalent route is to deduce the dispersion equation from the wave field equations and then utilize the generalized Hamilton equation to derive the ray equations. To obtain the generalized Hamilton equation for lossy media conveniently, one can make use of Sommerfeld-Runge's law of refraction and Whitham's conservation law of wave crests. Upon obtaining the ray equations, Fermat's principle can be deduced thereby in turn [A5].

Taking the specialties of acoustics into account, it is more appropriate to adopt the later treatment.

7.4A.1 The generalized dispersion equation

As mentioned above, it is firstly necessary to derive the dispersion equation for lossy media. Calling back to section 5.1 in Chap. V, the fundamental equations that describe the behavior of viscous, heat-conducting and non-ionized gases are the equation of continuity, the Stokes-Navier equation and the equation of heat conduction, which can be rewritten in the following forms

$$D_t\rho + \rho\nabla\cdot\boldsymbol{v} = 0 \tag{A1}$$

$$\rho D_t v + \nabla p + \rho g = \partial_{x_k}\left[\mu(\partial_{x_k}v_i + \partial_{x_i}v_k - \frac{2}{3}\delta_{ik}\partial_{x_l}v_l)\right] + \partial_{x_i}(\eta\partial_{x_l}v_l) \tag{A2}$$

$$\rho T D_t S = \mathcal{D} + \nabla\cdot(\kappa\nabla T) \tag{A3}$$

where ρ, v, p, S and T are the density, velocity of motion, pressure, equilibrium value of entropy per unit mass and temperature of the atmosphere, respectively. The acceleration due to gravity is g, while μ and η are the coefficients of viscosity and of bulk viscosity, respectively. The thermal conductivity is κ. $\mathcal{D}$ is the viscous loss rate, which is a second-order quantity of v [31]. For convenience of calculation, the right-hand side of Eq. (A2) has been written in tensorial form, where the subscripts i, j, k take the values 1,2,3 all over and δ_{ik} is the unit tensor or Kronecker's $\delta(\delta_{ik} = 1$ when $i = k$, $\delta_{ik} = 0$ when $i \neq k)$.

In the general case when μ and η are functions of coordinates, if we consider irrotational motions only (either the wind velocity $\boldsymbol{v}_0$ or the particle velocity $\boldsymbol{v}_1$ satisfies this condition), then, after developing and reuniting the right-hand side and rewriting it in vectorial form, we obtain for Eq. (A2)

$$\rho D_t\boldsymbol{v} + \nabla p + \rho\boldsymbol{g} - [\eta + (4/3)\mu]\nabla^2\boldsymbol{v} - (\nabla\cdot\boldsymbol{v})\nabla[\eta - (2/3)\mu] - 2(\nabla\mu\nabla)\boldsymbol{v} = 0 \tag{A2'}$$

We now suppose that the atmosphere has been perturbed, and each parameter characterizing its state deflects from its equilibrium value (denoted with the subscript 0) by a perturbed value (denoted with the subscript 1), i.e., the substitutions $\rho \to \rho_0 + \rho_1, p \to p_0 + p_1, \boldsymbol{v} \to \boldsymbol{v}_0 + \boldsymbol{v}_1$ etc. take places in Eqs. (A1), (A2′) and (A3). Assume further that the atmosphere is a stratified medium, i.e., the parameters characterizing its state are functions of the vertical coordinate z only. These functions, especially $\boldsymbol{v}_0$, μ, η and κ, are slowly varying functions of z, thus their derivatives with respect to z are small quantities of the first order, while μ, η and κ should be considered as quantities of the "zeroth order". At the same time, we assume that the wind velocity is steady, i.e., $\partial_t v_0 = 0$. Abandoning the corresponding equations for the quantities of the zeroth order, and neglecting the small quantities of the second order, we obtain an equations for the quantities of the first order

$$\frac{D_0}{Dt}\rho_1 + \boldsymbol{v}_1 \cdot \nabla\rho_0 + \rho_0\nabla\cdot\boldsymbol{v}_1 = 0 \tag{A4}$$

$$\frac{D_0}{Dt}\boldsymbol{v}_1 + \frac{1}{\rho_0}\nabla p_1 - \zeta\nabla^2\boldsymbol{v}_1 - \frac{\rho_1}{\rho_0}g = 0 \tag{A5}$$

$$\rho_0 T_0\frac{D_0 S_1}{Dt} - \kappa\nabla^2 T_1 = 0 \tag{A6}$$

where $\frac{D_0}{Dt} \equiv \left(\frac{\partial}{\partial t}\right) + v_0 \cdot V$, and

$$\zeta \equiv \frac{1}{\rho_0}\left(\frac{4}{3}\mu + \eta\right) \tag{A7}$$

Now there are seven variables (ρ_1, p_1, T_1, S_1 and three components of $\boldsymbol{v}_1$), but only five equations; therefore two additional equations are required. From thermodynamics and the equation of state for perfect gases, they are assumed to be valid for atmospheric air,

$$\frac{p}{\rho} = RT \tag{A8}$$

where R is the gas constant, we get

$$\rho_1 = \left(\frac{\partial\rho_0}{\partial p_0}\right)_{T_0} p_1 + \left(\frac{\partial\rho_0}{\partial T_0}\right)_{p_0} T_1 = \frac{\rho_0}{p_0}p_1 - \frac{\rho_0}{T_0}T_1 \tag{A9}$$

$$S_1 = \left(\frac{\partial S_0}{\partial p_0}\right)_{T_0} p_1 + \left(\frac{\partial S_0}{\partial T_0}\right)_{p_0} T_1 = c_p\left(\frac{T_1}{T_0} - \frac{\gamma-1}{\gamma}\frac{p_1}{p_0}\right) \tag{A10}$$

where c_p is the specific heat of air at constant pressure and γ is the ratio of specific heats.

Eliminating ρ_1 from Eqs. (A4) and (A9) and S_1 from Eqs. (A6) and (A10), we obtain

$$\frac{1}{p_0}\frac{D_0 p_1}{Dt} - \frac{1}{T_0}\frac{D_0 T_1}{Dt} + \frac{1}{\rho_0}\boldsymbol{v}_1 \cdot \nabla\rho_0 + \nabla\cdot\boldsymbol{v}_1 = 0 \tag{A11}$$

$$\rho_0 c_p\frac{D_0 T_1}{Dt} - \frac{D_0 p_1}{Dt} - \kappa\nabla^2 T_1 = 0 \tag{A12}$$

Where the famous relationship in thermodynamics

$$\frac{\gamma-1}{\gamma} = \frac{p_0}{T_0\rho_0 c_p}, \quad \text{for a perfect gas} \tag{A13}$$

has been utilized when obtaining Eq. (A12).

Substituting (A9) into Eq. (A5) and writing it in component form in rectangular coordinates (noting that the direction of g is opposite to that of the z-axis), we combine it with (A11) and (A12) as a set of simultaneous equations, where the subscript 1 is omitted from now on

$$\left(\frac{D_0}{Dt} - \zeta\nabla^2\right) v_x + \frac{1}{\rho_0}\frac{\partial p}{\partial x} = 0 \tag{A14}$$

$$\left(\frac{D_0}{Dt} - \zeta\nabla^2\right) v_y + \frac{1}{\rho_0}\frac{\partial p}{\partial y} = 0 \tag{A15}$$

$$\left(\frac{D_0}{Dt} - \zeta\nabla^2\right) v_z + \left(\frac{1}{\rho_0}\frac{\partial}{\partial z} + \frac{g}{p_0}\right) p - \frac{g}{T_0} T = 0 \tag{A16}$$

$$\frac{\partial v_x}{\partial x} + \frac{\partial v_y}{\partial y} + \left(\frac{\partial}{\partial z} - \frac{1}{H}\right) v_z + \frac{1}{p_0}\frac{D_0 p}{Dt}\frac{1}{T_0}\frac{D_0 T}{Dt} = 0 \tag{A17}$$

$$\frac{D_0 p}{Dt} + \left(\kappa\nabla^2 - \rho_0 c_p \frac{D_0}{Dt}\right) T = 0 \tag{A18}$$

where $H = -\rho_0(\mathrm{d}\rho_0/\mathrm{d}z)^{-1}$ is the scalar height of the atmosphere defined in Eq. (1.16).

Since all of the coefficients in the above equations are local constants (the variations are small within a wavelength – WKB approximation) within a certain range of the atmosphere according to our assumption, a plane wave solution of the following type is available

$$v, p, T \propto \exp[\mathrm{i}(\boldsymbol{k} \cdot \boldsymbol{r} - \omega t)] = \exp[\mathrm{i}\psi(\boldsymbol{r}, t)] \tag{A19}$$

where $\boldsymbol{k}$ is the wave number vector, ω the angular frequency, $\boldsymbol{r}$ the radius vector and ψ is the eikonal, such that

$$\boldsymbol{k} = \nabla\psi, \quad \omega = -\frac{\partial\psi}{\partial t} \tag{A20}$$

Substituting expression (A19) into Eqs. (A14)~(A18), we get

$$(\Omega + \mathrm{i}\zeta k^2) v_x - \frac{k_x}{\rho_0} p = 0 \tag{A21}$$

$$(\Omega + \mathrm{i}\zeta k^2) v_y - \frac{k_y}{\rho_0} p = 0 \tag{A22}$$

$$(\Omega + \mathrm{i}\zeta k^2) v_z - \left(\frac{k_z}{\rho_0} - \mathrm{i}\frac{g}{p_0}\right) p - \mathrm{i}\frac{g}{T_0} T = 0 \tag{A23}$$

$$k_x v_x + k_y v_y + \left(k_z + \frac{\mathrm{i}}{H}\right) v_z - \frac{\Omega}{p_0} p + \frac{\Omega}{T_0} T = 0 \tag{A24}$$

$$\frac{\Omega}{c_p} p - \left(\rho_0 \Omega + \mathrm{i}\frac{\kappa}{c_p} k^2\right) T = 0 \tag{A25}$$

where

$$\Omega \equiv \omega - v_0 \cdot k \tag{A26}$$

is the so-called intrinsic frequency discussed in sec. 3.5.2 and defined by Eq. (7.125).

Similar to that done in sec.6.1.1, equating the characteristic determinant of the above equation set (A21)~(A25) to zero gives

$$\begin{vmatrix} (\Omega + \mathrm{i}\zeta k^2) & 0 & 0 & -\dfrac{k_x}{\rho_0} & 0 \\ 0 & (\Omega + \mathrm{i}\zeta k^2) & 0 & -\dfrac{k_y}{\rho_0} & 0 \\ 0 & 0 & (\Omega + \mathrm{i}\zeta k^2) & -\left(\dfrac{k_z}{\rho_0} - \mathrm{i}\dfrac{g}{p_0}\right) & -\mathrm{i}\dfrac{g}{T_0} \\ k_x & k_y & \left(k_z + \dfrac{\mathrm{i}}{H}\right) & -\dfrac{\Omega}{p_0} & \dfrac{\Omega}{T_0} \\ 0 & 0 & 0 & \dfrac{\Omega}{c_p} & -(\rho_0\Omega + \mathrm{i}\dfrac{\kappa}{c_p}k^2) \end{vmatrix} = 0$$

we then obtain the corresponding local dispersion relation (as can be compared with Eq. (7.126))

$$\begin{aligned} G(\boldsymbol{r}, \boldsymbol{t}, \boldsymbol{k}, \omega) \equiv & \frac{\zeta\kappa}{p_0 c_p}\Omega k^4 + \left[\Omega + \frac{\kappa}{\rho_0 c_p}\left(\frac{\gamma g}{c^2} - \frac{1}{H}\right) k_z\right] k^2 + \frac{\Omega}{c^2}\left(\frac{g}{H} - \Omega^2\right) \\ & + \mathrm{i}\left\{\frac{\kappa}{\rho_0 c_p} k^4 + \left[\frac{\kappa}{\rho_0 c_p}\left(\frac{g}{H} - \Omega^2\right) - \frac{\Omega^2}{c^2}\zeta\right] k^2 + \left(\frac{1}{H} - \frac{g}{c^2}\right)\Omega k_z\right\} = 0 \end{aligned} \tag{A27}$$

In obtaining the above equation, we have made use of relation (A13) once again, and thus have

$$\frac{\rho_0}{p_0} - \frac{1}{T_0 c_p} = \frac{\rho_0}{\gamma p_0} = \frac{1}{c^2}$$

where c is the velocity of sound in air.

7.4A.2 The generalized Hamilton equation

In the case when the waves discussed are not plane waves but geometrical acoustics still holds, Eq. (A19) can be written as

$$v, p, T = (\breve{v}, \breve{p}, \breve{T}) \exp\left[\mathrm{i}\int \boldsymbol{k}\cdot\mathrm{d}\boldsymbol{r} - \mathrm{i}\int \omega \mathrm{d}t\right] \tag{A28}$$

where the amplitudes $\breve{v}, \breve{p}$ and $\breve{T}$ are slowly varying functions (compared to the exponential function) of space and time.

From the uniqueness condition of the eikonal approximation (A28), the Sommerfeld & Runge refraction law [A6]

$$\nabla \times \boldsymbol{k} = 0 \tag{A29}$$

and Whitham's conservation law of wave crest [A7]

$$\frac{\partial \boldsymbol{k}}{\partial t} + \nabla\omega = 0 \tag{A30}$$

can both be derived. These two partial differential equations (A29) and (A30), together with the algebraic equation (A27), combine as the basic equation system of our ray theory. Since this system is difficult to handle, the equivalent Hamilton's ordinary differential equations of geometrical optics are usually derived from it for substitution.

Following Suchy [A5], we derive the generalized Hamilton equations in the acoustic case as follows. From the rule of differentiation for implicit functions, the total derivative of $G(\boldsymbol{k}, \omega, \boldsymbol{r}, t)$ with respect to $\boldsymbol{k}$ is always equal to zero

$$\frac{\mathrm{d}G}{\mathrm{d}\boldsymbol{k}} = \frac{\partial G}{\partial \boldsymbol{k}} + \frac{\partial G}{\partial \omega}\frac{\partial \omega}{\partial \boldsymbol{k}} = 0$$

thus we have

$$\frac{\partial \omega}{\partial \boldsymbol{k}} = -\frac{\partial G}{\partial \boldsymbol{k}} \Big/ \frac{\partial G}{\partial \omega} \tag{A31}$$

Really, this just the group velocity along the ray and thus can be written as

$$\boldsymbol{v}_g \equiv \frac{\mathrm{d}\boldsymbol{r}_g}{\mathrm{d}t_g} = \left(\frac{\partial}{\partial t} + \frac{\partial \omega}{\partial \boldsymbol{k}} \times \nabla\right)\boldsymbol{r}_g = \frac{\partial \omega}{\partial \boldsymbol{k}} \tag{A32}$$

Since the differential operator $(\partial/\partial t) + (\partial\omega/\partial\boldsymbol{k}) \times \nabla$ is the group-centric time derivative, following the motion of the group with velocity $\partial\omega/\partial\boldsymbol{k}$, it can be called the "moving derivative". The situation here is similar to that for the Stokes' operator D/Dt (cf. Eq. (1.29)).

For absorbing media, $\partial\omega/\partial\boldsymbol{k}$ is a complex quantity which can be written as

$$\frac{\partial \omega}{\partial \boldsymbol{k}} = \boldsymbol{v}_g + \mathrm{i}u \tag{A33}$$

with

$$v_g = \mathrm{Re}\frac{\partial \omega}{\partial k}, \qquad u = \mathrm{Im}\frac{\partial \omega}{\partial k} \tag{A34}$$

the former can be regarded as the generalized group velocity. Now Eq. (A32) can be modified formally as

$$v_g + \mathrm{i}u = \frac{\mathrm{d}r_c}{\mathrm{d}t_c} \equiv \left(\frac{\partial}{\partial t} + \frac{\mathrm{d}r_c}{\mathrm{d}t_c} \times \nabla\right) r_c = \frac{\partial \omega}{\partial k} \tag{A32$'$}$$

Similarly, the total derivatives of G with respect to r and t would each be equal to zero also, hence

$$\frac{\mathrm{d}G}{\mathrm{d}r} = \frac{\partial G}{\partial r} + \left(\frac{\partial}{\partial r}k\right) \cdot \frac{\partial G}{\partial k} + \frac{\partial \omega}{\partial r}\frac{\partial G}{\partial \omega} = 0$$

Because of the refraction law (A29), the anti-symmetric part of the tensor $(\partial/\partial r)k$ vanishes, and this tensor can thus be substituted by the transposed tensor $k(\partial/\partial r)$. And on the other hand, from Whitham's law (A30), $\partial\omega/\partial r$ can be replaced by $-\partial k/\partial t$, therefore we obtain finally

$$\frac{\mathrm{d}G}{\mathrm{d}r} = \frac{\partial G}{\partial r} + \frac{\partial G}{\partial k} \cdot \frac{\partial}{\partial r} k - \frac{\partial G}{\partial \omega} \frac{\partial k}{\partial t} = 0$$

Dividing the above equation by $\partial G/\partial \omega$, and making use of Eq. (A33), we get

$$\frac{\mathrm{d}k}{\mathrm{d}t_c} \equiv \left[\frac{\partial}{\partial t} + (v_g + \mathrm{i}\omega) \times \frac{\partial}{\partial r}\right] k = \frac{\partial G/\partial r}{\partial G/\partial \omega} \tag{A35}$$

We now take the total temporal derivative of $G = 0$

$$\frac{\mathrm{d}G}{\mathrm{d}t} = \frac{\partial G}{\partial t} + \frac{\partial k}{\partial t} \cdot \frac{\partial G}{\partial k} + \frac{\partial \omega}{\partial t} \frac{\partial G}{\partial \omega} = 0$$

Making use of Eq. (A30) once again, we obtain

$$\frac{\mathrm{d}\omega}{\mathrm{d}t_c} \equiv \left[\frac{\partial}{\partial t} + (v_g + \mathrm{i}u) \cdot \frac{\partial}{\partial r}\right] \omega = -\frac{\partial G/\partial t}{\partial G/\partial \omega} \tag{A36}$$

Substituting Eq. (A31) into the right-hand side of Eq. (A32′), and coupling the resulting equation with Eq. (A35), we obtain the generalized Hamilton equations for absorbing media, written in symmetric form both in space and time

$$\frac{\mathrm{d}r_c}{\mathrm{d}t_c} = -\frac{\partial G/\partial k}{\partial G/\partial \omega}, \qquad \frac{\mathrm{d}k}{\mathrm{d}t_c} = \frac{\partial G/\partial r}{\partial G/\partial \omega} \tag{A37}$$

They are formally similar to the corresponding equations for non-absorbing media (compared with that described in Sec. 7.4), except in the present case where r_c, t_c and k are complex quantities and $G = 0$ is a complex equation as can be seen clearly from Eq. (A27). This causes an essential distinction between these two cases: The absorption gives rise to two additional equations that involve the variation of the imaginary part Im $\boldsymbol{k}$ of $\boldsymbol{k}$, and additional directional derivatives are gained due to the existence of the imaginary part $\boldsymbol{u}$ of $(\partial\omega/\partial k)$. Because of these additional directional derivatives, the wave vector $\boldsymbol{k}$ cannot be traced solely along the ray path as in the case of non-absorbing media, but a certain vicinity of this path must be included in the tracing. Owing to the coupling of the real and imaginary parts in Eq. (A37) for Re$\boldsymbol{k}$ and Im$\boldsymbol{k}$, the corresponding problem for ray tracing must be solved via approximate methods. Two such methods have been proposed in [A5].

7.4A.3 The generalized ray equations and fermat's principle

Substituting Eq. (A27) into Eq. (A37) and considering the case of a horizontal wind, i.e. $v_{0z} = 0$, only, we obtain an equation that is satisfied by the three components (all complex) of r_c in the form

$$\frac{\mathrm{d}x_c}{\mathrm{d}t_c} = v_{0x} - \frac{2\Phi}{K^2} k_x \tag{A38}$$

$$\frac{\mathrm{d}y_c}{\mathrm{d}t_c} = v_{0y} - \frac{2\Phi}{K^2}k_y \tag{A39}$$

$$\frac{\mathrm{d}z_c}{\mathrm{d}t_c} = \frac{1}{K^2}\left[-2\Phi k_z + \frac{\kappa}{\rho_0 c_p}\frac{T_0'}{T_0}k^2 - \mathrm{i}\frac{\omega_\mathrm{B}^2}{g}\Omega\right] \tag{A40}$$

where

$$\Phi \equiv 2\frac{\zeta\kappa}{\rho_0 c_p}\Omega k^2 + \Omega - \frac{\kappa}{\rho_0 c_p}\frac{T_0'}{T_0}k_z + \mathrm{i}\left[\frac{2\kappa}{\rho_0 c_p}k^2 + \frac{\kappa}{\rho_0 c_p}\left(\frac{g}{H} - \Omega^2\right) - \frac{\Omega^2}{c^2}\zeta\right] \tag{A41}$$

$$K^2 \equiv \frac{\zeta\kappa}{\rho_0 c_p}k^4 + k^2 - 3\frac{\Omega^2}{c^2} + \frac{g}{c^2 H} - \mathrm{i}\left[2\left(\frac{\kappa}{\rho_0 c_p} + \frac{\zeta}{c^2}\right)\Omega k^2 - \frac{\omega_\mathrm{B}^2}{g}k_z\right] \tag{A42}$$

in which the following well-known relations (cf. the related discussions in Sec. 1.3.3) are applied

$$\frac{1}{H} - \frac{g}{c^2} = \frac{\omega_\mathrm{B}^2}{g}, \qquad \frac{1}{H} - \frac{\gamma g}{c^2} = \frac{2c'}{c} = \frac{T'}{T}$$

When deducing the second relation, the "hydrostatic approximation" $p_0 = -\rho_0 g$ was used. In all of the above equations, primes denote derivatives with respect to z, and ω_B is the Väisälä-Brunt (V-B) frequency [Eq. (1.28)].

Because ρ_0, T_0, v_0, H and ζ, κ, etc., can be considered as functions of z only and c_p, γ and g as constants (the dependences of c_p and γ on ρ_0 and T_0 and the variation of g with the altitude z can be regarded as small quantities of the second order within the range that interests us) in general and because

$$\frac{p_0'}{p_0} = \frac{\rho_0'}{\rho_0}\frac{T_0'}{T_0}$$

from Eq. (A8), we obtain the parametric equations for determining various components of the wave vector

$$\frac{\mathrm{d}k_x}{\mathrm{d}t_c} = 0, \qquad \frac{\mathrm{d}k_y}{\mathrm{d}t_c} = 0 \tag{A43}$$

$$\begin{aligned} K^2\frac{\mathrm{d}k_z}{\mathrm{d}t_c} \\ &= \left[-\frac{\zeta\kappa}{\rho_0 c_p}k^4 - k^2 + \frac{1}{c^2}\left(3\Omega^2 - \frac{g}{H}\right)\right](k_x v_{0x}' + k_y v_{0y}') \\ &\quad - \frac{\kappa k^2}{\rho_0 c_p}\left[\zeta\Omega k^2 + \left(g - \frac{c^2}{\gamma H}\right)k_z\right]\frac{\rho_0'}{\rho_0} \\ &\quad - \left[\frac{\kappa k^2}{\rho_0 c_p}(\zeta\Omega k^2 + g k_z) - \frac{\Omega}{c^2}\left(\Omega^2 - \frac{g}{H}\right)\right]\frac{T_0'}{T_0} + \left[\frac{\kappa k^2}{\rho_0 c_p}k_z - \frac{g\Omega}{c^2}\right]\frac{H'}{H^2} + \frac{\kappa\Omega}{\rho_0 c_p}k^4\zeta' \end{aligned}$$

$$- \frac{k^2}{\rho_0 c_p} \left[\zeta \Omega k^2 + \left(g - \frac{c^2}{\gamma H} \right) k_z \right] \kappa'$$

$$+ \mathrm{i} \left\{ \left[2\Omega \left(\frac{\kappa}{\rho_0 c_p} + \frac{\zeta}{c^2} \right) k^2 - \frac{N^2}{g} k_z \right] (k_x v'_{0x} + k_y v'_{0y}) \right.$$

$$- \frac{\kappa k^2}{\rho_0 c_p} \left(\frac{k^2}{\gamma c^2} + \frac{g}{H} - \Omega^2 \right) \frac{\rho'_0}{\rho_0} + \left[\left(\frac{\kappa}{\rho_0 c_p} \left(\Omega^2 - \frac{g}{H} \right) + \frac{\zeta}{c^2} \Omega^2 \right) k^2 + \frac{g}{c^2} \Omega k_z \right] \frac{T'_0}{T_0}$$

$$\left. - \left[\frac{\kappa g}{\rho_0 c_p} k^2 - \Omega k_z \right] \frac{H'}{H^2} - \frac{\Omega^2}{c^2} k^2 \zeta' + \frac{k^2}{\rho_0 c_p} \left(\frac{k^2}{\gamma c^2} + \frac{g}{H} - \Omega^2 \right) \kappa \right\} \tag{A44}$$

The system of differential equations (A38)~(A44) is the Ray equation system for a lossy atmosphere. After including the atmospheric model, i.e., the given profiles of ρ_0, T_0, v_0, H and ζ, κ, etc., these equations may be integrated numerically and thus the complex rays may be traced out. The ray-tracing problem in this case, however, would be much more complicated than what we have done in the previous section.

Among many possible paths of the wave, it is necessary to pick out the path which makes the maximum contribution to the sound field. Usually, the criterion for picking this path is just Fermat's principle, where the generalized form for absorbing media can be written as [A5]

$$\delta \int_{r_1}^{r_2} \mathrm{d}\boldsymbol{r}_c \cdot \boldsymbol{k} = 0 \tag{A45}$$

where both δr_1 and δr_2 are equal to zero, and ω is a constant. In the case of real variables, the basis of this criterion is wave interference: waves within the distances of a wavelength to the determined path tend to constructive interference, whereas waves far from this path tend to destructive interference. In the case of complex variables, Eq. (A45) really gives two conditions for determining the ray path, both of which require the interference and the attenuation of the waves to be minima along the determined ray path. For discrete stratified media, the application of Fermat's principle leads to Snell's law of refraction, while for continuous stratified media, it leads to the Hamilton-Haselgrove equations that determine the ray's path. Therefore these two are equivalent. Hence the ray path, as a solution of equations (A38)~(A44), should satisfy Eq. (A45), but it is not certain to be a path that connects the source to the observation point. The path always starts from the source, and this determines the initial values for r_c. Requiring the path to end at the given observation point adds six conditions on the ray path, i.e., conditions concerning the real and imaginary parts of three spatial coordinates at the observation point, of which the imaginary parts will always be zero. The ray will end at the chosen observation point only if a proper choice of the remaining conditions is made.

Supplementary references for this section

[A1] Yang X R (Yang H J). Ray theory for an absorbing atmosphere, Journal of low frequency noise & vibration, 1986, 5(2): 78–85; having been read out on the 3rd International meeting on low frequency noise and vibration held in London, Sept. 1985.
[A2] Epstein P S. Geometrical optics in absorbing media. Proc. Nat. Acad. Sci. U. S., 1930, 16: 37–45.
[A3] Jones R M. Ray theory for lossy media. Radio Sci., 1970, 5: 793–801.
[A4] Bennett J A. Complex rays for radio waves in an absorbing ionosphere. Proc. IEEE, 1974, 62: 1577–1585.
[A5] Suchy K. Ray tracing in an anisotropic absorbing medium. J. Plasma Physics, 1972, 8, Part I: 53–65.
[A6] Poeverlein H. Sommerfeld-Runge law in three and four dimensions. Phys. Rev., 1962, 128: 956–964.
[A7] Whitham G B. Linear and Nonlinear Waves. New York: John Wiley & Sons, 1974.

7.5 Gaussian beam (GB[8]) approach

The method of the superposition of a Gaussian beam can be considered as a kind of modified tracing approach developed over the past twenty years. Initially, it was introduced in seismology by Cerveny et al [205], probably at the beginning of the 1980s, and has subsequently been applied in other fields, such as electromagnetic wave propagation and ocean acoustics [206] etc. The solution was originally obtained in the case of inhomogeneous media in one or two dimensions, but has subsequently been extended to three dimensions [207].

The basic concept of the theory is to launch a fan of beams from the source and to trace the propagation of these beams through the medium. The wave equation is solved in the immediate vicinity of each ray and the acoustic pressure at the receiver is obtained by summing the contribution of each of the individual beams. The GB method is limited by the high-frequency restrictions inherent in ray tracing. The theory assumes that the variations in the medium vary sufficiently slowly over a wavelength such that the interactions between the inhomogeneities can be neglected and that the wave equation can be approximated by the PE method.

In more detail, the GB method includes three steps as follows. Firstly, rays are launched from the source and propagate through the medium via numerically solving the ray equations. Secondly, the wave equation is solved in a coordinate system centered along each ray using the parabolic approximation. This leads to a system of first-order differential equations that can be solved step-by-step along each of the rays. The solution to the wave equation is expressed in the form of a Gaussian beam. Thirdly,

8 Abbreviating “Gaussian Beam” to “GB” is author’s coinage to echo with the widely adopted “FFP” (Fast Field Program) and “PE” (Parabolic Equation).

the final solution is obtained via the superposition of all the beams that intercept at the receiver.

This procedure overcomes the traditional difficulties of conventional ray methods: the solution is uniform everywhere and thus removes the singularities at caustics and sharp transitions into shadows. The procedure is not sensitive to the exact details of the medium since, contrary to ray methods, the GB method provides a local average. It is not necessary to find the rays that exactly intercept the receiver, thus reducing computation time.

Following [205], [208] and [209], the basic computational steps of the GB method will be briefly reviewed within the limitation of a two-dimensional case.

The first step is to solve the classical ray equations to obtain the central ray of the beam. These equations are (cf. section 6.2 of chapter II)

$$\mathrm{d}_t \boldsymbol{r} = \boldsymbol{v} + (\boldsymbol{k}/k)c$$

$$\mathrm{d}_t \boldsymbol{k} = -k\boldsymbol{\nabla} c - \sum k_i \nabla v \tag{7.135}$$

where $\boldsymbol{v}$ is the wind speed-vector, and $\boldsymbol{k}$ is the wave vector. These above-mentioned equations are usually solved using standard numerical techniques. Alternatively, they can be solved analytically if the sound-speed profile is linear, or it can be decomposed in layers using linear velocity variations, as mentioned in chapter 2.

The second step is to use the parabolic approximation to solve the wave equation locally in the ray-centered coordinate system, which is an orthogonal curvilinear system that follows a particular ray. In a two-dimensional medium, the ray-centered coordinates are specified by the unit vector $\boldsymbol{t}$ that is tangential to the ray in the (x, z) plane, and the unit vector $\boldsymbol{n}$ is normal to the ray as shown in Fig. 52.

The solution to the Helmholtz equation can be written as

$$u(s, n, \omega) = \left[\frac{c(0)q(0)}{c(s)q(s)}\right]^{1/2} \exp\left[-i\omega\left\{\tau(s) + \frac{1}{2}M(s)n^2\right\}\right] \tag{7.136}$$

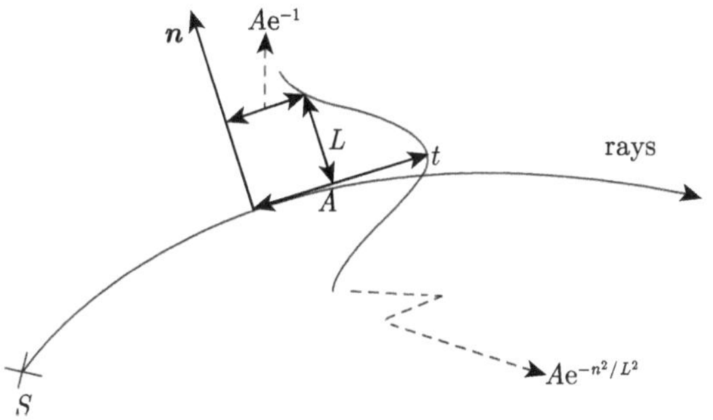

Fig. 52: Amplitude of the GB along the trajectory of a ray and the corresponding ray-centered coordinate system.

where s is the distance along the central ray from the source, n represents the length in a direction perpendicular to the ray at s, $c(s)$ is the sound speed and $\tau(s)$ is the travel time along the central ray. The function M is defined as

$$M(s) = p(s)/q(s) \tag{7.137}$$

where

$$p(s) = \varepsilon p_1(s) + p_2(s)$$
$$q(s) = \varepsilon q_1(s) + q_2(s) \tag{7.138}$$

and

$$\varepsilon = \varepsilon_1 + \mathrm{i}\varepsilon_2, \qquad \varepsilon_2 \geqslant 0 \tag{7.139}$$

The complex number ε is called the "GB parameter" and p_i, q_i $(i = 1, 2)$ are the two real independent solutions to the dynamic ray-tracing system given by

$$\mathrm{d}_s p_i = -[c_{nn}/c^2(s)]q_i(s)$$
$$\mathrm{d}_s q_i = c(s)p_i(s) \tag{7.140}$$

where $c_{nn} = \partial_n^2 c(s)$ denotes the second normal derivative of the sound speed $c(s)$. The condition that $\varepsilon_2 \geqslant 0$ in Eq. (7.139) guarantees that the energy is localized to the vicinity of the central ray.

Eq. (7.136) shows that the amplitude of each beam decreases with distance from the ray, which follows a Gaussian distribution – hence the term "Gaussian beam" (see Fig. 52). The set of Eq. (7.140) is solved step-by-step along each ray using standard numerical techniques in the same way as the ray equations. In the case of a linear sound speed gradient, this set gets simplified since $c_{nn} = 0$.

Following [210], the initial conditions at $s = 0$ are

$$p_1(0) = 0, \quad p_2 = 1\,\mathrm{sm}^{-1}$$
$$q_1(0) = 1\,\mathrm{m}, \quad q_2(0) = 0$$

We should note that the beam solutions are valid only when the ray-centered coordinates are well defined and regular in some region near the central ray.

Separating the real and imaginary parts of the ratio p/q, Eq. (7.136) can be written as

$$u(s, n, \omega) = \left[\frac{c(0)q(0)}{c(s)q(s)}\right]^{1/2} \exp\{-\mathrm{i}\omega\tau + \mathrm{i}[\omega/2c(s)]K(s)n^2 - [n/L(s, \omega)]^2\} \tag{7.136'}$$

where $L(s, \omega)$ controls the width of the beam and is given by

$$L^2(s, \omega) = -(2/\omega)\{\mathrm{Im}[p(s)/q(s)]\}^{-1} \tag{7.141}$$

and the phase-front curvature $K(s)$ is defined by

$$K(s) = -c(s)\mathrm{Re}[p(s)/q(s)] \tag{7.142}$$

The complex GB parameter defined in Eq. (7.139) is the quantity by which the properties of the beam, mainly its beam width $L(s, \omega)$ and the phase-front curvature $K(s)$, can be shaped. This complex parameter can be chosen to make the beam solution valid by defining the size of the region near the central ray. The beam width $L(s, \omega)$ is defined as the distance from the central ray at which the amplitude of the GB is $1/e$ times the amplitude on the central ray (see Fig. 51). $L(s, \omega)$ is inversely proportional to the square-root of the frequency, i.e., beams with higher frequencies are narrower than those with lower frequencies. The rate of change of phase is related to the local curvature, and provides the basis for the interpretation of $K(s)$ as a curvature [206].

In conventional ray theory, p and q are real and $q(s)/c_0$ is interpreted as a spreading loss. Ray theory fails at a caustic because the spreading becomes zero and Eq. (7.136) diverges. In the GB method this problem does not arise because q is nonzero.

The third step of the GB method is a superposition of all Gaussian beams passing in the neighborhood of the receiver R. We designate α as the launch angle of a ray with respect to an arbitrary axis passing through the source. The total field at a point located at the receiver R is therefore

$$U(R, \omega) = \int \varphi(\alpha, \omega) u(s, n, \omega) d\alpha \tag{7.143}$$

where the integration is over all of the beams. The solution $u(s, n, \omega)$ is calculated from Eq. (7.136) and φ (α, ω) is called the weight function, which is determined by expanding the wave field at the source and matches the high-frequency asymptotic behavior of the integral in Eq. (7.143) to the exact solution for a source in a homogeneous medium. It can be shown that [210]

$$\varphi(\alpha, \omega) = [\omega / 2\pi i c_0]^{1/2} F(\omega) G(\alpha) \tag{7.144}$$

where c_0 is the sound speed at the source, $F(\omega)$ is the source frequency spectrum and $G(\alpha)$ describes the radiation characteristics of the source. In practice, Eq. (7.143) is evaluated numerically by discretizing the integral.

The quality of the results obtained when using the GB method is critically sensitive to the choice of the GB parameter ε. The selection of ε has been discussed at length in [204], in which several options were studied. We deem that the best one is to control the beam-width L such that it does not become too large at the receiver, and the energy is therefore concentrated to the vicinity of the central ray. In the case of sound propagation above a ground surface, the source and receiver heights are typically between 1 and 5 m. If the beam width L is too wide, the bottom ends of the beams associated with the central rays that graze the ground will pass below the surface and spurious truncation occurs in the summation of Eq. (7.143).

In the case of propagation during downwards refraction, the GB tracing produces satisfactory results if the receiver is not too close to the ground. When the receiver is within a few wavelengths above the ground, the above-mentioned truncation will occur also in the beam summation. In the case of propagation during upwards refraction, the GB tracing only shows satisfactory results in the "insonified region" and near the shadow boundary. In the shadow region, the GE solution deviates significantly, leading to predictions of much lower sound levels. This is because of the narrow beam that must be used to avoid truncation due to the beams grazing the ground.

Chapter 8
Acoustic Remote Sensing for the Atmosphere

The phrase “remote sensing” implies a technique for detecting and/or measuring the location, shape, features and state(s) of an object using a specific tool. The term “acoustic remote sensing for the atmosphere” defines a technique for detecting the state(s) of an atmosphere, and various phenomena existing within it, by using sound waves as a tool. It is interesting to note that the word “sound” means “acoustic waves” when used as a noun, and has the meaning of “to examine” or “to probe” when used as a transitive verb. As such, we can see a close relationship between “acoustic waves” and “probing”.

The reason why acoustic waves are a powerful tool in atmospheric remote sensing lies in that the propagating acoustic waves are strongly affected by both macroscopic and microscopic structures within the atmosphere. As we have seen in the previous chapters, macroscopic structures, as well as temperature- and wind-profiles, induce wave reflection and refraction, while microscopic structures such as turbulence, etc., induce scattering. By utilizing corresponding relationships between these structures and the traveling path or the intensity of acoustic waves, one can “sense” the states of the atmosphere and the phenomena within it. In this manner, the problems discussed in this chapter may be regarded as “inverse problems” to atmospheric acoustics (i.e., the principal problems discussed in the previous chapters).

The remote-sensing technique includes two types: active and passive. The former refers to man-made sources of acoustic energy under control of the observer, as well as back-waves that are “acted” upon by the atmosphere that are received at the same place or in the vicinity of the detector; thus the state of the atmosphere can be judged by examining the travel time and the changes in wave amplitude and phase etc. The latter type refers to the direct measurement of sound waves, both man-made and natural, that exist objectively in the atmosphere, where the corresponding phenomena can be judged accordingly. In general, both types use ground-based instrumentation.

In correspondence with the point of view mentioned in chapter I that atmospheric acoustics may be divided into two categories of “classical” and “modern”, the acoustic remote sensing technique, as its inverse problem, can be accordingly divided into two similar categories as lower atmospheric- (tropospheric-) and upper atmospheric-remote sensing. The physical basis of the former is based on scattering due to microscopic structures and uses active remote sensing of audible sound is dominant. The latter is based on refraction by macroscopic structure, where passive remote sensing of infrasound is dominant.

Part One Acoustic remote sensing for the lower atmosphere (troposphere)

8I.1 Probing system

The main object of lower-atmospheric remote sensing is measuring variations of temperature and wind with altitude (profiles). The main instruments used are the echosonde, invented by McAllister in 1968 and patented in 1972; the "SODAR", which improved later and is a more complex form of the Doppler echosonde. The introduction of these kinds of instruments had added powerful tools to the arsenal of atmospheric researchers. Some authors even deemed that a "new era" in atmospheric acoustics had been ushered in via this introduction.

Over the past ten years, the SODAR-probing technique and computer science have developed synchronistically. As such rapid progress has been made in processing Doppler signals and analyses of probing results. New effective methods have been developed to separate SODAR back-wave signals from environmental noise, as well as ground–surface diffuse reflection.

Small SODARs of different configurations including multi-frequency systems that were developed and produced for commercial purposes [211]. Now they have been widely adopted for studying the lower atmospheric boundary layer (ABL). The design of antennae has been improved, and the development of computer-controlled phased array antenna has made using three-wave beam or five-wave beam probing possible [212].

Methods for estimating parameters of ABL turbulence, and parameters to standard test the similarity has been developed. By repeatedly comparing SODAR data with field measurement, the accuracy of average wind speed profiles and wind direction with modern Doppler SODAR and traditional meteorological standards is similar. However, achievements from using SODAR to measure vertical temperature profiles are rare, and there are many researchers still studying this difficult task. As in a whole system, using radio probing system to determine ABL temperature is obviously reasonable. The method of using SODAR to measure rainfall has been brought into practice [213].

Owing SODAR's characteristics of being accurate, cheap and reliable equipment, they have been used in environmental detecting systems and are capable of being used for remote-sensing the ABL boundary parameters better than other methods [214].

8I.1.1 Monostatic configuration

The main feature of McAllister's invention is the combination of an improved antenna design and the use of a facsimile recorder, as patented by Marti in 1919. The antenna

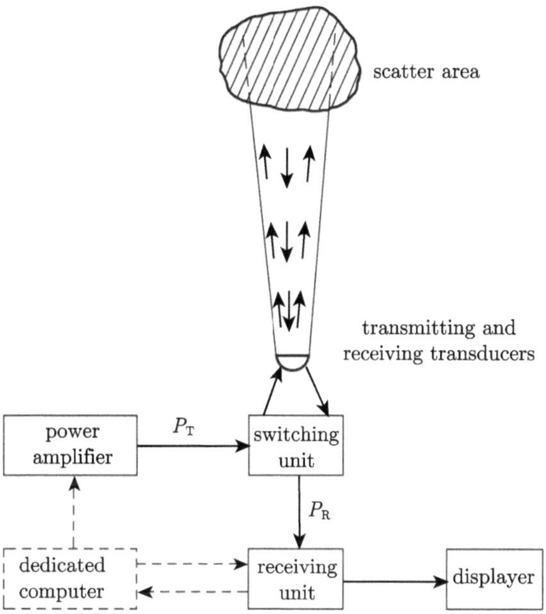

Fig. 53: Simplified block diagram of a monostatic echosonde.

he used was a monostatic configuration,[1] in which the transmission and reception transducers are identical or, at least, nearly collocated. The simplified block diagram shown in Fig. 53 describes the basics of the equipment McAllister used in 1968 [215] (the dash-lined part shows later supplements). In this set-up, an amplified acoustic tone burst P_T is sent through a switching device to the transmission transducer. After a suitable delay, the transducer is switched to the receiver mode, in which the echo signal P_R is passed to the receiver amplifier and then to a display or recording device, usually a facsimile recorder, though similar designs using an A-scope cathode ray tube (CRT) display or a z-axis modulated CRT display have also been employed. The most important characteristic of this equipment lies in the display, which is a real-time facsimile machine that can graphically display the correlation and structure of atmospheric scattering. The former "monostatic transducer" that he first used was an electro-dynamic driver and horn at the focus of a large dish, while a later set-up was an array composed of 196 loudspeakers, each being 20 cm in diameter. The peak pulse power of the array was over 500 W of electrical input.

Typical antennae achieve an angular beam width of $8^\circ \sim 10^\circ$. The reason for using the same antenna for reception rather than a more sensitive microphone lies

1 The term is adopted from radio technique. The meaning of the word "static" is only to tell the difference between radar based on the ground and the radar loaded on an airplane.

in this narrow beam width and the resulting much greater signal-to-noise (SNR) ratio.

The diagram shown in Fig. 53 is still representative of most of the more advanced systems used to date. The main improvement is the addition of a dedicated computer (the dash-lined part in the figure), by which the transmitted frequency, the timing signals of the control of sounder transmission and reception, including the pulse length and pulse repetition frequency, are derived. A passive diode gate is employed to protect the sensitive receiver circuits from the high-power transmission signal. The computer also operates an active gate which opens after the "dead time" that results from the antenna ringing after transmission.

The received signal is amplified and filtered to remove out-of-band noise components, and then is processed by a precision rectifier with a dynamic range of about 60 dB. The output of the rectifier can be recorded for quantitative analysis. A time-varying gain is used to compensate for the loss of amplitude resulting from spherical divergence of the acoustic echoes as they are received from progressively longer ranges.

In order to produce detectable echo signals, a high acoustic power transmission is desirable. This can be achieved by supplying a high electrical power to the antenna, which implies a high-power handling capability and high antenna efficiency. For a high-power reception, it is necessary to use an antenna with a large collecting area and it should operate at frequencies where atmospheric absorption is small. If the antenna is too large, however, the beam width becomes too narrow, and then significant sound energy will be lost via wind refraction.

It is usually possible to construct amplifiers with a noise level well below the environmental noise contribution caused by wind, traffic, aircraft, etc. In very quiet environments, however, the thermal noise in the receiving transducer may impose the limit of sensitivity. The intrusion of environmental noise can be minimized by clever antenna design, e.g., low side lobes and screening. A low-noise pre-amplifier is often used in antennae to eliminate problems caused by electrical pick-up when long cables (50∼200 m) are used to connect the antenna to the receiver. The pre-amplifier incorporates a step-up transformer to match the antenna impedance of a few ohms to the input impedance of the receiver, a few kilo-ohms.

Along with the rapid development of advanced integrated circuit designs, antenna elements, pre-amplifier receiving circuits as well as power amplifiers in more recent systems are put together in an almost hybrid form. This significantly improves the electronic performance and the portability of the system, while the actual cost is reduced [216–218].

Most conventional acoustic sounders use a carrier frequency in the range of 1 to 2 kHz, with maximum useful ranges from a few hundred meters to a few kilometers, and with a vertical resolution of 5∼20 m. For example, the detectable height of a commercial SODAR can reach 750 m when the frequency is 1400 Hz and with vertical resolution of 20 m. Sounds operating at frequencies greater than 5 kHz have a number of advantages [219]. For example, the acoustic output of some external noise sources such as traffic are often quite very little at these frequencies, and thus the system can be operated in a noisy town environment with minimal acoustic screening. Small antennae, which have the obvious advantage of being easily portable, also have short reverberation times, i.e. a short “dead time” after transmission, which allows sounding down to a few meters above the ground. These advantages, however, are at the expense of reduced range owing to the greater atmospheric acoustic attenuation at such high frequencies. Recent commercial and investigative systems tend to use high frequencies since, for a number of applications, the effective range of the acoustic radar is of minor importance.

The main limitations in the performance of an acoustic sounder depend on the characteristics of the antenna to a great extend, since present-day electronic techniques are capable of achieving almost ideal performance in a sounding system. In summary, the basic requirements for an effective monostatic antenna are: ① Narrow main beam to enable probing of a well-defined atmospheric volume; ② Low-side lobe levels to minimize reflections from nearby solid objects and to reduce reception from external noise sources; ③ High transmission efficiency to allow high acoustic outputs by using practically acceptable electrical input powers; ④ High receiver efficiency (equals to transmitting efficiency in many cases) to avoid degradation of the received SNR ratio by receiver noise.

8I.1.2 Bistatic configuration, Doppler echosonde

In bistatic systems, the receiver antenna and the transmission antenna are no longer “two combine into one” (or being located nearby), but are separately disposed far from each other. Obviously, the flexibility and facility in usage (where special consideration must be taken into account that such equipment are always being used in the open) of such systems are vastly inferior to monostatic systems, and are superior in other applications (especially in, for example, probing wind velocities). However, such superiorities recently have been overruled by the more advanced “tristatic” systems. Therefore the bistatic system is on the verge of disuse nowadays, and can scarcely be found in the market.

In earlier systems of this type, linear arrays of horns designed to exhibit a fan beam shape were suggested [220], however experience showed that extreme care must be taken to limit the beam width of such receiver arrays or, at least, to significantly decrease the gain in the horizontal direction, since in this direction it is generally too noisy to allow for the efficient operation of such units. Of course, if the noise nuisance is acceptable, the simultaneous usage of a fan beam transmitting antenna and a vertically directed receiving beam may be considered [221, 222].

The greatest value of the bistatic configuration lies in measuring the vertical profiles of mean winds on the basis of Doppler shift. Fig. 54 outlines the geometry

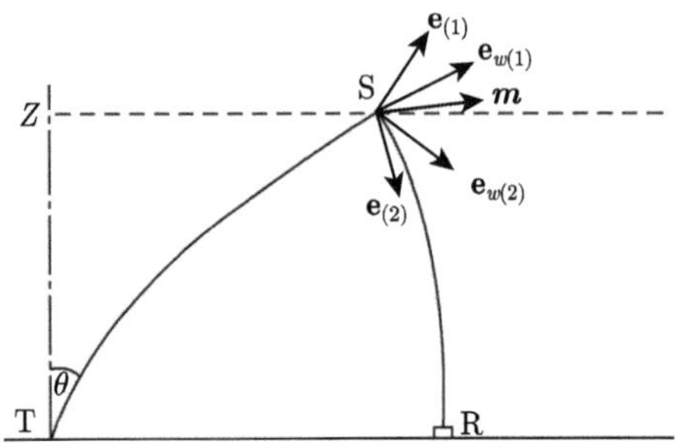

Fig. 54: Geometry for obtaining Doppler wind measurements. T: transmitter, R: receiver, S: scattering volume center at height z, $\boldsymbol{m}$: Mach vector of the wind velocity, $\boldsymbol{e}$: unit vectors in the corresponding directions.

of one "leg" of such a system. With two horizontal legs at right angles to each other and the addition of a vertical Doppler measurement, one can map the entire three-dimensional wind field at all heights within the echosonde range. There exist a number of different arrangements with various advantages and drawbacks. The following describes a particular design that seems one of the more flexible and dependable and has the advantage that the vertical narrow receiver (i.e. a "pencil beam") rejects much of the noise coming in from low-elevation paths from, e.g., road traffic, distant aircraft, residential areas, etc. The typical transmission antenna uses a fan beam geometry, one that sprays an initial pulse over a range of angles. Time-gating of the received signal then gives a Doppler value at each desired height.

The wind velocity is denoted by a vector $\boldsymbol{w}$ or by a Mach vector defined as $\boldsymbol{m} = \boldsymbol{w}/c$. Under simplified assumptions, such as horizontal stratification and negligible vertical wind or cross wind,[2] the horizontal component of the Mach vector m_x can be related to the Doppler shift Δf as follows [223]

$$\Delta f/f = -m_x \sin\theta + (m_x \sin\theta)^2$$

For a monostatic Doppler echosonde, when $\boldsymbol{m}$ is directed along the beam with no refraction, the frequency shift Δf satisfies

$$\Delta f/f = \pm 2m$$

where the plus sign holds when the wind blows toward, and the minus sign away from, the antenna.

Various methods for determining the frequency shift f from raw data have been proposed, especially ones to cope with the almost ubiquitous presence of noise. One of the common methods is to use FFT (fast Fourier transform) in the frequency domain,

2 In general, if we take the average value within a considerable period (for example 20 min), the influence of any vertical wind should be eliminated. On the other hand, the whole set of the sounding system will face the following case: For a certain "leg" of the system, the horizontal wind will become important, or other simplifying assumptions will not be tenable. Under this condition, more complicated results may be obtained, refer to [216].

while simultaneously measuring/estimating the spectral contribution of the noise. However, the FFT method requires a fairly powerful computer, whereas the CXCV (complex covariance) method in the time domain provides good data with a much less expensive micro-processor. The CXCV method requires careful choosing of the sampling frequency and filter bandpasses, where the choice of sampling frequency such as the Nyquist frequency (highest frequency allowed in a digitally sampled signal) needs to remain near the filter half-power point in order to minimize aliasing. A third process uses a "comb filter" made essentially from a number of narrow bandpass filters; typically, for a carrier frequency near 2 kHz, each "tooth" of the comb has a bandwidth of about 4 Hz.

With the increasing amount of inexpensive computer memory becoming available, echosonde designers are now the first choice for coherent signal processing, for example, averaging the signal in both amplitude and phase, and thus reducing random noise to a negligible level. The noise of rain hitting the receiver antenna makes echosonde observations impossible during a storm. In contrast, the so-called "echometer", patented by E. H. Brown in 1984, can provide vertical profiles of mean temperatures by virtue of the (necessary) coherent processing, which operates successfully through a rain squall, producing data in close agreement with those from direct measurements. Therefore, similar coherent processing of echosonde signals is quite effective for obtaining a large increase in the SNR.

Now, the above mentioned "tristatic system" will be introduced briefly. Such a system consists of three monostatic configurations placed at the same location, but pointing at three different directions, which are electronically synchronized. This system allows greater simplicity in the hardware but produces greater errors when the atmosphere lacks horizontal homogeneity because of separations between the three different scattering volumes.

8I.2 The physical foundations of acoustic sounding

Just as pointed out at the beginning of this chapter, the physical foundation of acoustic remote sensing in the troposphere is the scattering of acoustic waves off of non-homogeneities in the atmosphere. In this section, further discussions regarding the acoustic scattering off of non-homogeneities combined with practical applications of acoustic sensing, may be regarded as a supplement and extension to section 2 in chapter IV.

8I.2.1 The principle of pulse-echo sounding the atmospheric non-homogeneities

Assume that at time $t = 0$ the transmitter sends out a pulse of duration τ and of nearly constant frequency f, such that $\tau f \gg 1$. Moreover, the distance between the antenna and the scatterer r_S is somewhat larger than $c\tau/2$ and is such that the scatterer is in the transmitter's far field. The acoustic pressure at sufficiently large distances from any source of finite extent (other than a point source) can be written as (see, e.g., pp 45–47 in [56])

$$p_i = (D/r)F(\theta, \varphi, t - r/c) \tag{8.1}$$

where the function F is nonzero only if $0 < t - r/c < \tau$ and oscillates with angular frequency $\omega = 2\pi f$ throughout the pulse interval. Its normalization is such that the time average of F^2 is 1 for the time interval and for the direction toward the scatterer, taken here a $\theta = 0$. φ and θ are the polar angle and azimuth angle in spherical coordinates, respectively, while the constant D is such that $(D^2/\rho c)/r_S^2$ is the incident wave's average intensity at the scatterer during the irradiation interval.

The scatterer's dimensions are regarded here as sufficiently small compared to r_S so that the incident wave appears locally planar, such that the results in chapter IV are directly available.

The scattered-wave intensity I_{SC} varies with direction and with radial distance r from the scatterer as $(\mathrm{d}_\Omega\sigma)I_i/r^2$, where the differential cross section $\mathrm{d}_\Omega\sigma$ can alternatively be expressed as $\sigma_{\mathrm{back}}/4\pi$ for the backscattered direction (see section 1.2 in chapter IV), thus the intensity scattered back to the transmitter becomes (during $0 < t - 2r_S/c < \tau$)

$$I_{\mathrm{back}} = [D^2 r_S^{-2}/(\rho c)][\sigma_{\mathrm{back}} r_S^{-2}/(4\pi)] \tag{8.2}$$

Because $r_S > c\tau/2$, the backscattered pulse does not overlap the incident pulse, and so the operation mode of the transducer can be switched to that for reception in the interval between the termination of the transmission and the first arrival of the echo. The overall delay time, when multiplied by c, yields 2 r_S, so that the additional measurement of the echo's intensity, in conjunction with Eq. (8.2), suffices to determine the backscattering cross section τ_{back}.

Now suppose that the Born approximation is valid, i.e., there is no multiple scattering. Thus, $\hat{p}(\mathbf{x}_S)$ in the integrand of Eq. (4.11) can be replaced by the complex amplitude of the incident wave. Doing such is just the same as solving the integral equation by iteration, where the first iteration is accepted as satisfactory. Although this requires, in general, that the scattered wave in the steady state be much weaker than the incident wave wherever dominant non-homogeneities occur, no simple criteria involving magnitudes of γ_κ and γ_ρ (for their definitions see Eq. (4.9)) can establish the upper limit of the approximation's validity. It should, however, yield a good estimate of the scattered field if $|\gamma_\kappa| \ll 1$ and $|\gamma_\rho| \ll 1$ and if the path integrals of both $k|\gamma_\kappa|$ and $k|\gamma_\rho|$ are small when compared with unity.

Modification of Eq. (4.11) to the case where the incident wave is the pulse in Eq. (8.1), with a subsequent application of the Born approximation, yields the scattered wave in the form (applicable for bistatic as well as monostatic configurations)

$$P_{SC} = k^2 D \iiint [\gamma_{\mathrm{eff}}(\mathbf{x}_S)/r_S R] F(\theta_S, \varphi_S, t - r_S/c - R/c)\mathrm{d}V_S \tag{8.3}$$

where $r_S = |\boldsymbol{x}_S|$ is the distance from the center of transmitter (taken as the origin) to the scattering point; $R = |\boldsymbol{x} - \boldsymbol{x}_S|$ is the distance from scattering point $\boldsymbol{x}_S$ to reception point $\boldsymbol{x}$; and

$$\gamma_{\text{eff}}(\boldsymbol{x}_S) = \gamma_{\kappa}(\boldsymbol{x}_S) - \boldsymbol{e}_S \cdot \boldsymbol{e}_R \gamma_{\rho} \cdots (\boldsymbol{x}_S) \tag{8.4}$$

where unit vectors $\boldsymbol{e}_S$ and $\boldsymbol{e}_R$ point from the origin to $\boldsymbol{x}_S$, and from $\boldsymbol{x}_S$ to $\boldsymbol{x}$ respectively, (compare with the similar unit vectors $\boldsymbol{e}_k$ and $\boldsymbol{e}_r$ introduced in section 1 of chapter IV).

For the monostatic configuration, $\boldsymbol{e}_S.\boldsymbol{e}_R \sim -1$, so that Eq. (4.9) yields

$$\gamma_{\text{eff}}(\boldsymbol{x}) \sim 2\delta(\rho c)/\rho c \qquad \text{backscatter} \tag{8.5}$$

where $\delta\,(\rho c)$ is the deviation of the characteristic impedance of the atmosphere from its normal value ρc.

For a single clustered non-homogeneity of dimensions much smaller than a wavelength, the Born approximation leads to a scattered wave (8.3) which agrees with Eq. (4.4) obtained previously, where the backscattering cross section is consequently

$$\sigma_{\text{back}} = (k^4/\pi)[(1/\rho c) \iiint \delta(\rho c) \mathrm{d}V]^2 \tag{8.6}$$

8I.2.2 Scattering volumes delimited by electro-acoustic transducers

The transmitter and receiver in a sounding system are electro-acoustic transducers, and the transmission is characterized by function $i_{\text{tr}}\,(t)$ (loudspeaker excitation current) while the reception is characterized by function $e_{\text{erc}}\,(t)$ (microphone open-circuit voltage). Analogous quantities can be defined for "mechano-acoustic transducers": a rigid piston oscillating in an infinite baffle is characterized by a normal velocity $v_n\,(t)$, while one acting as a receiver is characterized by the force exerted on the piston face (being held virtually motionless) by the impinging sound wave. These two time-dependent functions are assumed to be linearly related to the transmitted and incident acoustic fields, respectively.

When driven at constant frequency ω by a current of complex amplitude $\hat{\imath}_{\text{tr}}$, the transmission transducer produces a far-field radiated acoustic pressure (see, e.g., section 4.10 in [56])

$$\hat{p} = -(\mathrm{i}\omega\rho/4\pi) M_{\text{tr}} \hat{F}_{\text{tr}}(\theta, \varphi)(\mathrm{e}^{\mathrm{i}kr}/r)\hat{\imath}_{\text{tr}} \tag{8.7}$$

in the direction with angular coordinates (θ, φ). Here the function $\hat{F}_{\text{tr}}\,(\theta, \varphi)$, whose phase is of minor interest, is normalized so that the transmitter radiation pattern $|\hat{F}_{\text{tr}}|^2$ is 1 when $\theta = 0$; while the constant factor $\omega\rho M_{\text{tr}}/4\pi$ is determined by the ratio $r|\hat{p}|/|\hat{\imath}_{\text{tr}}|$ along that axis ($\theta = 0$). The quantity M_{tr} is a convenient description of the transducer's ability to transform electrical current into far-field pressure.

The analogous description of a receiving transducer sets

$$\hat{e}_{\text{rec}} = M_{\text{rec}} \hat{F}_{\text{rec}}(\theta, \varphi)\hat{p} \tag{8.8}$$

to describe the voltage caused by a plane wave nominally having amplitude $\hat{p}$ at the transducer face and arriving from direction (θ, φ). Here the receiver directivity function $|\hat{F}_{\mathrm{rec}}|^2$ is normalized also to 1 at $\theta = 0$. The constant M_{rec} is the microphone response at normal incidence (in V / Pa). Equivalently, if a point source of volume velocity amplitude (source strength) $\hat{U}$ is located a great distance away at a point with coordinates (r, θ, φ), so that the $\hat{p}$ in Eq. (8.8) is $-(\mathrm{i}\omega\rho/4\pi)\hat{U}(\mathrm{e}^{\mathrm{i}kr}/r)$, that equation becomes

$$\hat{e}_{\mathrm{rec}} = -(\mathrm{i}\omega\rho/4\pi)M_{\mathrm{rec}}\hat{F}_{\mathrm{rec}}(\theta, \varphi)(\mathrm{e}^{\mathrm{i}kr}/r)\hat{u} \tag{8.8a}$$

Comparison of the above equation with Eq. (8.7) and reference to the reciprocity theorems indicate that if a transducer is a reciprocal transducer, then

$$M_{\mathrm{tr}} = M_{\mathrm{rec}}, \quad \hat{F}_{\mathrm{tr}} = \pm\hat{F}_{\mathrm{rec}} \tag{8.9}$$

As can be seen from Eq. (8.3), the scattered wave originates from a distributed source with strength density (volume velocity per unit volume)

$$\mathrm{d}U_{\mathrm{s}}/\mathrm{d}V_{\mathrm{s}} = [\gamma_{\mathrm{eff}}(\boldsymbol{x}_{\mathrm{s}})/\rho c^2]\partial_t p_i(\boldsymbol{x}_{\mathrm{s}}, t) \tag{8.10}$$

The receiver voltage is the superposition of the incremental contributions (Eq. (8.8a)) from each elemental volume, while the incident pressure is as given in Eq. (8.7). An appropriate relabeling and juxtaposition of the coordinate system consequently yields

$$\hat{e}_{\mathrm{rec}} = (\mathrm{i}\omega\rho k^2/16\pi^2)M_{\mathrm{rec}}M_{\mathrm{tr}}\hat{i}_{\mathrm{tr}} \iiint \hat{F}_{\mathrm{rec}}\hat{F}_{\mathrm{tr}}\gamma_{\mathrm{eff}}[\mathrm{e}^{\mathrm{i}k(R+r_{\mathrm{s}})}/r_{\mathrm{s}}R]\mathrm{d}V_{\mathrm{s}} \tag{8.11}$$

which may be rewritten as

$$e_{\mathrm{rec}}(t) = -(\rho k^2/16\pi^2)M_{\mathrm{rec}}M_{\mathrm{tr}} \iiint |\hat{F}_{\mathrm{rec}}\hat{F}_{\mathrm{tr}}|(\gamma_{\mathrm{eff}}/\boldsymbol{r}_{\mathrm{s}}R)\mathrm{d}_t i_{\mathrm{tr}}\mathrm{d}V_{\mathrm{s}} \tag{8.11a}$$

if the time dependence is explicitly inserted, in which $\mathrm{d}_t i_{\mathrm{tr}}$ is evaluated at $t - R/c - r_{\mathrm{s}}/c - \varepsilon/\omega$, where ε represents the position-dependent phase of $\hat{F}_{\mathrm{rec}}(\boldsymbol{e}_{\mathrm{R}})\hat{F}_{\mathrm{tr}}(\boldsymbol{e}_{\mathrm{s}})$.

Although both the inversions (8.11) and (8.11a) are derived for constant-frequency propagation, the latter should also apply to pulse propagation, whereby $i_{\mathrm{tr}}(t)$ is of nearly constant frequency in the interval $0 < t < \tau$ and is zero, or nearly zero, outside of that interval. The voltage output recorded during any small interval centered at t depends primarily on scattering within a volume between ellipsoids $t = (R + r_{\mathrm{s}})/c$ and $t = \tau + (R + r_{\mathrm{s}})/c$ (see Fig. 55). The volume is further restricted if the transmitter and receiver patterns are narrow-beam and if (for the case of bistatic systems) the beams are directed to intersect at a localized region centered at a point $\bar{x}_S$ and at distances $\bar{r}_S$ and $\bar{R}$ from the transmitter and receiver. For the monostatic case, we consider the beams to be coaxial and rely on the finite pulse duration to delimit the scattering to a finite volume.

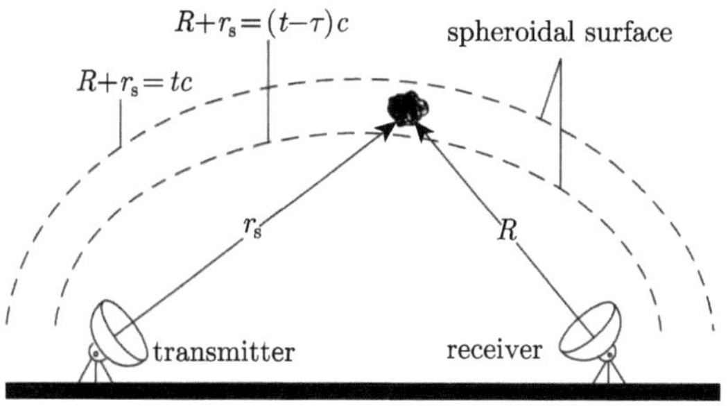

Fig. 55: Concentric spheroids delimiting region of possible scattering locations for bistatic sounding.

Since the scattering that reaches a receiver in the bistatic configuration comes from a finite volume regardless of whether or not the pulse duration is short, for simplicity we will first discuss bistatic sounding assuming constant-frequency transmission. Since $|\hat{F}_{\text{tr}}|$ and $|\hat{F}_{\text{rec}}|$ are 1 for the direction $\bar{e}_{\text{S}}$ from the origin to $\bar{x}_{\text{S}}$ and for the direction $\bar{e}_R$ from $\bar{x}_{\text{S}}$ to the receiver center, the scattering volume consists primarily of all points where $|\hat{F}_{\text{tr}}| \cdot |\hat{F}_{\text{rec}}|$ is greater than, say, 1/ 2. An estimate of its size is

$$\Delta V_{\text{S}} = \iiint |\hat{F}_{\text{tr}}|^2 |\hat{F}_{\text{rec}}|^2 \text{d}V_s \tag{8.12}$$

as will be explained below, in the derivation of Eq. (8.19).

The assumption that the scattering volume has dimensions much smaller than $\bar{r}_{\text{S}}$ and $\hat{R}$ allows us to replace r_{S} and R in denominator of the integrand in Eq. (8.11) by $\bar{r}_{\text{S}}$ and $\hat{R}$. Additional substitutions from Eqs. (8.7) and (8.8) consequently yield

$$\hat{p}_{\text{sc,ap}} = [\hat{p}_i/(4\pi)^{1/2}](\text{e}^{\text{i}k\bar{R}}/\bar{R})\psi \tag{8.13}$$

$$\psi = -k^2/(4\pi)^{1/2} \iiint |\hat{F}_{\text{tr}}\hat{F}_{\text{rec}}|\text{e}^{\text{i}\varepsilon}\gamma_{\text{eff}}\text{e}^{\text{i}k(R-\bar{R}+r-r)}\text{d}V_s \tag{8.14}$$

where $\hat{p}_{\text{sc,ap}}$ is the apparent pressure impinging upon the receiver from the direction of the scattering volume center, and $\hat{p}_i$ is the incident acoustic pressure at the volume's center $\bar{x}_{\text{S}}$. The distinction between $\hat{p}_{\text{sc,ap}}$ and $\hat{p}_{\text{sc}}$ arises due to the receiver weighs pressure contributions associated with different arrival directions differently.

8I.2.3 Acoustic radar equation

The above formulation extends readily to monostatic sounding with a reciprocal transducer from a single localized scatterer at a point with coordinates $(\bar{r}_{\text{S}}, \bar{\theta}_{\text{S}}, \bar{\varphi}_{\text{S}})$. The quantity ψ in Eq. (8.14) is replaced by one such that

$$|\psi|^2 = |\hat{F}(\bar{\theta}_{\text{S}}, \bar{\varphi}_{\text{S}})|^4 \sigma_{\text{back}} \tag{8.15}$$

where σ_{back} is given in Eq. (8.6) for a small weak non-homogeneity. Eq. (8.13) then yields the acoustic radar equation[3]

$$[I_{\text{sc}}/(4\pi r^2 - I_i)_0][(e^2_{\text{rec}})_{\text{av}}/(e^2_{\text{rec}})_{\text{av},0} = (\sigma_{\text{back}}/16\pi^2\bar{r}_s^4)|\hat{F}(\bar{\theta}_s, \bar{\varphi}_s)|^4 \tag{8.16}$$

where

$$(e^2_{\text{rec}})_{\text{av}}/(e^2_{\text{rec}})_{\text{av},0} = I_{\text{sc,ap}}/I_{\text{sc}}$$

is the ratio of the mean-squared voltage recorded to what would have been recorded if a signal had been of the same intensity incident from $\theta = 0$. Here I_{sc} is the actual acoustic intensity returning to the transducer, while $I_{\text{sc,ap}}$ is its apparent value when the returning wave is regarded as having come from the $\theta = 0$ direction. The quantity $(4\pi\, r^2 I_i)_0$ is equal to $4\pi r^2$ times the transmitted intensity in the $\theta = 0$ direction at a far-field distance r, which can be regarded as the acoustic power output times the directive gain associated with that direction. Similarly, $(4\pi r^2 I_i)_0|\hat{F}(\bar{\theta}_s, \bar{\varphi}_s)|^2$ is the power output times the directive gain associated with the direction $(\bar{\theta}_s, \bar{\varphi}_s)$.

8I.2.4 Incoherent scattering: bistatic acoustic sounding equation

If the non-homogeneities causing the scattering are dispersed throughout the scattering volume, one must take the relative phases of contributions from the different volume elements in Eq. (8.14) into account, and this can be done approximately via the substitution

$$R - \bar{R} + r_s - \bar{r}_s \approx (\bar{e}_s - \bar{e}_R)\cdot\xi \tag{8.17}$$

which results from a truncated power-series expansion in the components of $\boldsymbol{\xi} = \mathbf{x}_s - \bar{\mathbf{x}}_s$. Letting $\Delta\boldsymbol{k}$ to represent the change $(\boldsymbol{e}_R - \boldsymbol{e}_s)k$ in wave number vector undergone during the scattering, Eq. (8.14) yields

$$|\psi|^2 = \frac{k^4}{4\pi}\int\cdots\int \Phi(\xi)\Phi^*(\xi')\gamma_{\text{eff}}(\xi)\gamma_{\text{eff}}(\xi')\mathrm{e}^{\mathrm{i}\Delta k\cdot(\xi'-\xi)}\mathrm{d}V_\xi \mathrm{d}V_{\xi'} \tag{8.18}$$

where $\Phi(\boldsymbol{\xi})$ represents $|\hat{F}_{\text{tr}}\hat{F}_{\text{rec}}|\mathrm{e}^{\mathrm{i}\varepsilon}$ as evaluated at the position $\overline{x_s} + \xi$.

If $\gamma_{\text{eff}}(\boldsymbol{\xi})$ in different regions appears to be statistically indistinguishable, the idealization of a random medium is appropriate. The notion of a statistically homogeneous random process (whose correlation disappears over a relatively short distance) allows $\gamma_{\text{eff}}(\boldsymbol{\xi})\,\gamma_{\text{eff}}(\boldsymbol{\xi}')$ to be replaced by its ensemble average (or, equivalently, by the local spatial average of $\gamma_{\text{eff}}(\boldsymbol{\xi})\,\gamma_{\text{eff}}(\boldsymbol{\xi} + \Delta\boldsymbol{\xi})$); this average is the spatial auto-correlation function $R(\Delta\boldsymbol{\xi}; \gamma_{\text{eff}})$. When the auto-correlation function is negligibly small for any $\Delta\boldsymbol{\xi}$

3 The reason for its name is because it is the acoustic counterpart of the "free-space radar transmission equation" (widely referred to as "radar equation") for an electro-magnetic waves.

whose magnitude is comparable to or larger than a characteristic length over which $\Phi(\boldsymbol{\xi})$ changes appreciably, the incoherent scattering model results, whereby acoustic power that is scattered by moderately distant non-homogeneities are additive. Such assumptions reduce Eq. (8.18) to

$$|\psi|^2 = \eta(k, \Delta\boldsymbol{k})\Delta V_\mathrm{s} \tag{8.19}$$

where ΔV_s is as defined in Eq. (8.12) and where

$$\eta(k, \Delta\boldsymbol{k}) = (k^4/4\pi)\iiint R(\Delta\boldsymbol{\xi}; \gamma_\mathrm{eff})\mathrm{e}^{\mathrm{i}\Delta k\cdot\Delta\xi}\mathrm{d}(\Delta\xi_x)\mathrm{d}(\Delta\xi_y)\mathrm{d}(\Delta\xi_z) \tag{8.20}$$

$$= \frac{1}{4}\pi^2 k^4 S(\Delta\boldsymbol{k}; \gamma_\mathrm{eff}) \tag{8.20a}$$

Here $S(\Delta\boldsymbol{k}; \gamma_\mathrm{eff})$ is recognized as the spectral density of $\gamma_\mathrm{eff}(\boldsymbol{\xi})$ in wave number space.[4] The normalization adopted is such that

$$\langle\gamma_\mathrm{eff}^2\rangle = \int\int_0^\infty\int S(\Delta\boldsymbol{k}; \gamma_\mathrm{eff})\mathrm{d}(\Delta k_x)\mathrm{d}(\Delta k_y)\mathrm{d}(\Delta k_z) \tag{8.21}$$

gives the mean-squared value of $\gamma_\mathrm{eff}(\xi)$. Eq. (8.19), in conjunction with (8.13), which leads to the bistatic acoustic sounding equation

$$I_{\mathrm{sc,ap}}/(4\pi r^2 I_i)_\mathrm{O} = \eta\Delta V_\mathrm{s}/16\pi^2 r_\mathrm{s}^2\bar{R}^2 \tag{8.22}$$

where η is identified as the apparent bistatic cross-section per unit volume. The implication here is that the scattered intensity is directly proportional to the scattering volume, which is the distinguishing feature of incoherent scattering and requires that non-homogeneities causing the scattering be randomly dispersed and that any correlation length associated with the non-homogeneities is much smaller as compared with the scattering volume's dimensions. In contrast, if the scattering volume is very small in terms of a correlation length, the far-field acoustic-pressure contributions scattered by different volume elements are in phase and reinforce each other; then the scattering is considered to be coherent, and the apparent bistatic cross section is directly proportional to the square of the scattering volume.

8I.2.5 Echosonde equation

The incoherent-scattering idealization allows for a lucid interpretation of pulse-echo measurements of scattering from an inhomogeneous atmosphere. Eq. (8.13), via such an idealization, implies that for the monostatic case,

$$\delta(E/A)_{\mathrm{sc,ap}} = [(4\pi r^2 I_i)_\mathrm{O}\delta t/16\pi^2][\eta|\hat{F}|^4\delta V_\mathrm{s}/r_\mathrm{s}^4] \tag{8.23}$$

4 This is according to the so-called Wiener-Khintchine theorem: refer to N. Wiener, *Generalized harmonic analysis*. Acta Math., 1930, 55:117∼258.

is the apparent backscattering energy received per unit area due to scattering during the time interval δt from volume element δV_S at a distant point $(r_S, \theta_S, \varphi_S)$. The quantity $(4\pi r^2 I_i)_O$ is a representation of the power the transmitter radiates at time $t - 2r_S/c$. Hence the total apparent backscattered energy per unit area received up to time t is

$$(E/A)_{\mathrm{sc,ap}} = (1/16\pi^2) \iiint \left(\eta |\hat{F}|^4 / r_S^4\right) \left[\int_{-\infty}^{t-2r/c} (4\pi^2 I_i)_O \mathrm{d}t' \right] \mathrm{d}V_S \tag{8.24}$$

where $(4\pi r^2 I_i)_O$ is zero up to $t' = 0$ and is zero for $t' > \tau$. Taking the time derivative and subsequently transforming the r_S integration into one over $t' = t - 2r_S/c$ yields

$$I_{\mathrm{sc,ap}} = (c/32\pi^2) \int_0^{\tau} [\iint (\eta |\hat{F}|^4 / r_S^2 \mathrm{d}\Omega_S](4\pi^2 I_i)_O \mathrm{d}t' \tag{8.25}$$

which is the apparent backscattered intensity. The quantity in square brackets here is understood to be evaluated at $r_S = (t - t')c/2$. Consideration is limited to reception times t greater than the pulse duration τ. If the time t is further taken to be much greater than τ, Eq. (8.25) approximates to

$$I_{\mathrm{sc,ap}} \sim (c\tau/32\pi^2)[\eta \Delta\Omega_S / \bar{r}_S^2](4\pi r^2 I_i)_O \tag{8.25$'$}$$

where

$$\Delta\Omega_S = \iint |\hat{F}|^4 \mathrm{d}\Omega_S \tag{8.26}$$

should be interpreted as the solid angle being probed. The quantity $(4\pi r^2 I_i)_O$ now represents the time averaged over the pulse duration τ of transmitted power times directive gain of the transmitter. The radial distance $\bar{r}_S$ is approximately $ct/2$, which represents the average distance to the scattering volume (extending from $r_S = (t - \tau)c/2$ to $r_S = tc/2$). The quantity η is the average backscattering cross section per unit volume (Eq. (8.20a) with $\Delta\boldsymbol{k} = 2k\boldsymbol{e}_z$, which points along the beam's axis) for a spherical shell of solid angle $\Delta\Omega_S$. As before, $I_{\mathrm{sc,ap}}$ is the apparent acoustic intensity of the backscattered wave at the transducer, which accounts for the directional response characteristics during reception.

The applicability of the incoherent-scattering assumption to the derivation of Eq. (8.25′) requires $c\tau$ and $\bar{r}_S(\Delta\Omega\sigma)^{1/2}$ be large compared with the correlation length of the non-homogeneities. If this is not so, but the scattering atmosphere is nevertheless random, the prediction (8.25′) is an assembled average of all of the possible outcomes.

The generalization of the above considerations to pulse-echo sounding with the bistatic configuration yields

$$I_{\mathrm{sc,ap}} = (\eta/16\pi^2)(4\pi r^2 I_i)_O \iiint'' [|\hat{F}_{\mathrm{tr}}||\hat{F}_{\mathrm{rec}}|/r_S R]^2 \mathrm{d}V_S \tag{8.27}$$

where the double prime on the integral indicates that the region of integration is restricted to that lying between the prolate spheroids $r_s + R = tc$ and $r_s + R = (t - \tau)c$.

Now we introduce a dimensionless "aspect factor" $\tilde{A}$,

$$\tilde{A} = \{\bar{R}^2/[(c\tau/2)\Delta\Omega_{tr}]\} \iiint'' [|\hat{F}_{tr}||\hat{F}_{rec}|/r_s R]^2 \mathrm{d}V_s \tag{8.28}$$

where

$$\Delta\Omega_{tr} = \iint |\hat{F}_{tr}|2\mathrm{d}\Omega_s \tag{8.29}$$

is the apparent beam width in steradians of the transmitted beam and $\bar{R}$ is the distance from the receiver to the intersection of the transmitted beam's axis with the spheroid $R + r_s = ct$. The similarities of Eq. (8.27) with (8.25′) are more evidently emphasized here. Insertion of (8.28) into (8.27) yields the *echosonde equation* [224]

$$I_{sc,ap} = (c\tau/32\pi^2)[\eta\Delta\Omega_{tr}\tilde{A}/\bar{R}^2](4\pi r^2 I_i)_0 \tag{8.30}$$

For the monostatic configuration, the aspect factor $\tilde{A}$ becomes $\Delta\Omega_s/\Delta\Omega_{tr}$. When a reciprocal transducer is used to receive as well as transmit, $\tilde{A}$ must be less than 1, but approaches 1 for a sharp-edged beam. If $|\hat{F}|^2$ varies with angle as $\exp(-\alpha\theta^2)$, where α is somewhat larger than 1, then $\tilde{A}$ is 1/2; if it varies as $1/(1 + \alpha\theta^2)^2$, then $\tilde{A}$ is 1/3.

8I.3 Outputs of the acoustic sounder [225]

Acoustic sounder systems provide both qualitative and quantitative information of the thermal and mechanical structure of the lower atmosphere. The former is mainly achieved by the exploitation of acoustic echoes recorded by means of a facsimile recorder while the latter by a data logger, which normally refers to the Doppler shift record of the returned echoes and to the intensity of the acoustic signals that represent the state of the thermal structure of the atmosphere. Examples of both respects will be given to cover a wide range of applications in the following.

8I.3.1 Thermal plume detection

Thermal plumes are features present in convective conditions that result from the conductive random heating of air near the ground, which itself is heated by solar radiation – such randomness may result from variations in surface albedo and also from the presence of clouds. They originate in the super-adiabatic region near the ground and normally extend to the top of the convective boundary layer. Direct observations show that they are of great importance in the diurnal development of

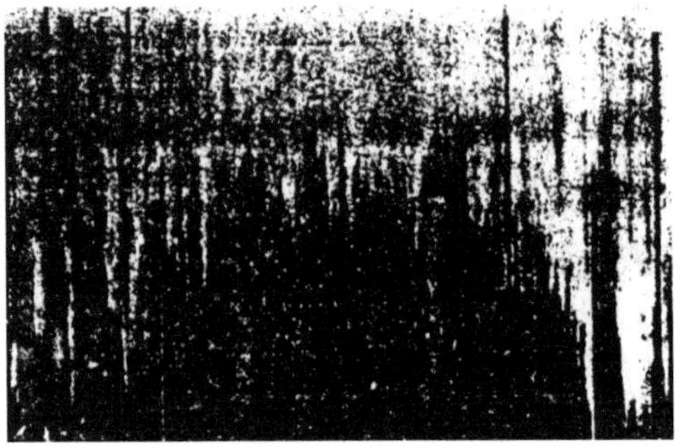

Fig. 56: Digital record showing typical thermal plume activity recorded with a high-resolution acoustic sounder (originally in color).

the boundary layer over land and are significantly responsible for carrying significant heat and moisture aloft from the ground. The onset time of convection and the change rate of convective layer depth are particularly important in atmospheric pollution studies. Similarly, in the evenings when convection dies down, turbulent mixing reduces dramatically and therefore the ability of the atmosphere to disperse pollutants is greatly affected. Such considerations illustrate that a system that can remotely record the state of convection over long periods of time is likely to be considerably usefuln.

Thermal plumes can be recorded by conventional high-frequency acoustic radar, but more recently facsimile recorders earned their place to the digital print-out, which in many cases provide better potential by incorporating different colors. A typical example is shown in Fig. 56, which was recorded with a high-resolution acoustic sounder operating under clear weather conditions and flat terrain. The potential of the new record against the facsimile recorder is quite obvious since it reveals details of the thermal structure that could not be detected previously.

8I.3.2 Monitoring of Inversions

One of the most useful applications of acoustic sounding may well prove to be the monitoring of inversions within the lowest few kilometers of the atmosphere. In such inversions, temperature increases with altitude, contrary to typical cases (where temperature decreases with increasing altitude) in the troposphere.

Inversions are of great meteorological significance since their presence may indicate preferred regions for strong wind shear that can affect safety in aviation. Sometimes they can also limit the growth of the convective boundary layer and hence influence many parameters such as maximum daytime temperature, visibility, air-pollutant concentration, etc. In some cases the height of the lowest inversion may be

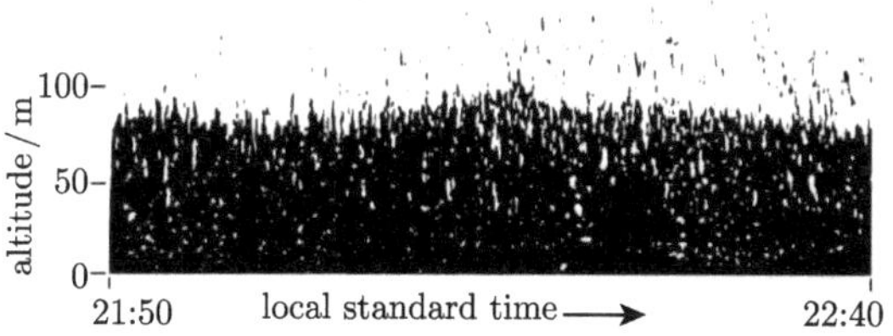

Fig. 57: A typical example of a temperature-inversion layer recorded with a conventional echo sounder.

taken as the depth of the convective boundary layer, which in turn is an important scaling parameter that describes turbulence in the lower atmosphere and is relevant to the dispersal of pollutants. Further, they can effectively impose a barrier to the transfer of heat, moisture and pollutants from the ground to the free atmosphere, and in this way may critically affect the formation of cloud and the air quality of the near-surface layer. A typical example of temperature inversion record is shown in Fig. 57, which was recorded using a conventional acoustic sounder equipped with a facsimile recorder. Similar records can be easily obtained during clear and stable weather conditions over all types of terrains.

81.3.3 Stable conditions and waves

With a stable stratification the atmosphere can support wave motion, and the interpretation of atmospheric data may well be complicated by the co-existence of waves and turbulence. This topic is particularly relevant to studies of the nocturnal boundary layer. Acoustic sounding offers an excellent method for remotely estimating (assessing) the significance of wave motion in particular situations [226]. Furthermore, the waves in many instances may become unstable and form patches of turbulence which may be the major contributors to turbulent transfer on these occasions. This is particular true for areas that are nearly continuously stable, such as the polar-regions in winter.

To illustrate the potential of acoustic sounding in this field, two examples that show a boundary layer in various states are provided. Fig. 58(a) presents an example of a stable boundary layer that is associated with a series of oscillating patterns on the top. It is considered that these were generated by breaking gravity waves [227]. Fig. 58(b) illustrates a quite different pattern of behavior in the nocturnal boundary layer. Here the acoustic echos are generated by a high level of mechanical mixing (wind shear) acting on the slightly stable temperature profile, which is quite common in many occasions. It should be noted here that the presented examples are only a small amount of the numerous "night type inversions" which vary from place to place according to weather conditions.

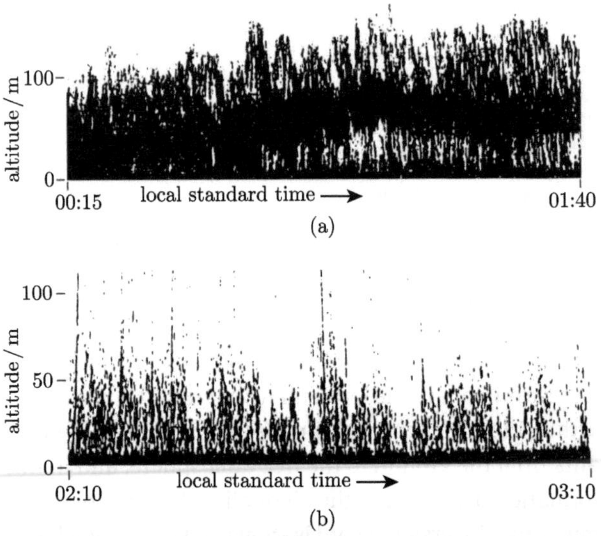

Fig. 58: Typical examples of boundary layer probing. (a) Stable boundary layer structure; (b) Slightly stable boundary layer.

8I.3.4 Quantitative comparisons

For properly evaluating acoustic sounders, it is necessary to make simultaneous direct measurements as close as possible to the scattering volume. Instrumentation supported by towers and tethered balloons are used for this purpose, although the distance between the direct sensor and the sampling volume can be fairly large [228]. The direct instrumentation attached to a balloon cable is capable of measuring wind components, absolute temperature, the temperature differential and atmospheric pressure etc [229]. From these basic quantities, estimates of the temperature-structure parameters C_T^2 and wind-structure parameters C_V^2, as well as their associated statistics, may be obtained.

The profile of C_T^2 can be obtained from echo intensities of a well-calibrated high-resolution monostatic acoustic sounder. Although the individual calculations represent a short time average, the site measurements highly correspond to the "remote estimations". Using the Doppler shift facility of the monostatic sounder, estimates of the axial component of the wind (usually vertical) can be obtained. A typical example of the comparison between these probing results and direct instrumentation is given in Fig. 59(a). By incorporating more complex operating configurations such as bistatic or triple monostatic ones, estimates of the mechanical turbulence and the three components of the wind can be obtained. A typical example of wind speed estimates using a high resolution tristatic acoustic sounder is given in Fig. 59(b). The estimated mean wind-speed profiles obtained by the sounder and observed by the meteorological

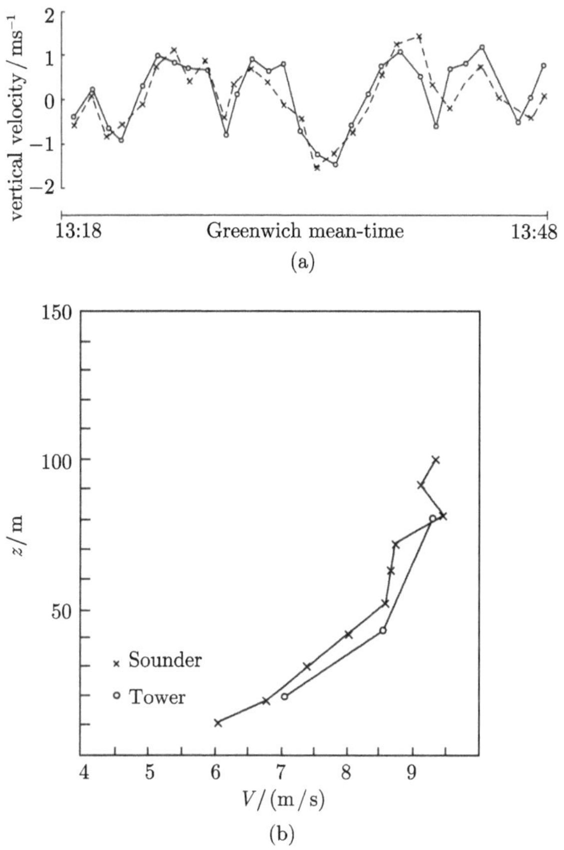

Fig. 59: Comparison between the results from acoustic sounding and from direct measurements. (a) Measurements of the vertical wind speed; (b) Measurements of the wind speed gradient.

tower agree quite well, however, of course, the basic differences between the two methods, namely, volume averaging and point measurements, should also be taken into account.

To sum up, the acoustic sounder is being increasingly used as a continuous monitor of the lower atmosphere since it gives a real-time record of atmospheric features which cannot be readily equated by in site sensors without the expense and complication of tethered balloons, radiosondes or meteorological towers. It is a useful instrument for boundary layer studies and promotes our understanding of the meteorological processes occurring in the lower atmosphere. In particular, it is a valuable aid for the interpretation of atmospheric in site instrument studies over a definite range of altitudes. The acoustic sounding system also provides quantitative details of the structure of the atmospheric boundary layer. In this way the sounder could therefore be used to obtain profiles of the structure parameters for temperature and wind velocity, and also for the rate of dissipation of turbulent kinetic energy. These quantities are useful for basic studies of the atmospheric boundary layer and are of significance when assessing the mixing capability of the lower atmosphere.

Doppler techniques are inherently more attractive since they do not require precise knowledge of the antenna characteristics or transmitter/receiver gains. A computer-controlled acoustic system can provide real-time profiles of mean wind speeds and turbulence velocity statistics to altitudes of several hundred meters. Such a system is useful for detailed studies of the atmospheric boundary layer, as well as for establishing the climatology of wind flow over various underlying surfaces.

It is worth noting for other possible area of acoustic sounders applications. This arises from the fact that the absorption of sound in air is a strong function of humidity, temperature and frequency of the sound itself. Suggestions have been made that multi-frequency absorption measurements using acoustic sounders could be used to obtain acoustic humidity and temperature profiles [230, 231]. This would have possible applications in the investigation of multi-path radio propagation effects caused by refractive index gradients, such as those associated with temperature inversions.

8I.4 Systematical algorithm for acquiring wind profiles from SODAR [232]

An inversion method to acquire wind profiles by means of an acoustic radar is introduced in the following paragraphs. This method consists of: determining the peak value of the Doppler frequency spectrum, and the smooth processing and use of the standard measured data to establish a confined linear retrieval operation. Each frequency spectrum peak value corresponds to one component of the wind speed and weighs those Doppler estimated values. Due to the limitations of the instruments, and the shear force present in the atmosphere, an estimation of wind in different layers will rely on each other. A spatial correlation is included in the retrieval process and the retrieval method can also estimate the degrees of freedom of the SNR at various heights.

8I.4.1 Doppler frequency spectrum acquired from SODAR

Consider a sound pulse signal transmitted from an acoustic radar: the initial moment of transmission is $t = 0$, the frequency is f and the pulse width is τ. When turbulence is caused at the height from z to $z+\mathrm{d}z$, it will reflect some of the energy of the signal to the receiver. As there are many factors such as antenna efficiency, wave-beam divergence, atmospheric absorption and turbulence scatter efficiency etc., the receiver can only record a small part $\sigma_S(z)\mathrm{d}z$ of the initial pulse energy, where $\sigma_S(z)$ is the linear density of the received energy. Assume that the sound velocity is c, among the received energy at the moment t, the energy of the front edge of the pulse is the part that reaches a height $ct/2$ and returns, where the energy of the back edge of the pulse is the part that reaches a height $c(t-\tau)/2$ and returns. If the envelope of transmitting sound power changes with time t is $p(t)$, and we consider the time delay of the coming-and-going path, then the power of the received signal without a Doppler frequency shift is

$$r(t) = \int_{c(t-\tau)/2}^{ct/2} \sigma_S(z) p\left(t - \frac{2z}{c}\right) dz \tag{8.31}$$

The spatial resolution of the received signal is $\Delta z_T = ct/2 - c(t-\tau)/2 = c\tau/2$. Each time t corresponds to a measurement interval and a corresponding dispersion power spectrum. The sampling number is N_S, and the sampling rate is f_S. The time spent acquiring each frequency spectrum is $T = N_S/f_S$ and the measurement center is at

$$z_i = \left(i - \frac{1}{2}\right)\Delta z_S, \quad (i = 1, 2, \dots, m) \tag{8.32}$$

where

$$\Delta z_S = \frac{cN_S}{2f_S} \tag{8.33}$$

By increasing the height we can obtain a series of measurement intervals until the distance $ct/2$ is large enough that the back-wave signal becomes so weak that it can be neglected. The value R_{ik} of the power spectrum in the measurement interval i corresponds to the frequency $f_{ik} (k = 1, 2, \dots, N_S)$, where

$$f_{ik} = \frac{f_S}{N_S}\left(k - \frac{N_S}{2}\right) \tag{8.34}$$

An example of a power spectrum is shown in Fig. 60, where the spectrum includes both signal and noise.

There are many methods that can be used to determine the location $\hat{f}_i$ that corresponds to the peak value of the wind component $\hat{u}_i$ of a frequency spectrum. For example, we can use the average frequency

$$\hat{f}_i = \sum_{k=1}^{N_S} f_{ik} R_{ik} / \sum_{k=1}^{N_S} R_{ik} \tag{8.35}$$

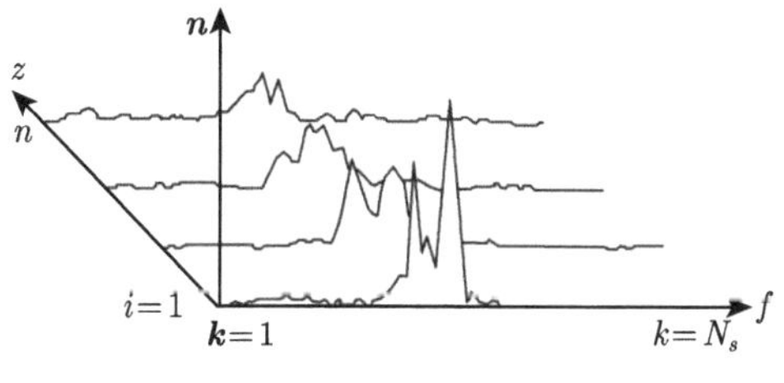

Fig. 60: A dispersing sound-power spectrum.

where f_{ik} can be regarded as a measured value to estimate $\hat{f}_i$ and as a function of the power-spectrum value R_{ik}.

8I.4.2 Spatial resolution of Doppler frequency spectrum

If we perform a Fourier transform over Eq. (8.31), we obtain

$$A(f) = P(f) \int_{-\infty}^{\infty} \sigma_S(z) \exp\left(2\pi i f \frac{2z}{c}\right) dz = P(f)S(k) \tag{8.36}$$

where $P(f)$ is the Fourier transform of $p(t)$, $S(k)$ is the spatial transform of $\sigma_S(z)$ and $k = 4\pi f/c$ is the spatial wave number. The power spectrum is $R(f) = A(f)A^*(f)$. For a pulse with maintaining time τ and with a rectangular envelope, the bandwidth of $P(f)$ is $\Delta f \approx 2/\tau$. That means that only when the spatial length is $2\pi/k = \perp c\tau/2 = \perp \Delta z_\tau$ will the contribution of $S(k)$ to $R(f)$ be useful. Thus the spatial resolution is determined by the larger of the two values Δz_τ and Δz_S.

If the Doppler frequency drift is varies with height, the transformation of Eq. (8.36) is not accurate. For a pulse with sine wave frequency f, a maintaining time τ, and which has a simple rectangular envelope, the Doppler frequency spectrum after demodulation is

$$P(f') = \frac{\sin\left(\pi[f' - f_D]\frac{f}{f_0}\tau\right)}{\pi[f' - f_D]} \tag{8.37}$$

where f_D is the Doppler frequency drift, $f_0 = f + f_D$. The Doppler frequency generated by wind velocity u is:

$$f_D = \frac{2f}{c}u \tag{8.38}$$

If the sampling in one measurement interval includes the contribution of many Doppler components, the obtained frequency spectrum is composed of all of the spectra centralized at different f_D.

$$A(f') = \int_{z_i-\Delta z_i/2}^{z_i+\Delta z_i/2} \sigma_S(z) \frac{\sin\left\{\pi[f' - f_D(z)]\frac{f}{f_0}\tau\right\}}{\pi[f' - f_D(z)]} dz \tag{8.39}$$

The spatial resolution of the velocity frequency is given by the following convolution

$$\rho(z) = \frac{\sin\left(\pi\frac{z}{\Delta z_\tau}\right)}{\frac{\pi}{\Delta z_\tau}} \times \frac{\sin\left(\pi\frac{z}{\Delta z_S}\right)}{\frac{\pi}{\Delta z_S}} \tag{8.40}$$

8I.4.3 Modeling of wind velocity profile

The above-mentioned method shows that the frequency spectrum value R_{ik} is determined at frequency f_{ik} and altitude z_i. But in fact, R_{ik} is also relevant to other spectral values at different heights. One way to process this kind of problems is to write the measuring values at different altitude $z_j (j = 1, 2, \ldots, n)$ as the sum of the weight composed of the unknown amplitude σ_j and Doppler frequency drift u_j as the following

$$R_{ik} = \sum_{j=1}^{n} \left\{ \sigma_j \frac{\sin\left(\pi\left[f_k - \frac{2f}{c}u_j\right]\tau\right)}{\pi\left[f_k - \frac{2f}{c}u_j\right]} \frac{\sin(2\pi[z_i - f_D(z)]/cT}{\pi[z_i - z_j]} \right\}^2 \tag{8.41}$$

and by inversion we can obtain the unknowns. The difficulty of the method lies in that the dependency of the weighed function on u_j has a very strong non-linearity. One of the adoptable methods is to take the frequency f_{ik} as a measurement value, and the accuracy in some respects depends on R_{ik}. For example, if we count the analyzing signals by frequency, then the largest R_{ik} will decide the measurement results of f_{ik}. The general model of f_{ik} relies on the estimated value of $u_j (j = 1, 2, \ldots, n)$ where the velocity component can be written as

$$\boldsymbol{f} = F(z, \boldsymbol{u}) + \boldsymbol{\varepsilon} \tag{8.42}$$

The measurement noise is indicated by vector $\boldsymbol{\varepsilon}$, and the function $F_k(z, u)$ describes a model for the change of the wind speed component on the vertical direction. A linear model is

$$\boldsymbol{f} = \boldsymbol{K}(z)\boldsymbol{u} + \boldsymbol{\varepsilon} \tag{8.43}$$

where $\boldsymbol{K}(z)$ is a weight function. Thus the equation of dispersion is

$$f_{ik} = \sum_{j=1}^{n} K_{ijk}u_j + \varepsilon_{ik} \quad (i = 1, 2, \ldots, m; k = 1, 2, \ldots, N) \tag{8.44}$$

If we assume that the detection of the spectrum peak is a single step to complete, and we then find each $\hat{f}_i$, then reversing equation (8.44) to determine the wind component $\boldsymbol{u}$ we just use the ordinary least-squares method. We then find an appropriate physical model $\boldsymbol{K}$ and an estimated value $\boldsymbol{u}$ of the velocity component that can exist physically. In examples of sound radar profiles, we can add binding factors (such as smoothness or restricting shear), then the problem becomes one of confined linear retrieval. For example, a common added binding factor can be used to formerly measure the value $\boldsymbol{u}_a$ of the velocity vector of wind profile using standard data. There will then be a covariance of relevant $n \times n$.

$$\boldsymbol{S}_\alpha = (\boldsymbol{u} - \boldsymbol{u}_\alpha)(\boldsymbol{u} - \boldsymbol{u}_\alpha)^{\mathrm{T}} \tag{8.45}$$

If the error of a measured speed vector value is $\pm\sigma_a$ and we assume that $\boldsymbol{S}_a$ is diagonal to element σ_a^2, then the covariance S_ε of the measured error in $mN_s \times mN_s$ will also be diagonal to element σ_{ik}^2. Thus the expected value of the velocity vector is

$$\hat{\boldsymbol{u}} = \boldsymbol{u}_a + \hat{\boldsymbol{S}}\boldsymbol{K}^T\boldsymbol{S}_\varepsilon^{-1}(\boldsymbol{f} - \boldsymbol{K}\boldsymbol{u}_a) \tag{8.46}$$

And the reversion of $n \times n$ covariance matrix is

$$\widehat{S}^{-1} = K^T S_\varepsilon^{-1} K + S_a^{-1} \tag{8.47}$$

By the n singular values of $\tilde{K} = S_\varepsilon^{-1/2} K S_a^{1/2}$, λ_i can be used to determine the SNR and the values of the independent velocity vectors in profile. A strangeness value of greater than 1 indicate that the measured value above the level of the noise. We then computing these values can give the number of independent velocity components one is able to measure above the noise level. The SNR is equal to λ_i at an altitude of z_i, while the number of degrees of freedom of the signal is

$$d_s = \sum_{i=1}^{m} \frac{\lambda_i^2}{1 + \lambda_i^2} \tag{8.48}$$

The number of degrees of freedom of the noise is

$$d_n = \sum_{i=1}^{m} \frac{1}{1 + \lambda_i^2} \tag{8.49}$$

From the above discussion we can see that the factors that determine the accuracy of velocity profile are the following: S_ε represents the noise of measured power spectrum value R_{ik}, S_a represents the uncertainty in former profile data, and K represents a model used for acquiring smooth profiles.

81.4.4 Weighted-function and covariance

The model represented by Eq. (8.43) does not include interrelations at the various altitudes. For example, the weighted-function K may be an interpolation function, so the wind velocity is estimated at different heights determined by the measuring intervals.

$$K_{ijk} = \frac{2fT}{c} \left\{ \frac{\sin(\pi[z_i - z_j]/\Delta z_s)}{\pi[z_i - z_j]/\Delta z_s} \right\}^2 \tag{8.50}$$

As mentioned above, the uncertainty of the actual velocity component represented by any f_{ik} is in some way relevant to R_{ik}. For example, we can use the following equation to establish the relation

$$\sigma_{ik}^2 = R_{ik}^2 \tag{8.51}$$

Rodgers [233] gave a covariance matrix that was based on the Markov process, which has actual physical meaning

$$S_{ajk} = \sigma_a^2 e^{-|\zeta - j|\delta z/h} \tag{8.52}$$

here δz is the interval of the layer orders of u in the vertical direction, and h is the vertical interrelated length of the velocity component. Eqs. (8.50)~(8.52) describe all of the interrelations caused by the atmosphere and the instrumental characteristics.

8I.4.5 Application examples

The following example is a practical application of this method. The related parameters are shown in Table IV. Here an assumption of standard measuring data is a zero-velocity profile. The measured frequency-spectrum data of the wind velocity by means of wound radar Aero Vironment 4000 are shown in Fig. 61. In this example, the Doppler velocity is a bit smaller – the typical velocity is 1 m s^{-1}. Fig. 62 is shown in grey scale in order to show the frequency-spectrum data of the same profile. The curve in the middle of the figure is the wind profile curve acquired by bassing on applying the assumption in Eq. (8.51) and rerersing from Fig. 61.

There are about 24 independent information that can acquired by using Eq. (6.51). The information is of great significance for interpreting sound radar data because

Tab. IV: Parameters used in reversion

c	sound velocity	340 m s^{-1}
F_s	sampling rate	960 Hz
f	transmitting frequency	4500 Hz
h	vertical related length of the wind velocity	10 m
m	number of measurement intervals	40
n	number of velocity components	30
N	number of sampling in one measurement interval	64
T	time of sampling in one measurement interval	67 ms
Δz_τ	resolution caused by the pulse length	8.5 m
Δz_s	resolution caused by the sampling length	11.3 m
δz	vertical interval of the velocity estimation	7.6 m
τ	pulse length	50 ms
σ_a	preset standard parameters	15 Hz

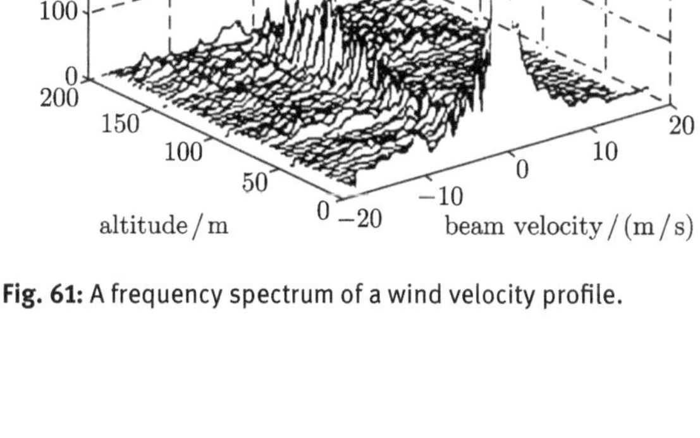

Fig. 61: A frequency spectrum of a wind velocity profile.

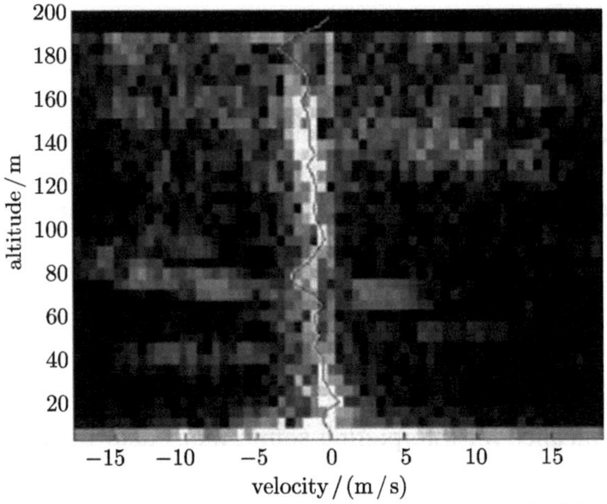

Fig. 62: Data from Fig. 61 expressed in grey scale. The curve in the middle of the figure is the wind velocity profile achieved by reversion

the method separates the factors of the instruments we used from the actual atmospheric behavior. Some assumptions have been made in this method, for example, the assumption of the dependency of linearity on wind velocity, and the assumption of the form and relevant length of the covariance matrix. But in this method we can certainly guarantee that the influence of these assumptions to the calculated errors can be accounted for.

8I.5 Passive remote sensing

As pointed out in the preceding part of this chapter, remote sensing of the lower atmosphere is dominated by active sensing, however, passive sensing certainly still exists, and the most important example should be that for thunder within the audible frequency band.

Thunder is a spectrum of sound generated by lightning in the atmosphere. The length of time stretched out depends on the differing transmission path length from the extended channel source. If an array of microphones is placed on the ground, it is possible to reconstruct the shape of the lightning channel by correlating the direction cosines and the arrival times of identifiable temporal segments of the thunder. This technique has been proven feasible in a continuing series of investigations [234]. Typically, 3 or 4 microphones are laid out in each array; a Y-shaped array with spacing of 30∼100 m between microphones has proved to be most effective. Owing to lots of kinking and bending natures of the lightning channel the acoustic signals vary with time violently [235]. Of course, sound that arrives from more distant portions of the channel will be attenuated at higher frequencies due to molecular absorption in the atmosphere. Typical power spectra that corresponds to differing arrival times have already been obtained [235].

When evaluating thunder data, it is important to know the temperature- and wind-profiles in order to reconstruct the direction of the sound rays back to their point of origin. The method of analysis and the expected errors in establishing the lightning channel geometry are discussed in [236] from data gathered by using a triangular array of 3 microphones placed 100 m apart. Typically, the horizontal extent of a lightning discharge will be three times the vertical extent. The accuracy with which the channels can be reconstructed has been confirmed by overlaying acoustic records onto photographs of visible portions of cloud-to-ground strokes. Typical positional errors seem to be within 4% when wind- and temperature-profiles are well known. When supportive wind measurements are not available, errors may be about 10%.

As a result of acoustic thunder research, we have learned that lightning is predominantly horizontal and is concentrated to a zone within clouds associated with freezing. A positive correlation between radar reflectivity values within the cloud and lightning activity has also been found. Thus this acoustic method has made an impact in an important area of cloud physics.

With regards to passive remote sensing that involve infrasonic waves that exist extensively in the atmosphere, these will be discussed in combination with the remote sensing for the upper atmosphere in the next part.

Part Two Acoustic remote sensing for the upper atmosphere

Owing to the fact that the atmospheric absorption of acoustic waves is directly proportional to the square of frequency, audible sound cannot reach the upper atmosphere (usually defined as above 50 km) from the ground with currently available and technically feasible power. Therefore the task of probing the upper atmosphere falls naturally to infrasonic waves. Also, due to limited technical levels, however, it is still difficult to emit infrasonic waves of sufficiently high power towards a given height in the atmosphere artificially, thus remote sensing of the upper atmosphere is almost always confined to passive sensing only. Fortunately, as it happens, there is an abundance of infrasonic waves arising from various sources in the atmosphere, therefore, a lot of information corresponding to these waves can be achieved according to passive sensing.

Remote sensing of the upper atmosphere of course has something in common with that discussed in Sec. 1 of this chapter (i.e. the lower atmosphere), but there also exists some differences in principle. The main difference lies in that the physical foundation is no longer scattering, but instead is in the refraction of acoustic waves. On the other hand, since the ionosphere is involved, in addition to measuring acoustic waves directly, radio sounders can also be used to detect any motions induced by acoustic waves via upper atmospheric ionization. Furthermore, acoustic waves and acoustic gravity waves can also affect the propagation of radio waves in the ionosphere through interactions with upper-atmospheric ionization. Of course, within the scope of this book, we consider only infrasound that passes through the upper atmosphere, which has been refracted back to the ground.

8II.1 Physical foundations of acoustic remote sensing for upper atmosphere

8II.1.1 Refraction

The physical foundation of remote sensing of the upper atmosphere is mainly via refraction induced by macroscopic structures within the atmosphere. The "abnormal" propagation phenomenon discussed in chapter III was attributed to just the bending (refraction) of sound waves back to the ground by upper atmospheric wind- and temperature-gradients. Because acoustic wavelengths (except for infrasound with extra low frequencies) are generally much smaller than the scales of variability within the atmosphere, sound propagation in the upper atmosphere can be treated from the viewpoint of ray acoustics and the refractive index structure (which corresponds to the macroscopic structure mentioned above) of the atmosphere.

The real part of the acoustic refractive index, which leads to the bending of acoustic rays, depends mainly on air temperature and secondarily on winds. Utilizing the horizontally stratified atmospheric model, a simple interpretation of refractive effects using Snell's law is permitted. Except for waves with frequencies small enough that the force of gravity must be taken into account in the equation of motion (as the condition considered in chapter VI), acoustic waves in the atmosphere are non-dispersive, i.e. the real part of their refractive index does not depend on their frequency. Therefore, the bending computed for one frequency will be applicable for all frequencies.

Atmospheric winds make the acoustic refractive index anisotropic, i.e. it depends on the direction of propagation. In fact, the real part of the acoustic refractive index can be written as follows

$$n = c_0/(c + \boldsymbol{v} \cdot \boldsymbol{k})$$

where c_0 is the reference sound speed, c is the local sound speed of air at rest, $\boldsymbol{v}$ is the wind velocity and $\boldsymbol{k}$ is a wave vector.

The case of an atmospheric model without winds is relatively easy to compute. The corresponding acoustic ray pattern can be obtained from acoustic sources at various heights, and thus can show how the temperature structure within the mesosphere and lower thermosphere bends rays back towards the ground, and even causes ducting at long distances [237].

The refraction caused by realistic atmospheric winds is more complicated, so that the propagating model must generally be done on a case-by-case basis. Cross winds deflect sound waves towards the azimuth, and winds along the propagation direction affect their travel time and vertical arrival angle. In [238], a model representing the average seasonal height variability of atmospheric winds was created and used to compute the bearing and travel time perturbations that they impose on long-range sound transmission. It was found that the typical bearing deviation was 5° and the group velocity variation was 50 m/s.

Acoustic ray-tracing computer programs such as HARPA (HAmitonian Ray-tracing Program for Acoustic-waves) [239] can now accept continuous three-dimensional models of atmospheric winds and temperature, as well as models of irregular reflecting terrain, and can compute acoustic ray paths by numerically integrating Hamilton's equations (differential expression of Fermat's principle). Such programs can account for horizontal refractive structure, which bends rays in the horizontal plane and affects the spatial coherence of waves [202].

Acoustic propagation to long distances at low frequencies is often more suitably modeled in terms of waveguide modes, which must include effects induced by upper atmospheric temperatures and winds [240]. Theory and observations of long-range ducted acoustic and internal-gravity wave modes in the atmosphere are reviewed in [241].

8II.1.2 Absorption

The imaginary part of the acoustic refractive index determines the rate at which the wave energy is absorbed by the atmosphere. Understanding how absorption affects low-frequency acoustic waves in the upper atmosphere is important, because it determines which frequencies can reach which heights.

In most cases, classical viscous dissipation and thermal conduction mainly dominate the absorption process. This is just the opposite of the case for high-frequency sound, in which the dominant absorption mechanism is molecular absorption (see chapter V). Of course, the largest non-classical contributions are still from rotational and vibration transitions in O_2 and N_2 [242], but these are normally considered only in precise estimates. Dissipation relations for frequencies in the acoustic gravity wave regime (wave periods longer than about 5 min) is derived in [243]. For purely acoustic waves in a non-turbulent atmosphere, however, reasonably accurate absorption estimates can be calculated using the formula for the amplitude attenuation coefficient a_{cl} (see Eq. (5.18)) for sound propagation in a perfect gas [244].

We see that the absorption, unlike the real part (the "bending part") of the refractive index, depends on the acoustic frequency: higher frequencies are absorbed more strongly than lower frequencies. This, together with the inverse dependence on atmospheric density, accounts for the fact that only acoustic waves with the lowest frequencies can reach the upper atmosphere.

As a rule-of-thumb example, the attenuation rate for an acoustic wave with a 1 Hz frequency at a 100 km altitude is about 0.1 dB/km [245].

8II.1.3 Inferring upper atmospheric properties from acoustic measurements

Snell's law is applicable to the atmosphere providing that the latter is horizontally stratified (see section 2, chapter III). Then, it is possible to deduce the sound speed at the apex (turning point) of an acoustic ray path by measuring the horizontal trace velocity, i.e. the speed at which an acoustic crest travels, along the ground. The vertical angle of arrival can then be measured equivalently. The remaining question is then: what height does the measurement apply to, i.e. at what height did the ray turn around? One way to answer this question is epeated trials, i.e. by assuming simple atmospheric temperature models and performing acoustic ray tracing accordingly [244].

Another way to recover temperature information is to perform an inversion using the Abel transform relation between the sound velocity profile and the dependence of the travel time (and thus the ground range) on the sound velocity at the ray turning point [239]. In principle, the upper atmospheric temperature profile can be recovered, but in practice, only a few points on the profile could be obtained from measurements at one range, because only two or three multi-path rays can, in general, only reach a given range. Measurements at multiple ranges can increase the profile resolution, although the effects of upper atmospheric winds make the measured sound velocity is "effective sound velocity", i.e. the sum of the sound velocity in the atmosphere at rest and the wind component in the propagation direction. Wind effects can be separated from those caused by temperature either via a horizontal ring of receivers, or by transmitting in reciprocal directions; however, both arrangements are probably too elaborate to be performed practically.

8II.2 Detecting systems for remote sensing

Acoustic-gravity wave sensors have historically been of the microbarograph type. Most modern microbarographs record changes in atmospheric pressure p relative to the pressure in a "reference volume" (a stable pressure chamber which is normally vented to the atmosphere through a "slowly leaking tube"). This leak provides a high-pass filtering action, which causes the output frequency of instrument to indicate $\mathrm{d}p/\mathrm{d}t$ which frequencies are below a "roll-off frequency" (determined by the time constant of the leaking tube) and p above that frequency. Another result of high-pass filtering is to "pre-whiten", or compensate for, background atmospheric pressure fluctuations (which increase in intensity at longer periods). In many operational systems, it is desirable to record $p(t)$ rather than dp/dt, so the microphone output should be filtered in such a way that the entire system response is relatively flat over some desired band.

Microbarographs are usually connected to "noise reducing pipes", which are typically about 300 m long with small openings every 1.5 m or so, which serve to average the sensor response over a specified spatial area, and thus filter out smaller-scale pressure fluctuations, which are presently regarded as "noise". Such filters also reduce the sensor response (above about 1 Hz) to ordinary acoustic noises. To be sure, the definitions of "signal" and "noise" are arbitrary and depend on where the attention will be focused at any given time; thus, spatial and temporal filters can be designed to allow certain fluctuations to pass and to reject those that are not of immediate interest, which are labeled as "noise". A simplified schematic diagram of a modern microbarograph system is shown in Fig. 63 [246].

During the passage of atmospheric waves, the atmospheric temperature, density and particle velocity, just the same as the pressure, produce fluctuations, but pressure is the quantity most wave sensors have been designed to sense. The reason seems to be partly historical, but also basically a question of instrument sensitivity. Pressure sensors have now been developed that can measure pressure changes of the order of 10^{-8}, or 0.01Pa under standard pressure. The accompanying temperature and density fluctuations are of the same order of magnitude. To match microbarograph sensitivity in terms of detectable sound waves, a temperature sensor would have to measure changes of the order of 8×10^{-7} K; a density sensor would have to measure fluctuations of about one part in 10^8; and a wind sensor would have to measure speeds of about 2×10^{-6} m/s, all well beyond the range of current field instruments. Microbarographs thus appear to be the most sensitive instruments presently available for the passive sensing of the higher-frequency portion of an acoustic gravity wave spectrum (whose periods between about 1 s and a few minutes).

At high (audible) frequencies, the ratios of wave-associated fluctuations in pressure, temperature, density and particle velocity remain constants. For example, the ratio of fluctuations in pressure and in particle velocity is just the characteristic impedance $\rho_0 c$ of the atmosphere, where the sound speed c is equal to the phase velocity ω/k. Just this fact allows us to make the above estimations of wave-associated fluctuations of various parameters without specifying the wave frequency, the wave vector magnitude or the direction. For longer wave periods, however, gravitational anisotropy strongly influences wave propagation, so that the phase velocity ω/k no longer equals the sound velocity c, but depends in a complicated way on wave frequency, wave vector magnitude and direction (see chapter VI). Therefore, the wave-variable ratios exhibit the same complicated behavior correspondingly. In general, the phase velocities of gravity waves are smaller than the sound velocity, and so too are wave-associated pressure fluctuations relative to velocity fluctuations. For example, in a particular asymptotic limit that corresponds to long-period gravity

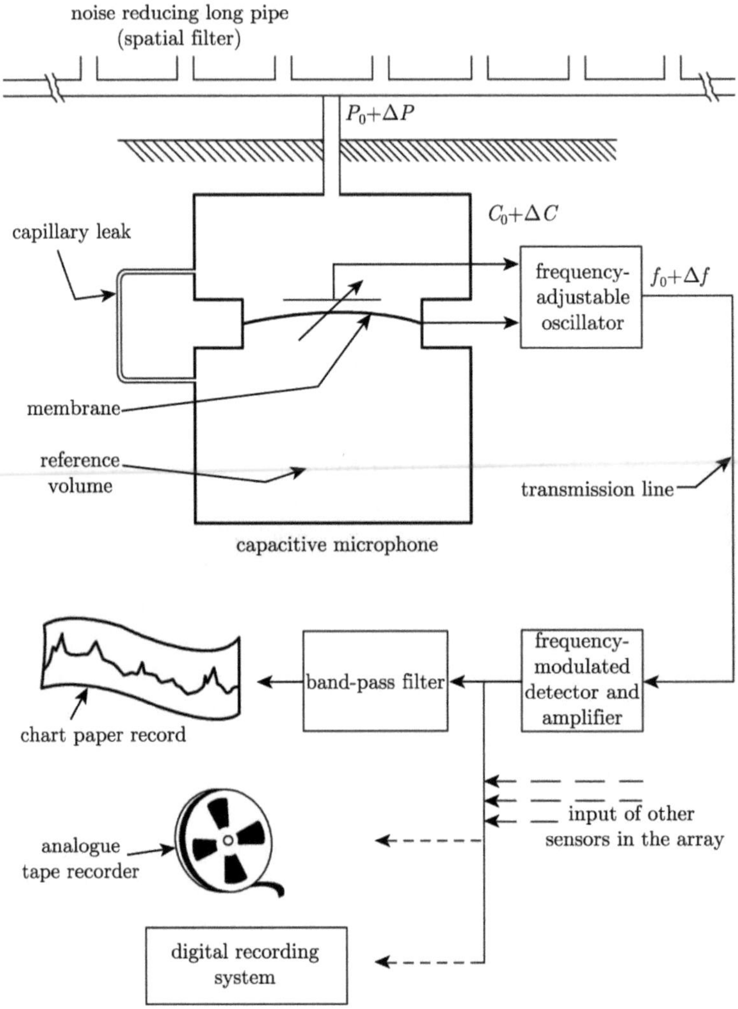

Fig. 63: A system for recording microbarometric pressure fluctuations.

waves whose vertical wavelengths λ_z are much smaller than the atmospheric scale height, $\omega \sim \omega_B(k_x/k_z)$, so $p/v_x = (1/2\pi)\rho_o\omega_B\lambda_z$ (where ω_B is the V-B frequency defined in Eq. (1.28)). This shows that pressure fluctuations (relative to velocity fluctuations) decrease as vertical wavelength becomes smaller. For a given horizontal wavelength (or horizontal phase-velocity), decreasing λ_z corresponds to lengthening wave periods.

Therefore, it would seem profitable to explore the capabilities of temperature, density or particle-velocity sensors when obtaining observations of internal gravity waves (such as atmospheric waves from natural sources, see section 10.2, chapter X). Particle-velocity sensors with sufficient sensitivity and frequency selectivity can provide information regarding the wave travel direction as well. It should be realized, of course, that the frequency range in which auxiliary sensors might provide increased wave observation capabilities is determined not only by instrument sensitivity but also by the levels of extraneous fluctuations in the quantities being measured. These "noise levels" are determined experimentally.

Because microbaro graphs are usually located at the earth's surface, the possible boundary-layer and other surface effects on the pressure-sensor response should be understood. It is clear that no vertical particle motions can exist at the surface. This fact alone requires that plane wave associated pressure perturbations (due to an incident and a reflected wave) double as compared to the magnitude of an incident plane wave alone. Near the surface, however, eddy viscosities play a complicated role in determining the wave-associated horizontal particle motions (and possibly the existence of boundary waves), and hence the pressure there. The details of this role, especially as it depends on surface roughness, are not yet as well understood as they should be if we are to understand exactly what surface-pressure sensors actually measure.

8II.3 Recognition of waves in the atmosphere

Not all pressure fluctuations in the atmosphere are "waves". For example, pressure fluctuations are also caused by non-propagating atmospheric motions such as advected "turbulence" are "non-waves". Clearly, such distinctions must rely on spatial as well as temporal pressure sampling, and criteria for assessing the degree of confidence for wave identification must be established.

To receive real waves, using only a single sensor is simply out of the question; instead an array must be used. In systems designed to measure pressure-wave properties, microbarographs are usually positioned on the ground to form a two-dimensional spatial array, in which individual sensors may be located from hundreds of meters to tens of kilometers apart, depending on the wavelengths and spatial coherence of the waves to be detected. Of course, if the sensors are located above the ground to form a three-dimensional array, valuable additional wave information will be obtained. However, the requirement of very high supporting structures and mechanical problems associated with wind noise reduction has retarded their development. The relative times of arrival of an identifiable pressure event at multiple sensors can then be used to determine the speed and direction of travel of that event past the array, where more sophisticated correlation techniques can give more accurate results.

Correlation of pressure signatures from spaced sensors does not, of course, automatically distinguish "wave" from "turbulence" or "noise". If relative arrival times of an event indicate a travel speed greater than the local sound speed, it is unlikely that the event could be anything but an acoustic wave, and such events are usually interpreted in terms of the "trace" of an acoustic wave arriving obliquely at the ground. But, in the case of more slowly moving events, the distinction between "wave" and "noise" is readily made only when spatial coherence is high enough and the waveform period is strong enough. As spatial coherence deteriorates, however, one is soon faced with conceptual and semantic problems related to the definitions of "wave" and "turbulence", and ultimately, with the task of imposing arbitrary criteria for distinguishing the two. Analytical tools that permit the choice of numerical criteria are provided by cross-correlation techniques, including cross-power spectral analysis.

Cross-correlation analysis [247] is perhaps the most commonly used technique for studying motion fields sampled at spaced sensors. The cross-correlation (sometimes also called the cross-covariance) function for two time series $x(t)$ and $y(t)$ is

$$\Phi_{xy}(\tau) = \lim_{T\to\infty} \frac{1}{2T} \int_{-T}^{T} x(t)y(t+\tau)\mathrm{d}t$$

Here the two functions are multiplied and averaged over a sample length $2T$, while the time shift t is varied. The higher the degree of correspondence of the two functions, the larger their averaged product Φ_{xy} will be. If $x(t)$ and $y(t)$ are data from spaced sensors, the value of τ for which $\Phi_{xy}(\tau)$ is maximized is taken as the time displacement, or the signal travel time between the two sensors. The value of τ for each pair of sensors in a network can be interpreted in terms of the trace velocity of a signal along the line joining the pair. When each trace velocity is plotted as a vector from a common origin and the tips of all the vectors align, then the analysis yields a single-plane-wave solution with a high degree of confidence (see Fig. 64).

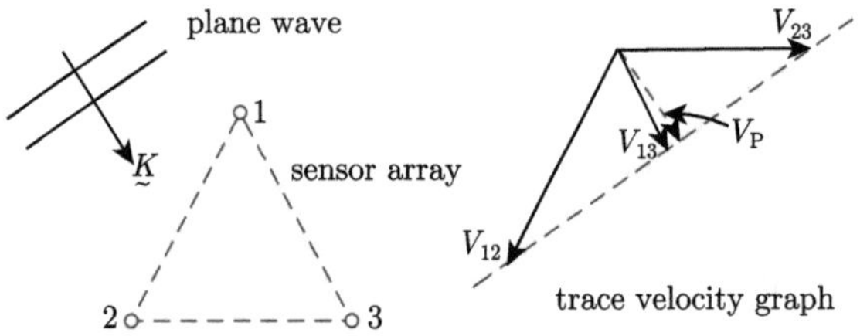

Fig. 64: Construction of the wave phase velocity vector from trace-velocity vectors measured across sensor pairs.

However, when records contain several frequency components of differing velocities, using the cross-correlation analysis alone may yield grossly misleading information about the patterns of movement [248]. The various frequency components may be separated by taking a Fourier transform of the cross-correlation function, which is complex if the cross-correlation function is asymmetrical about $\tau = 0$. This new function is called the cross-power spectral density, or simply the cross spectrum

$$\Phi_{xy}(\omega) \equiv C_{xy}(\omega) + \mathrm{i}Q_{xy}(\omega) = (1/\pi) \int_{-\infty}^{\infty} \Phi_{xy}(\tau)\mathrm{e}^{\mathrm{i}\omega\tau}\mathrm{d}\tau$$

where $C_{xy}(\omega)$ and $Q_{xy}(\omega)$ represent the in-phase (co-spectrum) and quadrature (phase difference equals to 90°) components of the complex spectrum of $\Phi_{xy}(\tau)$, respectively.

The quantity "coherence" is simply the normalized magnitude of the squared cross-spectrum

$$\mathrm{coh}(\omega) \equiv (C^2 + Q^2)/E_x E_y$$

where E_x and E_y are the power spectra of the x and y functions above. The phase angle of the cross-spectrum yields the time lag τ_0 between the two records as a function of ω

$$\omega\tau_0 = \arctan[Q_{xy}(\omega)/C_{xy}(\omega)]$$

Which in turn permits the computation of the velocity and the angle of arrival as a function of frequency.

Put as simply as possible, the squared magnitude of the cross-spectrum (the coherence) describes the accuracy with which a given set of multi-sensor data can be represented by the superposition of a number of pure plane waves. The phase angle of the cross-spectrum permits the computation of the velocity and the angle of arrival of each spectral component.

Analyzing the cross spectra from sensor pairs can, in principle, help distinguish between fluctuations caused by superimposed wave motions from those due to drifting or wind-carried eddies [248]. If the fluctuations are entirely due to wave motions, one would expect to observe different frequency components in the records moving with different speeds and in different directions. On the other hand, if eddies are "frozen in" the mean flow, one expects that all spectral components will have the same velocity, i.e., that of the mean flow. In the case of atmospheric gravity waves, however, wind speeds may often approach (or even exceed) wave propagation speeds, so that observed wave propagation speeds and directions may be closely related to the local wind conditions, and therefore, may be interpreted as advected turbulence. It is not certain whether such waves can be distinguished from other fluctuations; it may be possible to utilize the theoretically predicted phase relationships among pressure, density, temperature and wave-associated particle velocity to identify them.

8II.4 Passive remote sensing of infrasonic waves existing objectively in atmosphere

Waves of every hue exist almost ubiquitously in the atmosphere. Besides audible sound waves in smaller scopes, most of which are regarded as "noise" under the situations considered here, infrasonic waves, including infrasound in the general sense, acoustic gravity waves, internal gravity waves etc., are dominant. Owing to their weak attenuation in the atmosphere, these waves can exist at larger spatial scopes (up to the earth's whole atmosphere) and longer durations. The sources and properties of these waves are their distinguishing features, and are classified as either artificial and natural types via their association with definite mechanisms and phenomena. By means of their acceptance, location and analysis, one can obtain information related to their corresponding phenomena. For those associated with disastrous phenomena,

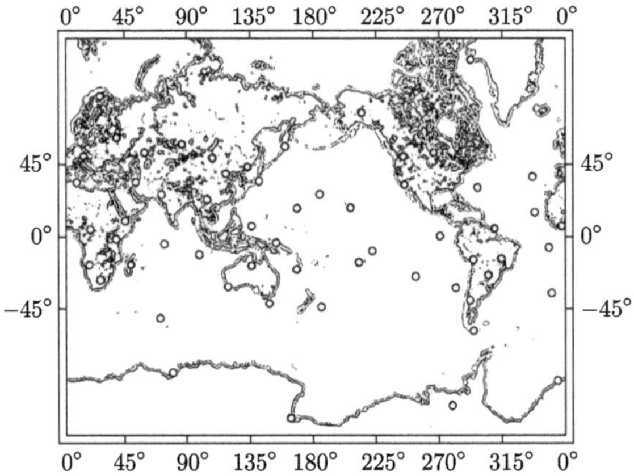

Fig. 65: Global infrasonic monitoring network. The small circles represent the measuring stations.

which make up the greater part, a definite value for prediction exists. Natural sources include atmospheric turbulence, bolides, aurorae, earthquakes, volcanic eruptions, ocean storms, waves related to mountain ranges, and snow avalanches. Artificial sources include supersonic aircraft or rockets, and industrial or nuclear explosions. The various atmospheric sound sources will be discussed in detail in chapter 10.

8II.4.1 Global infrasonic monitoring network [249]

There may be many reasons to generate infrasonic waves in the atmosphere. Among them, since the 1960s, infrasonic waves caused by nuclear explosions (in recent years along with the prohibition to atmospheric nuclear test, the attention to it has been divided) has received world-wide attention. By monitoring infrasonic waves, a nuclear explosive test can be discovered. In order to effectively find evidence of a nuclear explosive test, many countries in the world have united to establish an international infrasonic monitoring system (IMS). The infrasonic monitoring network includes about 60 monitoring arrays that are well-distributed around the world, as shown in Fig. 66. The measuring range of each array is about 1∼3 km, including 4∼8 continuously working monitoring stations. Each monitoring station can also detect speed and the direction of wind except for monitoring infrasonic signals. Infrasonic signals are detected via a microbarometer. The sampling rate of infrasonic waves is 20 pressure data per second, and the sampling rate of wind is one data per second. The infrasonic signals are processed with a spatial filter to enhance the SNR. The data sampled by each monitoring station of the arrays are transmitted in real-time to a central processor, then via satellites they are transmitted to the international data processing center at Vienna, Austria. At the international data processing center, large volumes of data are need to distinguish between the properties of infrasonic source and to determine the location of each infrasonic source.

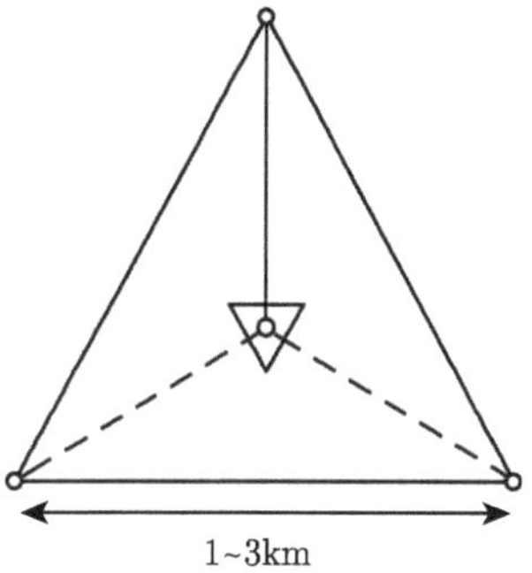

Fig. 66: A typical infrasonic monitoring array. The real form of an array is not regular, and not must be triangular.

Global infrasonic monitoring is a very complicated task. Many thorny problems associated with technique and research are faced by acoustic scientists each day. These problems were discussed in detail in the previous chapters. In order to effectively monitoring infrasonic waves, three problems need to be solved: ① whether infrasonic sources are generated by nuclear or non-nuclear sound sources; ② the propagation of acoustic waves in the turbulent atmosphere; ③ to extract weak signals of distant sources from noisy data. First, we must find convincing evidence from infrasonic data in order to prove whether a source event is explosive or not. Once we exclude the possibility that an infrasonic signal was generated by a nuclear explosive test, the monitoring task is thus finished; however the monitoring data are still of great scientific value. Subsequently, we need to determine the locations and parameters of the infrasonic sources, include geographical location, elevation and sustaining time. In order to determine the infrasonic location from received acoustic signals, we need to take advantage of an atmospheric model in the retrieval operation. Some of the currently accepted acoustic propagation models are empirical models for analogue acoustic propagation in an atmosphere. The above mentioned content includes, as discussed in previous chapters, the ray path method, the normal mode theorem, the two-dimensional parabolic equation (PE) method, etc.

Another problem is how to sample useful acoustic signals in noise. Currently, in the global infrasonic monitoring system, each typical monitoring station in the array is a spatial filter. Atmospheric turbulent noise is in the range 0.05~1 Hz. This frequency band covers the whole frequency range of interest in nuclear monitoring. We know that infrasonic signals at distances of 10~100 m are coherent, while infrasonic noise below 10 m that are generated by atmospheric turbulence near to the stations are incoherent. In the 1940s~1950s, Daniels proposed a method to decrease noise, whereby he superposed sampling signals from several sensors in a certain range upon the noise to increase the SNR. Increasing SNR in higher frequency range is a problem that still awaits a solution. Applying a part of operating network, we have successful monitoring results. For example, the large-scale explosion of a meteor shower that occurred somewhere between Hawaii and California, USA, on April 23 in 2001 was monitored by the infrasonic array, at a distance of 11000 km, in South Germany

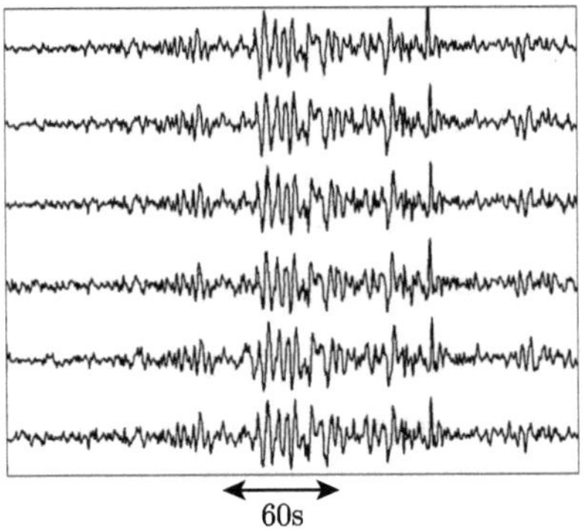

Fig. 67: The monitoring record in California, which shows high correlation. The burst of a meteor shower whose bursting point, which was 1800 km apart from the monitoring station in California, probed by a German monitoring station 11000 km beyond the burst point.

(see Fig. 67) [250]. The tropical storm "Barbara" in Pacific area in June, 2001 was monitored by the IMS infrasonic array in California, Hawaii and Alaska.

8II.4.2 Some prospects

Probing high-level atmospheric structure and dynamics using sound waves has matured technically over the years. However, to date most observational works have been purely phenomenological, while theoretical advances were mainly involved with "forward" propagating modeling. For future investigations, the practical tendency should be towards an improvement of low-frequency coherent sources, a more elaborate design of receiving arrays, and appropriate retrieval methods.

The concept of using RASS (Radio-Acoustic Sounding System) to survey atmospheric temperature has been developed in many countries [251]. Its theory involves radar waves that arise from resonant backward dissipation of periodic density fluctuations caused by upward acoustic waves. The temperature profiles can then be deduced from the Doppler frequency spectrum of the radar echo. The highest altitude required by this kind of measurements is currently about 20 km, although the utmost altitude limit allowed by RASS technique is still unknown. There is a successful start for studying the stratified imagery method of troposphere and mid-level atmosphere as well, which is advancing continuously. At the present stage, investigations of acoustic wave propagation and dissipation in the atmosphere related with sound remote-sensing probing mainly include two aspects: one is the study of the statistical properties of sound waves when traveling through the inhomogeneous anisotropic atmosphere; while the other is the study of the effect of the spatial construction of a wave field in the vertical plane to the propagation direction.

With respect to atmospheric research, these problems are mostly related to the coherent construction of the intermittence of atmospheric turbulence, and mid-scale anisotropy. The foundations of the study of these constructions is to determine their effects on diffraction, refraction and non-linear phenomena of acoustic waves [252]. In order to further improve and develop the study methods of atmospheric acoustics, the above-mentioned problems demand advance research both theoretically and practically.

Chapter 9
Non-linear Atmospheric Acoustics

So far, we have discussed only sound waves with small amplitudes; thus the related equations could be processed linearly and simply. However, these cases are just an approximation, where in reality sound waves are in essence non-linear. Under some conditions, very small non-linear terms in the fluid dynamics equations may cause entirely-new and very important phenomena. For example, the chief conduct of shock waves, owes to very small non-linear disturbances which accumulate continuously and develop into shock waves. Another example is radiation pressure, which is a very small but nonzero scalar value applied to a substance by non-linear effect, and its existence is excluded from linear models.

9.1 Non-linear effects in sound propagation [56]

9.1.1 Plane waves in homogeneous media

Let us start with the dynamic equations of an ideal gas (not considering viscosity and other lossy processes). In order to easily clarify the basic concepts, we limit ourselves to the one-dimensional case. Then, the continuity equation (1.31′) and motion equation (1.30a) become, respectively,

$$\partial_t \rho + \partial_x \rho v = 0 \tag{9.1}$$

$$\rho(\partial_t v + v\partial_x v) = -\partial_x p \tag{9.2}$$

As before, the specific entropy s is always regarded as constant, thus the density ρ can be treated as a function of the total pressure p

$$\rho = \rho(p, s) \tag{9.3}$$

$$x = \text{const} \tag{9.4}$$

If we specify v as a single-value function of p, we then have $\partial_t v = (\mathrm{d}_p v)\partial_t p$, and so on. A special solution to the above equations is similar to a plane wave propagating in the $+x$ or $-x$ direction. Substituting this supposition into (9.1) and (9.2), we get

$$\mathrm{d}_p \rho \partial_t p + \mathrm{d}_p(\rho v)\partial_x p = 0 \tag{9.5a}$$

$$\rho \mathrm{d}_p v \partial_t p + (\rho v \mathrm{d}_p v + 1)\partial_x p = 0 \tag{9.5b}$$

If the coefficient determinant is equal to zero, the above two equations are entirely identical; this condition (with $\mathrm{d}_\rho p = 1/c^2$) causes $\mathrm{d}_p v = \pm 1/\rho c$. By selecting the positive sign to correspond to propagation in the $+x$ direction, we can then derive (9.5a) or (9.5b) in the non-linear partial differential equation

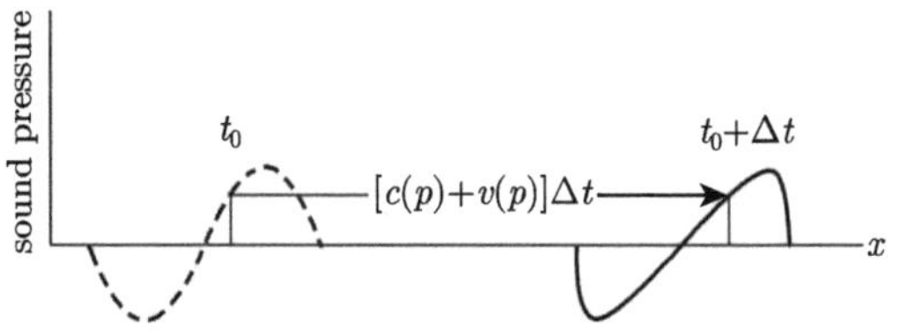

Fig. 68: The process of sound pressure waveform in a plane traveling wave. Each amplitude section traveling with specified speed $c(p) + v(p)$ relies on the amplitude

$$\partial_t p + (v + c)\partial_x p = 0 \tag{9.6}$$

Another processing method is as follows: Let us define $\eta(\rho) = \int\limits_{\rho_0}^{\rho} \frac{(\rho)}{\rho} \mathrm{d}\rho$, so we get $\partial\rho = (\rho/c)\partial_t\eta$, $\partial_x p = \rho c\partial_x\eta$ and so on, and Eq. (9.1) is derived into $\partial_t\eta + v\partial_x\eta + c\partial_x v = 0$, $\partial_t v + v\partial_x v + c\partial_x\eta = 0$. When $v = \eta$, we get the special solution (a simple wave)

$$\partial_t v + (v + c)\partial_x v = 0, \quad \partial_t p + (v + c)\partial_x p = 0$$

The result is the same as (9.6).[1]

The implication of (9.6) is that if we set $p(x_{\text{observe}}(t), t)$ to express the pressure at a moving observing point $x_{\text{observe}}(t)$, then p will be constant in time when $\mathrm{d}_t x_{\text{observe}} = v + c$. This time invariable can be obtained upon a comparison between equation $\mathrm{d}_t p(x_{\text{observe}}, t) = 0$ and (9.6). Because $v + c$ is a function of p, and p is a constant for the observer moving at speed $v + c$, so every point where the pressure amplitude p is fixed moves at a constant velocity (unrelated to time), although two points with different amplitudes will move with different speeds (see Fig. 68).

The parametric description of the solution is summed up to the details of $p(x, t)$ at time t. We set $p = p_0 + p'(x, t)$, while $p'(x, t) = f(x)$, then we have

$$p'(x, t) = f(\phi), \quad x = \phi + (v + c)t \tag{9.7}$$

$$p' = f(x - (v + c)t) \tag{9.7a}$$

where both v and c are taken values at $p_0 + f(\phi)$; at time t, the point on which p' is equal to $f(\phi)$ displaces a distance $(v + c)t$, beyond this distance x is ϕ.

For small amplitude sound waves, relations $\mathrm{d}_p v = 1/\rho c$ and $c = c(p)$ give

$$v \approx \frac{p'}{\rho_0 c_0}, \quad c \approx c_0 + (\partial_p c)_0 p' \tag{9.8}$$

---

1 See B. Riemann. *On the propagation of plain air waves of finite amplitude.* Abhandl. Ges. Wiss. Goettingen, 1860; This article was reprinted: The Collected Works of Bernhard Riemann, New York: Dover, 1953: 156~175.

Here we assume that the velocity of the surrounding fluid is zero. The differential quotient $(\partial_p c)_0$ (where the entropy is constant) is taken in the surrounding state circumstances, so it is a constant.

The two expressions in (9.8) can be combined into one

$$c + v \approx c_0 + \frac{\beta_0 p'}{\rho_0 c_0} \approx c_0 + \beta_0 v \tag{9.9}$$

where constant β_0 is

$$\beta_0 = 1 + (\rho c \partial_p c)_0 = \frac{1}{2}(\rho^3 c^4 \partial_{pp}^2 \rho^{-1})_0$$

(where the second form is derived from $\partial_p \rho^{-1} = -(\rho c)^{-2}$). Using another method, if p can be regarded as a function of s and ρ, then $\partial_p c^2$ is $\partial_\rho c^2/\partial_\rho p$ or $\partial_\rho^2 p/\partial_\rho p$. From this we can derive

$$\beta_0 = 1 + \frac{1}{2}\frac{B}{A}, \quad A = (\rho \partial_\rho p)_0, \quad B = (\rho^2 \partial_\rho^2 p)_0 \tag{9.10}$$

where A and B are the coefficients of $p(\rho, s)$ in the expansion with fixed s. For the two contributions of β_0, 1 and $\frac{1}{2}B/A$, should be linked with the difference between the fluid velocity and the sound speed, where the values correspond to their surrounding environment.

For an ideal gas, when entropy is fixed and p is directly proportional to ρ^γ, we get $A = \gamma p_0$ and $B = \gamma(\gamma - 1)p_0$, thus B/A is γ_{-1}, while β_0 is $(\gamma + 1)/2$. For air ($\gamma = 1.4$), we have $\beta_0 = 1.2$. For a liquid, the typical value of B/A is between 4 and 12, so from this view point we can see that liquid is more non-linear than gas.

Using the approximation (9.9), we can once again express (9.7) as follows

$$p(x,t) = f(\phi), \quad x = \phi + \left[c + \beta \frac{f(\phi)}{\rho c}\right] t \tag{9.11}$$

Here we conventionally omit the prime of p' and the subscript "0" of ρ_0, β_0 and c_0, where p now represents the sound pressure, and c represents the environmental sound speed. If the term $\beta f(\phi)/\rho$ in the second formula of (9.11) is omitted, then, returning to linear approximation, we get the well-known expression $p = f(x - ct)$ for a traveling plane wave.

Another presentation involves the detailed dependence of p on t when $x = 0$. Using the "wave delay" concept recommended in Chapter II, section 2.6, for a moving point with a fixed sound pressure amplitude p, its "wave delay" approximates to

$$\mathrm{d}_x t \approx \frac{1}{c + \beta p/\rho c} \approx \frac{1}{c} - \frac{\beta p}{\rho c^3} \tag{9.12}$$

then the proper correspondence of (9.11) is

$$p(x,t) = g(\Psi) t = \Psi + \frac{x}{c} - \frac{x}{c^2}\frac{\beta g}{\rho c} \tag{9.13}$$

where, when $x = 0$, $g(t)$ is $p(0,t)$ and t is Ψ. Here we assume $\beta p \ll \rho c^2$.

9.1.2 Synopsis of shock waves [253]

From (9.7a) and (9.9), we can see that the propagation velocity is a function of wave amplitude. A places where the amplitude is larger, the propagation is faster (see Fig. 68), and the form of a simple harmonic wave changes gradually into a saw-toothed wave. When the wave front suddenly steepens a shock wave appears, which is also called a blast wave. This phenomenon occurs when the state parameters in disturbed fluid are interrupted. For an ideal gas, by using this simple method we can estimate the required propagation distance when a harmonic wave changes into a shock wave. The propagation velocity of wave crest and wave trough are $c_0 + (1 + \gamma)V/2$ and $c_0 - (1 + \gamma)V/2$, respectively, where V is the particle velocity amplitude. Now suppose that after traveling a distance D the crest is just it to catch up with the front of it, then the required time is

$$t = \frac{D}{c_0 + \frac{1+\lambda}{2}V}$$

When catching up the trough, the distance difference traveled by the crest is

$$\left(c_0 + \frac{1+\gamma}{2}V\right)t - \left(c_0 - \frac{1+\gamma}{2}V\right)t = \lambda/2 \tag{9.14}$$

where λ is the wavelength. From the above two equations, we get the solution

$$D = \frac{\lambda}{2}\frac{1}{(1+\gamma)M} + \frac{\gamma}{4} \tag{9.15}$$

where M is the Mach number. When the frequency is 100 Hz, the wavelength approaches 0.3 m. If the intensity of the sound source is 100 dB, then $M = 2 \times 10^{-5}$, and a shock wave appears at a distance of $D = 3541$ m. When the intensity of the sound source is 180 dB, $M = 0.2$, and a shock wave appears only at a distance of $D = 0.27$ m. The above formula indicates that the distance for a shock wave appears, its distance is directly proportional to its wavelength, so at low frequencies, the distance needed for a shock wave to appearing must be greater. As mentioned above, because the basic equations of fluid dynamics are non-linear in essence, so too does the non-linearity of a wave also exist. In the absence of a disturbance, all simple harmonic waves can change into saw-toothed waves when their amplitudes are large enough, and eventually form shock waves.

Let uss now introduce some relations for a shock wave. In shock waves, the interruptions of state parameters is called a shock wave surface. The state parameters on both sides of the shock wave surface should satisfy certain conservative conditions. We mark the front side and rear side of shock wave surface by 1 and 2 respectively, and adopt the relative velocity as $u = U - v$, where U is the shock wave velocity, v is the particle velocity. These conditions are the three conservative laws mentioned in Chapter 1, section 4.3: mass conservation

$$\rho_1 u_1 = \rho_2 u_2 \tag{9.16}$$

momentum-flow conservation

$$p_1 + \rho_1 u_1^2 = p_2 + \rho_2 u_2^2 \tag{9.17}$$

and energy conservation

$$w_1 + \frac{u_1^2}{2} = w_2 + \frac{u_2^2}{2} \tag{9.18}$$

where $w = \frac{1}{\gamma - 1}\frac{p}{\rho} + \frac{p}{\rho}$ is the enthalpy.

Defining the Mach number when a shock wave flows forward as

$$M = \frac{U - v_1}{c_1}$$

we can get

$$\frac{v_2 - v_1}{c_1} = \frac{2(M^2 - 1)}{(\gamma + 1)M} \tag{9.19}$$

$$\frac{\rho_2}{\rho_1} = \frac{(\gamma + 1)M^2}{(\gamma + 1)M^2 + 2} \tag{9.20}$$

$$\frac{p_2 - p_1}{p_1} = \frac{2\gamma(M^2 - 1)}{\gamma + 1} \tag{9.21}$$

In the case of strong shock waves, we have

$$\frac{p_1 - p_1}{p_1} \to \infty, \quad M = \frac{U}{c_1} \gg 1$$

then the shock wave relations can be approximately expressed as

$$v_2 = \frac{2}{\gamma + 1}U, \quad \frac{\rho_2}{\rho_1} \approx \frac{\gamma + 1}{\gamma - 1}, \quad p_2 = \frac{2}{\gamma + 1}\rho_1 U^2 \tag{9.22}$$

In the following discussion, we will use the above relations to study shock waves propagating in different atmospheric conditions.

9.1.3 Generation of harmonic waves

One intension of Eq. (9.13) is that: along with an increase of propagation distance, high-order harmonics can be developed from a sound source with a fixed frequency. For example, if we assume the waveform at $x = 0$ is

$$g(\Psi) = P_0 \sin \omega\Psi$$

then Eq. (9.13) becomes

$$p = P_0 \sin \omega\Psi, \quad \omega t' = \omega\Psi - \sigma \sin \omega\Psi \tag{9.23}$$

where

$$\sigma = \frac{x}{\bar{x}}, \quad \bar{x} = \frac{\rho c^2}{\beta k p_0}, \quad t' = t - \frac{x}{c}$$

Here the forming distance of the shock wave $\bar{x}$ is the early value of x, thence $\omega\Psi$ is no longer a single-value function of $\omega t'$.

The wave form described by (9.13) is a periodic function of $\omega t'$ when σ is fixed; the reason is that as long as the second formula is satisfied, every time $\omega t'$ increases by 2π, $\omega\Psi$ is also forced to increase by 2π, however, when $\omega\Psi$ increases by 2π, p remains unchanged. Another deduction is that p must be an odd function of $\omega t'$. Therefore, the Fourier series expansion of p will have the following form

$$p = \sum_{n=1}^{\infty} p_{n,\mathrm{pk}}(\sigma) \sin n\omega t' \tag{9.24}$$

where $p_{n,\mathrm{pk}}(\sigma)$ is a Fourier coefficient. If we regard p as function of $\theta = \omega t'$ and σ, then

$$p_{n,\mathrm{pk}}(\sigma) = \frac{2}{\pi} \int_0^{\pi} p(\theta, \sigma) \sin n\theta \mathrm{d}\theta$$

If we transform the integral variable into $\xi = \omega\Psi$, then $\theta = \xi - \sigma \sin \xi$ while $p = p_0 \sin \xi$, so the above integral is derived into

$$\begin{aligned} p_{n,\mathrm{pk}}(\sigma) &= \frac{2P_0}{\pi} \int_0^{\pi} \sin \xi \sin[n(\xi - \sigma \sin \xi)](1 - \sigma \cos \xi) \mathrm{d}\xi \\ &= \frac{2P_0}{\pi} \int_0^{\pi} \cos[n(\xi - \sigma \sin \xi)] \cos \xi \mathrm{d}\xi \end{aligned}$$

The second form is obtained after partial integration. The above formula can also be written as

$$\begin{aligned} p_{n,\mathrm{pk}}(\sigma) &= \frac{2P_0}{\pi n \sigma} \int_0^{\pi} (\cos n\theta)[1 - (1 - \sigma \cos \xi)] \mathrm{d}\xi = \frac{2P_0}{\pi n \sigma} \int_0^{\pi} (\cos n\theta - \frac{1}{n} \mathrm{d}_{\xi} \sin n\theta) \mathrm{d}\xi \\ &= \frac{2P_0}{\pi n \sigma} \int_0^{\pi} \cos[n(\xi - \sigma \sin \xi)] \mathrm{d}\xi = \frac{2P_0}{n\sigma} J_n(n\sigma) \end{aligned}$$

where $J_n(n\sigma)$ is an nth order Bessel function.

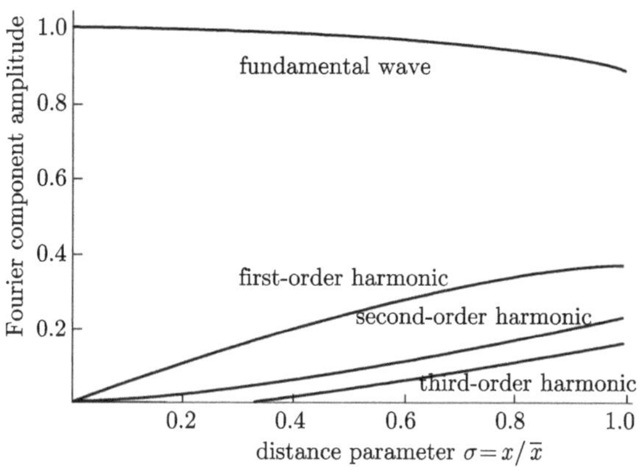

Fig. 69: The phase wave is a sinuous wave at $x = 0$, while the variation of amplitudes (unit is P_0) of each order harmonic with distance (unit is $\bar{x}$).

Substituting the above formula into (9.24), we obtain the Fubini-Chiron solution[2]

$$p = P_0 \sum_{n=1}^{\infty} \frac{2}{n\sigma} J_n(n\sigma) \sin\left[n\omega\left(t - \frac{x}{c}\right)\right] \tag{9.25}$$

But it must be emphasized that this formula is not applicable to the area beyond $\sigma = 1$.

Referring to the expansion of a power series of the Bessel function indicates

$$\frac{2}{n\sigma} J_n(n\sigma) \to \left(\frac{n\sigma}{2}\right)^{n-1} \frac{1}{n!}\left[1 - \frac{(n\sigma)^2}{4(n+1)} + \cdots\right]$$

So the amplitude of the fundamental frequency wave ($n = 1$) when σ is very small will decrease according to the following formula

$$p_{1,\mathrm{pk}}(\sigma) \approx P_0\left(1 - \frac{\sigma^2}{8}\right) \tag{9.26a}$$

while first-order harmonics ($n = 2$) will increase according to the following formula

$$p_{2,\mathrm{pk}}(\sigma) \approx P_0 \frac{\sigma}{2} \tag{9.26b}$$

so we can see as x changes linearly; higher order harmonics increase more slowly (see Fig. 69).

2 See E. Fubini-Chiron. *Anomalies in acoustic wave propagation of large amplitudes*, Alta Frep., 1935, 4: 530∼581; and D. T. Blackstock, *Propagation of plane sound waves of finite amplitude in non-dissipative fluids*, J. Acoust. Soc. Am., 1962, 34: 9∼30.

9.1.4 Nonlinear dissipative waves, Burger's equation

Now let us consider the absorption of media. For the sake of simplicity, we will only consider classical absorption, which is equivalent to the process of a fluid without relaxation, such as s monoatomic gas (or pure water). As mentioned in Chapter V, section 5.1, for this case, in the equations that describe fluid movement, Euler's equation should be replaced with the Navier-Stokes equation (5.5′). When considering heat conductivity, the equation of state ought to be modified. For this purpose, we first expand the pressure $p(\rho, s)$ into a power series of $\rho_1 = \rho - \rho_0$ and $s_1 = s - s_0$, reserving small quantities for the second-order.

$$p = p_0 + (\partial_\rho p)_s \rho_1 + \frac{1}{2}(\partial_\rho^2 p)_s \rho_1^2 + (\partial_s p)_\rho s_1$$

When considering formula (1.12), the second term is $c^2\rho_1$ by means of a thermodynamic relationship. We can then seek for any variations in the entropy s_1 caused by heat conductivity from formula (5.8), and substitute it into the above formula. Thus the resened equation of state can be rectified to include heat conductivity as

$$p = p_0 + c_0^2\rho_1 + \frac{1}{2}(\partial_\rho^2 p)_{\rho_0}\rho_1^2 - \frac{\kappa(\gamma - 1)}{c_p}\nabla \cdot \boldsymbol{v} \tag{9.27}$$

By substituting the sound wave disturbance formula (2.1) into the corresponding one-dimensional equations, and only reserving second-order terms, we get

$$(\rho_0 + \rho_1)\partial_t v + \rho_0 v \partial_x v = -c_0^2 \partial_x \rho_1 - \frac{(\gamma - 1)c_0^2}{\rho_0}\rho_1 \partial_x \rho_1 + \delta \partial_x^2 v \tag{9.28}$$

$$\partial_t \rho_1 + (\rho_0 + \rho_1) + \partial_x v_x + v_x \partial_x \rho_1 = 0 \tag{9.29}$$

where we used the rectified equation of state that includes heat conductivity, and δ is defined as $\delta \equiv \frac{4}{3}\mu + \frac{\gamma - 1}{c_p}\kappa$ in formula (5.18).

Now let us regard v as a function of x and $t' = t - x/c_0$. It is no longer a function of x and t: $v = v(x, t')$, we transform the coordinate system into the corresponding "first class accompanying coordinate system". Over a wavelength distance the distortion is very small, and every quantity is a slowly-varying function of x in the new coordinate system. If v is a first-order small quantity, then $\partial_x v$ is a second-order small quantity. So (9.28) and (9.29) become

$$\left(1 + \frac{\rho_1}{\rho_0} - \frac{v}{c_0}\right)\partial_{t'} v = \frac{\delta}{c_0^2 \rho_0}\partial_{t'}^2 v + \frac{c_0}{\rho_0}\left[1 + (\gamma - 1)\frac{\rho_1}{\rho_0}\right]\partial_{t'}\rho_1 - \frac{c_0^2}{\rho_0}\partial_x \rho_1 \tag{9.30}$$

$$\frac{1}{\rho_0}\left(1 - \frac{v}{c_0}\right)\partial_{t'}\rho_1 - \frac{1}{c_0}\left(1 + \frac{\rho_1}{\rho_0}\right)\partial_{t'} v + \partial_x v = 0 \tag{9.31}$$

From the two above formulae, if we eliminate the first-order terms then we will obtain an equation that consists of second-order terms. We multiply formula (9.31) by c_0 and add Eq. (9.30), we then replace ρ_1/ρ_0 with v/c_0, and the well-known Burgers equation is obtained

$$\partial_x v - \frac{\beta}{c_0^2} v \partial_{t'} v = \frac{\delta}{2c_0^2 \rho_0} \partial_t^2 v$$

where β is a "non-linear coefficient" defined in (9.10) (do not confuse it with the "thermal expansion coefficient" defined in (5.9)). In the limiting case of $\beta \to 0$, $\delta \to 0$, the solution to above formula can be reduced to the linear acoustic expression $v = f(x - ct)$ of plane waves in an ideal fluid.

If we regard v as a function of $x' = x - c_0 t$ and t: $v = v(x', t)$ is equivalent to the transformation of the coordinate system into a corresponding "second class accompanying coordinate system". Using an entirely analogical derivation, the corresponding Burgers equation can be obtained

$$\partial_t v + \beta v \partial_{x'} v = \delta \partial_{x'}^2 v \tag{9.32}$$

Using a non-linear transformation, we can simplify the Burgers equation to a linear equation. In formula (9.32), we set

$$v = -\frac{2\delta}{\beta \Psi} \partial_x \Psi \tag{9.33}$$

Then by eliminating the non-linear terms, the linear dissipation equation can be obtained

$$\partial_t \Psi = \frac{\partial}{\beta} \partial_x^2 \Psi$$

In this way, the problem of how to solve the non-linear Burgers equation becomes how to solve the linear dissipation equation. However, the applicability of this simple solution in practice is quite seldom, and most non-linear problems are very difficult to solve directly.

Next we will consider a common non-linear phenomenon: an N-wave. When an aircraft is flying in the air at supersonic speeds, two closely nearby and reverse-phased shock waves will be generated. Such the two shock waves joint together, and form a wave form that appears like the letter "N", and is hence named as such. We take the solution of the linear dissipation equation as

$$\Psi = 1 + \sqrt{\frac{c}{t}} \exp\left(-\frac{x^2}{2\mu t}\right)$$

where $\mu = \delta/\beta$. Using transformation (9.33), we can obtain the solution to the particle's vibrating velocity

$$v = -\frac{2\mu \partial_x \Psi}{\Psi} = \frac{x}{t} \frac{\sqrt{c/t} \exp(-x^2/4\mu t)}{1 + \sqrt{c/t} \exp(-x^2/4\mu t)}$$

by only considering cases where $t > 0$. At a given time t, the wave form v in space is shown in Fig. 70. From this we can see that the N-wave is one of the solutions to the

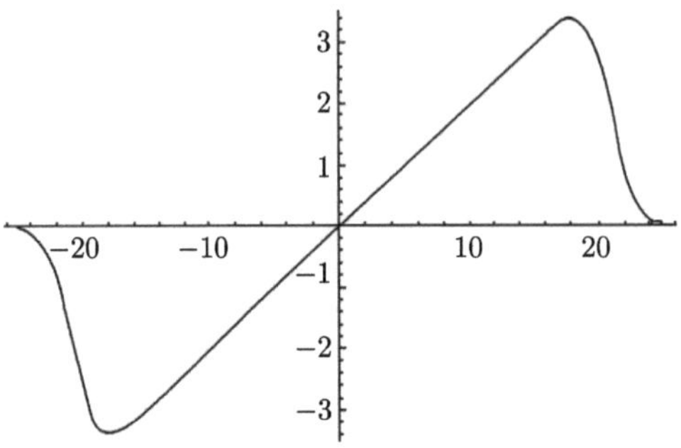

Fig. 70: An N-wave as a solution to the Burgers equation.

Burgers equation, which means we have mathematically proved that an N-wave can propagate in the atmosphere. The area under curve v is expressed as

$$A = \int_0^\infty v\mathrm{d}x = 2\mu \log\left(1 + \sqrt{\frac{c}{t}}\right)$$

Using the above formula, analogical to fluid mechanics, we define the Reynolds number as

$$R(t) = \frac{A}{2\mu} = \log\left(1 + \sqrt{\frac{c}{t}}\right)$$

Which gives the ratio between the influences of the non-linear term and the dissipative term, the ratio varies with time. When $R \gg 1$, the non-linear effect takes the dominant position, but when $t \to \infty$ the dissipative term will take the major one. Furthermore, the solution to the particle's velocity becomes

$$v = \frac{x}{t}\sqrt{\frac{c}{t}} \exp\left(-\frac{x^2}{2\mu t}\right)$$

This is the dipole solution of the heat-transfer equation. This solution is somewhat different to the present analogous numerical result, yet the main conclusion is consistent, i.e. after propagating a very long distance, the two steep wave fronts of the shock waves of the N-wave are no longer steep.

9.1.5 Nonlinear waves propagating in inhomogeneous media

Let us consider one-dimensional problems. We will let the z coordinate to be point vertically. For an atmosphere in an equilibrium state, $p = p_0(z)$, $\rho = \rho_0(z)$ particle velocity $v = 0$. Owing to gravitational action, the atmosphere remains pressure gradient, and the density and pressure must satisfy Eq. (1.13). Under the isothermal condition, the density is expressed by formula (1.15). Under the non-linear condition, the basic equations of fluid mechanics can be written as follows (see Chapter 1, section 1.4).

$$\partial_t \rho + v\partial_z \rho + \rho\partial_z v = 0$$
$$\partial_t v + v\partial_z v + \frac{I}{\rho}\partial_z p = -g$$
$$\partial_t p + v\partial_z p - c^2(\partial_t \rho + v\partial_z \rho) = 0$$

To solve the above equations by characteristic line method. The differential form of the characteristic relationship is expressed as

$$\mathrm{d}p + \rho c \mathrm{d}v - \frac{\rho c g}{v + c}\mathrm{d}z = 0, \quad \mathrm{d}_t z = v + c$$

Applying the result for to shock waves, we get

$$\mathrm{d}_z p + \rho c \mathrm{d}_z v - \frac{\rho c g}{v + c} = 0 \tag{9.34}$$

To solve this sort of problem numerical integration is necessary. For strong shock waves, some approximate results can be obtained. Strong shock waves have the following relationship

$$u = \frac{2}{\gamma + 1}U, \quad p = \frac{2}{\gamma + 1}\rho_0 U^2, \quad \rho = \frac{\gamma + 1}{\gamma - 1}\rho_0$$

where U is the shock wave velocity. Under the limiting condition, U is very large, so the third term of Eq. (9.34) can be neglected as comparing with the other two terms. Eq. (9.34) can then be simplified as

$$\frac{1}{U}\mathrm{d}_x U + \frac{\beta}{\rho_0}\mathrm{d}_x \rho_0 = 0$$

where

$$\beta = 1/\left(2 + \sqrt{\frac{2\gamma}{\gamma - 1}}\right)$$

Using the above equations, we can finally get

$$U \propto \rho_0^{-\beta}, \quad p \propto \rho_0^{1-2\beta}$$

This result is obtained under the condition that the mass density is attenuating exponentially. According to experimental atmospheric mass density data, the result needs to be modified appropriately.

9.2 Sonic boom

A sonic boom is a phenomenon of non-linear atmospheric acoustics generated by an aircraft flying at supersonic speeds. When the aircraft flies at supersonic speeds, the air in front of the aircraft will be extruded to a very strong high-pressure; while the air behind the aircraft will be dispelled to a very steep negative pressure. The high-pressure area and the negative-pressure area will join together to form a wave form as pointed out on p 329. the ground that is the sonic boom to be heard.

The fundamental theory of sonic booms was established and testified in the early 1950s and 1960s, and was applied to practical models in the 1970s. Over the past several decades, the production of more advanced supersonic aircrafts as well as estimates of the influence on the environment from noise generated when launching space aircrafts or when flying military aircrafts, have promoted understanding of the physics of sonic boom problems. Some important phenomena, such as the thicknesses of shock waves and their propagation in turbulence, have been researched. But due to the complexity of the problem, we must use a numerical analogue method to process the details of the physical phenomena Owing to advancements in computer science, sonic boom theory has being applied to numerical analogs and aerodynamic designs.

9.2.1 Fundamental theory of sonic boom [253]

Sonic boom analysis can be divided into three fields: "near field", "middle field" and "far field". Studies of these three fields correspond to the generation, propagation and development of a sonic boom. The boundary between the near, middle and far fields depends on concrete environmental conditions. Near field is an adequately small field (generally only one part of the entire length of an object), so that the atmospheric gradient is negligible. For small aircrafts such as fighters, even the shortest range of the far-fields is beyond 1500 m. For large aircrafts, such as supersonic passenger planes, they are designed to be longer than 90 m, thus signals received on the ground beyond the airline of 15 km belong to the middle field classification.

An aircraft moving at supersonic speeds will generate disturbances in the atmosphere. In the near field, for slender aircrafts or projectiles, the disturbance is very small, so that its behavior can be processed using the acoustic linear approximation. Therefore, in traditional sonic boom theory of slender objects, the disturbance of aircrafts in the atmosphere can be described via the linear theory of supersonic movement.

Disturbances created by blunt flying objects (such as the space shuttle) in the atmosphere is non-linear in essence, so its near-field theory is more complicated than the theory of slender objects.

Disturbances generated by aircrafts propagating through an atmosphere towards the ground can be basically treated as an outward propagating cylindrical wave. Because atmospheric media are inhomogeneous, the acoustic impedance between the aircraft and the ground will vary, implying that the wave-front curvature may change as well. For weak shock waves or simple harmonic waves, by only using geometric method, we can clearly describe the propagation law in the atmosphere.

When the distance is long enough, the wave form of the signals will vary. The reason is that for even very weak signals, after propagating a long distance, the accumulated non-linear effects will become quite obvious. The propagation of high-pressure signals travel faster than sound waves in the atmosphere, while the parts with low pressure propagate slightly slower. This kind of deformation usually causes the signal to turn into a shock wave. For a far-field signal, there is a tendency for it to become an N-wave.

Signals in the middle field begin to appear obvious for non-linear distortions, but the effect of an aircraft's geometric characteristics is also quite obvious. Signals in the far-field approach to a gradually advanced appearance, generally it is an N-wave, i.e. a sonic boom.

In the following, let us analyze the acoustic pressure generated by a sonic boom. Let U be the aircraft velocity relative to the fluid (parallel with the x axis), the components of the disturbed velocity are $U(1+u)$ on the x axis and Uv on the radial coordinate axis r (perpendicular to the x axis), respectively. When considering the solution to a cylindrical wave, we get

$$u = -\frac{1}{2\pi}\int_0^{x-Br} \frac{S''(\eta)\mathrm{d}\eta}{\sqrt{(x-\eta)^2-(Br)^2}} \tag{9.35}$$

$$v = \frac{1}{2\pi r}\int_0^{x-Br} \frac{S'(\eta)\mathrm{d}\eta}{\sqrt{(x-\eta)^2-(Br)^2}} \tag{9.36}$$

where $B = \sqrt{M^2-1}$, and $S(x)$ is the section area from the top to x. The disturbance is confined to be within the Mach angle, as determined by equation $x - Br = 0$. Using the method proposed by Landau [254], the above formula can be processed non-linearly. It should be indicated that the solution processed non-linearly does not correspond to the exact solution of the non-linear equation, where a theoretical analysis indicates that this sort of processing can describe a weak non-linear effect. Let $x - Br = \xi(x, r)$ (in the following we will discuss how to choose ξ), formula (9.35) and (9.36) will become

$$u = -\frac{F(\xi)}{\sqrt{2Br}}, \quad v = -\frac{BF(\xi)}{\sqrt{2Br}} \tag{9.37}$$

where

$$F(\xi) = \frac{1}{2\pi}\int_0^{\xi} \frac{S''(\eta)\mathrm{d}\eta}{\sqrt{\xi-\eta}}$$

which is called a F-function, which represents the sound source intensity, and includes characteristics of the aircraft as a sound source. According to Bernoulli's equation, the relationship between pressure and flow velocity can be given, and using formula (9.37) we can obtain

$$\frac{p-p_0}{p_0} = \gamma M^2 \frac{F(\xi)}{\sqrt{2Br}} \tag{9.38}$$

where p is the disturbed pressure, and p_0 is the air pressure in equilibrium state. By means of the relationship between the acoustic pressure and sound speed, we get

$$\frac{c - c_0}{c_0} = \frac{\gamma - 1}{2} M^2 \frac{F(\xi)}{\sqrt{2Br}} \tag{9.39}$$

where c is the sound velocity of a non-linear wave, c_0 is the sound velocity of a linear wave. Since ξ = const, so we have

$$\mathrm{d}_r x = \cot(\mu + \theta) \tag{9.40}$$

where $\mu = \arcsin(c/U)$ is the Mach angle, and θ is the inclination between the wave front and the x axis. By taking a proper approximation, in first approximations, we can get the expansion

$$\partial_r x = \cot \mu_0 - (\mu - \mu_0 + \theta) \operatorname{cosec}^2 \mu_0 \tag{9.41}$$

at the same order of approximation, we have

$$\theta \approx v, \quad \mu - \mu_0 \approx \frac{c_0}{U}\left(\frac{c - c_0}{c} - u\right) \sec \mu_0 \tag{9.42}$$

By means of the above formula, we can get

$$\partial_r x = \frac{1}{B} - \frac{M^2}{B}\left(\frac{c - c_0}{c} - u\right) - M^2 v \tag{9.43}$$

Substituting Eqs. (9.37) and (9.39) into (9.43), we obtain

$$\partial_r x = \frac{1}{B} - \frac{(\gamma - 1)M^4}{(2B)^{3/2}} \frac{F(\xi)}{r^{1/2}} \tag{9.44}$$

Finally, according the above formula, we have

$$x = Br - kF(\xi)r^{1/2} + \xi \tag{9.45}$$

where $k = \dfrac{(\gamma + 1)M^4}{\sqrt{2}B^{3/2}}$. Eqs. (9.37)~(9.39) and (9.45) give the non-linear solution to a sonic boom in the region $\xi/Br \ll 1$.

Following the conditions for weak shock waves, we take the mean value of two shock waves, and get conditions to determine ξ as

$$\frac{1}{2} kF^2(\xi)\sqrt{r} = \int_0^{\xi} F(\alpha)\mathrm{d}\alpha \tag{9.46}$$

If the half-angle at the top of the flying object is ε, the section-area function is $S(x) = \pi\varepsilon^2 x^2$, and the F-function can be expressed as

$$F(\xi) = 2\varepsilon^2 \sqrt{\xi} \tag{9.47}$$

and from Eq. (9.46) we can get

$$\sqrt{\xi} = \frac{3}{2}k\varepsilon^2\sqrt{r}$$

From this we can get the shock wave equation (9.45) as

$$x = Br - \frac{3}{4}k^2\varepsilon^4 r \tag{9.48}$$

The inclined half-angle of the shock-wave wave front is

$$\mu_0 = \frac{3}{8}\frac{(\gamma+1)^2 M^6}{(M^2-1)^{3/2}}\varepsilon^4$$

and the shock-wave intensity is

$$\frac{p-p_0}{p_0} = \frac{3}{2}\frac{\gamma(\gamma+1)M^6}{(M^2-1)}\varepsilon^4 \tag{9.49}$$

The results of numerical examinations indicate that when the half-angle at top of the flying object is less than 10° and the Mach number is between 1.1∼3, the above shock-wave intensity formula can give very good results.

For the far field case with $r \gg 1$, the shock-wave intensity formula becomes

$$\frac{p-p_0}{p_0} \approx \frac{2^{1/4}\gamma(M^2-1)^{1/8}K}{r^{3/4}\sqrt{(\gamma+1)}} \tag{9.50}$$

where $K = \sqrt{\int_0^{\xi} F(\alpha)\mathrm{d}\alpha}$ is called a "shape factor". From the above formula we can see that the dependence of the ground acoustic pressure on the Mach number is relatively slight but where its dependence on altitude is $r^{3/4}$.

Formula (9.38) expresses the cylindrical wave propagation of a wave disturbance generated by a flying object in an homogeneous atmosphere. This equation is obtained for the condition of a near field, but a very good approximation for a far field can be obtained as well. The far-field approximation of (9.38) is (9.50), and this solution includes the known result, i.e. that the shock-wave intensity is attenuated as $r^{-3/4}$, but not as $r^{-1/2}$ as for a sound waves.

Eq. (9.38) is applicable to axisymmetric objects. The concept of a locally axisymmetric flow was introduced by supersonic local controllin theory. We set φ as the rotational azimuth so that $F(x)$ can be extended to $F(x, \varphi)$. This is a result for a far field, which is tenable when the radius is large enough to make the effect from the local cross flow be negligible. This is suitable for a sonic boom, where only its far field properties are of interest.

Eq. (9.38) is raised from the viewpoint of aerodynamics. In a coordinate system that moves together with the flying object, a sonic boom is regarded as a wave. Doing so is advantageous for deriving pressure disturbances generated by flying objects. When analyzing sound propagation in a real atmosphere, the propagation of a sonic boom in the atmospheric coordinate system can be regarded as a ray. For steady flying, any viewpoint is applicable; but for non-uniform flying, using an object-following coordinate system will face the problem of a non-Newtonian reference system. Adopting the viewpoint of sound rays is advantageous in order to analogously computing the modelling propagating process using geometric acoustic methods, and is applicable for analyzing sound waves that cross a gradient atmosphere, as well as those from moving sound sources.

From the viewpoint of acoustic rays, Eq. (9.38) can be rewritten as

$$\frac{p(\tau) - p_0}{p_0} = \frac{F_r(\tau)}{\sqrt{B_r}}(\tau) \tag{9.51}$$

where, $\tau = t - c_0/s$. t is the propagation time along the acoustic ray, s is the distance along the ray, and c_0 is environmental sound velocity. Here F_r and B_r are modifications to F and B in Eq. (9.38).

9.2.2 Focus of sonic boom

In linear acoustics, a inward-concaved wave front will create convergent phenomena. Acoustic rays perpendicular to the wave fronts intersect and to form a focus area that are confined to acoustic-caustic lines shaped by the rays' envelope. Using ray acoustics we can obtain the locus of a sonic boom traveling to the ground, and easily determine any convergent phenomena created by the rays. But according to shock-wave theory, sonic boom convergence cannot form a caustic such as those created by an optical lens. This is because that when sound intensity increases due to the effects from convergence, the sound speed will quicken and sound rays will spread out thus the focussing cannot be formed. Non-linear effect causes the maximum amplitude of focusing to be quite limited, which is consistent with results from experiments. Only the different components of wave can be focused to one point, and diffraction of a caustic line will decrease the focusing intensity. Except for its non-linear effect, diffraction will also influence amplitude. Diffraction effects rely on frequency, thus its influence on lower frequencies will be stronger than for higher frequencies.

The area far from caustic lines is interesting as well. In the area above caustic lines, sound rays criss-cross. There are two sorts of sonic booms; normal sonic booms that cross caustic lines and are not focused; and others that cross caustic lines and afterwards are focused. The sonic boom after focusing will substantially attenuate, and obtain the contour of a U-wave. The sonic boom after focusing can be processed approximately via a Hilbert transform. On one side of the shadow area of a caustic line, an attenuated wave will emerge.

9.2.3 Thickness of shock wave

The thickness of a shock wave is a very important parameter of a sonic boom, because it is directly related to the loudness of the sonic boom that we can hear outdoors. Its thickness also affects the acceptable extent to people for which the noise is generated by supersonic flying.

The fundamental theory of sonic booms makes use of the weak shock-wave theory. This approach does not consider the structure of shock waves, but processes it only as a thin fault. However, experimental results show that the shock-wave structure of a sonic boom is thicker than that expected from the classical absorption mechanism. The occurrence of this case is caused by two parts: atmospheric turbulence and molecular relaxation. The structure of a shock wave is the result from an interaction between molecular relaxation and non-linear steepening. This phenomenon is well understood. The influence of the molecular relaxation process on the generation time of a shock wave in the atmosphere has long been well-known.

Two methods can be used to compute shock-wave structure. The first method is called the Pestorius-Anderson algorithm [255]. This approach, in the frequency domain, considers the relaxation effect, while in the time domain one should adopt the Whitham method to compute non-linear steepening. The second method [256] is to start from the Burgers equation, and in the time domain to compute both the relaxation effect and non-linear steepening. These two computing methods have been programmed for special purposes, and used test the corresponding theory.

Sometimes, shock-wave thickness can be described by the arising time of acoustic pressure. Plotkin made use of experimental data from flying to compute the happening time of a shock wave to be 1ms-psf/ Δp, where Δp is the super-pressure of the shock wave [257].

9.2.4 Simulating programs of sonic boom

The earliest computer simulating program was written by Friedman et al. [259] in 1963, and was widely applied to sonic boom analysis of a Boeing 2707 SST plan in 1960s. The majority of the details of this program is correct, but there are some errors in the expression of the formula for an acoustic ray tube area of B_r.

In 1972, Thomas published another simulation program [260], where he could directly input the acoustic pressure Δp obtained from wind tunnel tests. His simulation had a simple form and structure, and by using the data given by a near field wind tunnel test, his numerical simulation gave very good results. However as mentioned before, the results from the near field cannot reflect the deformation generated by sonic boom propagation. Thomas' program fully applied the fundamental theory of sonic booms and was applicable to any object flying non-uniformly in any windy, horizontally stratified atmosphere. But this sort of program was too complicated and required an enormous amount of computer time.

Carlson developed a program [261] for practical use: to compute an N-wave sonic boom generated by steady flying in a standard atmosphere. The program he proposed was in the form of equation (9.51) used in a far field, but included several factors related to the propagation through a windless steady atmosphere. These factors included: the influence of the path refraction of acoustic rays on amplitude and sustained time; the influence of an acoustic impedance gradient on amplitude and sustained time; and the amplitude increment caused by ground reflection. The separated locus of propagation was included and was used to compute a sonic boom through the ground attenuating layer with a certain breadth. The integration of the F-function was expressed by a normalized form factor that was proportional to the size of the flying object. For many sorts of flying objects, charts of the form factors were drawn. The merit of Carlson's method lies not only in being able to rapidly compute an N-wave sonic boom generated by steady flying, but also lies in its ability to explicitly show the scaling effect of the parameters, e.g. the Mach number and the length of flying objects. Form factors were also be used to generate the F-function to be employed in full ray tracing.

Another simplified method was developed by Plotkin [257], which was applied to non-uniform flying in a windless atmosphere. In this program, for a given atmosphere, the form of the acoustic rays only relied on the flying altitude and their initial vertical angle. The area of the ray tube could be computed from the form of the rays and by the differential quotient of the form of rays with respect to the flying parameters of the transporters. Once these quantities were given, the computation of a sonic boom generated by a non-uniform flying object then becomes an algebraic problem, and the computation time needed is two orders of magnitude smaller than that needed for a full-ray tracing program.

9.3 Recent researches for sound waves in atmospheric turbulence [252]

Turbulence is a non-linear phenomenon in a fluid. In many cases, gaseous movement in the atmosphere possesses turbulence characteristics. The results of studies of the propagation and scattering of waves in turbulent media form an important basis of the foundation of the modern remote-sensing technique of atmospheric sounding. As pointed out in Chapter 8, due to the strong dependency of sound waves on atmospheric temperature and wind velocity, as an extraordinary attractive tool, sound waves can be used for the sounding of atmospheric turbulence by remote sensing.

The propagation and scattering of sound waves in atmospheric turbulence are similar to the corresponding behavior of electromagnetic waves in the atmosphere, and also similar to the corresponding behavior of sound waves in an ocean. Therefore many problems regarding sound-wave propagation in the atmosphere can be solved by the available theory of wave propagation in randomly moving media.

During the latter half of the 20th century, the main theoretical results of sound-wave propagation in turbulence were for turbulence that was locally homogeneous and locally isotropic. For a detailed discussion, see Chapter 4, section 4.3 and 4.4. However, quasi-ordered medium-sized non-homogeneities always exist in the atmosphere, for example, heat convection. When surveying meteorological parameters at single points, non-homogeneities are expressed as intermittences of small-scale disturbances and anisotropies. In this case, average turbulent parameters such as temperature and structural parameters of velocity C_T^2 and C_V^2 (see Chapter I, section 1.2.3), energy loss rate ε and other parameters all have obvious changes, and are hard to be regarded as constant parameters of a given air flow. Since the 1970s, the influence of these characteristics to real atmospheric turbulence (such as the anisotropy of turbulence, the intermittence of turbulence, and quasi-ordered middle scale non-homogeneities) on the parameters of sound wave signals has been obvious. Many researchers make use of SODAR to detect the existence of heat convection. Early in 1968, McAllister obtained data of heat convection when he performed studies of atmospheric acoustic sounding [215]. However the existence of middle-scale atmospheric structure has rarely been considered in wave propagation theory until now. Experimental atmospheric physicists refer to the quasi-ordered non-homogeneity of both wind velocity vortex in temperature field and atmospheric wave as “coherent structure”. These coherent structures have an obvious influence on the propagation of sound waves. When acoustic pulses and infrasonic sound waves are propagating, the influence of coherent structures especially merits attention. The study of turbulence itself is a thorny non-linear problem, and the mathematical description of turbulence is very complicated, which in turn makes the problem of sound propagation even more intricate. In most recent studies of sound waves in turbulence, experimental methods are mainly used. Numerical methods are only used for studying some idealized models. In the following, we will introduce some chief results of research.

9.3.1 Influences from intermittence

Some interesting phenomena have been noticed in atmospheric experiments, such as: small-scale intermittence of turbulence caused by temperature and velocity structure parameters C_T^2 and C_V^2; the probability density function (PDF) of scattering signals is not consistent with the logarithmic normal function; the probability of strengthening the return wave signals increases; an increase in scattering intensity

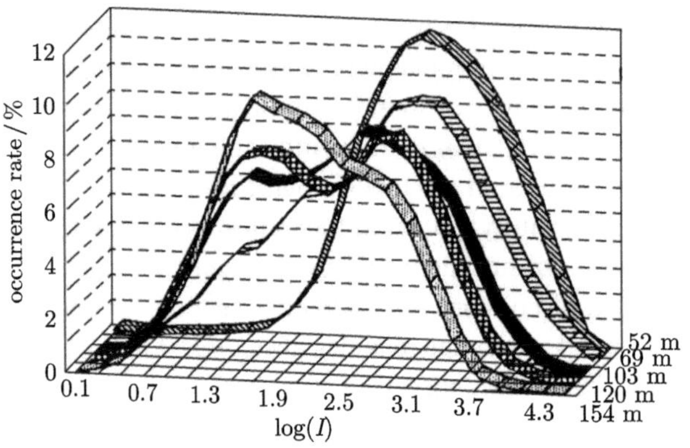

Fig. 71: The experimental columnar diagram for the probability density of the return signal scattered inversely from an altitude of 52 m to 154 m.

caused by turbulent intermittence, and so on. Petenko and Shurygin [262], by means of experimental work, studied the PDF of the inverse scattering of sound signals in a convective atmospheric boundary layer (ABL). Their result showed that when convective coherent structure appears, the PDF of the return wave signal intensity can be expressed as the sum of two logarithmically normal pdfs. The parameters of these pdfs depend on the altitude above the lower layer of the scattering volume and the degree of unsteadiness of the ABL. An experimental columnar diagram of an inverse scattering sound signal is shown in Fig. 71. From this result, it is possible to partly explain the inconsistency of the surveying results between C_T^2 by SODAR and C_T^2 measured on-the-spot in different ABLs. The present explanation for this intermittence is generally attributed to turbulence. However, until recently no theory was able to estimate the PDF parameters of real ABL return wave signals. The difficulties lie not only in that the effect of intermittence should be considered, but also in that the occurrence of a quasi-periodic irregularity must be deliberated.

9.3.2 Influences from anisotropy in small-sized turbulence

Experiments have proved that small-sized turbulence in ABL is anisotropic. This anisotropy is expressed by the sensitivity of inverse scattering of vertically launching sounding wave beams deviating from vertical direction. The sensitivity in this aspect was first observed by Neff (1975) [263]. If the non-homogeneity of the turbulence occurs in horizontal planes, it is possible that this phenomenon causes an increase in the inverse scattering intensity at vertical sounding. If the anisotropic structure is on the tilt, the intensity of the returned wave will be lower than the intensity caused by the isotropic non-homogeneity. Singal et al (1997) studied the influences of anisotropy

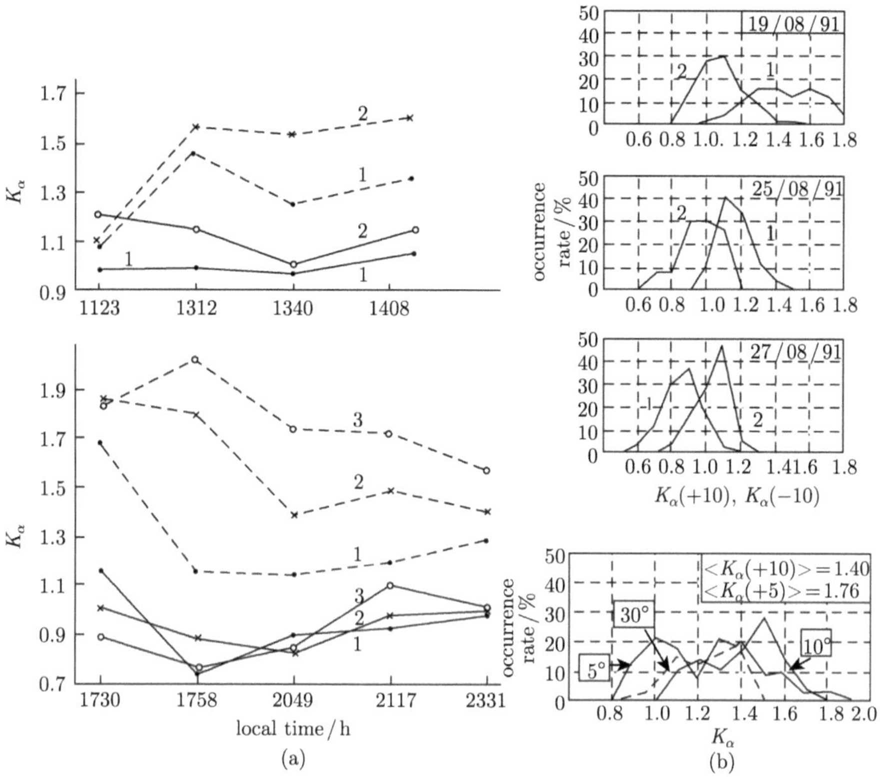

Fig. 72: Anisotropy of small-sized turbulence. (a) Time sequence of the anisotropic coefficient K_α at different altitudes; The upper figure corresponds to convective conditions, while the lower figure corresponds to temperature inversion conditions, solid line: $\alpha = 5°$, dotted line: $\alpha = 10°$; numerals beside the line represent different altitude ranges: 1—50~100 m; 2—100~150 m; 3—150~200 m. (b) In order from top to bottom the columnar diagrams of K_α under the temperature inversion condition, and K_α measured simultaneously at the opposite angles: 1 – $\alpha = 10°$; 2 – $\alpha = -10°$, the average value from four-night measurements: dotted line: $\alpha = 5°$; solid line: $\alpha = 10°$; dashed line: obtained from Neff's data (1975), in the higher inverted layer $\alpha = 30°$.

by SODAR [264]. The surveying results are shown in Fig. 72. The anisotropic coefficient of inverse scattering is defined as

$$K_\alpha = I_{\text{vert}}/I(\alpha)$$

where I_{vert} is the inverse scattering intensity in the vertical direction, and α is the angle of the wave beam deviating from the vertical direction.

From Fig. 72 we can see that under the inverted condition the value of the anisotropic coefficients are higher than what under the convective condition, and they will increase along with an increase of altitude of the scattering volume upon the lower surface. From Fig. 72(b), we can see that K_α also depends on the azimuth of the wave beam.

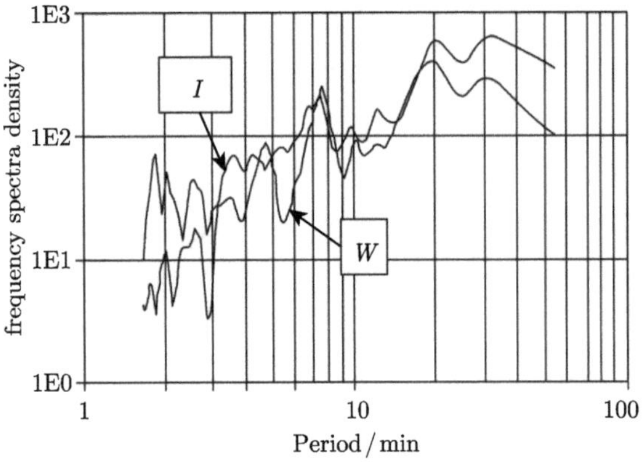

Fig. 73: An example for the periodic relationship between the power spectra of echo signal intensity and the vertical wind measured by Doppler radar and period. I: integral value of inverse scattering intensity over the whole altitude sounding range; W: the average value of the vertical wind velocity in the whole altitude sounding range.

9.3.3 Influence from quasi-periodic coherent structure of atmosphere boundary layer (ABL) on low-frequency power spectra of back-wave signals

Petenko & Bezverkhnii (1999) experimentally studied the influences of ABL coherent structure on the characteristics of low-frequency spectra of sound wave signals [265]. By means of the convective layer, they studied the reflection of infrasound waves within the frequency range $10^{-4}\sim10^{-2}$ Hz. For example, the low-frequency power spectra of a return-wave signal intensity and a Doppler frequency-shift in the frequency of SODAR in the vertical direction is shown in Fig. 73. The comb-shaped low-frequency spectra reminds us that the scattering turbulent structure is quasi-periodic. A typical extreme value corresponds to a period of 6∼9 min, which can be seen in the whole frequency spectra. The same observed period obtained also by some others previously. This period is close to the characteristic period of the Väisälä-Brunt frequency of a convection layer waveguide, so they proposed that the buoyancy wave in a convective layer influences the scattering characteristics of sound waves in the ABL.

9.3.4 Influences from coherent structure on the propagation of pulses in ABL

Many people have studied the influences of non-homogeneities of middle-scale wind velocities on the propagation of sound pulses and infrasound waves in the ABL [266]. Chunchuzov et al others experimentally studied the propagation of explosion sounds in a steady stratified ABL, where they measured the sustained time and propagating time of sound pulses. The waveform of sound wave signals received from an explosion

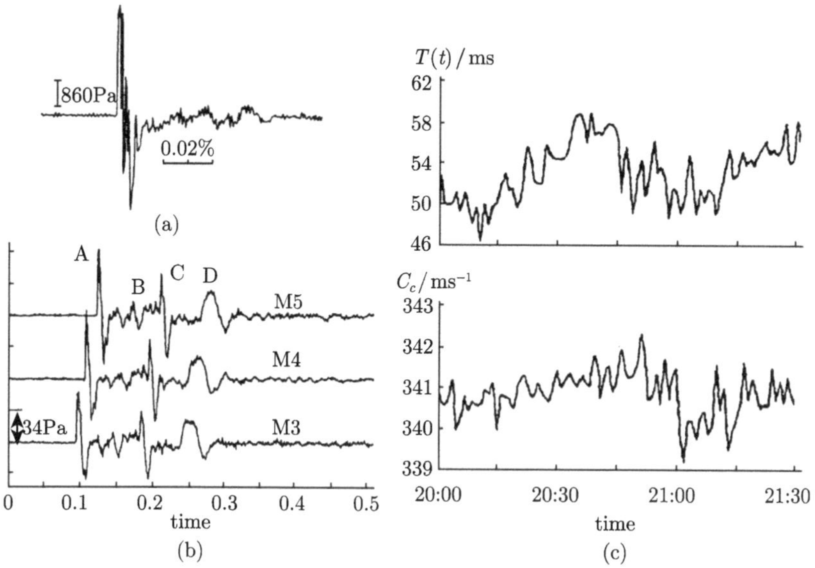

Fig. 74: Time fluctuations of sound pulses from a waveguide in the lower atmosphere. Measurement range of acoustic pressure and time are represented by the vertical or horizontal line segments beside the figure. (a) signals received from a nearby sound source; (b) identical signals received from the triangular antennae (M3, M4, M5) (r = 2.7 km); (c) the time variations of effective sound velocity measured by SODAR (lower figure) and the propagation time (upper figure).

source is given in Fig. 74: where Fig. 74(a), vepresents the signals detected near a nearby sound source; and Fig. 74(b) gives, the signals received from a triangular antenna at a distance of 2.7 km from the sound source, the sound waves propagating from different paths can be distinguished from the signals. Fig. 74(c) shows the propagation time of sound wave signals and the low-frequency fluctuation of wind velocity varying with time. The shapes of these two time sequences look very similar, and this is testimony to the feasibility of acoustic tomography for a steady stratified ABL. Using this method, we can estimate the three-dimensional space spectrum of wind velocity and temperature that are lacking in understanding of characteristic wave-numbers of buoyancy wave. In addition, these experiments monitored signals in the sound wave shadow area, which are obviously the results is due to the pulse scattering indused by mid-scale non-homogeneities which is comparable with the sound wavelength.

9.3.5 Sound scattering from anisotropy structure in mid-atmosphere

Tsuda (1986) found the mid-scaled long-term non-homogeneity of the stratosphere for the first time when surveying by radar [268]. Bush et al (1997) observed identical

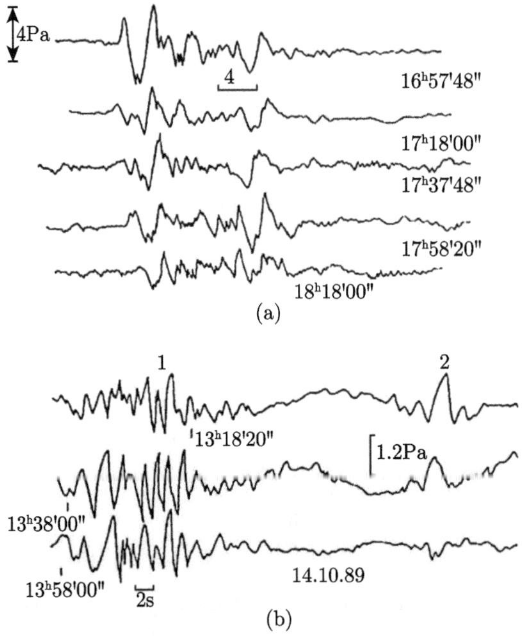

Fig. 75: Acoustic reflection and scattering generated by an anisotropic explosion in the mesosphere, received at a distance of r = 300 km from the sound source. (a) scattering in the stratosphere: on August 11, 1990, signals from the explosive device (equivalence 70 t, time interval 20 min) received in a sound shadow zeon; (b) reflection in the stratosphere (1) and mesosphere (2); on October 14, 1989, signals from an explosive device (equivalence 40 t, time interval 20 min) received within the audible domain.

structure when studying the long distance propagation of sound wave signals generated by a violent explosion [269]. The quasi-periodic fluctuations of refracted signals in stratospheric and the mid. atmospheric layer are shown in displayed in Fig. 75. The characteristics of these fluctuations remained unchanged for ten or more minutes. The acoustic topographic method Kulichkov (1998) sued to stud the stratospheric vertical structure, displayed that: the signal also has horizontal azimuth anisotropy in addition to vertical anisotropy [270].

9.3.6 Influences from turbulence on non-linear waves

Due to an increase of vibrating velocity of sound waves along with a decrease of atmospheric density, the propagation process of sound waves into the upper atmosphere is non-linear in essence; where the non-linear effect is the main reason for creating a form change and an increase in the sustained time of the sound pulse.

Atmospheric turbulence may influence the propagation of a sonic boom. On the condition that turbulence occurs, deformation will be generated by a sonic boom in two ways: one of them causes a rough microscopic structure after each shock wave; while the other thickens the shock wave. When a sonic boom crosses a turbulence, it will influence a non-linear deformation of the shock wave, and will mainly change its peak pressure and thickness. The results of experiments and numerical computation show that: the non-linear deformation of a sonic boom is weaker when atmospheric turbulence exists than when it does not exist. On an average, when turbulence exists, the peak pressure decreases and the thickness increases. The distribution of peak pressure and thickness are inhamogeneoes, the influence of a random velocity field is stronger than a random temperature field, and the result is consistent with practical measurements. But on the other hand, it is possible that atmospheric turbulence creates a convergent phenomenon of sonic boom wave fronts. The caustic surface will increase peak pressure and cause a variation in thickness.

The deformation of a theoretically stimulating explosion signal propagating to the upper atmosphere is shown in Fig. 76(a) [252]. Records of signals received from

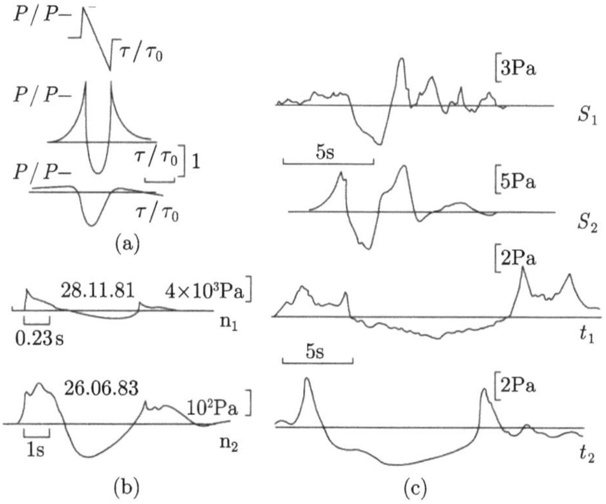

Fig. 76: Joint results of non-linear and mid-scaled turbulence in sound propagation. (a) Initial N-wave (upper figure), the sound pulse reflected by the upper atmosphere changes non-linearly from an N-wave into a U-wave: total reflection (middle figure); partial reflection (lower figure). Signals are represented by the relative pressure coordinate p/p_- and the relative time coordinate τ/τ_0. [p_- is the negative phase amplitude of pulses; $\tau = t - r/c$, r is the distance, $c(r)$ is the group velocity of the sound waves]; (b) signals received near the source: n_1 from a ground explosion (November 28, 1981, equivalence 260 t, distance of from the source 2 km); n_2 from an underground explosion (June 26, 1983, equivalence 2000 t, distance of from the source 4 km); (c) signals from two identical explosive sources received at a location of 240 km. S_1 and S_2 are the reflectance of the stratosphere, t_1 and t_2 are the reflectance from the thermosphere.

two explosions with different intensities and sustained time are shown in Fig. 76(b). Refraction in the stratosphere and the thermosphere of sound wave signals launched from these two explosions are shown in Fig. 76(c) and were recorded at distances of 200∼240 km apart from the sound source. As can be seen in Fig. 76(a), when no turbulence exists, the initial N-shaped waves become U-shaped waves. However, the forms of these waves are obviously deformed by turbulence. It is worth noticing that due to non-linear effects, although these two explosions are substantially different, the signals in Fig. 76(b) have almost identical sustained times and amplitudes.

9.4 Atmospheric solitary waves

9.4.1 Fundamental equations for atmospheric solitary waves

As introduced before, due to nonlinear effects, simple harmonic waves in propagating process will be gradually compressed and changed to saw-toothed waves. But for certain types of non-linear waves that have long propagation distances can retain their wave shapes, which are called solitary waves. The British scientist Russell early in 1844 observed that when a ship traveling along the canal suddenly stopped, a solitary water wave was formed and propagated far away. He reported this discovery at the British Science Congress. In 1895, Korteweg & de Vries derived a non-linear equation to describe these water waves, and afterwards it was named the KdV equation. Mathematically, solitary waves are a locally existing stable solution to the non-linear partial differential equations. Except for the KdV equation, other non-linear equations such as the Sine-Gordon equation also have solitary wave solutions. Solitary waves in atmosphere are mainly described by the KdV equation.

The reason for solitary waves being able to propagate for long distances results from interactions between frequency dispersion and non-linearity. The analysis in section 1.3 of this chapter indicates that an initial simple harmonic wave propagating non-linearly can generate high-order harmonic waves. When the frequency–dispersion effect exists, each order of harmonic waves have different frequencies so that their propagation velocities are different as well. Therefore the frequency–dispersion effect has the tendency to make wave forms disperse, while the non-linear effect has the tendency to make wave forms converge. These two tendencies cancel each other out, which allows solitary waves to maintain a stable wave form to propagate far away.

The original KdV equation only described a solitary waves on a water surface. When studying atmospheric physics, the KdV equation was extended to research of non-linear internal waves at the stratified atmospheric boundary [271]. Here we will now derive the generalized KdV equation. The basic model is that on both upper and sides of a rigid ground surface, two fluids with different characteristics are distributed, and the solitary waves exist at the horizontal boundary layer of these two fluids. Now, the adopted model for atmospheric solitary waves belongs only to the model's applicable to deep-water theory, and it requires a wavelength that is much shorter than the thickness of the upper fluid.

This is a two-dimensional problem: we set the x axis along the boundary layer and the y axis perpendicular to it; the horizontal and vertical components of the velocity are u and v, respectively. The governing equation of fluid mechanics are given as follows

$$\partial_t \rho + u\partial_x \rho + v \mathrm{d}_y \rho = 0 \tag{9.52}$$

$$\partial_\xi u_1 + \partial_y v_1 = 0 \tag{9.53}$$

$$\rho(\partial_t u + u\partial_x u + v\partial_y v) = -\partial_x p \tag{9.54}$$

$$\rho(\partial_t v + u\partial_x v + v\partial_y v) = -\partial_y p - \rho g \tag{9.55}$$

The derivation can be made in three steps: first we consider the solution to a lower fluid, then solve the movement of the upper fluid, and finally allow the solutions of the upper and lower fluids to satisfy the continuity condition in order to determine wave motion at the boundary. The lower fluid is confined to the interval $0 \leqslant y \leqslant h_0$. Let us adopt the following accompanying coordinate system

$$\xi = \varepsilon(x - c_0 t), \quad \tau = \varepsilon^2 t, \quad y = y$$

where ε is a small quantity. By expanding the velocity, pressure and density into:

$$u = \varepsilon^n u_n(\xi, y, \tau), \quad v = \varepsilon^{n+1} v_n(\xi, y, \tau)$$
$$p = p_0(y) + \varepsilon^n p_n(\xi, y, \tau), \quad \rho = \rho_0(y) + \varepsilon^n \rho_n(\xi, y, \tau)$$

Substituting these expansions into Eqs. (9.52)~(9.55), we can write the lowest-order equation as

$$\left.\begin{aligned} -c_0\partial_\xi \rho_1 + v_1\partial_y^3 \rho_0 &= 0 \\ \partial_\xi u_1 + \partial_y v_1 &= 0 \\ c_0\rho_0\partial_\xi u_1 &= \partial_\xi p_1 \\ \partial_y p_1 + \rho_1 g &= 0 \end{aligned}\right\} \tag{9.56}$$

After eliminating u_1, p_1, ρ_1, we get the following formula

$$\partial_y(\rho_0\partial_y v_t) - \frac{g}{c_0^2} v_1 \mathrm{d}_y \rho_0 = 0 \tag{9.57}$$

Transforming the velocity into

$$v_1 = -\phi(y)\partial_\xi f(\xi, \tau) \tag{9.58}$$

Eq. (9.57) then becomes

$$\mathrm{d}_y(\rho_0\partial_y \phi) - \frac{g}{c_0^2}\phi \mathrm{d}_y \rho_0 = 0 \tag{9.59}$$

By making a linear approximation, the other quantities can be expressed as follows

$$\begin{aligned} u_1 &= f(\xi,\tau)\mathrm{d}_y\phi(y) \\ p_1 &= -c_0\rho_0 f(\xi,\tau)\mathrm{d}_y\phi(y) \\ \rho_1 &= -\frac{1}{c_0}\rho_0\phi(y)\mathrm{d}_y\rho_0 \end{aligned}$$

Taking equations for order $O(\varepsilon^2)$

$$\begin{aligned} -c_0\partial_\xi\rho_2 + v_2\partial_y^3\rho_0 &= G_1 \\ \partial_\xi u_2 + \partial_y v_2 &= 0 \\ -c_0\rho_0\partial_\xi u_2 + \partial_\xi p_2 &= G_2 \\ \partial_y p_2 + \rho_2 g &= 0 \end{aligned}$$

where the non-homogeneous terms are

$$\begin{aligned} G_1 &= \frac{1}{c_0}\rho_0'\phi\partial_\tau f + \frac{1}{c_0}[\rho_0'\phi\phi' - \phi(\rho_0'\phi)']f\partial_\xi f \\ G_2 &= -\rho_0\phi'\partial_\tau f - [\rho_0(\phi')^2 - \rho_0\phi\phi'' + \rho_0'\phi\phi']f\partial_\xi f \end{aligned}$$

Eliminating u_2, p_2, ρ_2, we obtain the following formula

$$\partial_y(\rho_0\partial_y v_2) - \frac{g}{c_0^2}v_2\mathrm{d}_y\rho_0 = J(f,\phi) \tag{9.60}$$

and $J = -\frac{2}{c_0}(\rho_0\phi')'\partial_\tau f - \frac{1}{c_0}\{3[\rho_0(\phi')^2]' - 2\rho_0\phi'\phi'' - 2[\rho_0\phi\phi']'\}f\partial_\xi f$.

The condition that Eq. (9.60) has a non-singular solution demands

$$\int_0^{h_0}\phi\left[\partial_y(\rho_0\partial_y v_2) - \frac{g}{c_0^2}v_2\mathrm{d}_y\rho_0\right]\mathrm{d}y = \int_0^{h_0}\phi J(f,\phi)\mathrm{d}y \tag{9.61}$$

The boundary condition of the bottom fluid demands that

$$\phi = 0 \text{ and } v_2 = 0: \text{ at } y = 0. \tag{9.62}$$

Utilizing condition (9.62), Eq. (9.61) becomes

$$\begin{aligned} &c_0\rho_0[v_2(h_0)\mathrm{d}_y\phi(h_0) - \phi(h_0)\partial_y v_2(h_0)] \\ &= 2\left[\int_0^{h_0}(\rho_0\phi')\phi'\mathrm{d}_y\right]\partial_\tau f + \left\{\int_0^{h_0}[3(\rho_0\phi'^2)' - 2\rho_0\phi'\phi'' - 2(\rho_0\phi'\phi'')'\phi]\mathrm{d}y\right\}f\mathrm{d}_\xi f \end{aligned} \tag{9.63}$$

Now we will consider the upper fluid. Adopting the following accompanying coordinate system

$$X = \varepsilon(x - c_0 t), \quad \tau = \varepsilon^2 t, \quad y = y$$

in the interval $y \geqslant h_0$, Eqs. (9.52)~(9.55) can be expressed as

$$\left.\begin{aligned}\varepsilon \partial_y \rho - c_0 v \partial_x \rho + u \partial_x \rho + v \partial_y \rho = 0\\ \partial_\xi u + \partial_y v = 0\\ \varepsilon^2 \partial_y u - c_0 v \partial_x u + u \partial_x u + v \partial_y u = -\frac{1}{\rho_0}\partial_x p\\ \varepsilon^2 \partial_y v - c_0 \partial_x v + u \partial_x v + v \partial_y v = -\frac{1}{\rho_0}\partial_x p - g\end{aligned}\right\} \tag{9.64}$$

Expanding this to the particle velocity, pressure and density gives $O(\varepsilon^2)$

$$u = \varepsilon^2 U(\xi, y, \tau), \quad v = \varepsilon^2 V(\xi, y, \tau)$$
$$p = p_0(y) + \varepsilon^2 P(\xi, y, \tau), \quad \rho = \rho_0(y) + \varepsilon^4 R(\xi, y, \tau)$$

Substitution of the expansions into Eq. (9.64) and eliminating U, P, R, we get the Laplace equation of the order $O(\varepsilon^3)$

$$(\partial_x^2 + \partial_y^2)V = 0$$

The boundary condition is

$$V(x, y, \tau, \varepsilon) \to 0, \text{ when } y \to +\infty$$
$$V(x, y, \tau, \varepsilon) = V_0(x, \tau, \varepsilon), \text{ when } y = h_0$$

The solution to the Laplace equation is

$$V(x, y, \tau, \varepsilon) = \frac{1}{\pi} T \int_{-\infty}^{\infty} V_0(\chi, \tau, \varepsilon) \frac{y - h_0}{(y + h_0)^2 + (x - \chi)^2} \mathrm{d}\chi$$

T represents the main value of the integration. At $y = h_0$ we have

$$\partial_y V(x, y, \tau, \varepsilon) = -\frac{1}{\pi} \partial_x T \int_{-\infty}^{\infty} V_0(\chi, \tau, \varepsilon) \frac{1}{x - \chi} \mathrm{d}\chi$$

Finally, on the boundary layer, the vertical components of the velocity, and their first-order derivetives, are continuous, respectively.

$$\left.\begin{aligned}\varepsilon^2 v_1(\xi, h_0, \tau) + \varepsilon^3 v_2(\xi, h_0, \tau) = \varepsilon^2 V(\xi, h_0, \tau, \varepsilon)\\ \varepsilon^2 \partial_y v_1(\xi, h_0, \tau) + \varepsilon^3 \partial_y v_2(\xi, h_0, \tau) = \varepsilon^2 \partial_y V(\xi, h_0, \tau, \varepsilon)\end{aligned}\right\} \tag{9.65}$$

Without losing the generality, we can take

$$V_0(\xi, \tau, \varepsilon) = -\partial_\xi f.$$

To solve Eq. (9.65), we get

$$\left.\begin{aligned}\phi(h) = 0, \quad v_2(\xi, h, \tau) = 0\\ \mathrm{d}_y \phi(h) = 0, \quad \partial_y v_2(1) = T \int_{-\infty}^{\infty} \frac{f(\zeta, \tau)}{\xi - \zeta} \mathrm{d}\zeta\end{aligned}\right\} \tag{9.66}$$

Using Eq. (9.59) with conditions (9.63) and (9.66), ϕ can be solved, finally the equation for f can be given as

$$\partial_\tau f + \alpha f \partial_\xi f - \beta \partial_\xi^2 H(f) = 0 \tag{9.67}$$

where

$$\alpha = \left\{ \frac{3}{2} \int_0^{h_0} \rho_0(y) \phi'^3 \mathrm{d}y \right\} \Big/ \left\{ \int_0^{h_0} \rho_0(y) \phi'^2 \mathrm{d}y \right\}, \quad \beta = \rho_0 \Big/ \left\{ \frac{2}{c_0} \int_0^{h_0} \rho_0(y) \phi'^2 \mathrm{d}y \right\}$$

and H is the Hilbert transformation

$$H\{f(\xi)\} = \frac{1}{\pi} T \int_{-\infty}^{\infty} \frac{f(\zeta)}{\xi - \zeta} \mathrm{d}\zeta$$

Eq. (9.67) is just the generalized KdV equation. Assuming that f is a single-valued function of $\eta = \xi - \lambda\tau$, we can then get the following solution

$$f_s(\eta) = \alpha\delta^2 / (\eta^2 + \delta^2) \tag{9.68}$$

where $\lambda = \alpha\boldsymbol{\alpha}/4$, $|\delta| = 4\beta/(\alpha\boldsymbol{\alpha})$, so Eq. (9.68) is a solution that describes atmospheric solitary waves.

9.4.2 Detection of atmospheric solitary waves

So far, most solitary waves in atmospheric boundary layers were observed in the U.S. and Australia. For example, during spring in the Australian north region, the famous "morning glory" phenomenon can be seen [272]. These phenomena are like cirrus whose lengths often exceeds several hundreds of kilometers; they are the result of solitary waves that are displayed when the lower atmosphere is under the stably moist conditions. On the basis of estimation, the propagation velocity of these waves often exceeds 10 ms^{-1} with amplitude of several hundred meters and wavelengths of several kilometers. Doviak et al (1991) observed a solitary wave generated by a thunderstorm with an amplitude of 400 m and a wavelength of 2 km propagating through the inversion layer with thickness of 500 m at a velocity of 13 ms^{-1} at a distance of 100 km apart from its source in the central region of Oklahoma [273]. Ress & Rottman (1994) analyzed solitary waves in several cases observed in the Antarctic area, finding a range of phase velocities of 2~10 ms^{-1}, which were propagating from the dry land to the sty over the gradually inclining Antarctic coastal ice-shelf [274]. Now, it has been known that solitary waves with wavelengths of about 200 m and amplitudes less than 40 m can occur only at slower wind velocities.

Generally speaking, the amplitudes of solitary waves are generated by fluctuations of pressure and temperature measured on the ground surface or in the higher atmospheric. However, it is rather difficult to estimate the amplitude of a wave by using single-point sensors. As such, ground-base remote-sensing measuring systems are used, such as wind-profile instruments, RASS, laser radar and Doppler SODAR, which can provide images of these phenomena evolving in time and altitude. Among them, Doppler SODAR is regarded as a useful tool for surveying solitary waves at boundary layers. Sodar data can display images of variations of sound back wave intensity over time and altitude. These data provide direct information regarding the amplitudes of solitary waves propagating at altitudes below 1 km. Most SODARs can work well at altitudes about 1 km. The sodar characteristic signals of solitary waves are received by wave scattered by the boundary layer between the surrounding air and the returned air in side the wave. In SODAR data, the amplitude of solitary waves can be measured from the highest altitude of the returned wave. Christie et al. (1981)

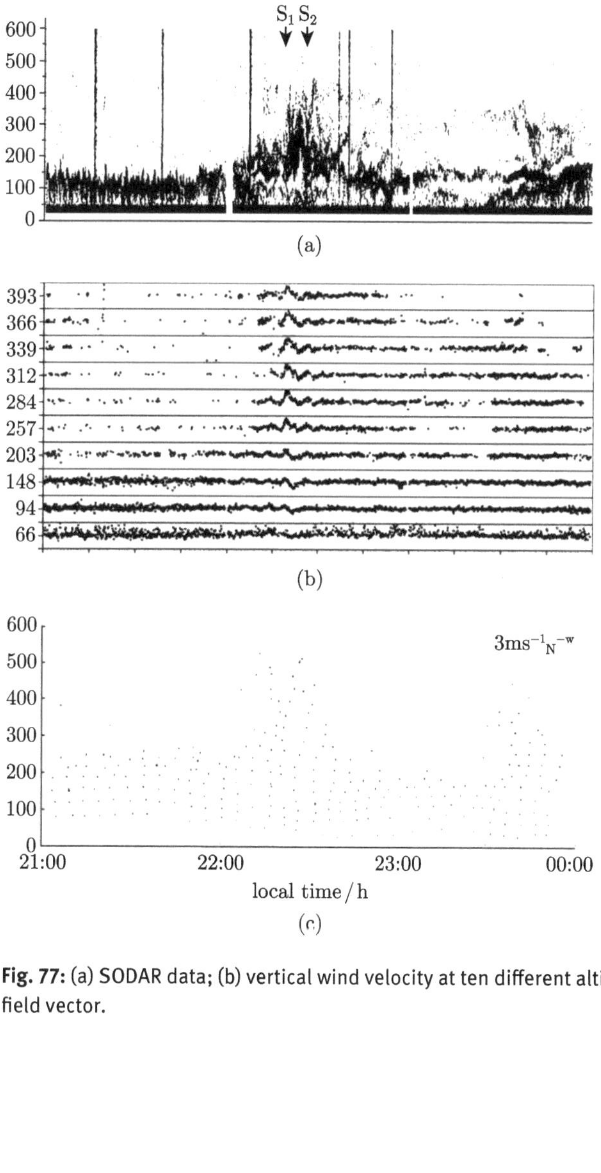

Fig. 77: (a) SODAR data; (b) vertical wind velocity at ten different altitudes; (c) horizontal wind field vector.

probed two cases of solitary waves by means of monostatic SODAR in Warramunga, Australia [275]. In each probing event, the number and amplitude of solitary waves was measured directly from the SODAR records. In the probing case, seven solitary waves were identified. Their amplitudes were between 200∼400 m and their wavelengths exceeded 1 km. Together, using the high-resolution SODAR with a largest measuring range of 100 m and a regular SODAR, Cheung & Little (1990) observed five events of solitary waves at night propagating in shallow stable boundary layers at Atmospheric Observatory in Boulder U.S., with amplitudes exceeding 500 m [276]. Moreover, satellites and other spatial image formation method, play auxiliary roles in surveying solitary waves.

A typical record of solitary waves propagating in a boundary layer in the sky over Rome City was made by Rao and Fiocco et al., as shown in Fig. 77. These solitary waves may have been generated by oceanic winds [277]. In the SODAR data, an area without an echo appeared, shown as S_1, S_2, and which indicate that returning air flow exists in solitary waves. Also shown in the figure are the characteristics of a solitary wave with a vertical velocity that monotonically increases with altitude. Furthermore, the amplitude-wavelength ratio they measured was well matched with the result deduced from theory. They continuous surveyed for over one year, however they observed solitary waves only three times. This fact denotes that solitary waves may not generally exist in the boundary layer in the sky over a city, and it may be also due to propagation conditions that can very easy destroyed them.

Chapter 10
Sound Sources in Atmosphere

10.1 Fundamental sound sources

The fundamental sound sources in an atmosphere can be categorized into three idealized types: monopole, dipole and quadrapole.

10.1.1 Monopole sources

A monopole source is a spherical source expanding and contracting uniformly. On the surface of a spherical source with radius a, the radial velocity of the vibration is given as

$$u = U\mathrm{e}^{\mathrm{i}\omega t} \tag{10.1}$$

where the amplitude U is a constant that represents the velocity amplitude. A spherically symmetric source will inevitably produce a spherical wave. The velocity potential of a particle satisfies the equation

$$\phi = \frac{A}{r} \exp\left(\mathrm{i}\omega t - \mathrm{i}kr\right) \tag{10.2}$$

Applying the boundary conditions, we get

$$A = \frac{a^2 U}{1 + \mathrm{i}ka} \exp(\mathrm{i}ka)$$

Sound pressure and particle velocity can be obtained from the equations given in chapter 2. The pressure solution that satisfies both the wave equation and the boundary conditions is

$$p = \rho \partial_t \phi = \frac{\mathrm{i}a^2 \omega \rho U}{r(1 + \mathrm{i}ka)} \exp(\mathrm{i}\omega t - \mathrm{i}kr + \mathrm{i}ka) \tag{10.3}$$

The amplitude of the sound pressure produced by a monopole is inversely proportional to the propagation distance.

The radiation power of a monopole source is expressed by its radiation resistance. The sound-field force acting on the sphere surface is

$$F = -Sp = \rho c \frac{(ka)^2 + \mathrm{i}ka}{1 + (ka)^2} Su = Z_r u \tag{10.4}$$

where S is the area of the sphere's surface. The above equation specifies that the force is proportional to the vibrating velocity. The proportion coefficient of the relation is just the radiation impedance

$$Z_r = \frac{F}{u} = \rho c \frac{(ka)^2 + \mathrm{i}ka}{1 + (ka)^2} S \tag{10.5}$$

If we separate the real part (radiation resistance R_r) and the imaginary part (radiation reactance X_r) we get

$$Z_r = \frac{(ka)^2 \rho c}{1+(ka)^2} S + \frac{ika\rho c}{1+(ka)^2} S = R_r + iX_r \tag{10.5'}$$

Where F can also be expressed as

$$F = -R_r u - M_r \dot{u} \tag{10.4'}$$

$M_r = \dfrac{X_r}{\omega}$ is the radiation mass. The sphere surface is subjected to a damping force and an inertial force caused by the existence of sound field. The average radiated power of the spherical sound source evaluated at the sphere's surface is

$$\bar{\omega} = \frac{S}{2} \mathrm{Re}\left[pv^*\right] = \mathrm{Re}\left[\frac{ka\rho c U^2 S}{2(i - ka)}\right] = \frac{1}{2} R_r U^2 \tag{10.6}$$

Where $v^* = U \exp(-i\omega t)$ is the complex conjugate of the particle velocity. It can be seen that the average radiated power depends on the radiation resistance, provided that the velocity of the sound source is known. In the case that the wavelength is much longer than the radius of the source, i.e., $ka \ll 1$, radiation impedance can be approximated as

$$R_r = (ka)^2 \rho c S \tag{10.7a}$$

Therefore, we can see that when the radiation area of the source is a constant, the sound radiation pressure will be lower at low frequencies. On the other hand, when the wavelength is much smaller than the radius of the sound source, i.e. $ka \gg 1$,

$$R_r = \rho c S \tag{10.7b}$$

Obviously, the radiation resistance is independent of frequency at high frequencies, and the larger the area of the sound source, the higher the sound radiation pressure becomes.

10.1.2 Dipole source

A dipole is formed by two monopoles with the same frequency and amplitude, but have opposite phases. Consider an acoustic dipole, whose two monopoles are separated by a distance d, and we choose the midpoint as the origin of the coordinate system. Let the distance between the observation point and the origin is r, the angle between the line connecting the observation point with the origin and the line connecting the two monopoles is θ. If the distance between the observation point and two origins are unequal, an acoustic path difference is given by

$$\Delta = \frac{d \cos \theta}{2}$$

will appear. According to the principle of linear superposition, the acoustic pressure at the observation point is the superposition of the acoustic pressure produced by the two monopoles. In atmospheric acoustics, we often care about the acoustic field far away from the sound source. However, we cannot neglect the effect of the phase difference caused by the path difference. The particle velocity potential in the acoustic field after superposition is

$$\phi = \frac{\mathrm{i}2A\sin(k\Delta)}{r}\exp(\mathrm{i}\omega t - \mathrm{i}kr) \tag{10.8}$$

When the wavelength is much longer than the distance between the two monopoles, the velocity potential can be rewritten as

$$\phi = \frac{Akd\mathrm{i}}{r}\cos\theta\exp(\mathrm{i}\omega t - \mathrm{i}kr) \tag{10.8'}$$

It can be seen that the radiation varies with direction and that the vibration amplitude decreases inversely proportional to the distance. The radiation directivity of the sound source can be expressed as

$$D(\theta) = \frac{p}{p\,|_{\theta=0}} = |\cos\theta| \tag{10.9}$$

The radiation intensity of the dipole is shown as

$$I = \mathrm{Re}\left(pv^*\right) = \frac{(|A|\,kd\cos\theta)^2}{2\rho cr^2} \tag{10.10}$$

Owing to the directivity of the dipole radiation, its average power through the spherical thce with radius r should be calculated by the following integration

$$\bar{\omega} = \int\!\!\int_S I\mathrm{d}s = \iint Ir^2\sin\theta\mathrm{d}\theta\mathrm{d}\varphi = \frac{2\pi}{3\rho c}(|A|\,kd)^2 \tag{10.11a}$$

The average radiation power through any sphere is a constant. In the case of a far field and long wavelength, the radiation directivity of monopoles and dipoles can be compared by evaluating their equivalent radiation resistances, which means that when $kd \ll 1$,the average sound power of the dipole radiation is

$$\bar{\omega} = \frac{2\pi}{3}\rho c(ka)^4 d^2 u_a^2 \tag{10.11b}$$

The dipole's equivalent radiation resistance can be obtained from the above equation

$$R_e = \frac{4\pi}{3}\rho c(ka)^4 d^2 \tag{10.12}$$

Hence, in case of a long wavelength, the radiation resistance of a dipole is directly proportional to k^4, while that of a monopole is directly proportional to k^2. Therefore, dipole radiation is smaller than a monopole at low frequencies.

10.1.3 Quadrupole sources

A quadrupole sound source consists of four identical monopoles located at the corners of an equilateral quadrilateral with side-length b, in which two of the adjacent monopoles vibrate in reverse phases. The acoustic pressure produced by a quadrupole at far distances can be obtained by the principle of superposition

$$p = \frac{iba^2 \rho ck^3 U \cos\theta \sin\theta}{2r} \exp(i\omega t - ikr) \tag{10.13}$$

The pressure produced by a quadrapole is inversely proportional to the distance between the source and the observation point. However, its radiation directivity is much more complicated than that of a dipole.

10.1.4 Piston sources

A piston with an infinite baffle (in practice, a baffle can be regarded as infinite when its length is much longer than the considered wavelengths) is another sound source that can better describe real situations than the above mentioned three ideal sound sources (monopoles, dipoles and quadrupoles). Let the surface of the piston source be a circular plane on which each point vibrates with the same velocity and moves perpendicular to the surface. The radiation is restricted in half space. The acoustic field of the piston source can be treated as the superposition of the acoustic field produced by an infinite number of monopoles. Considering a piston source with radius a, the vibrating velocity on the surface is specified in (10.1).

The mathematical description of the near field for a piston source is quite complex. Therefore, we will only consider acoustic radiation in the far field. We will use the spherical coordinate system (r, θ, φ), with its origin being at the middle of the piston. Let the radial coordinate on piston surface be σ, the infinitesimal area be $\mathrm{d}S = \sigma d\sigma d\varphi$ and the distance between the infinitesimal area and the observation point be h. The distance can be approximated as $h = r - \sigma \cos\varphi \sin\theta$ when h is much larger than the size of the sound source, with the solution being

$$p = \int_0^{2\pi}\int_0^{a} \frac{i\omega\sigma U}{2\pi r} \exp(i\omega t - ikr - i\sigma\cos\varphi\sin\theta)\, \sigma d\sigma d\varphi$$

Applying the property of Bessel functions, the above equation becomes

$$p = \frac{i\omega\sigma a^2 U}{2r} \frac{J_1(ka\sin\theta)}{ka\sin\theta} \exp(i\omega t - ikr) \tag{10.14}$$

This equation demonstrates that the radiation pressure of a piston source is inversely proportional to the distance. The directivity factor of the piston source can be written as

$$D(\theta) = \frac{p}{p_{\theta=0}} = \left|\frac{2J_1(ka\sin\theta)}{ka\sin\theta}\right| \tag{10.15}$$

For $x < 1, J_1(x) \approx \frac{x}{2}$, hence, if $ka \ll 1$, then $D(\theta) \approx 1$ which specifies that the radiation is almost non-directional when the wavelength is much larger than the size of the piston. When ka is larger than 3.83, the first root of $J_1(x)$ implies that there will be almost no radiation.

10.1.5 Fluid sources

Many sound waves are produced by violent flows of fluids, such as sea vibrations, wind and atmospheric, turbulences the concrete examples will be discussed in next section. Firstly let us derive the dominating equation for sound radiation of flowing fluids. Since the fluids in violent moving, the mass-flow continuity equation can be expressed as

$$\partial_t \rho + \nabla \cdot (\rho u) = q\rho \tag{10.16}$$

where q is flow rate of a fresh fluid volume entering into the specified space element. By utilizing the relationship between the change of mass density and the temperature change, we can obtain

$$\partial_t \rho \approx \gamma \rho \kappa \partial_t (p - \alpha \tau)$$

where γ is the adiabatic index, κ is the isentropic compressibility, α is the ratio of expansion coefficient a isothermal compressibility, and τ is the temperature perturbation. By applying the heat energy per unit mass and neglecting thermal conductivity, the following relation can be obtained

$$\partial_t (\rho - \alpha \tau) \approx \frac{1}{\gamma} \partial_t p - \alpha \varepsilon / C_p$$

The mass-flow continuity equation becomes

$$\rho \kappa \partial_t p + \nabla \cdot (\rho u) = q\rho + \frac{\alpha \gamma \rho \kappa \varepsilon}{C_p} \tag{10.17}$$

The momentum-flow continuity equation can be written as

$$\partial_t (\rho u) = F - \nabla P - \nabla T \tag{10.18}$$

where F is the force on a unit volume, P is the fluid pressure, and T is the total stress tensor excluding the fluid pressure. Taking time derivation of Eq. (10.17) and the divergence of Eq. (10.18), eliminating any terms that contain ρu in both equations and neglecting the thermal and viscous terms, we obtain the sound-pressure wave equation

$$\rho \kappa \partial_t^2 p - \nabla^2 p = \partial_{ij}^2 T_{ij} - \partial_i F_i + \partial_t (\rho s) \tag{10.19}$$

The non-linear terms on the right-hand side of Eq. (10.19) are usually known, which correspond to a quadripole, a dipole, and a monopole respectively. s in the last term of the right-hand side corresponds to the intensity of an effective monopole source.

Eq. (10.19) is the dominatiing equation that describes the sound waves radiated by flowing fluid, which will be applied to discussions of sea vibration, as well as to sound waves produced by turbulences.

10.2 Natural sound sources

A vast amount of infrasound exists in the atmosphere that are created by various natural sources, among which most are happened intermittently and occasionally, though some are near continuously, thus it is possible to use them for studying variability of the wind field in the upper atmosphere through their influences on long-distance sound propagation.

10.2.1 Ocean waves

Approximately 71% of the earth surface is covered by oceans. Ocean surfaces are often in a fluctuating state, which produces a huge amount of noise. However, not all ocean waves can create sound. Theoretical analyses have shown that they can produce sound waves, the so-called microbarom, only when standing waves are formed as single traveling waves cannot generate sound. Among the numerous observed natural infrasounds, microbaroms with periods of 2~8 s and amplitude around several μPa can be listed at the most typical ones. Those microbaroms can be often detected with an infrasound transmitter array with a spacing distance of 1 km. Such microbarom signals arrive from different directions, displaying their low spatial correlation; thus it turns out to be a surface source with big area which is now defined as coming from ocean waves, and they are tightly related to typhoons (Refer to chapter 7, section 7.4.2). Since storms can stimulate huge waves in the oceans, strong sound waves will occur during storms on the sea.

In general, an ocean-surface traveling wave cannot produce a sound wave because its phase velocity is much smaller than that of a sound wave. Ocean surface waves are dispersive, according to the dispersion relation for deep water gravity surface wave $\omega^2 = gk_S$ (subscript S refers to "surface"), where g is the acceleration due to gravity and k_S is the wave number of an ocean surface wave. The wavelength can be found to be 15~100 m. Using the deep water gravity wave speed formula $c_S = \sqrt{g/k_S}$, we can get the phase velocity, 5~12 m s^{-1}, which is much lower than that of a sound wave in the air. Sound waves and the ocean-surface waves satisfy the dispersion relation $\omega = ck$ as well as $\omega_S = c_S k_S$, respectively. If there is sound radiation, its frequency should be consistent with the frequency of the corresponding ocean waves, i.e. $\omega = \omega_S$, and the horizontal component of the waves the same is that for the water wave number for a sound wave, i.e. $k_x = k_S$. Applying the two dispersion relations we can compute the vertical component of the wave number of a sound wave, i.e. $k_y^2 = k_x^2 \left[(c_S/c)^2 - 1 \right]$. Since $c_S < c$, k_y is an imaginary number, implying that the sound wave is an attenuated wave that cannot travel very far.

On the other hand, non-linear interactions between ocean-surface waves also radiate sound waves. For example, the interaction between two ocean-surface waves propagating along the same straight line, but in opposite directions, can produce a sound wave whose frequency is the sum of the frequencies of the two ocean waves.

Some preliminary theory of ocean waves producing sound waves will be discussed below. The undulating ocean surface can be regarded as a moving lower boundary relative to the air above an ocean surface. The vibrating boundary causes a vibration of the air, and a sound wave is formed as the vibrations spread out. The radiation equation of the sound wave is the non-linear inhomogeneous wave equation, and the inhomogeneous term therein is just the sound source term. In general, sound sources include monopoles, dipoles and quadruples. They are produced by the rate of changes in time of air particle displacement, by pressure at the boundary and in the velocity potential on the boundary, respectively. All of the changes are caused by the vibration boundary. The ocean-surface acoustic radiation formula can be derived from Eq. (10.19)

$$\left(\partial_t^2 - c^2 \nabla^2 \right) (H(f)\rho') = \partial_{ij}^2 \left(H(f)\, T_{ij} \right) - \partial_j \left(P_{ij} \partial_j H(f) \right) + \partial_t \left(\rho_0 \mu_i \partial_i H(f) \right) \qquad (10.20)$$

where c is the velocity of sound in air, ρ' is the density change of air caused by sound waves, $T_{ij} = \rho u_i u_j + P_{ij}$ is the Lighthill stress tensor, and P_{ij} is the fluid pressure tensor. We take z as the vertical axis, and x and y as the horizontal axes, $f(x, y, z, t) = z - h(x, y, t) = 0$ as the axis along the sea surface, and $H(f)$ as the unit step function. The first, second and third terms on right-hand side of Eq. (10.20) represent quadrioples, dipoles and monopoles, respectively. This equation can be solved by using the Green function. Assuming that oceanic water is incompressible, and that the Lighthill stress tensor satisfies the following motion function, we find

$$\rho \partial_t u_i = -\partial_j T_{ij} \qquad (10.21)$$

Expressing the derivative of function H by δ function, the inhomogeneous term can be rewritten as:

$$\partial_i \left(\rho u_i u_j \partial_j f \delta(f) \right) - \rho \delta_t u_i \delta(f) + \rho \partial_t \left(u_i \partial_i f \delta(f) \right)$$

Since we are interested in the radiation of a transmittable sound wave, we will only consider the case of the far-field solution with enough long wavelengths. Thus, we can assume that the amplitude of surface wave h is a small quantity relative to the wavelength of the surface wave. To obtain the acoustic radiation, we keep the linear and the quadratic terms of the surface wave parameters, but neglect the of higher orders order terms. The general solution to the ocean-surface wave radiation can be obtained as

$$\rho' = -\frac{\rho_0}{8\pi c^3} \partial_t \int \frac{z}{R^2} u_z^2 \big|_{z'=0, \tau=t-R/c} \mathrm{d}x' \mathrm{d}y' \qquad (10.22)$$

where $R = \left((x-x')^2+(y-y')^2+z^2\right)^{1/2}$, and $u_z = \partial_t h$ is the vertical velocity of the surface of the water. As can be seen from Eq. (10.22), the solution contains the square of the vertical velocity of surface wave.

Now let us decompose Eq. (10.22). We will only consider the case when the ocean waves are harmonic. Expressing u_z in terms of harmonic vibration motion, the acoustic radiation field can be obtained by substituting it into Eq. (10.22). First, consider the simpler situation by assuming that there is only a pair of ocean-surface waves as in the two-dimensional case

$$h(x,y) = a_1 \cos(\omega_1 t - k_1 x + \phi_1) + a_2 \cos(\omega_2 t - k_2 x + \phi_2) \tag{10.23}$$

where ω and k_s satisfy the dispersion relation for deep ocean-surface waves, while $\omega = \sqrt{g|k_s|}$, a_1, a_2, ϕ_1, ϕ_2 are constants to be determined. The vertical velocity component of ocean-surface waves is then

$$\begin{aligned} u_z = &-\omega_1 a_1 \sin(\omega_1 t - k_1 x + \phi_1)\exp(-|k_1|z) \\ &+\omega_2 a_2 \sin(\omega_2 t - k_2 x + \phi_2)\exp(-|k_2|z) \end{aligned} \tag{10.24}$$

Basing on the above discussion, we can conclude that an individual surface wave cannot radiate a sound wave. Let us now consider the intersecting term composed by two waves

$$u_z^2 \to \omega_1\omega_2 a_1 a_2\left(-\cos\left(\omega^+ t - k^+ x + \phi^+\right) + \cos\left(\omega^- t - k^- x + \phi^-\right)\right) \tag{10.25}$$

where $k^\pm = k_1 \pm k_2$, $\omega^\pm = \omega_1 \pm \omega_2$, $\phi^\pm = \phi_1 \pm \phi_2$. Consider the case that $z = 0$. Substituting (10.25) into (10.22), we can get

$$p' = \frac{\rho_0\omega_1\omega_2 a_1 a_2}{4c^2}\left(\cos\left(\omega^+ t - k^+ x + \phi^+ - k_z^+ z\right) + \cos\left(\omega^- t - k^- x + \phi^- - k_z^- z\right)\right) \tag{10.26}$$

where

$$k_z^\pm = \sqrt{\left(\frac{\omega^\pm}{c}\right)^2 - (k^\pm)^2} \tag{10.27}$$

The expressions demonstrate that the non-linear interaction of two surface waves radiate two waves whose frequencies and horizontal wave numbers equal the sum and the difference of the frequencies and horizontal wave numbers of these two surface waves respectively. It can be proved that a wave whose frequency is the difference of those of the surface waves is an attenuation wave (it is essentially caused by a mismatch of phase velocities as mentioned above). Therefore, we only consider a wave whose frequency is the sum of the surface waves

$$p' = \frac{\rho_0\omega_1\omega_2 a_1 a_2}{4c^2}\cos\left(\omega^+ t - k^+ x - k_z^+ z + \phi^+\right) \tag{10.26a}$$

To ensure k_z^+ is a real number, the propagation directions of these two surface waves must travel, closely to a straight line and in opposite directions. To understand this intuitively, let us consider an ideal situation of an ocean standing waves. Let $a_1 = a_2 = a$, $\omega_1 = \omega_2 = \omega$, $k_1 = -k_2 = k_s$, $\phi_1 = \phi_2 = 0$, then

$$h(x, y) = a\cos(\omega t - k_s x) + a\cos(\omega t + k_s x) = 2a\cos(\omega t)\cos(k_s x) \tag{10.28}$$

In this situation, Eq. (10.26a) becomes

$$\rho' = \frac{\rho_0 \omega^2 a^2}{4c^2}\cos\left(2\omega t - k_z^+ z\right) \tag{10.29}$$

where $k_z^+ = 2\omega/c$ is a real number. This is a wave propagating vertically upward. It can be seen that the frequency of the sound wave is twice that of the ocean standing waves, and it is independent on the horizontal axis x. It is caused by a dependence of the wave-sound solution on the quadratic term in the equation of the ocean-wave velocity. The quadratic term in the sound-wave solution is

$$u_z^2 = a^2\omega^2\left[1 + \cos(2k_s x) - \cos(2\omega t)\cos(2k_s x) - \cos(2\omega t)\right] \tag{10.30}$$

Let us see the four terms in the brackets: the first two are unrelated to time so they cannot radiate sound waves. The third one can radiate attenuating waves only, as can be proved. The fourth one dependents on time with frequency 2ω, but is independent on the horizontal axis, which is consistent with the sound radiation described by Eq. (10.29). So only the fourth term can radiate sound. It can be seen from the above discussion that sound radiated by ocean standing waves is not the result of a simple piston moving on an ocean surface, but arises from a complex interaction between air vibrations above an ocean and ocean surface waves that stimulate this vibration.

Next we will consider sound radiation produced by interactions between ocean-surface waves within a certain range in the frequency domain. Let us still consider two-dimensional cases only. If the spectrum of an ocean-surface wave within a certain section has a continuous frequency spectrum denoted as $A(k_s)$, the disturbance of the ocean surface is obtained by a Fourier integral

$$h(x, t) = \int_{-\infty}^{+\infty} A(k_s)\cos(\omega t - k_s x + \phi(k_s))\,\mathrm{d}k_s \tag{10.31}$$

where $\omega = \sqrt{g|k_s|}$. Similarly, we define the spectrum of a radiation wave as $B(k_x, k_z)$, where k_x and k_z are the horizontal and vertical wave numbers, respectively, and $k = \sqrt{k_x^2 + k_y^2}$. The variation of mass density of a sound wave can be expressed as

$$\rho'(x, z, t) = \int_{-\infty}^{+\infty}\int_{0}^{+\infty} B(k_x, k_z)\cos(\omega t - k_x x - k_z z + \phi(k_x + k_z))\,\mathrm{d}k_z\mathrm{d}k_x \tag{10.32}$$

where $\omega = c\sqrt{k_x^2 + k_z^2}$. By combining Eq. (10.22) and (10.31), the following equation is obtained

$$\rho' = \frac{\rho_0}{4c^2} \int_{-\infty}^{+\infty}\int_{-\infty}^{+\infty} A(k_s)A(k_s')\omega\omega' \cos((\omega + \omega')t - (k_s + k_s')x - k_z z + (\phi + \phi'))\mathrm{d}k_s\mathrm{d}k_s' \tag{10.33}$$

where k_z is just k_z^+ in Eq. (10.26a). To obtain $B(k_x, k_z)$, we rewrite Eq. (10.33) in the form of Eq. (10.32). Using the transforms

$$k_x = k_s + k_s', \quad \xi = k_s - k_s' \tag{10.34}$$

The integral changes to

$$\rho' = \frac{\rho_0}{8c^2} \int_{-\infty}^{+\infty}\int_{-\infty}^{+\infty} A(k_s)\,A(k_s')\,\omega\omega' \cos((\omega + \omega')\,t - k_x x - k_z z + (\phi + \phi'))\mathrm{d}k_x\mathrm{d}\xi \tag{10.35}$$

where

$$k_z = \left(\frac{g}{2c^2}\,|k_x + \xi| + \frac{g}{2c^2}\,|k_x - \xi| + \frac{2g}{2c^2}\,|(k_x + \xi)\,(k_x - \xi)|^{1/2} - k_x^2\right)^{1/2} \tag{10.36}$$

To ensure k_z is a real number, there must be $|\xi| \gg |k_x|$, since g/c^2 is a small quantity. Then we can obtain the approximation

$$k_z \approx \left(\frac{2g}{c^2}\,|\xi| - k_x^2\right)^{1/2} \tag{10.37}$$

Substituting ξ (taking care that ξ is positive or negative) in (10.36) with k_z given in Eq. (10.37), we have

$$\rho' = \frac{\rho_0}{4c^2} \int_{-\infty}^{+\infty}\int_{0}^{+\infty} A(k_s)A(k_s')\omega\omega' \cos((\omega + \omega')t - k_x x - k_z z + (\phi + \phi'))k_z\mathrm{d}k_z\mathrm{d}k_x \tag{10.38}$$

where

$$k_s = \frac{1}{2}(k_x + \xi) = \frac{c^2}{4g}(k^2) + \frac{1}{2}k_x \tag{10.39a}$$

$$k_s' = \frac{1}{2}(k_x - \xi) = -\frac{c^2}{4g}\left(k^2\right) + \frac{1}{2}k_x \tag{10.39b}$$

$$\omega = (g\,|k_s|)^{1/2} \approx \frac{1}{2}ck + \frac{gk_x}{\sqrt{2}c}\frac{1}{k} \tag{10.40a}$$

$$\omega' = (g\,|k_s'|)^{1/2} \approx \frac{1}{2}ck - \frac{gk_x}{\sqrt{2}c}\frac{1}{k} \tag{10.40b}$$

The last terms in the previous four formulae are small quantities. Inserting these expressions into Eq. (10.38) and keeping all terms with the same order of g/c^2, the radiation acoustic field can be expressed as

$$p' = \frac{\rho_0 c^2}{16g} \int_{-\infty}^{+\infty} \int_{0}^{+\infty} A(k_s)A(k_s')k^2 k_z \cos((\omega + \omega')t - k_x x - k_z z + (\phi + \phi'))\mathrm{d}k_z \mathrm{d}k_x \tag{10.41}$$

Comparing Eq. (10.32) with (10.41) we can obtain the Fourier spectrum of the radiation acoustic field

$$\begin{aligned} B(k_x, k_z) &= \frac{\rho_0 c^2}{16g} k_z k^2 A(k_s) A(k_s') \\ &= \frac{\rho_0 c^2}{16g} \left[k_z k^2\right] A\left(\frac{c^2}{4g}k^2 + \frac{1}{2}k_x\right) A\left(-\frac{c^2}{4g}k^2 \frac{1}{2}k_z\right) \end{aligned} \tag{10.42}$$

Since c^2/g is a large quantity, if $A(k_s)$ varies slowly, Eq. (10.42) can be approximated to

$$B(k_x, k_z) = \frac{\rho_0 c^2}{16g} \left[k_z k^2\right] A\left(\frac{c^2}{4g}k^2\right) A\left(-\frac{c^2}{4g}k^2\right) \tag{10.42a}$$

Eq. (10.42a) demonstrates the relation between the spectra of an ocean-surface wave and that of the radiated sound wave. Some useful results can be obtained from this equation.

Let us now study the ratio g/c^2 in the above equation. The phase velocity of an ocean-surface wave is $\omega/k_s = \sqrt{g/k_s}$, and that of a sound wave is c. By equalizing and squaring them, we can get

$$k_s = \frac{g}{c^2} = K_{SA} \tag{10.43}$$

where K_{SA} is the angular wave number that makes the phase velocity of an ocean-surface wave equal to that of a sound wave. Converting to wavelength is obtained by using the values $c = 340$ m s^{-1} and $g = 9.8$ m s^{-2}

$$\lambda_{SA} = \frac{2\pi c^2}{g} \approx 74\,\text{km} \tag{10.44}$$

When the wavelength of a surface wave is smaller than $\lambda_{SA} \approx 74$ km, its phase velocity is smaller than that of a sound wave with the same wavelength. Since wavelengths of most of the ocean-surface waves are less than 74 km, most of them are slower than sound waves. In fact, this condition has already been used when obtaining the sound radiation solution.

The term bracketed in Eq. (10.42) can be regarded as the efficiency factor of a spectrum transferring for the sound waves radiated by the ocean surface waves. It increases with an increase of $|k|$, which means that ocean-surface waves with short wavelength will be more efficient for a radiating sound wave. In other words, ocean-surface waves with high frequencies can radiate more sonic energy, which is determined by its sound source property. This result shows that the spectrum peak of ocean-surface waves will shift to higher values in the spectrum of sound waves.

Eq. (10.42) can also be used to study the directivity of sound radiation. The sound-frequency spectrum, which is directly proportional to the vertical component of the sound wave number, can be expressed as

$$B(k_x, k_z) = \frac{k_z}{k} A'(k) = A'(k) \cos\theta$$

In which $A'(k)$ is determined by the shape of the ocean-surface wave spectrum. The equation also tells us that the sound radiation is proportional to the cosine of the polar angle, which has a maximum value in vertical direction and is zero in horizontal direction, and which gives the same polarity as a dipole whose polar axis is in the horizontal direction. The directivity is related to neither the wave number of sound wave, nor the shape of the ocean wave spectrum.

10.2.2 Heavy objects falling down into water

Heavy objects, such as rocks or ice, that fall into water can produce large sounds; the stimulated water waves can produce audible sound and long-distance propagating low-frequency infrasound that are able to propagate for large distances. If the object intense infrasound com be radiated by the surface water ware is very large, in the hundred-meter range, it can produce a vertical displacement of nearly the same scale as the size of the object on the water surface.

An object falling into a free-water surface can generate a series of complex effects, especially when the object is irregularly shaped. It is difficult to accurately describe produced the sound wave by applying a mathematical model. However, we can make some estimation theoretically for some idealized situations.

The basic assumption is that the water is an incompressible fluid, and sound radiation will not influence its motion in an obvious way. Consider a wave created by an ideal ball falling into water. The wave amplitude decreases sharply with the spreading of waves. Only consider a wave in the middle, formed by a circular ring with an inner radius a and an outer radius b. According to the volume-conservation law, the area of the inner circle and that of the concentric ring is the same. So we have $b = \sqrt{2}a$. The inner circle and the concentric ring form a reverse dipole, as shown in Fig. 78.

In the ideal model of the vibration of a piston with a baffle, the sound radiation is

$$p = -\frac{\exp(ikr)}{r} \frac{i\omega\rho U a^2}{2} \frac{2J_1(ka\sin\theta)}{ka\sin\theta} \tag{10.45}$$

where U is the velocity amplitude of the piston, a is the radius of the piston, θ is the angle between the line connecting observation point and sound source with the

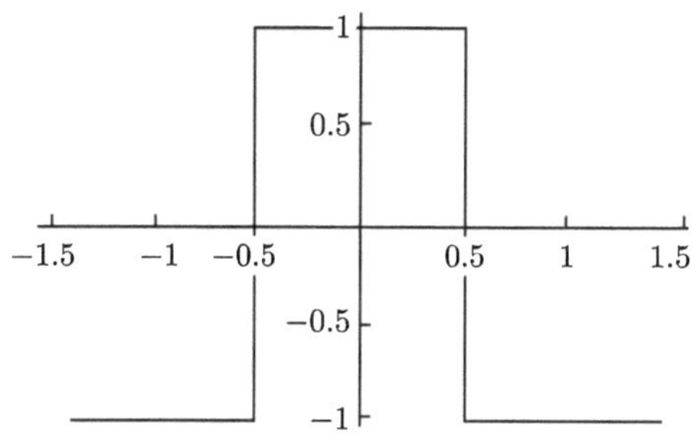

Fig. 78: Dipoles with reverse phase.

normal to water surface, ω is the angular frequency, and r is the distance between the sound source and the observation point far keyunt.

Radiation of the dipole with reverse phase can be seen as consisting of the difference of radiating acoustic waves formed by the two concentric circles with radius a and $\sqrt{2}a$, respectively. The sum of the radiations from the inner circle with reversed phase and the concentric ring is

$$p = -\frac{i\omega\rho U a^2}{r} D \tag{10.46}$$

where

$$D = \left[\frac{4J_1(ka\sin\theta)}{ka\sin\theta} - \frac{2J_1(ka\sin\theta)}{ka\sin\theta} \right] \tag{10.47}$$

D is a factor of that describes the radiated sound directivity. When $\theta = 0$, $D = 0$, this corresponds to dipole radiation. When $\theta = \pi/2$, the radiation reaches a maximum in the horizontal direction. Next let us estimate the radiation's amplitude. If the falling object is close to the water's surface, and falls slowly into water, the frequency of the radiated sound will be consistent with the vibration frequency of the surface wave. The frequency of the water wave can be computed by applying the following methods. Since the wavelength of the surface wave is a, we get $k_S a \approx 1$. Because $\omega \approx c_S k_S = \sqrt{g k_S} = \sqrt{g/a}$, when the amplitude of the surface displacement is h, the vibration velocity of the piston is $U = \omega h = h\sqrt{g/a}$. Substitution the expressions for U and ω into Eq. (10.46), we can get the amplitude of sound pressure

$$P \approx \frac{\rho g a h}{r} D(ka, \theta) \tag{10.48}$$

Assuming $h = 0.2a$, then we can get $k = \sqrt{g/a}c^{-1}$ and can obtain $k/k_S = c_S/c < 1$, which indicates that the wavelength of the sound wave is longer than that of the water wave. Take a numerical example: assuming that $a = 50$ m and $h = 10$ m, then $\omega - \sqrt{g/a} = 0.44$ rad/s, and the corresponding frequency is $f = 0.07$ Hz, $ka = 0.065$ and the air density is 1.3 kg/m^3. When $\theta = \pi/2$, then $D = 5.3 \times 10^4$, the signal 1 km apart is on the order of 0.003 Pa.

If an object falls into water from high altitude it will dosh against the water surface a high velocity V, The frequency of the surface wave cannot be calculated using the above method. Assume that the characteristic time T for one up-and-down of the water surface is determined by the time for the object to passing through the water surface. Thus, the characteristic frequency of the surface motion can be estimated by $\omega \approx 2\pi/T \approx 2\pi V/2a$. The velocity of surface displacement U is at is at the same order of magnitude with V. The maximum displacement is a. Under this circumstance, Eq. (10.48) changes to

$$P \approx \frac{\pi\rho a V^2}{r} D(ka, \theta) \tag{10.48'}$$

where $k = \omega/c = \pi V/ac$, $ka = \pi V/c$. Here is another classical numerical example: if $a = 50$ m, $V = 10$ m/s, and the corresponding frequency and is $f = 0.1$ Hz and $ka \approx 0.09$. When $\theta = \pi/2$, the maximum value of D is about 0.001. The amplitude of the acoustic pressure is about

$$P_{\text{max}} = \frac{20}{r_m} \text{Pa}$$

Therefore, we get the same magnitude order as the acoustic signal, which is 0.02 Pa at $r = 1$ km. Generally, a signal with an acoustic pressure larger than 0.01 Pa can be detected.

Another simplified model is a monopole, which may have a chance to occur at the initial stage when an object hits a water surface. The falling object produces an air-cave in the water, driving by the buoyant force it rises fast to form a convex on the surface. It we get rid of the outer ring in the previously mentioned dipole model, it will change to a monopole model. Since there is no counteraction from the outer ring, the acoustic pressure produced by the monopole would be larger.

10.2.3 Violent firing

A large fire is also another important sound source. Intense combustion processes can radiate infrasound. Actual measurements show that the frequency of the produced sound wave is related to the size of fire fields, where an increase and decrease of the detected acoustic pressure is consistent with the change of the fire's intensity. The process of fire burning is very unstable. In a fire field, frames in different spots have unstable periodicity, wake the ambient air pressure to oscilate, thus producing a sound waves. An empirical formula for the frequency and size of a fire field can be concluded from vast measurement data of fire scenes

$$f = \frac{1.5}{\sqrt{2r}} \tag{10.49}$$

where f is the frequency of the radiated sound wave, and r is the radius of the fire scene. This formula can be applied to fire scenes with radii ranging from 0.01 to 100 m.

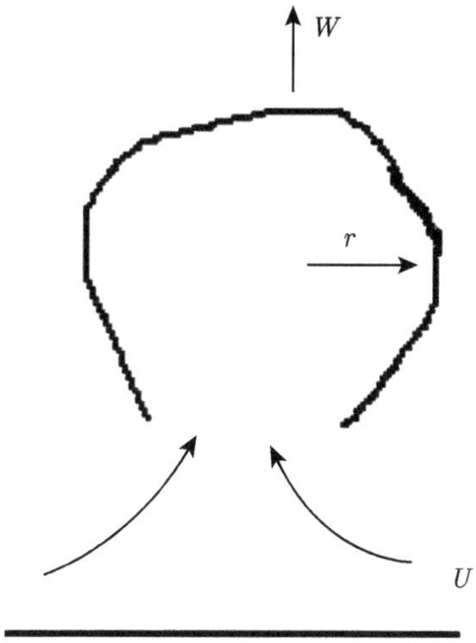

Fig. 79: Schematic diagram for the air flows in flame.

Studying flames is a very interesting and useful subject. It was been found in early studies that the flame of a gas emitted from tubes does not continuously burn. Instead, the fire has an on and off characteristic. Similar situations are also found in fire disasters. Getegen & Amged's model treats a fire scene as being formed by a series of sub-heating sources that alternately appear and disappear. Sound is produced in this process. How the sub-heat sources are formed and how they interact to produce periodicity remains a topic warranting further study. Next let us perform some simple theoretical analysis of the empirical formula. Heated gases rise in fires, and ambient gases replace them. The schematic diagram Fig. 79 defines some related variables. in this W is the rising velocity of the heated air flow and U is the velocity of the air flowing into the fire scene, they are approximately the same. The Froude number has different forms for different scenarios. The Froude number F that is related to high temperature can be defined as

$$F = \left(\frac{T}{\Delta T}\frac{U}{gr}\right)^{1/2} \tag{10.50}$$

where ΔT is the temperature of the flame and T is the environmental temperature, the characteristic time of the fire scene is $\tau = r/U$ and the frequency f of the sound wave is the reciprocal of the characteristic time

$$f = 1/\tau = U/r$$

utilizing to the definition of the Froude number given in (10.50), the frequency of the sound wave can be written as

$$f = \frac{F}{r}(gr)^{1/2} = F\left(\frac{\Delta T}{T}\frac{g}{r}\right)^{1/2} \tag{10.51}$$

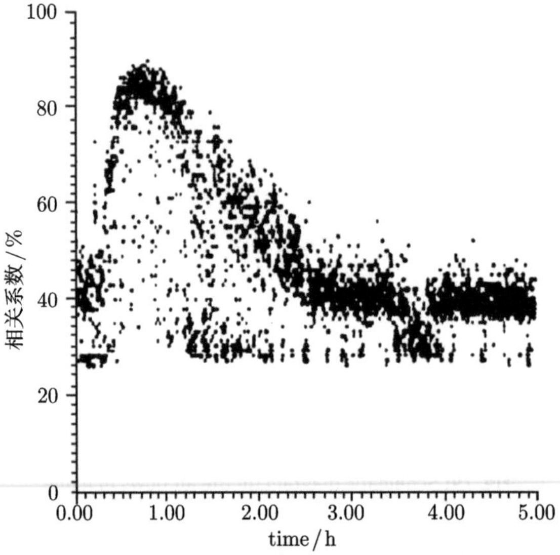

Fig. 80: Temporal change of correlation coefficient of the infrasonic waves generated from the conflagration occurred in a wood-up workshop.

When we consider a classical example where $g = 9.8\ \mathrm{m/s^2}$, $\Delta T = 2100$ K, $T = 300$ K, the following relation is obtained

$$f = \sqrt{70}\frac{F}{\sqrt{r}} \approx 8.4\frac{F}{\sqrt{r}}$$

Compared with the empirical formula Eq. (10.50), $f = 1.5/\sqrt{2r} \approx 1.04/\sqrt{r}$, we can calculate the Froude number, which is about $F \sim 0.124$. This result specifies that the buoyancy force in the fire is about ten times larger than the inertia force.

Some infrasound monitoring arrays have been installed to monitor infrasound produced in practical fire scenes. The data obtained can be used to compare with the main frequencies predicted by the empirical formulae.

In 2001, a conflagration happened in the storehouse of a factory producing regenerated wood in the USA. This fire produced an obvious sound signal that lasted 3 h, as shown in Fig. 80. When the intensity of the fire was strongest, the correlation coefficient of the measured data reached 0.9. The monitoring station also measured the azimuth angle as a function of time. When the fire spread to near the store house area, the measured azimuth angle changed with time. The main frequency of the produced sound wave when the fire was strongest was 1~2 Hz, while in the follow-up stage, it was 1~4 Hz.

In 2001, another conflagration happened in a natural gas station due to a leakage issue. The fire produced obvious infrasound. The infrasound monitoring station recorded that the fire lasted about 6 h. The recorded signal is shown in Fig. 81. The picture

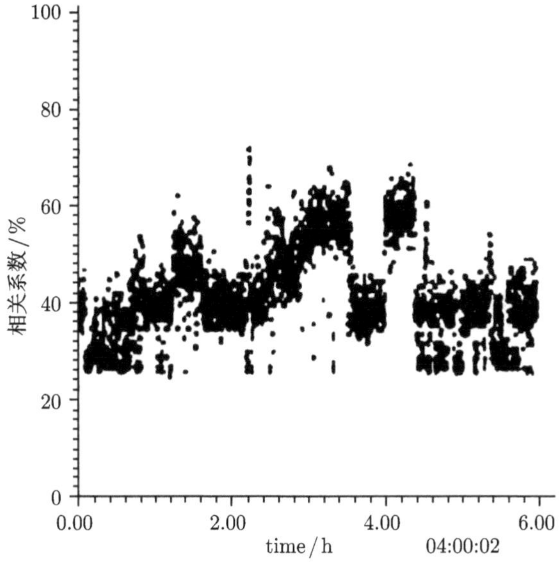

Fig. 81: Temporal change of correlation coefficient of the infrasonic waves generated from the contlagation occarred in a netural gas ceater.

clearly demonstrates that infrasound was produced by the explosion of the burning gas. Since the circumstances of these two contlagrations are known, to depose result with the empirical formula the monitored data.

10.2.4 Strong wind

(1) Mount ain-lee waves

Infrasound produced in the condition of a certain wind flow and certain mountain geometry is a called mount-back wave, whose typical period is about 40 s and whose typical amplitude is around 0.5 Pa. When the air is saturated enough to form cloud, mount-back waves can introduce spectacular and unique cloud images that are easily identified by eye. Because it is an important reason that creates fierce clear-air turbulences (CAT), its latent threat to airplanes is undoubted now.

Massive infrasound sources can be monitored near some hilly regions in winter. The results indicate that the topography characteristics of some hilly regions are particularly favorable for producing infrasound. The sensor array of the atmosphere monitoring station in Boulder (where the National Oceanic and Atmospheric Administration (NOAA) locates), near Lake Erie in Colorado, U.S., gives an observed example of a mountain-lee wave. This station recorded two types of sound signals that differed greatly in their levels and species, and both of which were from the mountain range

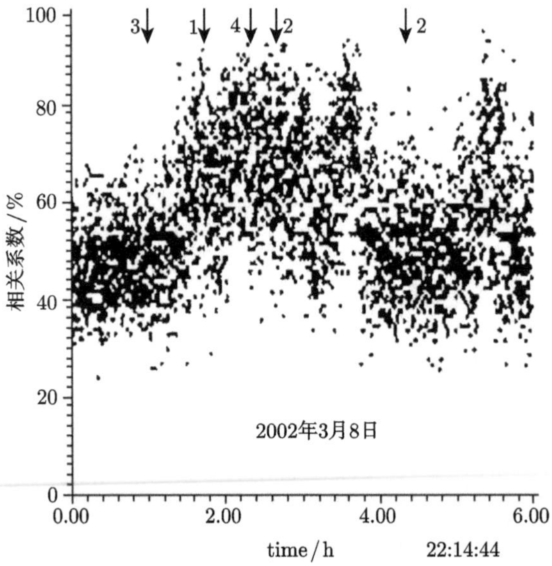

Fig. 82: Correlation coefficient as a fanotion of time.

direction. One type has frequencies in the range of 0.5∼1.0 Hz, and have a good correlation with signals, with a correlation coefficient larger than 0.7. The occurrence of the signals is consistent with bad weather conditions reported by high altitude pilots. This kind of sound signal is thought to be produced by turbulence at high altitudes. The other type of sound signal has frequencies in the range of 1∼5 Hz, also with a direction pointing to the mountain range. It interacts with low altitude currents and terrain, apparently, which is typical of mountain-lee waves. This kind of sound has low correlation coefficients, normally smaller than 0.6. Fig. 82 shows the functional relation of the correlation coefficients of the observed signals with time. The arrows in the picture specify the position where the turbulence occurred that was reported by the pilots. The numbers represent the degree of strength: 1 for slight, 2 for slight to moderate, 3 for moderate, 4 for moderate to severe, and 5 for severe. We must notice that just because no turbulence is reported does not mean that no turbulence is present as turbulence can quickly subside once the planes pass them.

(2) Storm and other bad weather

The intensity of infrasound produced by fierce convective storms is strong and has a long period, around 0.2∼2 min, which can be detected at distances of several hundreds to 1 million meters. The detected infrasound has a strong correlation with electromagnetic waves due to the ionosphere wave motion detected by Doppler. Therefore, it can be seen that those two phenomena are produced by processes from same source but are of different forms [290]. After analyses of several possible sound creation mechanisms, the most possible one is the vortex radiating mechanism, namely a typhoon cyclone. Since infrasound passes through the atmosphere very quickly, it is not possible to use them to used to predict storms. However, they can be a useful auxiliary tool for the current radar net warning system.

Besides typhoons and storms, many other types of bad weather such as thunderstorms, normal strong winds, hails and continuous rains can all produce infrasound with their own characteristics [291].

10.2.5 Earthquake

Earthquakes can produce audible sound, as well as infrasound. The infrasound produced can be caused by the vibration of ground surfaces and the vibration of mountains. The infrasound generated by the earth's surface can be divided into two categories: the seismic waves propagating to a certain regicn induce the vibration of ground sartace in that region, and thus radiate sound waves into the atmosphere in fundamentaly vertical direction – this is just the local intrasonic wdoe, which is somewhat similar to the sound wave produced by ackan surface, its period is identical with that for seisin ware on earth surface, there are some recorded infrasound corresponds to the wave-group consisting of longitudinal wave, transversal wave and surface wave, the periods are ten-odd See approximately Another hiud is the ground surface epicenter infrasound, which is radiated by the fierce vibration of the epicenter. It propagates through the atmosphere with a direction nearly horizontal to regions far beyond the epicenter. When propagating in wave guides formed in the atmosphere, the horizontal velocity of epicenter infrasound is slightly smaller than the sound velocity.

Infrasound generated by the vibration of mountains is called diffracted infrasound. It is created by terrain surface vibrations caused by seismic waves propagating through highlands or mountain ranges and other complex surroundings. Its propagation direction depends on the terrain, and as such, this kind of infrasound can propagate in different directions.

Infrasound generated by earthquakes can propagate to the ionosphere two hundred kilometers above the ground, and induce the electronic disturbance. The International Monitoring Center network has obtained electronic disturbance data from the rocket launchings carried out by the ionosphere during the Turkey earthquake in 1999, and data of the Kennedy Rocket Launch Center between 1998 and 2000 [292]. The results show that whatever whether a sound wave is caused by earthquakes or the launching of rockets, the electronic disturbances have features of shock waves with a period around 200~360 s. The shock wave signal is 2~5-fold stronger than the background noise. For earthquakes, the created sound sources are located at approximately the same places as the epicenters, while for rocket launching the corresponding sound source is at the spot in its orbit and at least 500~1000 m beyond the launching point with a flying height no less than 100 km.

10.2.6 Volcano eruption and meteorite fall

Volcanic eruptions bear vast energy. Erupted substances stimulate strong disturbances in the air, thus generates audible sound and infrasound. Volcanic eruptions are of different types. Volcanic eruptions with lava flowing slowly generates sound waves with wide frequency bands, but their fundamental and harmonics can be observed. Explosive volcanic eruptions can generate explosions, shock waves, and infrasound with very low frequencies. The strongest natural infrasound recorded in history is the volcanic eruption of Krakatoa in Indonesia on August 27th, 1883. The generated infrasound ran around the earth several times, propagated several hundred thousand kilometers. Its period was several hundred seconds and its amplitude was several hundred mPa it spots ten thousand kilometers beyond. The infrasound was so strong that it was detected by a simple barometer.

Infrasound is also generated by meteorite explosions. A famous example is the meteorite explosion in Tonguska, Siberia. The created infrasound had amplitude above 100 μPa and a period of 1 min.

10.2.7 Aurora

The mechanism of infrasound generated by aurora depends on types and activities of the aurora. When the ionosphere moves as a whole, aurorae will be produced who temperature in the ionosphere, where aurora is appoared, rises owing to irradiation of solar particles, and the warmer ionosphere rises up also. It will get cold after rising to a certain altitade and thus falls dow. This up-and-down movement radiates infrasound. Because the area of the ionosphere is very large, the period of the up and down motion is quite long, hence the period of the radiated infrasound is also long, around 1 min.

Another hind of infrasound in the ionosphere has a period of several minutes and a velocity of 500~1200 m s^{-1}, i.e. faster than the velocity of sound. This high-speed infrasound is generated by aurora arcs moving at supersonic velocities. The conducting particles in aurora will move as a whole with suppersonic speed under the action of the geomagnetic field, and At the same time, the charged particles will generated disturbances generate heat and disturbances in the ionosphere. Those disturbances radiate infrasound. This kind of infrasound accompanies the motion of aurorae, and thus they also propagate supersonically.

Aurorae can also generate infrasound wave with shorter periods of only a few Hz. and a period of a half or one minute. This may be produced two ways: firstly, via an instantaneous increase of the optical conductivity in aurorae, and secondly, by a pressure disturbance generated by a local aurora electron spray moving perpendicular through the ionosphere.

Tab. V: The observed infrasonic waves from natural sources

sources	meteorological phenomena microbarom	evil weather	frontal passage	torrent	mountainlee wave	volcanic meteorite	aurora	earthquake	solar eclipse
formatin mechanis	radiating into atmosphere from ocean waves formed by typhoon	—	pressure fluctuations; vibration of buoyancy force	unstability of shear	aircurrent passing through mountain range	volcanic eruption	aurora arc; supersonic motion in electric efflux pulse radiation	temperature decrease in local atmosphere causing by the moon shadow	
period (sec)	5→7	12→60	180→1500	240→7200	20→70	1→300	10→400	12→25	600→1800
amplitude (Pa)	0.01→0.1	0.05→0.3	5→200	5→20	0.05→0.3<500	0.1→0.5	0.1→0.2	3→10	
lasting time	several hours → several days	several minutes → several hours	several minutes → several hours	several hours → several days	several hours → several days	<1 hour	several seconds → several hours	<1 hour	several hours
horizontal trace velocity (m/s)	340	330–360	2–20	15–75	340	320–500	400–1200	340	un measured
Transverse correlation distance (multiple of wave number)	−2	−14	—	—	−14	−14	—	—	—
obvious characteristics	close to single-frequency wave, beadstring shaped wave envelope; lowcorrelation	azimuthvary with time; reaching peak value when spring changing into summer	large amplitude; usually in wave shape	low phase velocity	fixed azimuth lasting for several days; easily generating in winter locally	longperiod wave ahead of short-period wave	highspeed short pulse ahead of long-period wave; can be observed only at magneticlatitude >30°	longperiod wave ahead of shortperiod wave	longperiod wave ahead of short-period wave with larger amplitude

10.2.8 Others[1]

Many geophysical and astronomical phenomena can generate infrasound. Besides the important sources discussed above, some others such like frontal passage, torrent, and eclipse, etc can be mentioned.

In conclusion, some important characteristics of several major natural infrasound sources detected up to now are listed in Tab. V.

10.3 Artificial sound sources

10.3.1 Airplanes

Airplane engines can generate loud sounds when flying. While planes fly supersonically at a height r of several km above the ground, the sound heard on the ground is not generated by the planes directly. When planes fly supersonically, the air in front is squeezed which creates a high-pressure region, while the air behind is devoid and forms a steep negative pressure. The high positive and high negative pressure regions combine to form waves that are shaped like the letter "N", which propagates to the ground and are heard as large sounds, called a sonic boom. The theoretical analysis of sonic boom has been discussed in detail in section 9.2 chapter 9.

If the plane is large enough, the N-wave contains infrasound. A classical example is the sonic boom generated by the supersonic Concorde [293]. The concorde was an airliner proposed and produced in 1962 by British and French governments with a speed at two-fold sound speed. Because of Concorde's huge sonic boom, it was restricted to flying over ocean areas at heights at least aisote 18 km, such that it can keep away from other commercial airplanes and bad weather. Observation of French Flers' infrasound array system shows that a clear infrasound signal appears several times a day, which is related to regular flights between North America (New York) and Europe (London and Paris) of French and English Concorde airliners.

10.3.2 Rockets

Infrasound is created by air dynamic sound at 10$\sim$30 km above the ground when launching rockets in and reentrying the atmosphere. A strong N-wave will be generated when rockets fly at supersonic speeds. The wave is able to propagate through

1 There are many references regarding this category of phenomena. Here we only choose one of each kind in different views, but are not representativeness. The discussion on various wave shapes can be seen in [202].

several thousand kilometers via atmosphere duct. The amplitude of the infrasound generated by rockets can reach the order of twenty μPa and the signal can last several to ten minutes. When the size of the rockets is smaller, the energy of the infrasound generated is also weaker and the corresponding propagation distance is shorter. A rockets' shell can radiate sound waves that contain a lot of infrasound. The infrasound is strong near the launching field which is harmful to buildings and nearby devices.

10.3.3 Explosions in upper atmosphere

Early in the 1950s, the measurement of sound propagation of explosion sources was used to infer the temperature and the wind in the stratosphere [294, 295]. In these noteworthy controllable experiments, bombs weighing 1800 kg were dropped from B-29 aircraft and a circular sensor array located 225∼450 km beyond was used to record the sound. The high-altitude wind can be deduced by the delay time, elevation angle, azimuth angle and the deviation between arriving direction and the known source direction.

10.3.4 Nuclear tests in atmosphere

In the 1950s and the 1960s, numerous attempts were attempted in order to find long distance propagating models of low-frequency sound waves and the acoustic gravity wave in the atmosphere. The stimalation derived from detecting, localizing and equivalence-estimating of the atmaspheric nuclear explosions far beyond. Because part of the acoustic energy radiated by the sound source passes through the high-altitude atmosphere, it is needed to understand how high-altitude atmosphere affects the received signal. It was quickly realized that the received signal could be used to study the high-altitude atmosphere itself. The earliest trial was to compare the dispersion of acoustic gravity wave modes of the received signals with those calculated from the stratified atmospheric model [296]. In [297], it is mentioned that high-altitude wind affects the dispersion model of a hydrogen explosion pressure wave. [298] attempted to explain the abnormal time delay of the explosion signal by stating that the ray path is refracted and split by the horizontal wind and the temperature structure. Some advanced models that include season effects are used to improve the estimation of the direction of explosion signals. All of the considered models have undergone many careful improvements. It is still a pity that there have been no major advancements for extracting useful information of the atmosphere from nuclear explosion data.

There are many articles about the effects of atmospheric nuclear explosions on the ionosphere. For example, a description of the F-layer can be regarded as a significant wave disturbance of the guided wave mode of the sound wave and the gravity wave [299, 300]. However, attention was paid to the positioning of and estimating the magnitude of explosions has been emphasized repeatedly over the analysis of the atmosphere.

10.3.5 Explosion of U.S. space shuttle "Challenger" [141]

The tragic explosion of the U.S. space shuttle "Challenger" on January 28th, 1986 was a rare occurrence in human history in both its height and the manner of the explosion. However, it is lucky that the opportunity was caught and the corresponding infrasound was recorded, which became the only material in the world available for this type of event.

The Beijing infrasound station possesses a set of high sensitivity capacity micro-barometers placed on ground and multi recorders to monitor various infrasound in the atmosphere in long terms. 12 h and 53 min after the explosion, the Beijing infrasound station received signal waves, as shown in Fig. 83 (the cooperative unit, Guizhou Institute of Meteorological Sciences, also received a similar signal 13 h and 29 min after the explosion). Based on the analysis of the four strongest wave crests a, b, c, and d, the corresponding wave properties are given in Tab. VI.

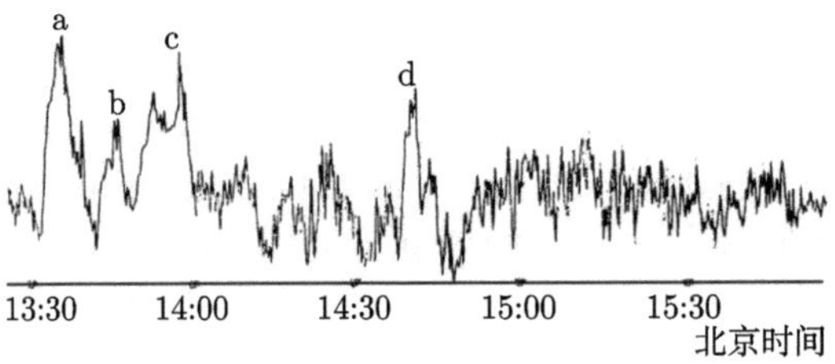

Fig. 83: Simplified block diagram of a monostatic echosonde.

Tab. VI: Parameters of infrasonic waves generated by the explosion of "Challenger"

serial number of the waveform	a	b	c	d
Arrival time (Beijing time)	13:31	13:42	13:48	14:38
Period (sec)	648	396	744	480
Amplitude (mV)	146	82	116	130
Amplitude for converted sound wave (Pa)	32.5	12.5	29.0	23.5
Corresponding wave velocity (m/s)	302	298	295	278

References

[1] Brown E H, Hall F F. Jr, Advances in atmospheric acoustics. Review of Geophysics and Space, 1978, 16(1): 47.
[2] Reynolds O. On the refraction of sound in the atmosphere, Phil. Trans. Roy. Soc. London, Ser. A, 1876, 166: 315.
[3] Stokes G G. On the Effect of Wind on the Intensity of Sound, Report, 22, Brit Ass, Dublin; Mathematical and Physical Papers of G. G. Stokes, 1857, 4: 110, New York: Cambridge Univ. Press. 1904.
[4] Tyndall J. On the atmosphere as a vehicle of sound, Phil. Trans. Roy. Soc. London, Ser. A, 1874, 164: 183.
[5] Teisserenc de Bort L P. Variations de la temperature de l'air libredans la zone comprise entre 8 km et 13 km k'altitude, C. R. Acad, Sci., 1902, 134: 987.
[6] van den Borne G. Über die schallverbreitung bei explosions katastrophen, Phys. Z., 1910, 11: 483.
[7] Emden R. Beiträge zur thermodynamik der atmosphäre, Meteorol. Z., 1916, 33: 351.
[8] Whipple F J W. The high temperatures of the upper atmosphere as an explanation of zones of audibility, Nature, 1923, 111: 187.
[9] Gowan E H. Low frequency sound waves and the upper atmosphere, Nature, 1929, 124: 452.
[10] Gutenberg B. The velocity of sound-waves from gun-fire in southern California, Eos Trans. AGU, 1938, 19: 156.
[11] Stroud W G, Nordberg W. J. Walsh. Atmospheric temperatures and winds between 30 and 80 km, J. Geophys. Res., 1956, 61: 45.
[12] Taylor G I. Eddy motion in the atmosphere, Phil. Trans. Roy. Soc. London, Ser. A, 1915, 215: 1.
[13] Obuhov A M. Orassяniĭ zouka v trubulentnom potoke, Dokl. ANSSSR, 1941, 30: 611.
[14] Bergmann P G. The wave equation in a medium with a variable index of refraction, J. Acoust. Soc. Amer., 1946, 17: 329.
[15] Bergmann P G. Propagation of radiation in a medium with random inhomogeneities, Phy. Rev., 1946, 70: 486.
[16] Blokhinzev D. The propagation of sound in an inhomogeneous and moving medium, J. Acoust. Soc. Amer., 1946, 18: 322.
[17] Kraichnan R H. The scattering of sound in a turbulent medium, J. Acoust. Soc. Amer., 1953, 25: 1096.
[18] Lighthill M J. On the energy scattered from the interaction of turbulence with sound or shock waves, Proc. Cambridge Phil. Soc., 1953, 49: 531.
[19] Batchelor G K. Wave scattering due to turbulence, in Symposium on Naval Hydraulics (ed. Sherman F S), 409–, Washington D. C.: National Academy of Sciences. 1957.
[20] Tatarski V. I. Wave Propagation in a Turbulent Medium. Translated by R. A. Silverman. McGraw-Hill, New York, 1961.
[21] Monin A S. Characteristics of the scattering of sound in a turbulent atmosphere, Sov. Phys. Acoust., 1962, 7: 370.
[22] Gilman G W, Coxhead H B, Willis F H. Reflection of sound signals in the troposphere, J. Acoust. Soc. Amer., 1946, 18: 274.
[23] McAllister L G. Acoustic sounding of the lower troposphere, J. Atmos. Terr. Phys., 1968, 30: 1439.
[24] KolmoGorv A N. The local structure of turbulence in an incompressible viscous fluid for very large Reynolds numbers, Dokl. AN SSSR, 1941, 30: 301.

[25] Monin A S, Yaglom A M. Statistical Fluid Mechanics, MIT Press, Cambridge, Mass., 1974, I; 1974, II.
[26] Wesely M L. The combined effect of temperature and humidity fluctuations on refractive index., J. Appl. Meteorol., 1976, 15: 43.
[27] Brown E H, Ckuffird S F. On the attenuation of sound by turbulence, J. Acoust. Soc. Amer., 1976, 60: 788.
[28] Lindzen R S. Internal gravity waves in an atmosphere with realistic dissipation and temperature, Geophys. Fluid Dyn., 1970, 1: 303.
[29] Georges T M. Ionospheric effects of atmospheric waves, ESSA Tech. Report IER 57 ITSA 54, U. S. Govt. Printing Office, Washington, D. C. 1967.
[30] Beer T. Atmospheric Waves, London: Adam Hilger. 1975.
[31] Morse P M, Ingard K U. Theoretical Acoustics. McGraw-Hill, New York, 1968.
[32] Bergmann P G. The wave equation in medium with a variable index of refraction, J. Acoust. Soc. Amer, 1946, 17: 329.
[33] Wallot J. Der senkrechte Durchgang elekromagntischer Wellen durch eine Schicht räumlich veränderlicher Dielectrizitätskonstante, Ann. Phys. (Leipzig), 1919, 60: 734.
[34] Försterling K. Über die Ausbreitung des Lichtes in inhomogenen Medien, Ann. Phys. (Leipzig), 1931, 11: 1.
[35] Försterling K, Wuster O. On reflection in inhomogeneous media. Ann. Phys. (Leipzig), 1950, 8: 129.
[36] Rayleigh Lord. Theory of Sound, 2nd ed. Dover. 1945.
[37] Rytov S M. i F. S. Àdkeviq, Ob otrænii zlektromagnitnys Bodn ot sloя s otrica-telnoĭ dizlektriqeskoĭ postoяnnoĭ TF, 1945, 10: 285.
[38] Whittaker E T., Watson G. N. Modern Analysis, 4th ed. Cambridge 1962.
[39] Muskat M, Meres M. W. Reflection and transmission coefficients for plane waves in elastic media Geophys., 1940, 5: 115.
[40] Iamada R. On the radio wave propagation in a stratified atmosphere. J. Phys. Soc. Japan, 1955, 10: 71.
[41] Epstein P. Reflection of waves in an inhomogeneous absorbing medium. Proc. Natl. Acad. Sci. U. S. A., 1930, 16: 627.
[42] Rawer K. Elektrische Wellen in einem geschichteten Medium. Ann. Phys. (Leipzig), 1939, 35: 385.
[43] Burman R, Gould R N. The reflection of waves in a generalized Epstein profile. Can. J. Phys., 1965, 43: 921.
[44] Elias G I. Das Verhalten elektromagnetischer Wellen bei räumlich veränderlichen elektrischen Eigenschaften. Elecktr. Nachr. Tech., 1931, 8: 4.
[45] Potler P S, Murphy S R. Solution of the wave equation in a medium with a particular velocity variation. J. Acoust. Soc. Amer., 1962, 34: 963.
[46] Maserov E P. K voprosu o volnovodnom rasprostranenii zvuka v sloistoneodnorodnyh sredah, Akust., 1959, 5(3): 332.
[47] ManalovsknĭV P. K integrnrovann nvkotoryh odnorodnyh dineĭnyh dnfferenqialnyh urabneniĭ vtordka, Izd Harov Univ, 1959.
[48] Miller S C. Good R H. A W K B type approximation to the Schrödinger equation. Phys. Rev., 1953, 91: 174.
[49] Fröman N, Fröman P O. JWKB Approximation Contributions to the Theory. North-Holland Publishing Co. 1965.
[50] Abramowitz M. Stegun I A (eds.). Handbook of Mathematical Functions. Dover, N. Y. 1965.
[51] Fok V A. Tablncy Funkciĭ Zĭrn, M. G. T. T. I. 1946.
[52] Heading J. An Introduction to Phase-integral Methods. Methuen; J. Wiley. 1962.

[53] Dorognicyn A A. Asnmptotiqeskie zakony raspredeleni sobstvennyh znaqennī dl nekotoryh osbyh vidov differencialnyh uravneniī vtorogo pordka, Uspehi Mat. Nauk, 1952, 7: vyp. 6.
[54] Murphy E L. Ray representation of diffraction effects in the split beam sound field. J. Acoust. Soc. Amer., 1949, 43(3): 610.
[55] Brehovsknh L M. Opole toqeqnogo istoqinnika vsloisto-neodnoī srede, Izv AN SSSR. ser. fizin. 1949, 13(5): 505.
[56] Pierce A D. Acoustics, An Introduction to Its Physical Principles and Applications. New York: McGraw-Hill. 1981; Amer. Inst. of Physics, New York, 1991.
[57] Cagniard L. Réflexion et réfraction des ondes séismiques progressive. Gauthier-Villars, Paris, 1939.
[58] Milne E A. Sound Waves in the atmosphere. Phil. Mag., 1921, 6(42): 96.
[59] Jaeger G. On the propagation of sound in moving fluid, Sitzungsber. Kais. Akad. Wiss. (Vienna), Math-Naturwiss. Kl. sec. IIa, 105: 1040.
[60] Barton E H. On the refraction of sound by wind. Phil. Mag., 1901, 6(1): 159.
[61] Fujiwhara S. On the abnormal propagation of sound waves in atmosphere. Bull. Cent. Meteorol. Obs. Japan, 1912, 1(2); 1916, 4(2).
[62] Hamming R W. Numerical solution of ordinary differential equations, in Digital Computer User's Handbook (ed. Klerer M and Korn G A), New York: McGraw-Hill. 1967.
[63] Moler C B, Solomon L P. Use of splines and numerical integration in geometrical acoustics. J. Acoast. Soc. Amer., 1970, 48: 739.
[64] Whitham G B. Group velocity and energy propagation for three-dimensional waves. Common. Pure Appl. Math., 1961, 14: 675.
[65] Engelke R. Ray trace acoustics in unsteady inhomogeneous flow. J. Acoust. Soc. Amer., 1974, 56: 1291.
[66] Ugincius P. Ray acoustics and Fermat s principle in a moving inhomogeneous medium. J. Acorst. Soc. Amer., 1972, 51: 1759.
[67] Tedrick R N. Meteorological focusing of acoustic energy. Sound: Its Uses and Control, 1963, 2(6): 24.
[68] Hayes W D. Round table discussion on sonic boom problems. Aircraft Engine Noise and Sonic Boom, AGARD Conf. Proc., 1969, 42: 36.
[69] Hall S-L. V. Distortion of the sonic boom pressure signature by high speed jets. J. Acoust. Soc. Amer., 1978, 63: 1749.
[70] Gutenberg B. Propagation of sound waves in the atmosphere. J. Acoust. Soc. Amer., 1942, 14: 151.
[71] Cook R K. Strange sounds in the atmosphere, Part 1 and Part 2 (with J. M. Young), Sound: Its Uses and Control, 1962, 1: 12–16, 25–33.
[72] Cox E F. Abnormal audibility zones in long distance propagation through the atmosphere. J. Acoust. Soc. Amer., 1949, 21: 6–16, 501.
[73] Gutenberg B. In Compendium of Meteorology. American Meteorological Society, Boston. (ed. T. F. Malone) 1951.
[74] Hayes W D. Energy invariant for geometric acoustics in a moving medium. Phys. Fluids, 1968, 11: 1654.
[75] Blohincev D. Akustika Neodnorodoī Dviuxeīs Sredy, ogizgostehizdat, 1946.
[76] Atvars J, Schubert L K, Ribner H S. Refraction of sound from a point source placed in an air jet. J. Acoust. Soc. Amer., 1965, 37: 168.
[77] Schubert L K. Numerical study of sound refraction by a jet flow, I: Ray acoustics. J. Acoust. Soc. Amer., 1972, 51: 434.
[78] Bateman H. The influence of meteorological conditions on the propagation of sound. Mon. Weather Rev., 1914, 42: 258.

[79] Attenborough K, Hayek S I, Lawther J M. Propagation of sound above a porous half-space. J. Acoust. Soc. Amer., 1980, 68: 1493.

[80] Sabatier J M, Bass H E, Bolen L N K. Attenborough and V. V. S. S. Sastry, The interaction of airborne sound with the porous ground the theoretical formulation, J. Acoust. Soc. Amer., 1986, 79: 1345.

[81] Paul D I. Acoustical radiation from a point source in presence of two media. J. Acoust. Soc. Amer., 1957, 29: 1102.

[82] Paul D I. Wave Propagation in acoustics using the saddle-point method. J. Math. Phys., 1959, 38: 1.

[83] Piercy J E, Embleton T F W, Sutherland L C. Review of noise propagation in the atmosphere. J. Acoust. Soc. Amer., 1977, 61: 1403.

[84] Parkin P H, Scholes W E. The horizontal propagation of sound from a jet engine close to the ground, at Radlett. J. Sound Vib., 1964, 1: 1.

[85] Attenborough K. Acoustical impedance models for outdoor ground surfaces. J. Sound Vib., 1985, 99: 521.

[86] Delany M E, Bazley E N. A Note on the effect of ground absorption in the measurement of aircraft noise. 1971, J. Sound Vib., 16: 315.

[87] Chessel C I. Propagation of noise along a finite impedance boundary. J. Acoust. Soc. Amer., 1977, 62: 825.

[88] Morse P M. Feshbach H. Methods of Theoretical Physics, McGraw-Hill, 1953.

[89] Isei T, Embleton T F W, Piercy J E. Noise reduction by barriers on finite impedance ground. J. Acoust. Soc. Amer., 1980, 67: 46.

[90] Daigle G A. Diffraction of sound by a noise barrier in presence of atmospheric turbulence. J. Acoust. Soc. Amer., 1982, 71: 847.

[91] Price M A. Attenborough K., Heap N. W. Sound attenuation through trees: Measurements and models. J. Acoust. Soc. Amer., 1988, 84: 1836.

[92] Senior T B A. Low frequency scattering, J. Acoust. Soc. Amer., 1973, 53: 742.

[93] Clay C S, Medwin H. Acoustical Oceanography: Principles and Applications. New York: Wiley. 1977.

[94] Tatarskii V I. The Effects of the Turbulent Atmosphere on Wave Propagation. Israel Program for Scientific Translations, Jerusalem (1971).

[95] Brown E H. Acoustic Doppler radar scattering equation and general solution. J. Acoust. Soc. Amer., 1972, 52: 1391.

[96] Brown E H, Clifford S F. Spectral broadening of an acoustic pulse propagating through turbulence. J. Acoust. Soc. Amer., 1973, 54: 36.

[97] Clifford S F, Brown E H. Acoustic scattering from a moving turbulent medium. J. Acoust. Soc. Amer., 1974, 55: 929.

[98] Huang M N. Sound scattering from atmospheric turbulence. Progr. Astronaut. Aeronaut., 1976, 46: 35.

[99] Ford G W, Meecham W C. Scattering of sound by isotropic turbulence of large Reynolds number. J. Acoust. Soc. Amer., 1960, 32: 1668.

[100] Brown E H. Turbulent spectral broadening of backscattered acoustic pulses. J. Acoust. Soc. Amer., 1974, 56: 1398.

[101] Lewis R M, Bleistein N, Ludwig D. Uniform asymptotic theory of creeping waves. Commun. Pure Appl. Math., 1967, 20: 295.

[102] Friedlander F G. Sound Pulses. Cambridge, 1958.

[103] Pridmore-Brown D C. Sound propagation in a temperature and wind-stratified medium. J. Acoust. Soc. Amer. 1962, 34: 438.

[104] Watson G N. Theory of Bessel Functions, 2nd ed., Cambridge, 1952.

[105] Miller J C P. British Association Mathematical Tables. Part-vol. B: The Airy Integral, Cambrideg, 1946.

[106] Bauer H. J. Influences of transport mechanisms on sound propagation in gases. Adv. Mol. Relaxation Processes, 1972, 2: 319.

[107] Evans L. B., Sutherland L. C. Absorption of sound in air. Wyle Res. Rep. No. WR 70-14. for U. S. Army Res. Off. 1970.

[108] Hirschfelder K. F., Curtiss C. F., Bird R. B. Molecular Theory of Gases and Liquids. New York: Wiley. 1954.

[109] Cottell T L, McCoubrey J C. Molecular Energy Transfer in Gasses. London: Butherworth. 1961.

[110] Bass H E, Sutherland L C. Piercy J. Evans L. Absorption of sound by the atmosphere. Physical Acoustics, Vol. XVII, 145–232, Academic Press, 1984.

[111] Bhatia A B. Ultrasonic Absorption. London: Oxford Univ. Press. 1967.

[112] Raff L M, Winter T G. Origin of the temperature dependence of the “rotational relaxation” time. J. Chem. Phys., 1968, 48: 3992.

[113] Bauer H-J, Bass H E. On acoustic measurements of rotational energy transfer. J. Chem. Phys., 1972, 57: 1763.

[114] Greenspan M J. Rotational relaxation in nitrogen, oxygen and air. J. Acoust. Soc. Amer., 1959, 31: 155.

[115] Bauer H-J. Phenomenological theory of the relaxation phenomena in gases. Physical, Acoastics. Vol. II A, 47–131, Academic Press. 1965.

[116] Tempest W, Parbrook H O. The absorption of sound in argon, nitrogen and oxygen. Acustica, 1957, 7: 354.

[117] Bass H E, Keeton R G. Ultrasonic absorption in air elevated temperatures. J. Acoust. Soc. Amer., 1975, 58: 110.

[118] Bass H E, Sutherland L C. On the rotational collision number for air at elevated temperatures. J. Acoust. Soc. Amer., 1976, 59: 1317.

[119] Evans L B, Sutherland L C. Absorption of sound in air. J. Acoust., Soc. Amer., 1971, 49: 110(A).

[120] Evans L B, Bass H E, Sutherland L C. Atmospheric absorption of sound: Theoretical predictions. J. Acoust. Soc. Amer., 1972, 51: 1565.

[121] Bauer H J, Shields F D, Bass H E. Multimode vibrational relaxation in polyatomic molecules. J. Chem. Phys., 1972, 57: 4624.

[122] Shields F D. On obtaining transition rates from sound absorption and dispersion curves. J. Acoust. Soc. Amer., 1970, 47: 1262.

[123] Bass H E, Bauer H J, Evans L B. Atmospheric absorption of sound: Analytical expressions J. Acoust. Soc. Amer., 1972, 52: 821.

[124] Holman J P. Thermodynamics. New York: McGraw-Hill. 1969.

[125] U. S. Standard Atmosphere, Washington D. C: U. S. Gov. Print. Off., 1962.

[126] Gray D E. (ed.). American Institute of Physics Handbook. 3rd ed. New York: McGraw-Hill. 1972.

[127] Henderson M C, Herzfeld K F. Effect of water vapor on the Napier frequency of oxygen and air. J. Acoust. Soc. Amer., 1965, 37: 986.

[128] Monk R G. Thermal relaxation in humid air. J. Acoust. Soc. Amer., 1969, 46: 580.

[129] Bass H E, Shields F D. Vibrational relaxation and absorption in O_2/H_2O mixtures. J. Acoust. Soc. Amer., 1974, 56: 856.

[130] Schwartz R N, Slawsky Z I, Herzfeld K. F. Calculation of vibrational relaxation times in gases. J. Chem. Phy., 1952, 20: 1591.

[131] Evans L B. Vibrational relaxation in moist nitrogen. J. Acoust. Soc. Amer., 1972, 51: 409.

[132] Zuckerwar A J, Meredith R W. Sound absorption in N_2H_2O gas mixtures at elevated temperatures. J. Acoust. Soc. Amer., 1980, 68: S107.

[133] Bass H E, Hottman S D, Bauer H J. Laser induced fluorescence study of the deactivation of N_2 by H_2O, H_2S and CH_4. J. Chem. Phys., 1980, 72: 2113.
[134] Bass H E, Shields F D. Absorption of sound in air: High frequency measurements. J. Acoust. Soc. Amer., 1977, 61: 571.
[135] Sutherland L C. Review of experimental data in a support of a proposed new method for computing atmospheric absorption loss. U. S. Dep. Transp. Rep. 1975, No. TST-75-87.
[136] Goff J A. Saturation pressure of water on the new Kelvin scale, Fundamentals and Standards of Humidity and Moisture: Measurement and Control in Science and Industry, V 3, 1965, Chap. 7, New York: Reinhold. 1965.
[137] List R J (ed.). Smithsonian Meteorological Tables, 6th Rev. ed. Smithsonian Washington D. C.: Inst. Press. 1958.
[138] Jaeschke M, Heller W J, Meier G E A. Acoustic damping in a gas mixture with suspended submicroscopic droplets. J. Sound Vib., 1975, 43: 467.
[139] Wei R J, Tian Y R, Lu Q J. Absorption of sound in water fog composed of submicron droplets. J. Acoust. Soc. Amer., 1987, 81(5): 1350.
[140] Marble F E, Wooten D C. Sound attenuation in a condensing vapor. Phys. Fluids, 1970, 13: 2657.
[141] Cole J E III, Dobbins R A. Propagation of sound through atmospheric fog. J. Atmos. Sci., 1970, 27: 426.
[142] Cole J E III, Dobbins R A. Measurements of the attenuation of sound by a warm air fog. J. Atmos. Sci., 1971, 28: 202.
[143] Davidson G A. Sound propagation in fog. J. Atmos. Sci., 1975, 32: 2201.
[144] Davidson G A. Barger's equation for finite amplitude acoustics in fogs. J. Sound Vib., 1976, 45: 473.
[145] Davidson G A. Propagation of audible sound through air water fog. J. Acoust. Soc. Amer., 1977, 62: 497.
[146] Knudsen V O, Wilson J V, Anderson N S. The attenuation of audible sound in fog and smoke. J. Acoust. Soc. Amer., 1948, 20: 849.
[147] Wei R J, Wu J R. Absorption of sound in water fog. J. Acoust. Soc. Amer., 1981, 70: 1213.
[148] Hines C O. Internal atmospheric gravity waves at ionospheric heights. Can. J. Phys., 1960, 38: 1441.
[149] Lamb H. Hydrodynamics, 6th ed. New York: Dover. 1945.
[150] Martyn D F. Cellular atmospheric waves in the ionosphere and troposphere. Proc. Roy. Soc., A, 1950, 201: 216.
[151] Tolstoy I. The theory of waves in stratified fluids including the effects of gravity and rotation. Rev. Mod. Phys., 1963, 35: 207.
[152] Jones W L. Energy momentum tensor for linearized waves in material media. Rev. Geophys. Space Sci., 1971, 9: 917.
[153] Beer T. Lamb waves and the conservation of energy Am. J. Phys., 1972, 40: 774.
[154] Midgley J E, Liemohn H B. Gravity waves in a realistic atmosphere. J. Geophys. Res., 1966, 71: 3729.
[155] Eckart C. Hydrodynamics of Oceans and Atmospheres, New York: Pergamon Press. 1960.
[156] Rossby C G. Planetary flow pattern in the atmosphere. Q. J. Roy. Met. Soc. Supplement, 66: 68 (1940).
[157] Platzman G. The Rossby wave. Q. J. Roy. Met. Soc., 1968, 94: 225.
[158] Garrett C J R. Atmospheric edge waves. Q. J. Roy. Met. Soc., 1969, 95: 731.
[159] Pitteway M L V, Hines C O. The reflection and ducting of atmospheric acoustic gravity waves. Can. J. Phys., 1965, 43: 2222.
[160] Johnston T W. Atmospheric gravity wave instability. J. Geophys. Res., 1967, 72: 2972.

[161] Clemmow P C, Dougherty J P. Electrodynamics of Particles and Plasmas. Addison Wesley, Reading, Mass. 1969.
[162] Scover R S. Gravity waves in the atmosphere. Arch. Wet. Geophys. Biokl., A, 1953, 4: 176.
[163] Fogle B, Haurwitz B. Noctilucent clouds. Space Sci. Rev., 1966, 6: 279.
[164] Jones D, Maude A D. Evidence for wave motions in the E region in the ionosphere. Nature, 1965, 206: 177.
[165] Thome G D. A study of large scale traveling disturbances in the ionosphere using the Arecibo UHF radar. Report CRSR 236. Cornell Univ., Ithaca, 1966.
[166] Lindzen R S, Blake D. Lamb waves in the presence of realistic distributions of temperature and dissipation. J. Geophys. Res., 1972, 77: 2166.
[167] Bretherton F P. Lamb waves in a nearly isothermal. atmosphere. Q. J. Roy. Met. Soc., 1969, 95: 754.
[168] Yang XunRen, Xie JinLai. Detection and analysis of the infrasonic waves attached to the tragic explosion of U. S. space shuttle "Challenger". J. Low Freq. Noise Vib., 1986, 5: 100.
[169] Phillips N A. The equations of motion for a shallow rotating atmosphere and the traditional approximation. J. Atmos. Sci., 1966, 23: 626.
[170] Chapman S, Lindzen R S. Atmospheric Tides. D. Reidel, Dordrecht 1970.
[171] Salomons E M. Computational Atmospheric Acoustics. Kluwer London: Acad. Publ. 2001.
[172] Nobile M A, Hayek S I. Acoustic propagation over an impedance plane. J. Acoust. Soc. Amer., 1985, 78: 1325.
[173] Raspet R, Franke S J. Residue series solution of impulse noise propagation into a shadow zone. J. Acoust. Soc. Amer., 1988, 83: 1964.
[174] Hallberg B, Larsson C, Israelsson S. Numerical ray tracing in the atmospheric surface layer. J. Acoust. Soc. Amer., 1988, 83: 2059.
[175] Bass H E. Atmospheric acoustics. In Encyclopedia of Applied Physics, VCH Publishers Inc. 1991, 2: 145.
[176] Jensen F B, Kuperman W A. Porter M. B., Schmidt H. Computational Ocean Acoustics. New York: Amer. Inst. of Physics. 1994.
[177] Walkden F, West M. Prediction of enhancement factor for small explosive soures in a stratified moving atmosphere. J. Acoust. Soc. Amer., 1988, 84(1): 321.
[178] Raspet R., Lee S. W., Keuster E., Chang D. C., Richards W. F., Gilbert R. N Bong. A Fast Field Program for sound propagation in a layered atmosphere above impedance ground. J. Acoust. Soc. Amer., 1985, 77: 345.
[179] Lee S W, Bong N, Richards W F, Raspet R. Impedance formulation of the Fast Field Program for acoustic wave propagation in the atmosphere. J. Acoust. Soc. Amer., 1986, 79: 628.
[180] West M, Sack R A, Walkden F. The Fast Field Program (FFP). A second tutorial: application to long range sound propagation in the atmosphere. Applied Acoustics, 1991, 33: 199.
[181] Raspet R, Lee S W, Kuester E, Chang D C, Richards W F, Gilbert R, Bong N. A fast-field program for sound propagation in a layered atmosphere above an impedance ground. J. Acoust. Soc. Amer., 1985, 77: 345.
[182] Lee S W, Bong N, Richards W F, Raspet R. Impedance formulation of the fast field program for acoustic wave propagation in the atmosphere. J. Acoust. Soc. Amer., 79: 628-(1986). J. Acoust. Soc. Amer., 1986. 79: 628.
[183] Nijs L. Wapenaar C P A. The influence of wind and temperature gradients on sound propagation, calculated with the two-way wave equation. J. Acoust. Soc. Amer., 1990, 87: 1987; J. Acoust. Soc. Amer., 1992, 91: 498.
[184] Wilson D K. Sound field computations in a stratified moving medium. J. Acoust. Soc. Amer., 1993, 94: 400.

[185] Wilson D K. Propagation in a stratified atmosphere and solution with the FFP-L program. Internal Riport. The Pennsylvania State Univ. 1990.
[186] West M, Sack R A, Walkden F. The fast field program (FFP). A second tutorial: application to long range sound propagation in the atmosphere. Applied Acoustics, 1991. 33: 199.
[187] Filippi P, Habault D, Lefebvre J P, Bergassoli A. Acoustics. Basic physics. theory and methods. San Diego: Academic. 1999.
[188] Myers M K, McAninch G L. Parabolic approximation for sound propagation in the atmosphere. AIAAJ., 1978, 16: 836.
[189] Tappert F D, Lee D A range refraction parabolic equation. J. Acoust. Soc. Amer., 1984, 76: 1797.
[190] Gilbert K E, White M J. Application of the parabolic equation to sound propagation in a refraction atmosphere. J. Acoust. Soc. Amer., 1989, 85: 630.
[191] Di X, Gilbert K E. Wave propagation in a 3-D turbulent atmosphere: horizontal coherence. Proc. Eighth Symposium on Long-Range Sound Propagation, Pennsylvania: The Pennsylvania State Univ. 1998: 169.
[192] Chen C, Lin Y, Lee D. A three dimensional azimuthal wide-angle model for the parabolic wave equation. J. Comp/Acoust., 1999, 7: 269.
[193] Davis J A, White D, Cavanaugh R C. In NORDA parabolic equation workshop (ed. J. A. Davis, D. White and R. C. Cavanaugh). NORDA TN-143, Naval Ocean Research and Development Activity, Bay St. Louis, MS, 1982.
[194] Claerbout J F. Fundamentals of Geophysical Data Processing, New York: McGraw-Hill. 1976; Oxford: Blackwell. 1985.
[195] Knightly G H, Lee D, Mary D F St. A higher order parabolic wave equation. J. Acoust. Soc. Amer., 1987, 82: 580.
[196] West M, Gilbert K, Sack R A. A tutorial: on the parabolic equation (PE) model used for long range sound propagation in the atmosphere. Applied Acoustics, 1992, 37: 31.
[197] Press W H, Flannery B P, Teukolsky S A, Vetterling W T. Numerical Recipes. The Art of Scientific Computing. Cambridge: Cambridge U. P. 1986.
[198] Salmons E M. Improved Green's function parabolic equation method for atmospheric sound propagation. J. Acoust. Soc. Amer., 1998, 104: 100.
[199] Berkhout A J. Applied Seismic Wave Theory. Advances in Exploration Geophysics, Amsterdam: Elsevier. 1987.
[200] Gilbert K E, Di X. A fast Green's function method for one-way sound propagation in the atmosphere. J. Acoust. Soc. Amer., 1993, 94: 2343.
[201] Yang Xunren. Ray tracing for infrasonic waves generated by typhoon. Chinese J. Acoust., 1984, 3(2): 183.
[202] Gossard E E, Hooke W H. Waves in the Atmosphere. Elsevier Scient. Publ. Co. 1975.
[203] Georges T. M. 3-D ray tracing for acoustic gravity waves. Proc. Conf on Effects of Acoustic Gravity Waves on Electromagnetic Wave Propagation. AGRD Proc. 1972, 115: 2-1 to 2-8.
[204] Pierce A D, Posey J W. Theoretical predictions of acoustic-gravity pressure waveforms generated by large explosions in the atmosphere. Air Force Camb. Res. Lab., AFCRL-70-0134, 1970.
[205] Cerveny V, Popov M, Pšencik I. Computation of wave fields in inhomogeneous media-Gaussian beam approach. Geophys. J. R. Astron. Soc., 1982, 70: 109.
[206] Porter M B, Bucker H P. Gaussian beam tracing for computing ocean acoustic fields. J. Acoust. Soc. Amer., 1987, 82: 1349.
[207] Porter M B, Bucker H P. Applications of Gaussian beam tracing to 2 and 3D problems in ocean acoustics, Proc. of the 12th World Congress on Scientific Computation (ed. W. F. Ames and C. Brezinski), (Pairs:MACS), 1988: 227.

[208] Weber M. Computation of body-wave seismograms in absorbing 2-D media using the Gaussian beam method: Comparison with exact methods. Geophys. J., 1988, 92: 9.

[209] Gabillet Y, Schroeder H, Daigle G A, L'Espérance A. Application of the Gaussian beam approach to sound propagation in the atmosphere: Theory and experiments. J. Acoust. Soc. Amer., 1993, 93(6): 3105.

[210] Muller G Efficient calculation of Gaussian beam seismograms for two dimensional inhomogeneous media. J. R. Astron. Soc., 1984, 79: 153.

[211] Coulter R L, Kallistratova M A. Two decades of progress in SODAR techniques: a review of 11 ISARS proceedings. Meteorology and Atmospheric Physics. Austria: Springer-Verlag. 2004, 85:(1–3): 3–19.

[212] Ulyanov Y N, Butakova S V. Sonic antennas for acoustic and radio-acoustic remote sensing. 4th International Conference on Antenna Theory and Techniques. 2003, 2(2): 564–569.

[213] Bradley S, Webb T. Use of an ultrasonic sodar to sense raindrop size distributions. Source: Journal of Atmospheric and Oceanic Technology, 2002, 19(8): 1203–1207.

[214] Kallistratova M A, Coulter R L. Application of SODARs in the study and monitoring of the environment. Meteorology and Atmospheric Physics. Austria: Springer-Verlag. 2004, 85(1–3): 21–37.

[215] McAllister L G. Acoustic sounding of the lower troposhere. J. Atmos. Terr. Phys., 1968, 30: 1439.

[216] Asimakopoulos D N, Stefanou G, Helmis C D. Atmospheric acoustic minisounder. J. Atmospheric Oceanic Tech., 1987, 4: 345.

[217] Coulter R L, Martin T J, Hart R L. A battery-powered minisodar for long-and short-term operations. International J. Remote Sensing, 1992, 15: 245.

[218] Mursch-Radlgruber E, Gregg D W, King C W, Ruiffieux D, Sharp K A H, Wolfe D E, Neef W D. NOAA's protable high frequency minisodar-design and first results. International J, Remote sensing., 1992, 15: 325.

[219] Moulsley T J, Cole R S. High frequency atmospheric acoustic sounders. Atmospheric Environment, 1979, 13: 347.

[220] Carsey F A. A linear zipper array for probing the wind and structure profile of the lower atmosphere (abstract). J. Acoust. Soc. Amer., 1972, 52: 1309.

[221] Davey R F. A coherent acoustic Doppler radar for real-time wind measurement. In Preprints of the 17th Radar Meteorology Conference, Boston: Amer. Meteor. Soc., 1976: 270–275.

[222] Kaimal J C, Wescott J W. An acoustic Doppler sounder for measuring wind profiles in the lower boundary layer. In Preprints of the 17th Radar Meteorology Conference, 290–296, Boston, Mass: Amer. Meteor. Soc. 1976.

[223] Georges T M, Clifford S. F. Acoustic sounding in a refracting atmosphere. J. Acoust. Soc. Amer., 1972, 52: 1397.

[224] Neff W D. Quantitative evaluation of acoustic echos from the planetary boundary layer. Tech. Rep. ERL. 332-WPL, 38: 34 NOAA, Boulder, Colo. 1975.

[225] Asimakopoulos D N, Helmis C G. Recent advances on atmospheric acoustic sounding, International J. Remote Sensing, 1994, 15: No. 2, 223.

[226] Eymard L. Ondes de gravite dans la couche limite planetaire etude experimenta par sondage acoustique. Tech. Note No. 54, France: Centre National d'etudes des Telecommunications, 1978.

[227] Caughey S J, Readings C J. An observation of waves and turbulence in the earth's boundary layer. Boundary layer Meteorology, 1975, 9: 279.

[228] Caughey S J. The Cardington turbulence instrumentation. Unpublished UK Meteorological Office Turbulence and Diffusion Note, No. 82(1977).

[229] Haugen D A, Kaimal J C, Readings C J, Rayment R. A comparison of ballon-borne and tower-mounted instrumentation for probing the atmospheric boundary-layer. J. Appl Meteor., 1975, 14: 540.

[230] Little C G. Acoustic methods for the remote probing of the lower atmosphere. Proc. IEEE, 1969, 57: 571.

[231] Gething J T, Jensen D. Measurement of temperature and humidity by acoustic echo sounding. Nature (Phys. Sci.), 1971, 231: 198.

[232] Bradley S G, Hunerbein S. V. A systematic algorithm for obtaining SODAR wind profiles. Meteorology and Atmospheric Physics. Austria: Springer-Verlag. v 85, 2004, 85(1–3): 77–84.

[233] Rodgers C D. Inverse methods for atmospheric sounding, Theory and practice. London: World Scientific, 2000.

[234] Few A A. Lightning channel reconstruction from thunder measurements. J. Geophys, Res., 1970, 75: 7517.

[235] Few A A. Power spectrum of thunder. J. Geophys, Res., 1969, 74: 6926.

[236] Teer T. L., Few A. A.. Horizontal lightning. J. Geophys Res., 1974, 79: 3436.

[237] Barry G. Ray tracings of acoustic waves in the upper atmosphere. J. Atmos. Terr. Phys., 1963, 25: 621.

[238] Georges T M, Beasley W H. Refraction of infrasound by upper atmospheric winds. J. Acoust. Soc. Amer., 1977, 61: 28.

[239] Jones R M, Riley J P, Georges T M. APPA-a versatile three dimensional Hamiltonian raytracing program for acoustic waves in the atmosphere above irregular terrain. NOAA Report, Boulder, Colo. 1986.

[240] Pierce A D. Propagation of acoustic-gravity waves in a temperature and wind-stratified atmosphere. J. Acoust. Soc, Amer., 1965, 37: 218.

[241] Francis S H. Global propagation of atmospheric gravity waves: a review. J. Atmos. Terr. Phys., 1975, 37: 1011.

[242] Jones R M, Georges T M. Infrasound from convective storms: III. Propagation to the ionosphere. J. Acoust. Soc. Amer., 1976, 59: 765.

[243] Francis S H. Acoustic gravity modes and large-scale traveling ionospheric disturbances of a realistic dissipative atmosphere. J. Geophys. Res. 1973, 78: 2278.

[244] Donn W L, Rind D. Microbaroms and the temperature and wind of the upper atmosphere. J. Atmos, Sci., 1972, 29: 156.

[245] Georges T M. HF Doppler studies of traveling ionospheric disturbances. J. Atmos. Terr. Phys., 1968, 30: 735.

[246] Derr V E (ed.). Remote Sensing of tine Troposphere. Chap, 21, U. S. Govt. Printing Off. 1972.

[247] Briggs B. H., Phillips G. S., Shinn D. H. The analysis of observations on spaced receivers of fading radio signals. Proc. Phys. Soc. London, B, 1950, 63: 106.

[248] Gossard E E. The apparent movement of the spectral components in fading records of ionospherically reflected radio waves. J. Geophys. Res., 1967, 73: 1563.

[249] Hedlin M A H. Global monitoring of subaudible sound in Earth's atmosphere. Oceans., 2003, 1(1): 55–60.

[250] Garces M, Hetzer C, Lindquist K, Hansen R, Drob D, Picone M. Stratospheric arrivals in the upstream wind direction: case study of the April 23, 2001, bolide, The Infrasound Technology Workshop, Kona, Hawaii.

[251] Takahashi K, Masuda Y, Matura N, Kato S, Fukao S, Tsuda T, Sato T. Analysis of acoustic wave fronts in the atmosphere to profile the temperature and wind with a radio acoustic sounding system. J. Acoust. Soc. Amer., 1988, 84: 1061.

[252] Kallistratova M A. Acoustic waves in the turbulent atmosphere: a review. Journal of Atmospheric and Oceanic Technology, 2002, 19(8): 1139.

[253] Whitham G B. Linear and Nonlinear Waves, New York: John Wiley & Sons. 1974.
[254] Landau L D. On shock waves at large distances from the place of their origin. Soviet J. Physics, 1945, 9: 496.
[255] Anderson M O. The propagation of the spherical N wave in an absorbing medium and its diffraction by a circular aperture Tech Report ARL 74-25, Univ. Texas. 1974.
[256] Kang J. Nonlinear acoustic propagation of shock waves through the atmosphere with molecular relaxation, Ph. D thesis. Pennsylvania State Univ. 1991.
[257] Plotkin K J. PCBoom3 sonic boom prediction model version 1.0e, Wyle Research Report WR 95-22E, 1998.
[258] Plotkin K J. State of the art of sonic boom modeling. J Acoustical Society of America. 2002, 111(1): 530.
[259] Friedman M P, Kane E J, Sigalla A. Effects of atmosphere and aircraft motion on the location and intensity of a sonic boom. AIAAJ. 1963, 1(6): 1327.
[260] Thomas C L. Extrapolation of sonic boom pressure signatures by the waveform parameter metho. NASA, TN. D-6832.
[261] Carlson H W. Simplified sonic boom prediction. NASA Technical Paper 1122, 1978.
[262] Petenko I V, Shurygin E A. A two-regime model for the probability density function of the temperature structure parameter in the convective boundary layer. Bound.-Layer Meteor., 1999, 93: 381–394.
[263] Neff W. Quantitative evaluation of acoustic echoes from the planetary boundary layer. NOAA Tech. Rep. ERL. 322-WPL, 34, 1975.
[264] Singal S P, Gera B S, Kallistratova M. A., Petenko I. V. Sodar aspect sensitivity studies in the convective boundary layer. Int. J. Remote Sens., 1997, 18: 1809–1819.
[265] Petenko I V, Bezverkhnii V A. Temporal scales of convective coherent structures derived from sodar data. Meteor. Atmos. Phys., 1999, 71: 105–116.
[266] Wilson D K. Scattering of acoustic waves by intermittent temperature and velocity fluctuations. J. Acoust. Soc. Amer., 1997, 101: 2980–2982.
[267] Chunchuzov I P, Otrezov A I, Sound propagation in the atmospheric boundary layer. Proc. Eighth Int. Symp. on Acoustic Remote Sensing, Moscow, Russia, ISARS, 1996, 1.1–1.12.
[268] Tsuda T, Sato N, Hirose K, Kato S. MU Radar observations of the aspect sensitivity of backscattered VHF echo power in the troposphere and lower stratosphere. Radio Sci., 1986, 21: 971–980.
[269] Bush G. A. Kulichkov S. N., Svertilov A. I. Some results of experiments on sound scattering by anisotropic inhomogeneities in the middle atmosphere. Izv. Russ. Acad. Sci., Atmos. Oceanic Phys., 1997, 33: 445–452.
[270] Kulichkov S N. Acoustic sounding of inhomogeneous structures in the middle atmosphere. Izv. Russ. Acad. Sci., Atmos. Oceanic Phys., 1998, 34: 5–11.
[271] Ono H. 'Algebraic Solitary Waves in Stratified Fluids', J. Phys. Soc. Japan, 1975, 39: 1082–1091.
[272] Menhofer A, Smith R K, Reeder M J, Christie D R. 'Morning-Glory Disturbances and the Environment in which They Propagate', J. Atmos. Sci. 1997, 54: 1712–1725.
[273] Doviak R J, Chen S S, Christie D R. 'A thunderstorm generated solitary wave observation compared with theory for nonlinear waves in a sheared atmosphere'. J. Atmos. Sci, 1991, 48: 87–111.
[274] Rees J M, Rottman J W. Analysis of solitary disturbances over an antarctic ice shelf. Boundary-Layer Meteorol, 1994, 69: 285–310.
[275] Christie D R, Muirhead K J, Clarke R H. 'Solitary waves in the lower atmosphere'. Nature 1981, 293: 46–49.
[276] Cheung T K, Little C G. Meteorological tower, microbarograph array, and sodar observations of solitary-like waves in the nocturnal boundary layer. J. Tmos. Sci., 1990, 47: 2516–2536.

[277] Rao M P Castracane P, Casadio S, Fua D, Fiocco G. Observations of atmospheric solitary waves in the urban boundary layer. Boundary-Layer Meteorology, 2004, 111(1): 85.

[278] Arendt S, Fritts D C. Acoustic radiation by ocean surface waves. Journal of Fluid Mechanics, 2000, 415 (25): 1.

[279] Posmentier E. A theory of microbaroms. Geophys. J. R. Astrn. Soc, 1967, 13: 487–501.

[280] Donn W L, Posmentier E. Infrasonic waves from the marine storm of April. 7 1966. J. Geophys. Res., 1967, 72: 2053–2061.

[281] Rind D. Microseisms at Palisades 3. Microseims and microbaroms. J. Geophys. Res., 1980, 85: 4854–4862.

[282] Dowling A P, Ffowcs Williams J E. Sound and Sources of Sound. Ellis Horwood: 1983.

[283] Ostrovsky L A. On generation of infrasound by large objects falling into water. IEEE International Geoscience and Remote Sensing Symposium., 2002, 22: 873–875.

[284] Cetegen B M, Ahmed T A. Experiments on the periodic instability of buoyant plumes and pool fires Combustion and Flame, 1993, 93: 157–184.

[285] Bedard A, Nishiyama R T. Infrasound generation by large fires: experimental results and a review of an analytical model predicting dominant frequencies. IEEE International Geoscience and Remote Sensing Symposium., 2002, 2: 876–878.

[286] Gaydon A G, Wolfhard H G. Flames, their structure, radiation and temperature, 4th ed. London: Chapman and Hall. 1979; Toeppler A. Ann. Phys., Leipzig, 1866, 128: 126.

[287] Drysdale D. An Introduction to Fire Dynamics. Chichester: A Wiley Interscience. 1985.

[288] Bedard Jr. Atmospheric turbulence aloft: A review of possible methods for detection, warning and validation of prediction models. AIAA 93–0847, 31st Aerospace Sci. Meeting & Exhibit. 1993.

[289] Nishiyama R T, Bedard A J, Kirschner Jr, Strong A L. Winds over mountains and infrasound: possible applications for detecting regions related to aircraft turbulence reports. 2002 IEEE International Geoscience and Remote Sensing Symposium, 2002, 2(2): 879–881.

[290] Kessler E. Thunderstorms. U. S. Govt. Printing Off. 1982, 3: 117.

[291] Xie Jin-lai, Yang Xun-ren, Li Ying-bo, Li Qi-tai. On the correlativity between atmospheric gravity waves and various weather phenomena. J. Low Freq. Noise Vib., 1988, 7(4): 142.

[292] Afraimovich E L, Kosogorov E A, Plotnikov A V. Shock-acoustic waves generated during rocket launches and earthquakes. Cosmic Research, 2002, 40(3): 241.

[293] Balachandran N K, Donn W L. Concorde sonic booms as an atmospheric probe. Science, 1977, 197:47.

[294] Crary A P. Stratosphere winds and temperatures from acoustical propagation studies. J. Meteorol., 1950, 7: 233.

[295] Crary A P, Bushnell V C. Determination of high-altiude winds and temperatures in the Rocky Mountain area by acoustic soundings, October 1951, J. Meteorol., 1955, 12: 463.

[296] Donn W L, Shaw D M. Exploring the atmosphere with nuclear explosions. Rev. Geophys., 1967, 5: 53.

[297] MacKinnon R F. Effects of winds on atmospheric pressure waves produced by hydrogen bombs. J. Meteorol. Soc. Japan, 1968, 46: 45.

[298] Meecham W C. Effects of atmospheric wind structure on shorter-period nuclear-generated infrasound. J. Geophys. Res., 1968, 73: 377.

[299] Row R V. A coustic-gravity waves in the upper atmosphere due to a nuclear detonation and an earthquake, J. Geophys. Res, 1967, 72: 1599.

[300] Baker D M, Davies K. Waves in the ionosphere produced by nuclear explosions, J. Geophys. Res, 1968, 73: 448.

Bei Fragen zur Produktsicherheit wenden Sie sich bitte an:
If you have any questions regarding product safety,
please contact:

Walter de Gruyter GmbH
Genthiner Straße 13
10785 Berlin
productsafety@degruyterbrill.com